Brain Electricity

Robert W. Baloh

Brain Electricity

The Interwoven History of Electricity and Neuroscience

Springer

Robert W. Baloh
UCLA
Los Angeles, CA, USA

ISBN 978-3-031-62993-8 ISBN 978-3-031-62994-5 (eBook)
https://doi.org/10.1007/978-3-031-62994-5

This Springer imprint is published by the registered company Springer Nature Switzerland AG
The registered company address is: Gewerbestrasse 11, 6330 Cham, Switzerland

To my father, “the electric man”

Preface

> By rapid averaging and other procedures, electronic computers are helping investigators to extract much new information from the complex recordings of the electrical activity of the brain.
>
> Mary "Mollie" Brazier, June 1, 1962, *Scientific American*

Electricity and neuroscience are topics that have fascinated science historians since ancient times. Each topic has a rich literary tradition with breathtaking discoveries made by colorful characters that changed the evolution of human culture. Yet, it has only been in recent times that medical science historians have appreciated the close interrelationship of these two topics. In England, Iwan Rhys Morus focused on the Victorian era and its fascination with the application of electricity to the body and brains of living beings from electrotherapy to electrocution. In America, Stanley Finger linked the application of electricity to the birth of neuroscience, and in France, Céline Cherici emphasized how the history of electricity was at the crossroads of biology and medicine.

As a researcher and practicing neurologist, I naturally have a different perspective than these eminent medical science historians, but I am indebted to them for their insights. I have had a lifelong interest in electricity dating back to my father showing me how to design an electric circuit as a child and reading articles about electricity in *Scientific America* as a teenager. I read about Thomas Edison and Nikola Tesla, and having grown up near Pittsburgh, I was fascinated by the battle of the electric currents between Edison and George Westinghouse. It was years later, however, before I became aware of the role these and other pioneers in electricity played in the developing fields of neurophysiology and neurology. During my neurology residency at UCLA, I met Mollie Brazier and became enamored with her stories and books on the history of neurophysiology.

Early researchers on static electricity had no way of measuring electricity, so they used their body as a rough indicator of the magnitude of electricity. A strong shock indicated more electricity than a weak shock. One of these investigators even noted a taste sensation after a shock to the tongue, a hint that nerves were activated by electricity. The discoveries of Luigi Galvani and Alessandro Volta marked a major turning point in the history of both electricity and neuroscience. Galvani convincingly showed that electricity could activate nerves and muscles and Volta's bimetal electrochemical battery provided a continuous source of electricity for physical and medical experiments. The discovery that electricity and magnetism were tightly interrelated soon followed.

Michael Faraday's lines of force theory of electromagnetic waves marked the beginning of the modern era of electricity, and his induction coil provided a new and better way to stimulate nerves (faradization). This in turn inspired Michael Todd's pioneering work on epilepsy and brain electrical stimulation. Emil du Bois-Reymond and Herman Helmholtz identified and measured the speed of a nerve action potential, and their student Julius Bernstein was the first to record an action potential. David Ferrier electrically stimulated the cerebral cortex in primates documenting localized function, and Charles Sherrington stimulated the spinal cord demonstrating the synapse and the integrative function of the nervous system.

James Clerk Maxwell was a pivotal figure in the fields of electromagnetism and physics. He converted Faraday's lines of force theory into mathematical equations that provided the foundation of modern physics. The equations showed how electromagnetic waves pass through space as vibrations in the all-pervasive ether. This led to discovery of wireless telegraphy and ultimately radio transmission. Curiously, it also led to belief that brain waves could travel in the ether making it possible to read minds and even communicate with the dead. Two of the most famous scientists of the latter nineteenth century, Oliver Lodge and William Crookes, were spiritualists who actively participated in psychic research along with their scientific research. In fact, Crookes attributed development of his radiometer and cathode ray tube to his studies in spiritualism.

The role of electricity in the developing medical specialty of Neurology in the nineteenth century cannot be overemphasized. Nearly all the early neurologists in Europe and America were electrotherapists. Since electricity activated nerves, nerve cells, and muscle, it seemed only natural that electrotherapy was best used for treating nerve, brain, and muscle diseases. Indeed, electrotherapy was most effective for treating two of the most common neurological diagnoses of the time, hysteria and neurasthenia. We now know that was because electrotherapy is a powerful placebo.

Throughout history, there has been a close interrelationship between new discoveries in electronics and advances in neuroscience. Vacuum tubes developed to improve wireless transmission led to oscillographs for recording nerve impulses and polygraphs for recording electroencephalography (EEG). The invention of transistors and long-lasting batteries made it possible to make implantable electric neurostimulators and advances in microelectrode technology led to intraneuronal recordings and deep brain stimulation. The discovery of nuclear magnetic resonance (NMR) made it possible to image the brain in unbelievable detail.

This book traces the interrelationship between electricity and neuroscience from ancient times to the late twentieth century, focusing on the key scientists and their contributions. I aim for a wide scientific audience but particularly students of neuroscience. The historical context of discovery provides an ideal framework for understanding modern concepts of electromagnetism and neuroscience. Furthermore, the stories of these pioneering researchers can be inspirational to those beginning their career in neuroscience.

Las Vegas, NV, USA Robert W. Baloh

Contents

About the Author

Robert W. Baloh is well known as a leader in the field of neurotology. After completing his neurological training at UCLA in 1972, he joined the faculty at the UCLA School of Medicine with a joint appointment in Neurology and Head and Neck Surgery in 1973. Baloh became Professor in 1982, Distinguished Professor in 2002, and Emeritus professor in 2021. Author of 14 books, over 350 articles in peer-reviewed scientific journals, and over 100 book chapters, he is a pioneer in the study of the vestibular system: the inner ear/brain system which helps people to maintain their sense of balance and spatial awareness. He has trained more than 40 fellows who are now leaders in the field of vestibular research. Baloh also has a longstanding interest in the history of neurology and has written extensively on the subject. Among his many honors, Baloh received the Hallpike/Nylen Prize at the Bárány Society Meeting in Prague, Czechoslovakia, in 1992 and had an international Dizziness and Balance Symposium in his honor at the 2014 American Academy of Neurology meeting in Philadelphia. His book *Clinical Neurophysiology of the Vestibular System*, written with Vicente Honrubia, is currently in the fourth edition and is the standard in the field. His other books include *Dizziness, Hearing Loss and Tinnitus: The Essentials of Neurotology* (1984), *Neurotology* (1994), *Dizziness: Why Do You Feel Dizzy and What Will Help You Feel Better* (with Greg Whitman, 2016), and *Vertigo: Five Physician Scientists and the Quest for a Cure* (2017). Recently, he became interested in the boundary between neurology and psychology with the publication of *Havana Syndrome: Mass Psychogenic Illness and the Real Story Behind the Embassy Mystery and Hysteria* (with Robert Bartholomew, 2020) and *Medically Unexplained Symptoms: A Brain-Centered Approach* (2021). Most recently he published a book about exercise and brain health in 2022.

Chapter 1
The Mystery of Electromagnetism

> *A lodestone attracts only magnetic bodies, electrics attract everything. A lodestone lifts great weights; a strong one weighing two ounces lifts half an ounce or an ounce, Electrics attract only light weights: e.g., a piece of amber three ounces in weight lifts only one-fourth of a barleycorn weight.*
>
> *William Gilbert, 1600*

Abstract To ancient civilizations, electricity and magnetism had mystical qualities suggesting some type of supernatural force. Not surprisingly, ancient physicians tried to harness this force to treat a variety of illnesses. In the mid-sixteenth century, William Gilbert made the first clear distinction between electricity and magnetism and his contemporary Roger Bacon proposed a general theory of magnetism as a force acting at a distance. Their work marked the beginning of modern scientific methodology. To deal with the conflict between religion and the new science, Rene Descartes proposed a two-universe solution, one that obeyed physical laws and the other that only God could understand. Descartes' theory on brain function led to the bizarre conclusion that animals could not feel pain because they lacked a soul localized in the pineal gland. Several early scientists including Roger Bacon saw a parallel between magnetism and gravity although Isaac Newton saw them as two different forces acting at a distance. Edmond Halley was the first to suggest that the Earth's magnetism came from forces deep below the Earth's surface.

Keywords Magnetism · Compass · Electrics · Anelectrics

When ancient people first witnessed electromagnetism, they were in awe. Lodestone attracted metal objects and amber when rubbed with fur attracted light objects. These inanimate objects seemed to possess supernatural powers. Lodestone was stronger than amber since it attracted heavier objects. Not surprisingly, physicians and faith healers at the time tried to harness these mysterious powers for treating a variety of diseases.

R. W. Baloh, *Brain Electricity*, https://doi.org/10.1007/978-3-031-62994-5_1

Lodestone

According to Greek mythology, a shepherd noticed that his feet were sticking to the ground while herding his sheep on Mount Ida in a region of northern Greece called Magnesia [1]. It took great effort to lift his feet and he noticed that the bottom of his walking stick also stuck to the ground. Curious for an explanation, he dug into the ground and found a stone that attracted the nails in his shoes and the metal bottom of his walking stick. The phenomena became known as magnetism after Magnesia and the stone, lodestone, was composed of the mineral magnetite.

As far back as 2000 BCE, ancient cultures including Greeks, Egyptians, Hindus, Chinese and even possibly American Indians were aware of the magnetic properties of lodestone. The Chinese were the first to recognize that the Earth had a natural magnetism and that a piece of lodestone would align itself with the poles of the earth.

In his writings in the fourth century BCE, Wang Xu noted that a "south-pointer" allowed people to always know their position on the planet. The south pointer, also called a south-pointing fish, was a large spoon in the shape of a ladle made from lodestone (Fig. 1.1). When the south pointer was placed on the ground or on any flat surface so that it balanced on the bottom back half of the spoon the handle would point toward the south. This primitive compass was initially used exclusively to orient the construction of buildings and roads using the geometric principles of feng shui. These early Chinese compasses deviated up to 15 degrees from the true south depending on location which explains slightly different street orientations in different towns and villages [1].

It wasn't until around 800 CE that the Chinese began using a splinter of lodestone shaped like a tadpole floating on water as a compass. This primitive compass

Fig. 1.1 Ancient Chinese south pointer or south pointing fish. Spoon in the shape of a ladle made of lodestone. When placed on a flat surface the handle pointed south. (From Needham J. *Science and Civilization in China*. Cambridge, Cambridge University Press, 1962, volume 4, figure 344 with permission)

was an intermediate between the lodestone spoons and iron needles of later compasses. A few hundred years later, the Chinese scientist Shen Kuo described how he magnetized a needle by rubbing it against a lodestone and then hanging the needle with a single strand of silk attached at the center to produce a compass [2]. Kuo noted that sometimes the needle pointed south and sometimes north.

Although Marco Polo is often credited with bringing the compass to Europe after visiting China in the late thirteenth century, European mariners were already using the compass in the late twelfth century. One of the best early descriptions of a European compass was provided in a satirical poem by the French poet and monk Guyot de Provins around 1205. *"…there is an art which the sailors have, which cannot deceive. They take an ugly brown stone, the magnet, to which iron willingly attaches itself, and touching a needle with it, they fix the needle in straw, and float it on the surface of water, whereupon it turns infallibly to the Pole star"* [3].

At the time, the spherical Earth was thought to be the center of the Universe with the moon, planets, sun, and heavenly bodies all revolving around the Earth daily. The Earth's poles represented the axis around which all the bodies revolved. One star, the pole star was observed to be relatively fixed in the northern sky thought to be located on or near the axis of celestial rotation. To explain why the compass needle pointed toward the pole star, it was assumed that there were mountains of lodestone at the Earth's poles, the only material known to attract a compass needle at the time. Stories were told of passing ships being destroyed as their iron nails were pulled out by lodestone mountains.

Interestingly, Chinese compass needles have always been made with the prime end pointing to the south, whereas European compass needles have always been made to point to the north. This supports the likelihood that they were developed independently of each other.

Amber

Amber represents the world's first precious gemstone. Long before the silk and salt roads existed, the amber road was distributing amber for processing throughout the ancient civilizations on the European, Asian, and African continents. Amber that washed up onto the shores of the Baltic Sea was harvested and transported to jewelry workshops around the Mediterranean beginning more than 10,000 years BCE. Around 1000 BCE the Greeks began navigating the Black Sea and via the river Dnieper opened a trade route to the Baltic region.

Early theories on the origin of amber varied from the secretions of whales or giant ants to crystalized bird tears. According to Greek mythology, there once were two suns in the sky but the gods banished one into the ocean. When the sun hit the bottom of the ocean it broke up into thousands of pieces that washed ashore as amber.

We now know that amber is fossilized tree resin from pine trees of forests that covered the area prior to the formation of the Baltic Sea. The thick, sticky, resin oozed down the tree trunks, trapping tiny animals and vegetation in its path. The hardened resin ultimately made it to the developing sea by rains and flooding. Amber occurs in several colors from a common yellow to a rare bluish tinge. Red amber indicates the fossilized resin was exposed to the sun for a long time before washing into the sea. Black amber is filled with impurities and is fragile to work with but is most valuable when carved by an experienced artisan.

Around 600 BCE the first great Greek philosopher, Thales of Miletus, made the remarkable observation that when he rubbed amber with animal fur the amber developed the peculiar property of attracting nearby light objects such as thread or small pieces of cloth [4]. It could even provide a slight shock to the person holding the amber. As a party trick, women in ancient Greece would rub their amber jewelry to provide a shock to a frog. Thales felt that amber had a "soul" that accounted for its ability to attract other objects.

The Greek word for amber is "elektron." The materials that had these strange attractive properties were called "elektricks" and their attraction, "elektrick force." These terms are the forerunners of the words—electron, electricity, electric current, and electronics.

Early Theories on Electromagnetism

Much of what we know about ancient Greek and Roman scientific knowledge can be traced to the writings of the Roman scholar and historian, Pliny the Elder who spent a great deal of his life writing more than a million words in a 37-volume encyclopedia entitled *Naturalis Historia* (Natural History) in the first century CE. Pliny was enthralled with magnetism. *"What phenomenon is more astonishing? Where has nature shown greater audacity? For iron, the tamer of all substances is drawn to the magnet, following some intangible attraction, and, as it comes nearer, leaps to meet the magnet"* [5].

The Greeks were aware that lodestone not only attracted pieces of iron and steel but that the metals themselves became magnetized so that they would attract other pieces of iron and steel. A whole chain of metal rings could be lifted and held together by a single lodestone. Pliny felt that there were two types of magnetism, one that occurred naturally such as with lodestones and another that occurred when stones like amber were rubbed with fur. The naturally occurring type was much stronger than the one that required rubbing.

The early Christian church had a different way of dealing with the mysteries of electromagnetism. Declare them a miracle. Saint Augustine summarized the current knowledge on magnetism in his masterwork *De Civitate Dei* (*The City of God*), published in 428 CE. He wrote that he was "thunderstruck" when he first saw a lodestone lift a chain of iron rings and iron flakes moved around a silver plate by a lodestone moving beneath the plate. But he was puzzled why a lodestone, unlike amber rubbed with fur, could not move even tiny pieces of straw. He concluded that man should not concern himself with such a conundrum. It was enough for a

Christian to attribute such mysteries to the *"goodness of the Creator, who is the one true God"* [5].

The first effort to explain magnetism in terms of natural phenomena is found in the poem *De Rerum Natura* (*On the Nature of Things*), written in the first century BCE by the Roman poet and philosopher Tutus Lucretius Carus [6]. The story of how this poem was rediscovered and copied by an Italian book hunter in the fifteenth century is the theme of a recent Pulitzer Prize–winning book, entitled *The Swerve* by Stephen Greenblatt. In 1417, a former papal secretary, Poggio Bracciolini discovered a copy of Lucretius' poem in a remote German monastery and reintroduced it to Western culture [7].

Lucretius was a disciple of the famous Greek philosopher, Epicurus, who in the fourth century BCE elaborated on the theory of the atom. Epicurus theorized that the universe is composed of tiny indivisible particles called atoms that are in constant motion, joining together to create objects including all living beings and eventually breaking up and returning to individual atoms. There was nothing in the Universe other than atoms and void.

In his epic poem, Lucretius emphasized that human beings are no different from any other type of matter present throughout the Universe, composed of groupings of atoms that eventually will revert to individual atoms. He "reassured" us that we should not fear death since there is nothing but atoms and void, no supernatural beings. Of course, this theory was not well received by religions either at the time or in subsequent years.

Lucretius wrote that atoms could only be set in motion by their own weight or by being propelled by some external force. Regarding magnetism, he speculated that the lodestone emitted a dense stream of atoms that bombarded and displaced the atoms in the space between the lodestone and the iron. This emission of atoms resulted in a void in the space so that the atoms of the iron slid into the void and atoms in the back on the other side of the iron pushed it toward the lodestone.

To explain why only select materials were attracted by lodestone, Lucretius suggested that some such as gold were so dense and heavy that they could not be moved while others such as wood had large spaces between the atoms that they were made of so that the atoms emitted by the lodestone passed right through. But why didn't other substances emit atoms? What was unique about lodestone?

Ancient Medical Uses of Lodestone and Amber

Ancient physicians reasoned that since lodestone had the power to attract iron, it also might have the power to attract evil spirits and illnesses from the body. As far back as 1200 BCE, Chinese physicians used lodestones on acupuncture sites to control the flow of Qi (chi) through the body and around 600 BCE Hindu physicians recommended lodestone to treat a wide range of ailments. The Greek physician Hippocrates recommended the use of loadstone to attract and remove pain associated with a variety of diseases. He advised patients with headaches to lay with their heads on a lodestone. Egyptian physicians recommended the use of lodestone bracelets to relieve the pain of arthritis. Cleopatra wore a small lodestone on her forehead to preserve youth and beauty. Many people thought that lodestone had strong aphrodisiac potency.

Ancient physicians often applied magnetite externally in a pulverized form compounded with other ingredients called Emplastrum Magneticum (magnetic plaster) to treat a variety of diseases such as arthritis, poisoning, and gout. The magnetic plaster was applied directly to an affected body part. Pliny the Elder recommended a magnetic plaster for treating burns and Galen of Pergamum compared the magnetic plaster to cathartic drugs and suggested that the plaster was useful for attracting bile and phlegm and for removing poisons.

Although amber was very popular for jewelry and decorative ornaments, it was also popular in ancient medicine, particularly for making amulets to protect the wearer from evil spirits and a wide range of diseases. A well-known Roman physician, Calistratus felt that amber amulets protected against madness and a powder of amber mixed with honey could cure ailments of the throat, stomach, eye, and ear. In Japan, red amber was thought to purify the blood and in China, a mixture of amber powder and opium was used as a tranquillizer. Amber powder has long been used as a salve to treat skin conditions and in modern times it is still used in cosmetic treatments to enhance skin elasticity and remove wrinkles.

We now know that the attractive property of amber rubbed with fur is due to static electricity, but the use of static electricity for treating medical conditions awaited the development of better techniques for generating it in the seventeenth century. Ancient healers and physicians, however, did have a source of natural electricity generated by animals. Electric rays, eels, and catfish were known to produce electricity as far back as 3000 BCE. The Egyptians called the electric catfish the "thunder of the Nile." They used the electric catfish to treat headaches and nerve pain.

Pliny the Elder described the numbing effects of the shocks of electric rays and catfish in *Naturalis Historia*. This numbing effect was used to treat a variety of painful ailments. The personal physician of the Roman Emperor Claudius, Scribonius Largus, provided a detailed medical text for treating gout affecting the feet called podagria: "*For any sort of podagria when the pain comes on, it is good for one to put a living black torpedo-fish under his feet while standing on a beach (not dry but one on which the sea washes), until he feels that his whole foot and shank are numb just up to the knees. This will both relieve the current pain and alleviate future occurrences*" [8]. The use of electric fish to treat pain became known as ichthyo-electroanalgesia and is still in use in some countries.

Magnetism and the Dark Ages

For about the next thousand years after the fall of the Roman Empire (often called the Dark Ages), there was little interest in magnetism apart from the compass for navigation. The average person's understanding of magnetism was based on ancient lore and religious teachings. For example, garlic and diamonds were thought to weaken magnets according to Pliny the Elder. Sailors who looked after the compass on board ships were warned against eating garlic or onions well into the sixteenth century. Saint Augustine used magnetism to defend the concept of miracles to skeptics who questioned their validity. If magnetic attraction can't be explained, why should we need to explain miracles.

To describe magnetism, poets often used human attributes such as love and virtue. The twelfth-century French troubadour poet Bernart de Ventadorn wrote: *"the fair lady draws me toward her like a magnet"* [9]. Guido Guinizzelli, the thirteenth-century Italian poet and originator of the dolce stil novo (sweet new style), wrote:

In what strange regions near the polar star
May the great hills of massy lodestone rise,
Virtue imparting to the ambient air
To draw the stubborn iron; while afar
From the same stone, the hidden virtue flies
To turn quivering needle to the Bear
In splendor blazing in the northern skies [10].

Replace virtue with magnetic field and one has a modern poetic description of magnetism.

In his canzone on love, Guinizzelli wrote: *"love has a home in a gently noble heart, like, in the same manner, adamas* [medieval word for magnet] *has home in an iron mine"* [9]. Medieval poets took advantage of the phonetic resemblance of the words for lover (amans) and magnet (adamas). William Shakespeare used *adamant,* another word for magnet, in Midsummer Night's Dream: *"You draw me, you hard-hearted adamant, but yet you draw not iron; for my heart is true as steel."*

Peter the Wanderer

Into the vacuum of critical thinking in the Dark Ages stepped a little-known French engineer Pierre Pelerin de Maricourt, better known as Petrus Peregrinus or Peter the Wanderer. He acquired the name Peregrinus or Pilgrim after visiting the Holy Land. His *Epistola de Magnete* (*Letter on the Magnet*) is considered one of Europe's earliest scientific reports [11]. A contemporary, the English Franciscan friar and philosopher Roger Bacon, called it the definitive work on magnets and lauded Peregrinus for his mathematical and experimental skills. The letter, dated August 8, 1269, was composed during the siege of Lucera in southern Italy and provided Peregrinus time to reflect on experiments he had carried out earlier. He was part of the army of Charles I, Duke of Anjou and King of Sicily, who was besieging Lucera at the behest of the Pope. Peregrinus' job was to design and build machines for heaving stones and fireballs into Lucera [9].

The letter consisted of two parts: a first part providing a description of the polarity of magnets, how to determine the north and south poles and the attraction and repulsion between poles. The second part described instruments that utilized magnetism, the floating compass and a new pivoted compass and ended with wild speculation about a magnetic clock and a perpetual motion machine.

Peregrinus was the first to make a clear distinction between the north and south poles of a magnet. In his experiments, he made a sphere of lodestone and observed that when a compass needle was placed at different points around the globe, there were two unique locations where the needle aligned itself perpendicular to the surface of the globe. When the compass needle was placed at other positions around the globe, the needle outlined a series of meridians all of which came together at the two opposite poles. Peregrinus recognized the similarity of his magnetized globe, which he called the "terrella" (little Earth), to the Earth's magnetism. He proposed that the Earth's north and south poles were on the axis around which all of Creation revolved.

Peregrinus made several other observations that had eluded the ancient philosophers. He showed that the north pole of a magnet attracted the south pole of another magnet, whereas the north poles of different magnets repelled each other. He also observed that breaking a lodestone into multiple smaller pieces resulted in each piece having its own north and south poles. "*You will also notice that the ends of the two stones which before their separation were together, after breaking will become one a north pole and the other a south pole. If now these same broken portions are brought near each other, one will attract the other, so that they will again be joined...*" [11].

Peregrinus then addressed the question of why the compass needle points to the north pole. He dismissed the prevailing notion that there were mountainous deposits of lodestone at the north pole attracting the needle since there were lodestone deposits at other locations around the Earth, yet these deposits had minimal effect on the direction of the compass needle. He also observed that the Pole Star was not fixed at the celestial north pole but rather slightly moved around it and that the compass needle did not follow the movement. Considering the symmetrical pattern of the compass needle around his small lodestone globe, he reasoned that the Earth's magnetism was also symmetrical, a basic property of the Universe unrelated to any local deposits of lodestone on the Earth or position of celestial bodies, "*...the poles of the lodestone derive their virtue from the poles of the heavens*" [11].

The notion that the Earth was the center of the Universe with all celestial bodies rotating about it was engrained in religious doctrine at the time. A staunch Catholic such as Peregrinus did not consider questioning the doctrine.

The ancient Greek astronomer, Aristarchus of Samos, proposed a heliocentric model of the solar system as far back as the third century BCE with the Earth revolving about the Sun once a year and rotating about its axis once a day. However, Aristotle, Plato, and Ptolemy favored the geocentric model of the solar system, and this model was accepted by the developing Christian religion and remained dominant throughout the Middle Ages. Even after Nicholas Copernicus, the German astronomer and Catholic Cannon provided a strong scientific defense of the heliocentric model in the sixteenth century, the Inquisition of 1615 concluded that the heliocentric model was absurd and heretical because it contradicted the Holy Scripture. Galileo Galilei, the Italian astronomer and polymath who championed

Copernicus' heliocentric model, was tried by the Inquisition and spent the remainder of his life under house arrest [12].

Peregrinus ended his *Letter* by proposing a perpetual motion machine consisting of a toothed wheel powered by a magnet. A small, rounded weight made of brass or silver was located between two of the teeth. Peregrinus indicated that the metallic ball "*seeks the center of the earth in virtue of its weight, thereby aiding the motion of the teeth*" [11]. Several versions of Peregrinus' perpetual motion machine were subsequently made but unsurprisingly none of them worked. Although the notion of perpetual motion had great appeal during the Middle Ages, modern thermodynamic theories show that it is impossible to produce such devices.

Magnetism in Medieval Medicine

Magnets and magnetite plasters and concoctions played a large role in medieval medicine [13]. The earliest use of lodestone in medicine was to aid surgeons in removing an iron arrow tip from the body. As far back as 600 BCE, the Hindu surgeon Sucruta described how he used a magnet to help extricate an arrow tip. He noted that the magnet worked best if the arrow tip did not have barbs and if there was a wide opening parallel to the fibers of the tissue. In the thirteenth century, the English physician Gilbertus Anglicus wrote in his *Compendium Medicinae* that a magnet could help the surgeon localize and extract iron concealed in the flesh. More bizarre were reports of successful treatment of hernias after having the patient swallow iron fillings and using a magnet to draw in and restore the protruding intestine. Others reported successful extraction of metal particles from the eye using a magnetic plaster spread over the eye.

Magnets applied to the head were common for treating a wide range of neurological and psychiatric disorders. In the sixth century, Aetius of Amida recommended the application of a magnet as the treatment of choice for hysteria and muscle spasms. Magnets were used to draw epileptic seizures from the brain. The twelfth-century nun and mystic Hildegard of Bingen wrote about religion and philosophy but also wrote a treatise on plants and minerals for medical purposes. Because of this, she is often considered the first German woman doctor. Magnetism was at the center of her healing process. She followed the usual practices of the time but also went a bit further recommending placing a lodestone in the patient's mouth to treat fits of rage and anger and to prevent lies and malicious gossip. She also felt that holding a lodestone in the mouth could make fasting more bearable.

Although most uses of magnetite in medieval medicine were external, it was also common to pulverize magnetite and mix it with milk for internal application. The eleventh-century Persian philosopher and physician Avicenna recommended this concoction for accidental poisoning with rusted iron suggesting that the poison would be attracted to the magnetite and speed up the excretion from the intestines. Others recommended the mixture for treating a variety of poisonings. This may have worked by inducing vomiting. The thirteenth-century German Dominican friar and philosopher Albertus Magnus, who was later canonized as a Catholic saint, recommended a milk/magnetite mixture for treating heart failure and edema in his book *Mineralia*.

Paracelsus, Doctor, and Mystic

The culmination of medieval fascination with magnets in medicine occurred with the enigmatic Theophrastus Bombast von Hohenheim, better known as Paracelsus, a physician and alchemist born in Switzerland in 1493. Paracelsus developed an elaborate theory on how magnets cure disease and introduced the concept of "animal magnetism," a forerunner of mesmerism and hypnosis.

Paracelsus was a paradoxical character who was either loved or hated by his medical peers. He regularly criticized his fellow physicians and apothecaries and displayed contempt for Galen and Avicenna, the two ancient physicians considered the pillars of medieval medicine. He famously burned Avicenna's massive medical text *Canon of Medicine* in the Basel Market Square in 1527. He wrote and taught in German rather than the traditional Latin, and he preached that the only way to learn the art of medicine was to practice it on patients. He was known for his abusive language and heavy drinking, and he ridiculed physicians with pompous titles.

Regarding Paracelsus' own titles, there are conflicting reports. Some questioned his descent from the Swabian noble family Bombast von Hohenheim suggesting that he invented the title and that his father, who was also a physician, was born a commoner. The title Paracelsus is even more controversial since it seems inconsistent with his distaste for Latinized names. In Latin, it means literally surpassing Celsus, the first-century Roman author of *De Medicina*, the best surviving treatise on ancient medicine. Some suggested that his circle of friends honored him with the name, while others suggested that his enemies used it as ridicule and that he later accepted it for spite. Regardless, he first used it in medical publications in 1536 and regularly thereafter.

Paracelsus despised the ancient "humoral theory," replacing it with an equally speculative theory based on astronomy and alchemy. The human body was a microcosm of the solar system. The sun was the heart, the moon the brain, Jupiter the liver, Saturn the spleen, and Venus the kidneys. From the constellation of the planet diseases arose. He described what he called a "life force" in the human body that when depleted resulted in disease. The "life force" could be replenished with herbs and foods and with chemicals such as sulfur and mercury. Magnets could energize the body's "life force" and thereby start the healing process.

Since magnets had the mysterious power to attract iron, Paracelsus felt that they could also attract diseases from the body. The magnetically extracted disease could be "inoculated" into the ground, a plant, or even another person. He described detailed procedures for transplanting diseases from the body using a magnet. For example, to treat epilepsy, which he felt was due to too much nervous fluid in the brain, he applied the repulsing pole of a magnet to the head and spine and the attracting pole to the abdomen.

Paracelsus also considered the human body to possess magnetic properties. *"Man possesses a magnetic power by which he can attract certain effluvia of a good or evil quality in the same manner as a magnet will attract particles of iron... A magnet may be prepared from iron that will attract iron, and a magnet may be prepared out of some vital substance* (for example urine or blood) *that will attract vitality... If such a magnet is applied to a part of the patient's body, it attracts and absorbs vitality from that part in the same manner as a sponge absorbs water, and it will thereby allay the inflammation existing in such a part, because it attracts the superabundance of magnetism carried to that place by the rush of the blood"* [14].

This "animal magnetism" not only had the power to influence disease but also human interactions. *"In this way diseases can be cured in one person and caused to appear in another; love between two persons of the opposite sex may thus be created, and magnetic links be established between persons living at distant places, because there is only one universal principle of life, and by it all beings are sympathetically connected together"* [14].

There is a spiritual quality to Paracelsus' writings and like most medieval physicians he was deeply religious. In the end, it was God's will that diseases occurred and only through God's will could physicians heal the patient. As with Saint Augustine, Paracelsus felt that ultimately only God could understand the complex nature of disease.

Problems with the Compass

Even though Peregrinus described what he considered a symmetrical magnetism aligned with the Earth's celestial north pole, navigators were already noticing that compass needles often deviated from the true north [1]. Initially, the discrepancies were attributed to defective needles or carelessness but after these possibilities were refuted, the phenomena became known as "declination," a tilting of the magnetic pole from the north pole. Systematic measurements of compass needle declination in Europe in the mid-fifteenth century identified an 11–12 degrees deviation to the east of the true north.

In the late fifteenth century, at the beginning of the great Portuguese and Spanish voyages of discovery, it became more and more apparent that the compasses being used could not provide reliable measurements of longitude at sea. During his voyage to the Americas, Christopher Columbus was alarmed that his compass needle deviated a good amount from the Pole Star. Columbus suggested to his crew that this was probably due to the previously undetected movement of the Pole Star. However, in the mid-sixteenth century, the Neapolitan scholar John Baptista Porta suggested that the declination of the compass needle changed as Columbus moved across the Atlantic Ocean, from about 10 degrees east of north in Europe to about

10 degrees west of north in the Caribbean islands. The meridian with approximately zero declination was at the Azure Islands.

Around this time, the Flemish cartographer, Gerardus Mercator, concluded that declination resulted from the fact that the magnetic axis of the Earth was tilted from the rotational axis. The magnetic meridians intersected at the magnetic poles, not the geographic poles. Like Porta, Mercator noted that declination was zero at the Azures. Declination explained many of the distortions of coastlines on contemporary maps. Mercator speculated on a map of the Artic region (since no one had ever been there) depicting both poles as magnetic islands (presumably adhering to the lodestone mountain theory of the Earth's magnetism). Remarkably, California recently claimed by Spain was located north of the Arctic Circle on his map.

Another phenomenon the navigators noticed as ships explored more and more of the Earth's seas was a property of the compass initially called "dip," later called "inclination." The compass needle balanced on a pivot would tilt up or down vertically depending on the latitude of the ship, the further north or south of the equator, the greater the inclination. When a compass designed for the northern hemisphere was used in the southern hemisphere, the north end of the needle was pulled upward, and the south end was pulled downward. Whereas a compass designed for the southern hemisphere showed the opposite effect.

Initially, this "bothersome" effect was corrected by counterbalancing with weight to make the needle swing primarily in the horizontal plane for its area of use. However, others saw inclination as a potentially useful way to estimate latitude. The English compass maker, Robert Norman, published a pamphlet in 1581, in which he described two versions of a compass for measuring inclination. The first version was reminiscent of the original floating compass, a magnetized needle in cork that was designed to have neutral buoyancy in a container of water. With this devise, he could measure both declination and inclination although not very accurately. For a more accurate measurement of inclination, he constructed a vertical compass with a central pivot so that the needle settled at the angle of inclination. With these devices, navigators observed that the angle of inclination gradually increased from almost zero at the equator to almost one as one approached the poles. Any theory on the Earth's magnetism had to account for declination and inclination.

William Gilbert and *De Magnete*

William Gilbert was born in Colchester, England in 1544, the eldest of 11 children [15, 16]. He was a brilliant student and entered St. Johns College, Cambridge to study mathematics at age 14. Over time his interests turned to medicine, and he graduated as a Doctor of Medicine in 1569. After touring the continent and visiting the best medical schools, Gilbert developed a medical practice in London with a reputation as an expert in maritime diseases. Having an inquiring mind, Gilbert expanded his interests beyond medical practice to a wide range of scientific inquiries. He began with experimental chemistry but was soon drawn to the strange phenomena of magnetic and electric attractions. Gilbert hosted small groups of amateur scientists in his house where they discussed scientific advancements and performed experiments on their own.

Over time, Gilbert's reputation as a physician and experimenter solidified leading to his selection as President of the Royal College of Physicians in 1600. This year was of particular significance since it was also the year that Gilbert published his monumental work on magnetism and static electricity, *De Magnete,* and the year he was called to court and appointed physician to Queen Elizabeth I. His service to the Queen was short-lived since she died in early 1603, and Gilbert died later in the same year during the great plague epidemic.

De Magnete, with the full translated title "*On the magnet, Magnetick Bodies also, and on the great magnet the Earth; a new Philosophy demonstrated by many arguments and experiments*" was the first book to establish magnetism and electricity as a science [17]. The work is often considered key in the evolution of science from the ancient to the modern. It was initially published in Latin with several editions in the early seventeenth century and was first translated into English in the late nineteenth century.

De Magnete was divided into six books and 115 chapters although many of the chapters were short, not much more than an outline. The focus of the book was on magnets, much of which reiterated and expanded the work of Peregrinus. Even though the title page listed the author as William Gilbert M. D.—Electrician, there was only a single chapter devoted to amber and static electricity. However, in many ways, this chapter was the most impactful of all the chapters.

Gilbert's work on magnetism can be divided into three broad areas, his work on lodestone and the Earth's magnetism, the problems with the compass and navigation. and the relevance of magnetism in the heliocentric model of the solar system. Like Peregrinus, Gilbert concluded that the Earth was a giant magnet but unlike Peregrinus, Gilbert had additional information including measurements of declination and inclination of the Earth's magnetic field and overwhelming evidence for the heliocentric model of the solar system.

In experiments very similar to those of Peregrinus, Gilbert made a lodestone sphere which he also called a terrella (little Earth) and he measured the orientation of small pieces of wire (~1 cm) placed at locations all over the terrella. He noted that the angles these tiny pieces of wire made with the surface of the terrella were the same as the measurements of inclination observed at different latitudes on Earth. He concluded that the magnetism of the terrella was due to the magnetized iron of the lodestone so the Earth's magnetism must be due to the magnetism within the planet itself and not due to mountains of lodestone at the poles, the Pole Star, or the celestial pole as proposed by Peregrinus.

Gilbert proposed that magnetism was spread through the whole substance of the Earth and it was symmetrical about the Earth's axis of rotation. To explain declination, he suggested that although the compass needle was mainly influenced by the magnetism of the globe it was also influenced to a lesser degree by magnetic rocks located on the continents. He speculated that in the middle of an ocean equally distant from land, the compass needle would point directly north. To test this hypothesis, he made terrellas with lumps of lodestone protruding from the surface to represent mountainous continents and gouged out areas to represent ocean basins. He reported that the models supported his hypothesis in that when a compass needle was near a lump it deviated from the pole of the terrella whereas a compass needle near a basin pointed toward the pole.

Although he believed that the operation of the Universe was outside of experimental proof, he had praise for Copernicus and his sun-centered solar system. He reasoned that the distance to the other stars must be enormous so that it would be impossible for them to rotate about the Earth daily. The stars near the equator would have to be moving at unimaginable speeds whereas those close to the poles would hardly move at all.

Gilbert speculated that magnetism was key to generating the Earth's rotation and the orbital rotation of other planets and stars. But his explanation for how magnetism achieved this effect borders on the bizarre. Despite forward thinking on the origin of the Earth's magnetism, he still believed in the ancient notion that magnets had souls and moved themselves. Furthermore, he objected to the term "magnetic attraction" preferring what he called magnetic "coition." He used this animistic term in its literal sense, meaning to mate or join. *"Coition, we say, not attraction for the term attraction has wrongfully crept into magnetic philosophy, through the ignorance of the Ancients"* [17].

There was no physical force acting between magnets and iron but rather a coming together like two people might approach one another. *"Thus, the magnetic coition is the act of the lodestone and of the iron, not one of them alone"* [17]. Since the Earth is a magnet, it can move itself, rotating on its axis and circle the sun. Other heavenly bodies moved themselves without any force acting between them, no action at a distance as some suggested.

Fortunately, Gilbert published his heliocentric theory in a reformed England. Expressing similar views in Catholic Europe would likely have led to being burned at the stake. In the same year (1600), the Italian philosopher and former priest Giordano Bruno was burned at the stake for supporting a heliocentric model of the solar system.

In his chapter on "electrics," Gilbert made a clear distinction between lodestone and amber for the first time. He demonstrated that the attractions of these two substances had entirely different properties. "*A lodestone attracts only magnetic bodies, electrics attract everything. A lodestone lifts great weights; a strong one weighing two ounces lifts half an ounce or an ounce, Electrics attract only light weights:* e.g., *a piece of amber three ounces in weight lifts only one-fourth of a barleycorn weight"* [17].

Gilbert made a device he called a versorium (Latin for "turn around"), a kind of electroscope to test for the presence of an electric charge. The versorium consisted of a thin light-weight needle balanced on a pivot (like a compass but the needle was not magnetized). It would move toward a nearby object if was electrically charged (Fig. 1.2). His versorium was much more sensitive to the weak forces of electric attraction than bits of thread or scraps of paper. With this more sensitive instrument, he found that many other materials exhibited an amber-like effect. A wide variety of

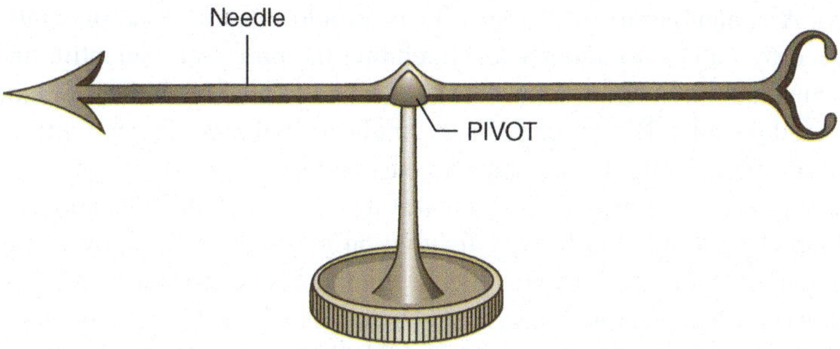

Fig. 1.2 Gilbert's versorium. Light-weight needle balanced on a pivot like a compass, but the needle was not magnetized. The needle turned toward a nearby object that was charged with electricity. (Redrawn from Gilbert W. *De Magnete*, 1600)

gemstones as well as fake gemstones made of glass or crystal attracted the needle when rubbed. Substances that attracted the needle were called "electrics" and those that did not anelectrics.

Several factors were unique to "electric" materials. Gilbert found that the attractive properties of electrics were particularly sensitive to moisture. Condensation of water on the surface of the material inhibited the attraction so the experiments were difficult to perform in humid weather. Some liquids such as alcohol inhibited attraction, whereas others such as olive oil had no effect. Extreme heat markedly inhibited electric attraction. None of these factors altered the attractive properties of lodestone.

Gilbert proposed that the attraction of electrics was caused by an "electric effluvium" (from the Latin effluere, meaning to flow out) that was emitted when the materials were rubbed. *"For as no action can be performed by matter save by contact, these electric bodies do not appear to touch, but of necessity something is given out from the one to the other to come into close contact therewith and be a cause of incitation to it"* [17]. The effluvia from the electric somehow reaches out to a nearby object, attaches itself to the object, and then pulls it back to the electric. The effluvia act like "material rods" attaching to nearby objects and holding on to them until the force is spent and the objects are released and fall way.

Although Gilbert understood that poles of a magnet repel other like poles while unlike poles attracted unlike poles, he was unaware that electric materials could also attract and repel each other. His electric effluvia only explained the attraction between electric materials. He did observe that objects fell off an electric once they had touched it, but he attributed that to a loss of the effluvia force.

Francis Bacon and Force Acting at a Distance

Another Englishman who took advantage of the relative safety of a reformed Elizabethan England at the turn of the seventeenth century was Francis Bacon, a natural philosopher and statesman who served for a period as Lord Chancellor of

England. A contemporary of Gilbert, Bacon would take the experimental method introduced by Gilbert to another level defining the process of scientific induction that became known as "Baconism." He is often considered the father of modern science. Paradoxically, Bacon, a staunch Anglican, believed in an Earth-centered Universe despite compelling evidence to the contrary.

Bacon entered Cambridge University at age 12 where his education was conducted largely in Latin and heavily focused on Aristotelian metaphysical philosophy. Although Bacon admired Aristotle, he felt that his methods were argumentative and arbitrary often leading to wrong conclusions. Like Gilbert, he saw a new method, the experimental method. Bacon felt that all knowledge is derived from inductive reasoning—a method of reasoning whereby a series of experimental observations are synthesized to come up with a general principle. *"Once these particulars have been gathered together, the interpretation of Nature proceeds by sorting them into a formal arrangement so that they may be presented to the understanding*" [18].

Bacon avidly read Gilbert's *De Magnete* and though he was impressed with his experimental observations he ridiculed Gilbert's explanations of magnetic force and gravity [19]. Unlike Gilbert, who believed that bodies could influence other bodies only through physical contact, Bacon had no problems with suggesting force acting at a distance. In Aphorism 45, Bacon wrote, *"Again, if there be any magnetic power which operates by consent between the Earth and heavy bodies, between the globe of the Moon and the seas (which seems very likely in the half-monthly tidal cycle), or between the stellar heavens and the planets by which the latter are summoned and drawn to their apogees, then all these operate at very long distances indeed"* [20]. Bacon pointed out that an obvious example of the magnetic force acting at a distance is magnetism deep within the Earth acting on a compass needle on the surface of the Earth.

Bacon felt that there was a definite link between the Earth's gravity and magnetism. The same magnetic force that draws iron to a magnet draws heavy bodies to the Earth. He noted that gravity and magnetism shared a common property—*"the closer heavy bodies get to the Earth, the more strongly and vigorously they are carried toward it, and the further away the more weakly and slowly (as happens in magnetic attractions)"* [21]. Magnetism not only explained attraction at a distance but also an attraction at small distances, such as when *"amber and jet attract straws; bubbles merge when they come together; certain purgative medicines draw humours down from above; and things of that kind"* [20].

Bacon considered magnetism the universal force that explained the movement of the heavenly bodies, the Earth's gravity, and electrical attraction. But what was this magnetic force? In his essay *Pan, or Nature,* Bacon described how the universal force might work. *"The body of Nature is most elegantly and truly represented as covered with hair; in allusion to the rays which all objects emit; for rays are like the hairs or bristles of nature; and there is scarcely anything which is not more or less radiant. This is very plainly seen in the power of vision, and not less so in all kinds of magnetic virtue, and in every effect which takes place at a distance. For whatever produces an effect at a distance may be truly said to emit rays. But Pan's hair is longest in the beard, because the rays of the celestial bodies operate and penetrate from a greater distance than any other"* [22].

In this remarkably prescient paragraph, Bacon suggests that there is a unifying concept in nature that can explain forces acting from afar and near and that these forces result from rays emitted by objects. Examples of the rays are light (vision) and magnetism. It would be many centuries before light rays were shown to be electromagnetic waves and scientists are still searching for a unifying concept of gravity and electromagnetism.

It is important to recognize that Bacon's idea of force acting at a distance was a paradigm shift in how scientists viewed the Universe. From the ancient philosophers through Gilbert, there was a consensus that physical bodies could influence other physical bodies only through direct contact. Of course, this is why magnets were assumed to have animate souls. There was a mystical quality to Bacon's explanation for the force acting at a distance, but he did suggest a possible mechanism—rays emitted by physical bodies.

Magnetic Cosmology

Although Gilbert and Bacon were the first to suggest that magnetism could explain the movement of the heavenly bodies, they each had critical flaws in their basic premises. Gilbert believed that magnetic bodies had a soul to explain their movement, whereas Bacon believed in an Earth-centered universe with all heavenly bodies circling the Earth. These religious beliefs dominated their scientific endeavors. A contemporary of Gilbert and Bacon, the German mathematician and astronomer, Johannes Kepler, also struggled with conflicts between religion and science but he reconciled the conflict by proposing that God created the world according to an intelligible plan and that plan should be accessible to man. Despite this arrangement, religious zealots hounded him through much of his scientific career.

After extensive measurements of planetary movement, Kepler confirmed the Copernican heliocentric model and concluded that the planets moved in an elliptical orbit rather than a circular orbit around the sun. He completely rejected the animistic model of Gilbert. Since the pull on the planets grew weaker the further they were from the sun, Kepler concluded that there must be a "corporeal" force between the planets and the sun acting at a distance [23].

Kepler speculated that the sun generated "rods" of magnetic force as it rotated on its axis and in turn these rods of magnetic force swept the planets around the sun. To explain the elliptical orbits of the planets, he proposed that the magnetic force of the sun alternately attracted and repelled the planets based on the alignment of the planetary poles with the sun's monopole. Unlike Bacon, Kepler made no attempt to explain the nature of the magnetic force acting at a distance but rather described the universe in spiritual terms with the sun corresponding to the Father, other heavenly bodies to the Son and the space in between to the Holy Spirit.

Christianity and Medieval Science

The early Christian church had a paradoxical relationship with the teachings of the ancient Greek philosophers Plato and Aristotle. Many early Christians followed the stoic philosophy as taught by Plato and Aristotle but there were fundamental aspects of stoicism that were incompatible with Christianity. Like the Christians, Stoics believed that material wealth and status were of no importance and that virtue was the key to happiness. But unlike the Christian God who had human qualities and cared for the human condition, the stoic God was an all-powerful and pervasive force in the Universe, equivalent to what we call "nature." The stoic God had little interest in the human condition possibly even unaware of it.

Although Plato influenced the early Christian church, it wasn't until the twelfth and thirteenth centuries that Aristotle became such an influential figure in Christian doctrine. At that time, his works on metaphysics and ethics were rediscovered and translated into Latin and thereby became available to medieval Christian scholars. Saint Thomas Aquinas, often considered the greatest medieval Christian thinker, played a key role in incorporating Aristotle's teachings into Christian theology. Aquinas was particularly influenced by Aristotle's notion that knowledge is not innate but rather is obtained through one's senses and from logical inference of self-evident truths.

Through Aquinas, Aristotle's teachings became deeply embedded in Christian theology. To doubt Aristotle was considered heretical. A good example is Aristotle's belief in an Earth-centered model of the Universe. Despite compelling evidence against the model that developed in the fifteenth to seventeenth centuries, the Christian hierarchy would have none of it.

The Jesuits, a Catholic order founded in 1534, nicely illustrates the paradoxical relationship of the Roman Catholic Church and science. The Jesuits were accepting of the scientific experimental method for understanding the workings of nature, but they rigidly upheld the belief of an immobile Earth at the center of the Universe.

Niccolo Cabeo was a Jesuit monk who read Gilbert's *De Magnete* with great interest and published his own experiments and critique entitled *Philosophia Magnetica* in 1629. Cabeo estimated that Gilbert's lodestone model of the Earth was about 10 cm in diameter and that the elevations of the mountains and depressions at the base of the oceans would be no more than a tenth of a millimeter if constructed to scale [24]. Using his own terrella (Little Earth) he showed that these tiny irregularities in the Earth's surface had little effect on a compass needle and thus could not explain declination. In his writing, however, Cabeo was more interested in defending the Earth-centered model of Creation than explaining why a compass needle might deviate from the true north.

Despite Gilbert's extensive experiments on static electricity, he described only the attractive forces and failed to mention repulsive forces. His effluvium theory accounted only for electrical attraction. Niccolo Cabeo, on the other hand, noted that when he rubbed amber, it would first attract the versorium needle and bits of paper but if they touched the amber, they would suddenly be repelled and fly away from the amber [25]. The amber could have both attractive and repulsive qualities. It would be more than a century before the American, Benjamin Franklin, described the positive and negative charges of static electricity.

Rene Descartes and the Dual Universe Solution

The French mathematician and philosopher, Rene Descartes, came up with a clever solution to the science/God problem—two separate universes, a physical universe and a spiritual universe. There was the nonthinking universe that followed straightforward physical laws and the thinking universe occupied by God and humans that was understandable only by God. This construct allowed him to speculate on the laws of the physical universe hopefully without offending the church censors.

Descartes, a mathematician and philosopher like the ancient philosophers Plato and Aristotle, relied on logic rather than experimentation to understand the physical universe. His use of logic is illustrated by the famous statement known as his most certain proposition—"*I think, therefore I am.*" Just as Gilbert and Bacon represented the new scientific approach, Descartes represented the old philosophical approach. Descartes' view of the physical universe was a version of the atom theory of Epicurus and Lucretius discussed earlier in the chapter. But Descartes believed that there was no empty space; the space between masses was filled with tiny atoms like a gas, that he called ether [aether in Latin]. Movement of the ether resulted in movement of masses present within the ether.

Descartes explained the orbit of the Earth and other planets around the sun as due to the Earth and planets being "pushed" by ether as it moved about the sun in a large vortex [26, 27]. The vortex of ether could move some matter to its outside while other matter could move toward its center explaining the Earth's gravity. All bodies must occupy some space and no two bodies could occupy the same space. However, a large body could contain spaces that smaller bodies or thin fluids could enter and pass through.

Recall that Lucretius explained the pull of iron toward a magnet by postulating that the magnet emitted streams of atoms that bombarded and displaced atoms in the space between the magnet and the iron. This produced a void and the iron slid into the void toward the magnet. There was no possibility of a void in Descartes' ether model, so he proposed an elaborate scheme of left- and right-handed corkscrew-shaped atoms that moved in the ether about a magnet and worked like a corkscrew wine bottle opener. These atoms entered tiny channels in the magnet and the iron attracted or repelled the iron from the magnet depending on the direction of the corkscrews.

Descartes' drawing of the corkscrew particles moving about a large sphere of lodestone (comparable to Gilbert's terrella) in his *Principia Philosophiae* published in 1644 is often described as the first illustration of the concept of a magnetic field (Fig. 1.3). On superficial inspection, this may seem the case, but on a closer look, there are major differences between Descartes' illustration and our current concept

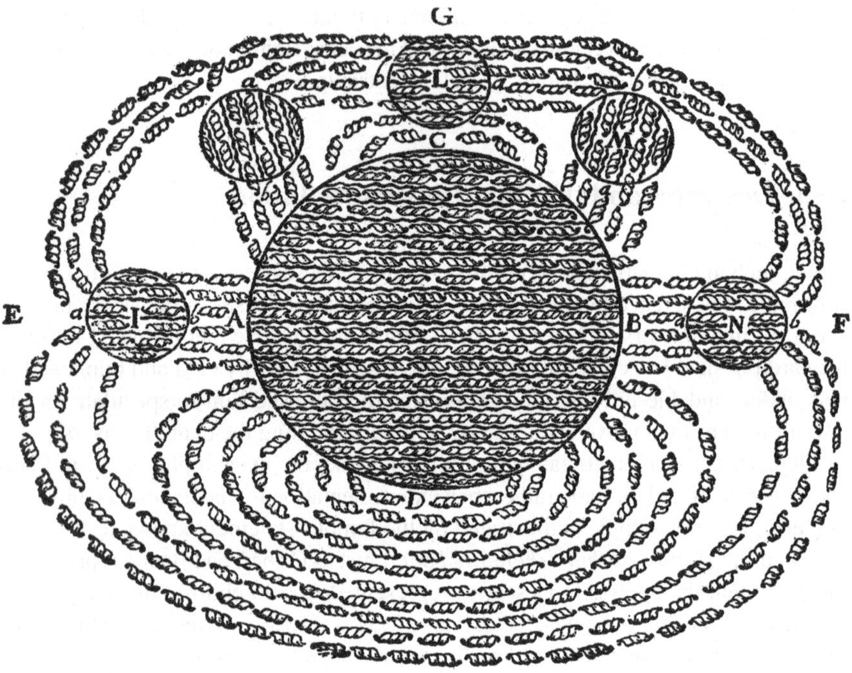

Fig. 1.3 Descartes theory of magnetism published in 1644. Left and right-handed corkscrew-shaped atoms moved in the ether around a magnet. The corkscrew atoms worked like a wine bottle opener pulling objects toward the magnet. (From Descartes R. *Principia Philosophiae*, 1644, p. 271, free Google eBook)

of a magnetic field. For one, the scheme requires that the channels in the lodestone and iron in the "magnetic field" be exactly aligned so that the corkscrew atoms can flow between the two structures. This is obviously not the case with a magnetic field. Descartes did not even attempt to explain the orientation of a compass needle with the magnetic field.

The physical universe as described by Descartes was a mechanical clockwork universe composed of matter with no empty space and no energy outside of the kinetic energy of matter in motion. On the other hand, his spiritual universe composed of God and human souls was immaterial and unable to be described in physical terms. The human brain was at a key crossroads in his dual model since it housed the machinery for control of the automatic workings of the body and the soul for thinking and self-awareness like God.

Descartes considered nerves to be hollow tubes that were connected to the fluid-filled compartments in the brain called ventricles. For reflex behavior, such as withdrawing from a painful stimulus, he postulated that thin filaments within each nerve tube controlled tiny valves in the ventricles of the brain that controlled the flow of animal spirits into the nerves. Painful pressure or heat against the skin would move the filaments (like pulling on a rope to ring a bell), open the valves, and release animal spirits from the ventricles into the nerve causing a reflex muscle contraction that pulled the extremity away from the painful stimulus [28].

For voluntary behavior, Descartes proposed that the rational immortal soul, which he localized in the pineal gland, perceived sensations such as pain and pleasure and controlled the flow of spirits by deciding when to open and close the tiny valves within the ventricle. An analogy was a spider sitting at the center of his web aware of any slight movement of a fiber in the web. Descartes picked the vestigial pineal gland because it was an unpaired structure (there was only one immortal soul) and was located between the lateral ventricles.

Descartes' theory became very popular primarily for its simple mechanical explanation of reflex behavior but there were unintended consequences. It raised an interesting question—Do animals feel pain? A heated debate occurred in the scientific and medical community. Since Catholic religious doctrine, faithfully followed by Descartes, taught that only humans have an immortal soul, it followed that non-human animals could not "feel" pain; they were automatons. They had animal spirits in the ventricles of the brain that controlled reflex behavior but lacked an immortal soul to sense the emotion of pain and pleasure. This may seem foolish since everyone has seen animals grimace and moan after injury suggesting that they are in pain. Descartes dismissed these reactions to injury as automatic reflexes that did not indicate that the animal "feels" pain. The result was that he and his many followers conducted experiments on animals without concern for sedation.

Descartes the "Physician"

Both of Descartes' paternal and maternal grandparents were physicians and he almost certainly spent some time studying medicine even though he did not officially practice medicine. He did spend much of his life providing medical advice to his friends and he proudly proclaimed that he acted as his own physician. "*I share the opinion of Tiberius, who was inclined to think that everyone over thirty had enough experience of what was harmful or beneficial to be his own doctor. Indeed, it seems to me that anybody who has any intelligence, and who is willing to pay a little attention to his health, can better observe what is beneficial to it than the most learned doctors*" [29]. Considering the medical knowledge and practice of medicine at the time, this statement was probably not far from the truth.

Descartes wrote on numerous occasions that the "preservation of health" was the goal of his studies. He resolved to devote his life *"to nothing other than trying to acquire some knowledge of nature from which we may derive rules of medicine which are more reliable than those we have had up till now"* [30]. He boasted that he spent more time dissecting human bodies than any physician, yet his understanding of human anatomy was as flawed as that of Galen. His recommendations for maintaining a healthy body and longevity were aligned with those of Hippocrates and Galen: Eat a healthy diet, exercise in moderation, and get adequate sleep. He warned against bloodletting, newfangled drugs, and extreme or bizarre treatments in general.

Psychosomatic symptoms were as common in the seventeenth century as they are in the twenty-first century. When Princess Elizabeth consulted Descartes for an upset stomach in 1644, he lectured her on the importance of the soul on the body: *"I know no thought more proper for preserving health than a strong conviction and firm belief that the architecture of our bodies is so thoroughly sound that when we are well we cannot easily fall ill except through extraordinary excess or infectious air or some other external"* [31]. On the other hand, he felt that if one expects to be ill there is a good chance they will become ill: *"people who are convinced by an astrologer or doctor that they must die at a certain time, and for this reason alone fall ill, and frequently even die"* [31]. We now call the benefits of positive expectation, the placebo effect, and the detriment of negative expectations, the nocebo effect. The nocebo effect plays an important role in generating most psychosomatic symptoms.

The prevention of aging and prolongation of life was a preoccupation with Descartes. *"It should not be doubted that human life could be prolonged if we knew the appropriate art. For since our knowledge of the appropriate art enables us to increase and prolong the life of plants and such like, why should it not be the same with man?"* [32]. It was common knowledge among the learned community of the seventeenth century that Descartes was working on ways to prolong life and there were rumors that he had already discovered the secret to prolonging life (the fountain of youth). Although Descartes modestly proclaimed that he could not promise to make man immortal he was sure it was possible to lengthen life by hundreds of years.

Consider the shock to his friends and colleagues when the news broke that Descartes had died at age 54 of pneumonia while in Sweden. He acted as his own physician right up to the end using tobacco-flavored wine to bring up the phlegm. He initially resisted being bled but near the end, he gave in, and he was bled a large quantity just before his death.

Robert Hooke and Isaac Newton

Two preeminent English scientists of the latter seventeenth century, Robert Hooke and Isaac Newton, were responsible for developing the universal principle of gravity, but they had differing opinions on the mechanism of planetary movement. Hooke was flamboyant and enjoyed the notoriety of his scientific reputation whereas Newton was somewhat of a recluse who despite his monumental discoveries was suspicious of fellow scientists and had only a few close associates.

A key observation that Hooke claimed to have communicated to Newton in 1769 was the inverse square rule of gravitational force—the magnitude of gravitational force decreased by the square of the distance between two bodies. The inverse square rule formed the basis of Newton's mathematical equations for planetary movements about the sun confirming Kepler's elliptical orbits. Newton later denied he received the information about the inverse square rule from Hooke suggesting it was common knowledge in the mid- seventeenth century [33].

Like many in his time, Hooke wondered why the planets moved around the sun since they were not tied to it by any visible attachments. Why did the planets take a curved path rather than a straight line like other bodies set in motion by a push? He concluded that the sun must *"have an attractive power upon all the bodies of the planets, and of the earth that move about it, and that each of those again have a respect answerable, whereby they may be said to attract the Sun in the same manner as the Load-stone hath to Iron, and the Iron hath to the Load-stone"* [34]. It took almost another century before the inverse square rule was shown to apply to magnetic force as well as gravity.

Isaac Newton, on the other hand, had little to say about magnetism [35]. He recognized the similarities between gravity and magnetism but felt that they were two separate entities. The laws of universal gravitation formulated using his new calculus explained not only the motion of objects on Earth and the heavenly bodies but also the moon's influence on tides and the trajectory of comets. Like many of his contemporaries, he saw gravity as a force acting at a distance but had little to say about the nature of the force.

Newton had a lifelong interest in light, having performed a series of experiments showing how white light broke up into a spectrum of colors when passed through a prism. It was generally assumed that some property of the prism created the colors. Newton showed that it wasn't the prism that created colors but rather that white light was composed of a spectrum of rainbow colors and that the prism refracted each color to a slightly different degree, separating the spectrum of colors within white light. He proved this point by passing one of the colors that had been separated by a prism through a second prism and showing that it remained the same without any change.

What exactly was light? Newton initially conceived of a "ray" of light, a stream of "corpuscles" in motion. Hooke disagreed suggesting that light was a wave-like sound. Newton responded to his nemesis by pointing out that light does not turn corners like sound. Descartes considered light a pressure in the ether and that color resulted from a rotation of the ether particles. Humans perceive light and color when pressure impinges upon the eye.

But was light a particle or a wave? Newton grappled with this dilemma for years without coming to a clear conclusion. To address the problem, he turned to Descartes' concept of ether. He saw ether as like air but subtle and more strongly elastic. Just as sound was due to vibrations in the air, light rays resulted from high-frequency vibrations in the ether. Colors represented vibrations at slightly different frequencies. Not surprisingly, Hooke responded by claiming that Newton had stolen his wave theory of light without giving him proper credit. As we will see later, light is now considered to be both a wave and a particle.

Newton became interested in electricity late in life suggesting that it played a critical role in the operation of nature. In the second edition of his famous book, *Principia* he describes electricity as *"a certain most subtle spirit which pervades and lies hid in all gross bodies*" [27]. This mysterious "spirit" accounted for attraction and repulsion that occur at short distances between most objects when they are rubbed and between particles that are close together. Electricity could also emit, reflect, and refract light. He also speculated that electricity was responsible for carrying sensations to the brain in "the solid filaments of the nerves" [36] and in turn relay the commands from the brain to the muscles to initiate movement.

Edmond Halley and the Earth's Magnetic Field

Although Edmond Halley is best known for his description of a comet that orbited the sun with a period of 76 years, he spent much of his life studying the phenomena of declination in the compass needle and speculating on the mechanism of the Earth's magnetic field [1]. He was a strong proponent of Newton's theory of gravity and was one of a few scientists that Newton trusted and could work with. When Halley was just 16 years of age in 1672, he measured the magnetic declination in London and noted that the compass needle pointed slightly west of the north pole, not east as had been described earlier in the century. Was the Earth's magnetic field slowly shifting from east to west?

In 1676 when he was studying at Oxford, Halley was given the chance to join an astrological expedition in the southern seas to map the location of stars in the southern hemisphere. He took advantage of the expedition to make systematic measurements of declination throughout the voyage. When he returned, he was determined to explain the declination measurements based on a "general rule."

Halley devised the idea of four magnetic poles, two in the north and two in the south. All four poles influenced a compass needle, but the greatest influence was from the nearest pole. For example, he estimated that the European North Pole was on a Meridian that ran along the edge of England about 7 degrees from the North pole and the American North Pole on a Meridian passing through the center of California and about 15 degrees from the North pole. All four poles were gradually shifting to the west to varying degrees. Halley admitted that this was largely speculation since no one had been able to visit the polar regions because of the extreme weather conditions and measurements of declination were available for only a small part of the globe.

As for the origin of the Earth's magnetism, Halley suggested that the Earth's interior consisted of a solid inner sphere and an outer surface shell separated by some type of fluid. Both the inner sphere and the outer shell were magnetized and rotated about a central axis but the rate of rotation of the central sphere lagged the rotation of the shell because of the viscosity of the intervening fluid. He estimated that it took about 700 years for the inner sphere to fall one rotation behind the outer shell. This model predicted not only a gradual shift of declination of the compass needle but also a complete reversal every 350 years. This was the first attempt to explain the Earth's magnetic field based on the Earth's interior and although there are some glaring inaccuracies some aspects were remarkably prescient.

In 1698, Halley convinced King William III to award him a Captain's position in the British Admiralty and to commission a nautical geographical expedition to measure magnetic declination throughout the Atlantic Ocean particularly the uncharted south. His initial voyage on the *Paramour* in October 1698 was cut short due to disputes with his second in command but a second voyage in 1699 resulted in more than 200 declination measurements throughout the Atlantic Ocean. On return, Halley introduced the concept of contour lines to generate a map of declination values that could be easily understood. He used smoothed lines to join places with equal declination values, so-called Halleyan lines. In 1702, Halley generated a magnetic declination contour map of the entire world based on all the declination values that had been reported up to that time.

In Brief

To ancient civilizations, electricity and magnetism had mystical qualities suggesting some type of supernatural force. Not surprisingly, ancient physicians tried to harness this force to treat a variety of illnesses. In the mid-sixteenth century, William Gilbert made the first clear distinction between electricity and magnetism and his contemporary Roger Bacon proposed a general theory of magnetism as a force acting at a distance. Their work marked the beginning of modern scientific methodology. To deal with the conflict between religion and the new science, Rene Descartes proposed a two-universe solution, one that obeyed physical laws and the other that only God could understand. Descartes' theory on brain function led to the bizarre conclusion that animals could not feel pain because they lacked a soul localized in the pineal gland. Several early scientists including Roger Bacon saw a parallel between magnetism and gravity although Isaac Newton saw them as two different forces acting at a distance. Edmond Halley was the first to suggest that the Earth's magnetism came from forces deep below the Earth's surface.

References

1. Turner G. North pole, south pole. The Experiment: New York; 2010.
2. Merrill RT, McElhinny MW. The Earth's magnetic field: its history, origin and planetary perspective. 2nd ed. ed. San Francisco: Academic Press; 1983.
3. Turner G. North pole, south pole. New York: The Experiment; 2010. p. 13.
4. Park B. A history of electricity. New York: Wiley; 1898.
5. Fowler M. Historical beginnings of theories of electricity and magnetism. http://galileoandeinstein.physics.virginia.edu/more_stuff/E&M_Hist.html.
6. Lucretius. On the nature of things. Trans. Rose WHD, rev. Smith MF. Cambridge MA: Harvard University Press; 1924, rev. 1975.
7. Greenblatt S. The swerve. New York: Norton and Company; 2011.
8. Keyser PT. The purpose of the Parthian galvanic cells: a first century A.D. electric battery used for analgesia. J Near East Stud. 1993;52:93–4.

9. Sparavigna AC. Petrus Peregrinus of Maricourt and the medieval magnetism. Mechanics. Mater Sci Eng. 2016; ISSN 2412-5954, https://arxiv.org/pdf/1512.02634.pdf.
10. Turner G. North pole, south pole. New York: The Experiment; 2010. p. 14.
11. Peregrinus P. The letter of Petrus Peregrinus on the magnet, A.D. 1269, Trans by Brother Arnold, with introductory notice by Brother Potamian. New York: McGraw; 1904.
12. Galilæus GL. His system of the world, the third dialogue. Available on line at: http://www.chlt.org/sandbox/lhl/Salusbury/page.376.php?.
13. Häfeli U. The history of magnetism in medicine. In: Magnetism in Medicine. Berlin: Wiley-VCH; 2006. p. 1–25.
14. Paracelcus. Volumen Medicinae Paramirum. Translated from the original German by Leideker KF. Baltimore: Johns Hopkins Press; 1949.
15. Turner G. North pole, south pole. New York: The Experiment; 2010. p. 32–47.
16. Bordeau SP. Volts to Hertz – the rise of electricity. Minneapolis: Burgess Publishing Company; 1982.
17. Gilbert W. De Magnete. Trans Mottelay PF. New York: Courier Dover Publications; 1958.
18. Turner HS. Francis Bacon's common notion. J Early Mod Cult Stud. 2013;13:7–32.
19. Wang X. Francis Bacon and magnetical cosmology. Isis. 2016;107:707–21.
20. Wang X. Francis Bacon and magnetical cosmology. Isis. 2016;107:714.
21. Wang X. Francis Bacon and magnetical cosmology. Isis. 2016;107:713.
22. Wang X. Francis Bacon and magnetical cosmology. Isis. 2016;107:715.
23. Kepler J. The secret of the universe. Trans. Duncan AM. New York: Abaris; 1981.
24. Turner G. North pole, south pole. New York: The Experiment; 2010. p. 46–7.
25. Turner G. North pole, south pole. New York: The Experiment; 2010. p. 65.
26. Rene Descartes – mechanical push universe theory. New-science-theory.com, 2022.
27. Spratt J-S. The Descartes-Newton paradox: clashing theories of planetary motion at the turn of the eighteenth century. Senior Capstone Projects. Paper 609, 2016.
28. Zimmer C. Soul made flesh: the discovery of the brain – and how it changed the world. New York: Free Press; 2004.
29. Shapin S. Descartes the doctor: rationalism and its therapies. Br J Hist Sci. 2000;33:139.
30. Shapin S. Descartes the doctor: rationalism and its therapies. Br J Hist Sci. 2000;33:136.
31. Shapin S. Descartes the doctor: rationalism and its therapies. Br J Hist Sci. 2000;33:146.
32. Shapin S. Descartes the doctor: rationalism and its therapies. Br J Hist Sci. 2000;33:137.
33. Bennett JA. Cosmology and magnetical philosophy, 1640-1680. JHA. 1981;12:165–77.
34. Wang X. Francis Bacon and magnetical cosmology. Isis. 2016;107:717.
35. Gleick J. Isaac Newton. New York: Pantheon Books; 2003.
36. Home RW. Newton on electicity and the aether. In: Bechler Z, editor. Contemporary Newtonian research. New York: D Reidel Publishing Co.; 1982. p. 191.

Chapter 2
The Early Age of Electrotherapy

> *As soon as any of the Thunder Clouds come over the Kite, the pointed Wire will draw the Electric Fire from them, and the Kite, with all the Twine, will be electrified, and the loose Filaments of the Twine will stand out every Way, and be attracted by an approaching Finger.*
>
> Benjamin Franklin, 1752

Abstract The early eighteenth century marked the beginning of the age of electricity. Francis Hauksbee built the first efficient static electricity machine and Stephan Gray demonstrated that electricity could be conducted over a distance. Gray's "flying boy" experiments started a trend of electrifying human subjects for public entertainment that ultimately led to early electrotherapy. The Leyden jar provided the first simple way to store electricity for both physical and medical experiments. Ben Franklin was a pivotal character in the history of electricity and neuroscience. His kite experiment proved that lightening was static electricity and his experiments on human subjects showed that electricity applied to the head caused loss of consciousness and retrograde amnesia. Early efforts at electrotherapy were promising although poorly documented and ripe with quackery. The dangers of electricity to the human body were appreciated for the first time.

Keywords Static electricity · Electric fluid · Capacitor · Lightning rod · Hysterical paralysis

By the end of the seventeenth century, many of the physical properties of static electricity had been described but there was little understanding of exactly what it represented. Gilbert had clearly separated electricity from magnetism but did not recognize that there could be attraction and repulsion between electrified objects. He demonstrated that static electricity was not unique to amber but could be produced by rubbing a wide variety of substances and he devised a primitive electrometer that allowed him to separate materials into *electric* and *nonelectric*. The Italian Jesuit, Niccolo Cabeo, noted that electrified objects could attract and repel each other like the poles of a magnet questioning Gilbert's "effluvia theory" that only explained electrical attraction.

R. W. Baloh, *Brain Electricity*, https://doi.org/10.1007/978-3-031-62994-5_2

Otto von Guericke and His Electricity Generator

Hand-rubbing objects to produce static electricity was largely restricted to party tricks that awed the audience. To study and understand static electricity, a better method of generating it was needed. In 1663, the German inventor and politician, Otto von Guericke, produced an apparatus for generating static electricity that provided the backbone for the research and medical uses that developed in the eighteenth century [1, 2].

Guericke, who attended the finest universities in Germany and the Netherlands, is best known for his invention of the vacuum pump in 1650. His pump consisted of a piston in an air gun cylinder along with two-way flaps that could pull air out of a vessel producing a partial vacuum. The notion that *nature abhors a vacuum* was engrained in ancient philosophy and strongly endorsed by Descartes, and no one had produced a vacuum. Guericke demonstrated the remarkable force of air pressure with his experiments using the vacuum pump.

In his most famous experiment conducted in 1657, Guericke made two 20-inch metal hemispheres, joined them together, and sucked the air out of the sphere using his vacuum pump. The air pressure outside of the sphere held the two halves together so that 16 horses, eight harnessed on each side, could not pull the hemispheres apart.

Guericke's electric generator consisted of a large sulfur globe at the end of a long shaft mounted on a wooden platform (Fig. 2.1). He heated sulfur powder inside a glass globe and the glass was later broken and removed leaving the sulfur globe. When the globe was rubbed with the hand or a piece of cloth, it generated sparks of static electricity. The shaft could be turned by hand (no crank) and the globe could be removed from the platform and used as a source of stored static electricity for experiments. With his apparatus, Guericke, like Cabeo, noted that his electrified globe could both attract and repel nearby small objects sometimes first attracting them and then repelling them.

With his electric generator, Guericke made a remarkable observation. When he removed a fully charged globe, the ability to attract and repulse could be passed along a several foot long strip of linen touching the globe. He had demonstrated electrical conduction although he did not fully appreciate the significance of the observation. He did point out that the finding was not compatible with the effluvia hypothesis for electrical attraction proposed by Gilbert.

St. Elmo's Fire

St Elmo's fire is luminous glow from a rod-like object such as the mast of a ship or a church spire that occurs during a lightning storm. It is often accompanied by a hissing or buzzing sound. Named after the patron saint of sailors, St. Erasmus of Formia (also known as St Elmo), sailors were in awe of the phenomenon, and many considered it a good omen. In their epic voyages, both Columbus and Magellan took

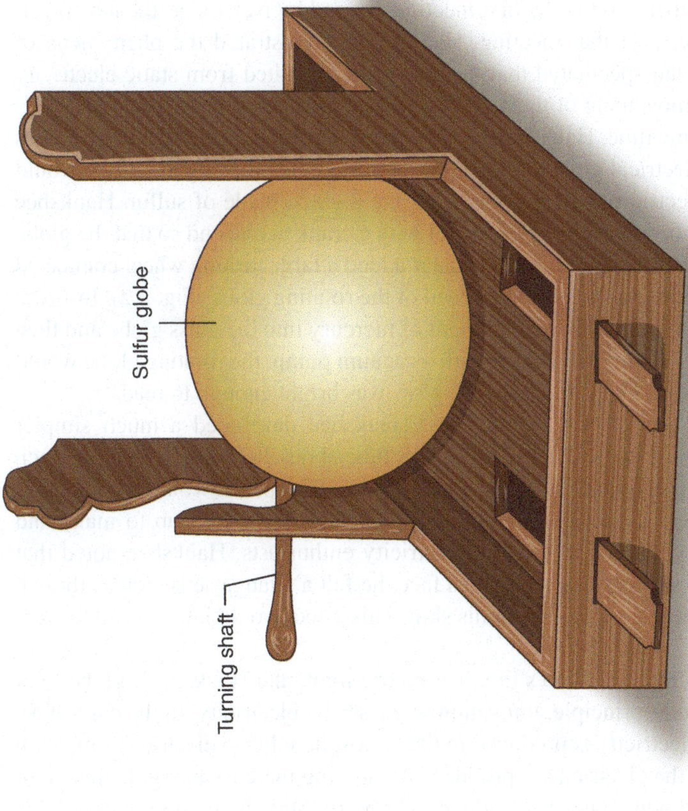

Fig. 2.1 Otto Von Guericke's static electricity generator. A large sulfur globe mounted on a platform with a shaft could be hand turned and removed from the platform for experiments

advantage of the sighting of St. Elmo's fire in the masts of their ships to lift the sagging spirits of their crews [3].

The French astronomer, Jean Picard, first described barometric light in 1675. Picard noticed a luminous glow of a mercury barometer tube after he shook it. What caused the glow was a mystery. In the early eighteenth century, Francis Hauksbee showed that a similar glow could be produced in a glass tube electrified by rubbing it with a cloth [4]. He called the effect "phosphorescence."

Hauksbee, the son of a draper, apprenticed as a draper and ran a draper shop in London before transitioning to making air pumps and pneumatic engines. His talent for making instruments led to his becoming an assistant in Isaac Newton's laboratory and a member of the Royal Society. He made his first presentation at the Royal Society in December 1703 at the first meeting chaired by Newton as the new president of the Society. At that meeting, Hauksbee demonstrated the phenomena of barometric light and speculated that the luminosity resulted from static electricity produced by the movement of the mercury against the glass tube.

Around the same time, Hauksbee began experimenting with electricity machines inspired by the electrical globe of von Guericke. He noted that a glass globe would generate static electricity when rubbed just like a globe made of sulfur. Hauksbee placed a glass globe at the center of an axil with a crank at one end so that the globe could be easily turned on its mount. He later added a large turning wheel connected to the axle by a pulley to increase the speed of the rotating globe (Fig. 2.2). In 1705, he found that if he injected a small amount of mercury into the glass globe and then created a partial vacuum in the globe with a vacuum pump, the rotating globe would glow when he placed his hand on it. The glow was bright enough to read.

In addition to his electrical machines, Hauksbee developed a much simpler device for generating static electricity, a glass tube about 30 inches long and 1 inch in diameter. When he vigorously rubbed the tube with paper, it would attract small pieces of brass leaf as far away as a foot. These tubes were cheap to make and became extremely popular with early electricity enthusiasts. Hauksbee noted that when he brought a charged tube near his face, he felt a strange sensation as though fine hairs were being pushed against his skin. This effect would later be called *electric wind.*

We now know that St. Elmo's fire, barometric light, and Hauksbee's globe light all work on a similar principle, ionization of gasses by electricity. In the case of St. Elmo's fire, the electricity is produced in the atmosphere by an electric storm. With barometric light, the electricity is produced by moving mercury along the inside of the glass tube and with Hauksbee's globe light by rubbing the rotating glass globe. In each case, electrons are released into the surrounding air (electric wind), where they collide with gas molecules causing the release of photons of light, the wavelength depending on the gas molecules present, oxygen, nitrogen, helium, etc. This is the same principle underlying modern gas-discharge lamps such as neon and mercury vapor lamps.

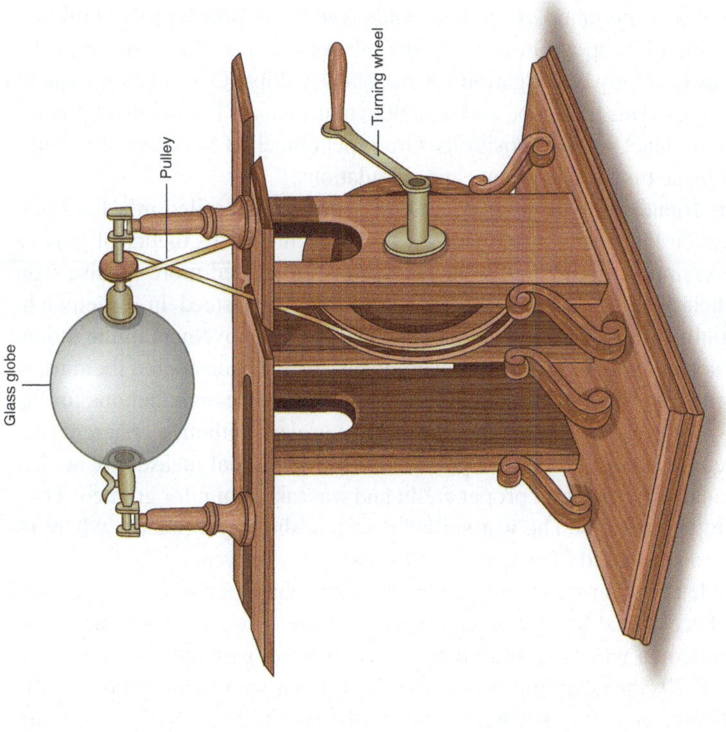

Fig. 2.2 Francis Hauksbee's static electricity generator. Glass globe at the center of the axil was connected to a large turning wheel by a pulley to increase the speed of rotation

Stephen Gray and His "Flying Boy"

The story of Stephen Gray and his experiments with static electricity is remarkable considering his lifelong struggle with poverty [5, 6]. Gray was born into a poor family in Canterbury, England, in 1666 and after minimal schooling he apprenticed first to his father and then an older brother in the dyeing trade. He had a natural curiosity for science and spent much of his free time reading books on science obtained from friends who had access to the libraries of wealthy residents living in Canterbury (the only source of such books at the time).

Gray's fortunes improved when he was introduced to John Flamsteed, a clergyman and self-taught astronomer who became England's first Astronomer Royal. In 1675, Flamsteed was appointed by the King and given the responsibility of building an observatory on a hilltop at Greenwich. His job was to chart the position of the heavenly bodies to improve navigation for the Navy's ships. Over the subsequent years, Flamsteed used his telescope and sextant to make tens of thousands of precise measurements of planet and star positions. Gray spent much of his spare time helping Flamsteed make his measurements and calculations.

Through his friendship with Flamsteed, Gray became embroiled in a feud that developed between Flamsteed and Isaac Newton. Working on his theory of gravity for a revised version of his *Principia* in 1694, Newton needed more precise data regarding planetary positions, so he decided to visit Flamsteed in Greenwich. Flamsteed initially provided him with measurements on the movement of the moon, but he was reluctant to provide other measurements since he considered them to be his personal property. Newton argued that Flamsteed would become famous for his measurements and that the measurements would be useless without a coherent theory to give them meaning. Flamsteed provided some additional measurements but felt that Newton did not give him proper credit and was taking him for granted. They soon became bitter enemies. The war spilled over into the Royal Society where as president, Newton excluded Flamsteed and his associates for years.

Gray was able to obtain some part-time work with John Desaguliers, a scientific consultant, and occasional Royal Society's demonstrator, who ran a boarding house for visiting gentlemen with scientific interests. He was not paid but received room and board. In 1720, the boarding house was torn down to provide room for the Westminster Bridge and Gray became homeless. His friend Flamsteed helped him obtain a pensioned position at the London Charterhouse. It was there that Gray performed his famous experiments and recruited a charity boy from the Charterhouse school for his initial "flying boy" experiments.

The London Charthouse is a historical complex of buildings that dates back originally to the fourteenth century but was bought in 1611 by Thomas Sutton and converted into an almshouse for older men and a school for young boys. Sutton, considered the wealthiest commoner in England, made his money when coal was discovered on land he had leased in Newcastle. In his will, which was contested but finally upheld in court, he established a home for 80 poor male pensioners and a boarding school for 40 boys. Just the almshouse remains open to this day.

While at the Charterhouse, Gray became interested in static electricity and mostly used Hauksbee's glass tube to generate electricity. On one occasion, he noted that a large cork he had placed in the end of the tube to keep out dust and dirt attracted small pieces of paper after he rubbed the tube just like the tube itself. The humidity had to be just right for this to work. This seemed odd so he stuck a small strip of fur in the end of the cork and observed that small pieces of paper were attracted to the end of the strip. He could even attach an ivory ball to the end of the strip and the ball attracted the pieces of paper. He then tried longer strips and finally a long piece of oily hemp twine attached to an ivory ball and the paper still was attracted to the ball. Gray realized that the "electric virtue" was not a fixed phenomenon but rather *fluid-like* and could *flow* over long distances.

Gray had no idea what was being conducted along his hemp twine but gradually began experimenting with other potential conductors. He found that if he suspended the electrified twine by silk threads the electricity would flow for long distances. On the other hand, if he suspended the twine by thin metal wires the electricity was rapidly dissipated. Based on these observations he divided materials into two types, those that held an electric charge such as glass, silk, and resin and those that dissipated an electric charge such as water and metals. He also observed that he could transfer electrification from one body to another nearby body even though there was no physical contact, later called electrification by induction.

Many of Gray's experiments on electrical conduction over a distance were performed at the manor house of the Reverend Granville Wheler, a wealthy relative of Flamsteed and a member of the Royal Society. The two men dropped electrified hemp twine from the house tower to the garden and across a paddock some 800 feet using garden stakes with short pieces of silk to keep the twine from touching the ground. If the twine touched the house or the ground the electric charge rapidly dissipated. Wheler discussed these experiments with his friends at the Royal Society and Gray reported the results in a letter to Desaguliers. It was Desaguiliers who first used the term "conductor" for materials such as metals that rapidly dissipated an electric charge and the term "insulator" for materials such as silk that maintained an electric charge.

Gray's experiments on electrical conduction extended to all sorts of material including live animals and even human subjects. His "flying boy" experiments became famous throughout Europe providing a small source of income to help support his scientific endeavors. He initially suspended a charity boy from the Charter school with silk cords attached to a horizontal bar supported by two vertical bars anchored to the floor (Fig. 2.3). When he rubbed his glass tube to generate an electric charge and placed it near the boy's feet without touching them, he noted that the boy's extended hands would attract small pieces of paper just like the electrified tube.

As word of his "flying boy" experiments spread through the public, Gray gradually expanded his demonstrations to scientific societies and private homes. He became a showman who learned to build tension and excitement in the viewing audience. Instead of his glass tube to electrify the boy, he used an electrical machine such as the one invented by Hauksbee so he could build up an even greater charge.

Fig. 2.3 Stephen Gray's "flying boy." Boy suspended with silk cords fastened to a horizontal bar attached to two vertical bars anchored to the floor. When Gray rubbed his glass tube to generate static electricity and placed it near the boy's legs, the boy attracted small pieces of paper or could turn pages of a book without touching them

He might use a child from a member of the scientific society, from the house of the exhibition or from the street anxious for a few coins. A favorite trick was to hold a book near the boy's hand and have him turn the pages without touching them.

Gray liked to get the audience involved in the demonstration selecting a few members to participate in the show. A chosen person was asked to stand on an insulating stand and hold the boy's hand. Despite the lack of any sensation, the participant acquired the boy's ability to turn the pages of a book without touching them. To add to the drama, Gray had a person from the audience stand on the floor next to the boy and after increasing the electric charge in the boy he dimmed the lights and asked the person to touch the boy's nose. As the audience member's finger neared the boy's nose, there was a loud crackling sound and a spark of lightning jumped from the boy's nose to the finger. Except for an occasional minor shock, the flying boys suffered no apparent ill effects.

By the time Gray performed his key experiments on electrical conduction, Isaac Newton had died and leadership in the Royal Society had changed hands. But Gray was too poor to pay dues, so he was not a member of the Society and electricity was not considered that important of a topic. In addition, the Society's proceedings were not published for a few years due to limited finances. Most members were familiar with Gray's findings; however, many attended Desagulier's routine demonstrations at the Society. The Royal Society did finally recognize Gray's scientific contributions with the first Copley Medal in 1731 for his work on electrical conduction and the second Copley Medal in 1732 for his work on electrical induction. As in life, Gray died in poverty in 1736. Although there is no marking, he is probably buried in a common grave in the area reserved for pauper pensioners from the London Charthouse.

Grays discovery of electrical conduction was an entirely new concept of electricity, a kind of "fluid" that could flow along an oily twine or metal wire. Rubbing an object with a cloth somehow caused the fluid to transfer from the cloth to the object. Although others had previously noted that electrified objects could both attract and repel each other, it wasn't until 1733 that the French chemist, Charles Francois du Fay studied the phenomena of electrical attraction and repulsion in detail. Du Fay made the simple observation that chaff that was attracted to an electrified amber rod was repelled by an electrified glass rod and vice versa. He made the apparent logical conclusion that there are two types of electrical fluid, resinous (amber-like) and vitreous (glass-like). Normally, objects had roughly equal amounts of the two fluids so that they were uncharged. When an amber rod was rubbed with fur, it gained an excess of resinous fluid, whereas when a glass rod was rubbed with silk, it gained an excess of vitreous fluid. Two bodies with excessive amounts of the same fluid repulsed each other while bodies with different fluids attracted each other.

DuFay tested a wide variety of substances and observed that all materials except metals and liquids could be electrified by rubbing. The problem was the electric charge would dissipate over time and the release could not be controlled. As machines for generating static electricity became larger and more efficient, the need for storage and controlled release became even more apparent.

Georg Matthias Bose and the Prime Conductor

Georg Matthias Bose was a professor of natural philosophy at the University of Wittenberg, Germany, where he conducted his famous experiments with electricity from 1742 to 1745 [6, 7]. He made a glass globe machine like Francis Hauksbee's machine and used it to conduct experiments and provide dramatic demonstrations for fellow scientists and the public. In the process, he discovered what he called "a prime conductor" that stored electrical charge generated by the machine.

Bose came up with the idea of his prime conductor when he had an assistant hold a metal bar in one hand while touching the rotating globe of his machine with the other hand. If the assistant stood on a block of resin to insulate him from the floor, the metal bar built up a charge over time. The electricity passed through the assistant's body and accumulated on the metal bar. This was surprising since at the time it was assumed that only insulators such as glass and resins could store static electricity. Conductors such as metals and water were known to rapidly dissipate a charge. Bose showed that a properly insulated metal object could store static electricity.

Bose is probably best known for his public demonstrations of the mystery and glamor of electricity. He was a true showman. His "electric kiss," an extension of

Gray's flying boy demonstrations, was immensely popular. Bose would select an attractive young woman from the audience and after she was properly charged with electricity while standing on an insulated platform, he would invite a young man to give her a kiss. Sparks would fly as the couple received a brief shock.

His most talked about exhibition became known as "beatification." This was an extension of his prime conductor experiments. He had an assistant dressed in a suit of armor with many sharp steel edges and spikes along with a metal helmet. When the assistant sat on an insulated chair and received an electric charge from the friction machine, his metal armor would spark and glow in the dark. The electrified helmet would shoot out plasma rays of light like those around the head of a canonized saint, thus the term beatification. The glowing light was another example of St. Elmo's fire. A colleague was so impressed that he published details of the demonstration in the Royal Society's *Philosophical Transactions.*

Another crowd favorite was Bose's fire-on-water demonstration. He covered a container of water with a thin film of alcohol and set the alcohol ablaze by transferring an electric charge from the friction machine to the water. Since ancient times, water and fire were thought to be opposites so the spectacle of burning water made a lasting impression on the audience. On the practical side, Bose demonstrated how electricity increased water flow in a metal siphon. Normally the flow would drop off after a time but if the siphon was electrified the water stream markedly increased. This observation set the stage for the later development of electrostatic spraying.

The Leyden Jar

As word of Bose's work on electricity began to spread, a German Lutheran Cleric and lawyer, Ewald Georg von Kleist performed a series of experiments that would ultimately lead to the development of a compact storage device for static electricity [8]. Von Kleist had studied at the University of Leyden in Holland where he developed an interest in electricity before later becoming dean of a cathedral in Pomerania, at the time part of the Kingdom of Prussia. His first invention was an efficient prime conductor based on Bose's work. Von Kleist suspended a large metal pipe on silk threads attached to the ceiling so that the pipe almost did not touch the globe of his friction machine. Although cumbersome, this turned out to be an excellent way to store an electric charge for further experiments with electricity.

Like many of his colleagues, von Kleist was impressed by Bose's fire-on-water experiment. He wondered whether electricity could be stored in the liquid of a glass container. He chose a medicine bottle filled with alcohol assuming the glass would act as an insulator and protect him from any electric shock. He fitted the bottle stem with a cork and passed a nail through the cork and into the alcohol. He then attached the outer end of the nail to his prime conductor pipe to charge the liquid while he held the bottle in one hand. Initially, he observed no effect but when he carried the bottle into a darkened room, he noticed that the bottle was glowing suggesting that the alcohol contained an electric charge, producing St. Elmo's fire. He then made a critical mistake. He touched the nail with his free hand. This caused a violent shock, throwing him across the room.

Although he clearly discovered a method of storing electricity in a bottle, von Kleist did not completely understand how the system worked. Making matters worse, when he attempted to repeat the experiment, it didn't work. Of course, on repeating the experiment he was careful to avoid holding the bottle in his hand while it was being charged for fear of another shock. He told colleagues about his remarkable experiment, but they also had difficulty repeating it. Like him, they were wary of holding the bottle after hearing of his violent shock. As it turns out, that was the critical step that was missing in attempts to repeat the experiment.

Based on his prior study at the University of Leyden, von Kleist maintained regular communication with colleagues at the university particularly a fellow lawyer, Andreas Cunnaeus, who happened to be working as an assistant to Physicist, Pieter Musschenbroek [9]. These two decided to figure out why von Kleist was having so much difficulty repeating his experiment. Instead of alcohol, they used water in a glass jar. After a series of attempts to store electricity in the water including suffering several shocks, they discovered the critical importance of an outside conductor on the jar (von Kleist's hand in his experiment). Without it, electrostatic back pressure developed markedly limiting the ability of the jar to store electric charge. They soon discovered that all that was required was an insulator such as glass with a conductor such as metal or water on each side of the insulator. The simplest device was a glass jar coated on the inside and outside with metal foil (Fig. 2.4). With this device, they could store large electrical charges and generate enormous sparks. They had invented the first electrical capacitor which later came to be known as the "Leyden jar."

The Leyden jar provided a valuable new tool for electrical experimentation, as we will see later in the chapter, particularly in the field of medicine. It also made it much easier to demonstrate electricity to the public. Jean-Antoine Nollet, a French clergyman and physicist who worked with Charles du Fey, organized one of the most talked about demonstrations for the King, in Paris in 1750. He had a group of Carthusian monks hold hands forming a 900-foot circle. When he discharged a Leyden jar into one of the monks, all the monks received a brief shock causing them to jump into the air in unison.

Benjamin Franklin and His Kite Experiment

In the early eighteenth century, America was still a developing nation well behind the European community in its scientific endeavors. There were a few cities, however, such as Philadelphia and Boston where fledgling scientific societies were flourishing and overall, there was broad public interest in the sensational scientific discoveries coming out of Europe. In 1743, a young Philadelphia publisher, inventor and avid experimenter, Benjamin Franklin, attended a public demonstration on static electricity in Boston that was like the demonstrations conducted by Gray and

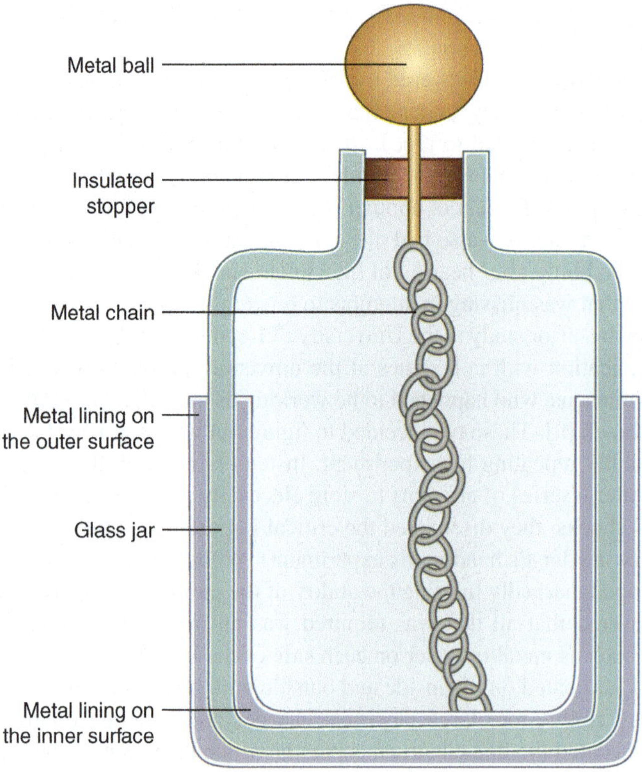

Fig. 2.4 Leyden jar. Glass jar coated on the inside and outside with metal foil. The inner foil was connected to a metal ball at the top by a metal chain and rod passed through an insulated stopper

Bose in Europe. Franklin was enthralled with the exhibit and decided to pursue his own experiments. He initially acquired an electrical rubbing tube from a colleague in London and later made his own electrical machine with a large wheel and an electrical rubbing glass globe.

One of Franklin's earliest discoveries was that "electrical fire" was drawn to sharp metal points [10]. This was not a new observation, but Franklin immediately saw the practical implications of the phenomenon. At the time, it was generally assumed that lightning was a giant electrical spark although for obvious reasons this was not so easy to prove. Franklin proposed that a sharp metal rod mounted vertically on a rooftop could dissipate the electrical charge of a lightning bolt harmlessly to the ground, a "lightning rod." Franklin first suggested an experiment to prove lightning was electricity in 1750. He recommended extending a sharp pointed metal conductor high into a cloud during a thunderstorm. If it contained electricity, the lightning rod would extract it and conduct it to the ground where it could be observed and collected. Since there were no structures in Philadelphia tall enough for the experiment at the time, Franklin was waiting for the steeple at Christ Church to be completed in 1754.

Franklin is probably best known for his famous "kite experiment" in which he is credited with demonstrating that lightning is electricity, although there is some controversy concerning how and when he performed the experiment. Franklin described how such an experiment might be performed in the *Pennsylvania Gazette* in October 1752. He began by describing how to make the kite with a pointed wire sticking out of the top to attract the electricity.

> To the End of the Twine, next the Hand, is to be tied a silk Ribbon, and where the Twine and the silk join, a Key may be fastened. This Kite is to be raised when a Thunder Gust appears to be coming on, and the Person who holds the String must stand within a Door, or Window, or under some Cover, so that the Silk Ribbon may not be wet; and Care must be taken that the Twine does not touch the Frame of the Door or Window. As soon as any of the Thunder Clouds come over the Kite, the pointed Wire will draw the Electric Fire from them, and the Kite, with all the Twine, will be electrified, and the loose Filaments of the Twine will stand out every Way, and be attracted by an approaching Finger. And when the Rain has wet the Kite and Twine, so that it can conduct the Electric Fire freely, you will find it stream out plentifully from the Key on the Approach of your Knuckle [11].

Franklin clearly implies that the kite experiment was successfully conducted in Philadelphia (thought to be in June 1752) but there are no surviving written documents in which Franklin describes the details of the actual experiment. In his book: *The History and Present State of Electricity, with Original Experiments,* published in London in 1767, Joseph Priestley best known for his discovery of oxygen, suggested that Franklin did not initially publicize his experiment because of concerns about ridicule should it be unsuccessful. He told no one except his son, William, who helped him raise the kite. Priestley provided the following description of the experiment that was presumably told to him by Franklin who was in London at the time Priestley wrote his book.

> The kite being raised, a considerable time elapsed before there was any appearance of its being electrified. One very promising cloud had passed over it without any effect; when, at length, just as he was beginning to despair of his contrivance, he observed some loose threads of the hempen string to stand erect, and to avoid one another, just as if they had been suspended on a common conductor. Struck with this promising appearance, he immediately presented his knucle to the key, and (let the reader judge of the exquisite pleasure he must have felt at that moment) the discovery was complete. He perceived a very evident electric spark. Others succeeded, even before the string was wet, so as to put the matter past all dispute, and when the rain had wet the string, he collected electric fire very copiously. This happened in June 1752, a month after the electricians in France had verified the same theory, but before he heard of any thing they had done [12].

Assuming Franklin performed the experiment in June of 1752 as Priestly reported, he was not the first to demonstrate that lightning was electricity. This honor goes to French physicist, Thomas-Francois Dalibard who followed Franklin's earlier suggestion to construct a tall lightning rod that could be extended into a low cloud during a thunderstorm [13]. In May 1752, Dalibard constructed a 40-foot-tall metal rod at Marly-la-Ville in France and placed it on an insulator. Dalibard had to return to Paris, so he arranged for a local man, Monsieur Coiffier, to do the experiment when a storm arose. After hearing thunder on May 10, Coiffier was able to generate sparks when he held an insulated brass wire near the lightning rod. The experiment became known as the "Philadelphia experiment" as news spread rapidly through Europe. Franklin later met and became friends with Dalibard. Of course, Monsieur Coiffier's heroics were largely forgotten.

One thing for sure is that Dalibard's metal rod and Franklin's kite were not struck by lightning. If they had been, Coiffier and Franklin would probably have been electrocuted. The rod and kite picked up electrical charge in the storm clouds.

The Fatal Experiment

Georg Wilhelm Richmann was a German physicist born in the Baltic city of Pernov in 1711. Pernov (now Pärnu in Estonia) became part of the Russian Empire after Peter the Great's Great Northern War. Richmann eventually gravitated to St Petersburg where he became director of physics at the Russian Academy of Sciences. With the invention of the Leyden jar in 1745, just about every physicist in Europe became interested in electricity. Richmann's work focused on developing an instrument to quantify electricity. Gilbert's versorium identified the presence of electricity but there still was no objective measure of the amount of electricity. It was known that when a piece of thread was attached to an electrified object at one end, the other end of the thread would flare away from the object because like electrical charges repel each other, the greater the charge, the greater the flare.

Richmann developed an instrument to measure the angle of the flare with a protractor-like structure behind the thread producing one of the first electrometers [14]. He published a description along with engravings of his instrument in the *Novi commentarii of the Academy in 1752.* Richmann was aware of Franklin's work on lightning rods and Dalibard's successful experiment with a lightning rod in France. He decided to repeat the experiment using his electrometer to measure the amount of atmospheric electric charge. He set up an insulated iron rod inside his house with the electrometer attached. He then connected a metal wire to the iron rod and ran the wire up to the roof with the end sticking upward into the sky.

On August 6, 1753, while he was at work at the Academy, Richmann observed a thunderstorm developing, so he rushed home with an engraver from the Academy to document the event [15]. As Richmann leaned over near the rod to monitor his electrometer, the engraver observed a ball of lightning jump from the iron bar to Richmann's head, knocking him unconscious. As the engraver ran for help, Mrs. Richmann rushed in and found her poor husband unresponsive. Despite heroic attempts to revive him including mouth-to-mouth respiration, he was pronounced dead.

A detailed postmortem examination revealed that the electricity had entered the body through the head and exited from a foot, shredding his shoe. There was discoloration of the skin where the electricity entered the head and bleeding into the lungs but no obvious damage to the brain or heart as the electricity passed from head to foot. When Benjamin Franklin heard about the tragic event, he emphasized the importance of grounding a lightning rod. If Richmann had only run a second wire from the iron rod to the ground the tragedy almost certainly would not have happened. Franklin couldn't resist adding that the Richmann tragedy did confirm his doctrine of lightning. In a later eulogy, Joseph Priestly who was a clergyman in addition to a scientist noted that not every scientist was fortunate to die so gloriously.

Franklin's One-Fluid Theory

Although Benjamin Franklin is best known for his experiments with atmospheric electricity, his most important and lasting contribution to the field was his "one fluid" theory [16]. Franklin suggested that "electrical fire" was present in all substances and that electrical charge which could be positive or negative was normally in a balanced or neutral state. Excess or deficiency of positive or negative charge could be produced by friction between substances producing an imbalance in the amount of electric fluid. For example, when a glass tube was rubbed with a silk cloth, the glass became charged with an excess of fluid or a positive charge. By contrast, the opposite occurred when amber was rubbed with fur, the amber developed a deficiency of fluid or a negative charge. In Franklin's one-fluid theory, a positive charge corresponded to an excess of Du Fay's vitreous fluid and a negative charge corresponded to an excess of his resinous fluid.

Franklin suggested that electrical fluid consisted of "extremely subtle" particles that repelled each other yet were attracted to other equally subtle particles with an opposite charge. Franklin assumed that electrical fluid would flow from a positively charged substance with excessive fluid to a negatively charged substance with a deficit of fluid. If an uncharged conductor encountered a positively charged glass tube, some of the excess fluid on the tube would flow onto the conductor, whereas, if the same conductor encountered a negatively charged amber rod, electrical fluid would flow from the conductor to the amber in an attempt to balance the amount of fluid. Thus, Franklin's electrical fluid moved from positive to neutral and from neutral to negative. This convention persists to this day even though electrons move in the opposite direction.

Based on his one-fluid theory, Franklin conducted a series of experiments designed to understand how a Leyden jar works. He designed a Leyden jar with removable inner and outer metal coatings and demonstrated that the charge must be stored on the glass in between. The inner or outer coatings showed no signs of electric charge when they were removed, but when reassembled the charge was restored (Fig. 2.5). He then sandwiched a plate of glass between two thin sheets of lead and charged the equivalent of a flat Leyden jar by connecting one lead sheet to the prime conductor of his electric machine and the other to the ground. If he then removed the lead sheets, he could feel prickling shock-like sensations when he touched the

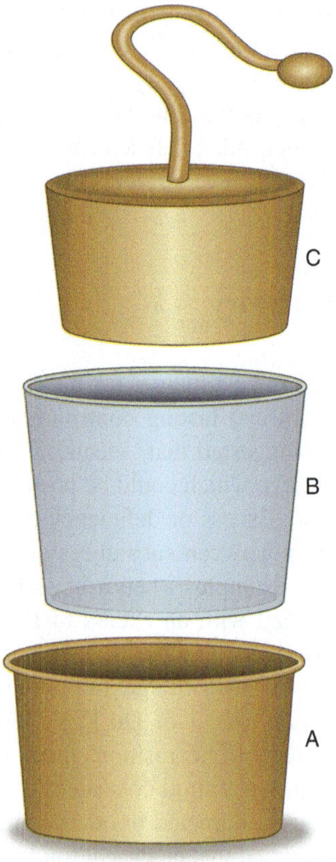

Fig. 2.5 Ben Franklin's Leyden jar with removable inner (**c**) and outer (**a**) metal cups. The charge was stored in the glass cup (**b**) in between

glass in various locations but nothing when touching the lead plates. When he reassembled the plates, the flat Leyden jar had its original charge showing that the conductors simply served to move charge to and from the glass. In the process, he had produced the first plate capacitor, versions of which are still in routine use in electrical circuits.

With his work with Leyden jars, Franklin also produced the first electrical battery. He noted that if he wired together a series of Leyden jars in parallel (positive to positive and negative to negative) he could generate a massive electric charge limited only by the number of Leyden jars. He chose the term "battery" based on the analogy of the enhanced discharge of the grouped Leyden jars to the discharge of an artillery battery. Soon, Leyden jar batteries became routinely used by public demonstrators and instrument makers included batteries along with their standard electrical machines.

Abbé Nollet and Medical Electricity

Some of the earliest suggestions that electricity generated from friction machines might have medical uses came in France in the early 1740s. Charles Francois du Fay, previously mentioned for his two-fluid theory of electricity, teamed up with Jean-Antoine Nollet and traveled through France staging electrical demonstrations for the public as Gray and Bose did in England and Germany. These included the "flying boy" and "electric kiss" that so captivated the audiences. In 1734 they traveled to England to exchange ideas with English physicists and in 1736 to the Netherlands where they met Pieter Musschenbroek.

Nollet, who studied theology and physics, was ordained a deacon in the Catholic Church in 1728 and assumed the title Abbé Nollet [17]. He later suspended his clerical career to focus on electricity research but maintained the title Abbé throughout his life. He built and sold electrical instruments beginning in 1735 to finance his own experiments. Nollet suggested the name "Leyden jar" for the storage device invented by the Dutch investigators and he invented one of the first electrometers in 1748.

From his experiences with electricity in human subjects, Nollet speculated that electricity might be able to revitalize a paralyzed human. He tested the notion in a few paralyzed patients and noted that he could trigger muscle contractions in a paralyzed limb and tingling in an arm that had not had any sensations for years. Based on an observation made by Bose that blood exiting from an opened human vein markedly increased if the person was electrified, Nollet suggested that electricity might be useful for treating patients with blood circulation problems. Although he was not a physician, Nollet played an important role in the rapid expansion of the use of electricity in medicine in the mid-eighteenth century.

The Beginnings of Electrotherapy

As the electrical craze spread throughout Europe more and more people were being electrified in the frequent public demonstrations. Just like traveling theaters, the electrical demonstrators would announce their arrival in a town with printed leaflets listing their experiments. They would search out local homeowners who would host the electrical soirées for a percentage of the entrance fees. The show would be repeated in the same house for a week or two and then the itinerant showmen would move on to the next town. Over time, more and more demonstrators incorporated medical healing into their presentation of sparks and shocks.

In Germany, in 1745, Johann Gottlob Krüger, a practicing physician and professor of medicine at the University of Halle noted that people who were repeatedly exposed to the "electric kiss" developed small red spots on their hands that slowly disappeared over a few hours [18]. His student, Christian Gottlieb Kratzenstein, measured the pulse of people before and after electrification and observed a consistent increase in pulse rate about a third of the resting pulse rate. He also noted that the people tended to perspire during the electrification. Like Abbé Nollet, Kratzenstein suggested that electricity might improve blood circulation.

Based on these observations, Kratzenstein began recommending electric treatments for a wide variety of diseases thought to be due to poor circulation. He also noted that electricity produced sedation and tiredness so it might be beneficial for treating hyperexcitability and sleep disorders. The fact that electrical stimulation could trigger involuntary muscle contractions suggested to both Krüger and Kratzenstein that electrification might be useful for treating a variety of neurological conditions including stroke and palsy. Kratzenstein described how gentle electrification allowed a woman with arthritis and contractures of her fingers to play her harpsichord again. Reports of dramatic cures with electricity began to appear in popular magazines and to a lesser extent in medical journals. Based on his pioneering work on electricity in humans, it has been speculated that Kratzenstein was the model for Doctor Frankenstein in Mary Shelley's famous book written several decades later (see Chap. 3) [19].

In Italy at the University of Padua, a professor of experimental philosophy, Giovanni Poleni was providing regular public presentations for prestigious visitors at his theater [18]. In 1745, Poleni heard of the popular demonstrations being presented in Venice using an electrical machine by the Army physician, Christian Xavier Wabst and Poleni sent his assistant to Venice to learn how to reproduce them. On return of his assistant, Poleni included electrical experiments in all his public demonstrations. He regularly measured the pulse of his noble spectators and himself before and after electrification. He confirmed the findings of Kratzenstein and word of the potential healing power of electricity rapidly spread throughout Italy.

In the Italian town of Vicenza, the marquis Luigi Sale maintained a large electrical machine in his villa which he used to treat the local population. Sale's most famous patient was his servant, Giambattista Negretti. The so-called "sleepwalker from Vicenza" was well known to the local medical community and several booklets were published regarding his case. Each year in springtime, Negretti would get out of bed though still sleeping and begin a search for food and drink even going to a local tavern without awakening. Doctors had tried a variety of techniques to awaken him including painful stimuli all without success. Sale decided to try electrification and the results were dramatic. After a few shocks, Negretti awoke and profusely thanked the marquis, praising the "blessed electrical machine." [18]

Pivati's Medicated Tubes

Not unlike many medical treatments in the mid-eighteenth century, quackery was rampant with electrical therapy. The sparks and shocks of electricity were ideal for impressing hopeful patients. Case reports of successful treatment were anecdotal without systematic follow-up and conflicts of interest for instrument makers and physicians with electrical machines were rampant.

One of the most publicized examples of electrical quackery in the mid-eighteenth century was the "medicated tube" developed by the Italian writer and self-made physicist, Gianfrancesco Pivati while living in Venice [20]. Pivati filled glass tubes with herbs and medicines and then electrified the glass. He claimed that the healing substances evaporated through the pores of the glass diffusing into the surrounding air so that patients need only breathe the air to be cured. With this method, there was no need to swallow medication. Purgatives could be given without the annoying gastrointestinal side effects. Pivati was a member and received support from the prestigious Institute of Sciences of Bologna.

Among the many people Pivati claimed to have healed was the 75-year-old Bishop Donadoni of Sebenico who suffered from long-standing severe arthritis. In the spring of 1747, having heard of Pivai's miracle cures, the Bishop visited him in Venice. When he arrived at Pivati's villa, he had to be helped from the gondola by clerics who accompanied him since his arthritis was so severe that he could not move his fingers, bend his knees, or walk without support. After examining the Bishop, Pivati filled a glass tube with medication, sealed it with corks, and rubbed the glass with a leather skin. He then brought the electrified "medicated tube" near the bishop's body generating sparks from all his involved joints.

Pivati reported an immediate response: *"the bishop freely opened both his hands, clenched them, and found he had so much strength that, having taken one of the cleric's by the arm, he held him so tightly that the latter had to beg to be released...Upon leaving, he got into his boat almost as if he were a vigorous boy, still extolling the prodigy that he had seen affected in himself to all and sundry."* [21]

Pivati described the Bishop's cure in a letter entitled *On Medical Electricity* sent to the Secretary of the Institute for Sciences of Bologna, Francesco Zanotti, who became a strong proponent of the new treatment. The letter was published and electricians throughout Europe began to try Pivati's miracle cure. Physicians in Turin and Balogna claimed dramatic successes after placing medications into the glass globes of their electrical machines claiming that the medication evaporated through the sealed glass just as through Pivati's electrified tubes. But from the rest of Europe, reports were less encouraging with many reporting complete failure.

Among those having difficulty repeating Pivati's dramatic cures was Abbé Nollet in France. Nollet decided to visit Italy and observe for himself the *"the new Wonder more difficult to be explained than any other"* [22]. He left Paris in April 1749 and toured the Italian Peninsula for 9 months meeting with the healers using the new technique and the patients claiming to be cured. When he returned, he wrote a summary of his experiences that was published by the Royal Academy of Sciences in Paris and the Royal Society in London.

Nollet reported that he was able to eye-witness the medical and electrical treatments only in Turin. He watched the treatments in a variety of people including himself and was not impressed with the results. Those receiving electrified purgatives showed no signs of being purged and he found no evidence for the testimonials of dramatic cures. When he visited Pivati in Venice he received only excuses for delays in treatments. It was too hot or there were too many people. Pivati did admit that the Bishop of Sebenico was now the same as before his medical electrical treatment. There was no cure. In private, Nollet considered Pivati a charlatan and an ignoramus.

Electrotherapy for Paralysis

One of the earliest observations made by traveling electricians in Europe was that application of electricity to an arm or leg triggered an involuntary muscle contraction even jerking of an extremity. At the time nerves were thought to cause muscle contraction by releasing *vital force*, a supernatural substance unique to living organisms, beyond the understanding of man. Could it be that vital force was nothing other than electricity? If so, could electricity cure paralysis?

As mentioned previously, Abbé Nollet was one of the first to attempt such treatments. He teamed up with physicians and surgeons at the Hôspital des Invalides in Paris to test the healing powers of electricity on a group of soldiers paralyzed from injuries in war. The results were not impressive, and he did not feel that they warranted publication.

The first report of a successful treatment of a paralyzed patient with electricity was by Jean Jallabert in Geneva, Switzerland. Jallabert was a mathematician and physicist who began his career as a Protestant cleric. His most important contribution to the field of electricity was the observation that if one approached an electrified object with a conductor that had a sharp pointed tip it would discharge the electricity from a greater distance and without a spark compared to a conductor with a blunt tip. Ben Benjamin Franklin was likely aware of Jallabert's discovery when he got the idea for a lightning rod.

Jallabert published a report of a successful "cure" of a 52-year-old man with a paralyzed arm in December, 1747 [18]. The man called Nogues was involved in an accident 15 years earlier leaving him with an atrophic right arm void of sensation. Jallabert carried out his treatment with several prominent physicians and surgeons as witnesses. For 45 days Jallabert sparked the muscles of the arm for 2 h in the morning and in the afternoon. These localized shocks were combined with generalized body shocks throughout the day.

Norgues first noticed some return of sensation and then some slow but definite movement of his fingers. Jallabert reported a gradual improvement in the color and muscle tone in the paralyzed arm and by the end of therapy, the muscles in Norgues' arm were visibly larger and firmer and he was able to lift small weights and place a hat on his head. Norgues was overjoyed with the success of his treatment and the only setback was a painful shock from an overcharged Leyden jar. Unfortunately, travel restrictions prevented his return to Jallabert so we have no information on the long-term benefits of the electrotherapy.

After receiving instructions on the electrification of paralyzed patients from Jallabert, Boissier de Sauvages, the Royal Professor of Medicine at the prestigious Montpellier Medical School in France claimed to have cured several patients with paralysis. For example, an elderly beggar who could not move his left arm or right leg had such a dramatic recovery that his wife called it a miracle. Describing his cures in a letter to Jallabert, De Sauvages wrote *"thanks to your instructions electricity has become fashionable, in this town everybody wants to be electrified"* [23]. The problem with these claims of cure was that paralysis is not a diagnosis but rather a sign of many different disorders some permanent such as spinal cord injury and stroke and others potentially reversible such as peripheral nerve injuries (probably in the case of Norgues) and hysterical paralysis (likely with the beggar). Other conditions such as severe arthritis and Parkinson's disease might present like a palsy and be transiently overcome with improved effort based on high expectations.

Reports of miraculous cures of paralysis with electricity in England led to a vigorous debate at meetings of the Royal Society [24]. One physician reported that electrical treatments helped a man with painful contractures of the hand due to severe arthritis but provided no benefit to a woman with a wasted paralytic arm. In December of 1757, the distinguished physician and member of the Royal Society, John Pringle described a woman cured of hysterical paralysis by electrical treatments. John Pringle was a close friend of Benjamin Franklin and Franklin had just arrived in London as a representative of the Pennsylvania Assembly. Pringle asked Franklin to comment for the Society on his results with electricity for treating paralysis in America. Franklin made his report on December 21, 1757, and it was published in the *Transactions* in 1758.

Franklin began his report by noting that for several years based on newspaper reports of dramatic cures in Italy and Germany, numerous "paralytics" were brought to him in Pennsylvania for electrification [25]. Most of the paralytics had suffered from a stroke. His technique was to use his electrical machine to provide a preliminary shock to all parts of the affected limb or side and then to build up a charge in two large Leyden jars and then discharge the electricity of the combined jars onto the affected limb or limbs three times a day.

According to Franklin: *"The first Thing observ'd was an immediate greater sensible Warmth in the lame Limbs that had receiv'd the Stroke than in the others; and the next Morning the Patients usually related that they had in the Night felt a prickling Sensation in the Flesh of the paralytic Limbs, and would sometimes shew a Number of small red Spots which they suppos'd were occasion'd by those Pricklings. The Limbs too were found more capable of voluntary Motion, and seem'd to receive Strength,"* [25]

Franklin noted that patients often showed signs of improvement over the first few days of treatment raising their spirits, but the improvement was not maintained, and they became discouraged by the end of the treatment session. He concluded: *"I never knew any Advantage from Electricity in Palsies that was permanent"* [26]. Furthermore, Franklin was aware of the potential placebo effect of electrical treatments. *"And how far the apparent temporary Advantage might arise from the Exercise in the Patients journal and coming daily to my house, or from the Spirits given by the Hope of Success, enabling them to exert more Strength in moving their Limbs, I will not pretend to say"* [26].

Electrotherapy for Hysteria

Hysteria has a rich tradition in medical history. The term is derived from the Greek word for the uterus, *hysterika.* In ancient times, hysteria was a medical syndrome characterized by nervous irritability, swooning, fits, paralysis, fatigue, and many other symptoms thought to occur uniquely in women. In fact, ancient Greek physicians believed the condition resulted from a *wandering uterus*, moving about the body producing different symptoms depending on where it happened to settle. When the uterus moves upward in the woman's body, forcing its way into crowded places at the center, it could cause the woman to have difficulty breathing, swoon, or faint and become speechless.

Some likened the uterus to a viscous animal that could invade other organs. Since they thought it was attracted by fragrant smells and repelled by fetid smells, physicians treated hysteria by rubbing the vagina with honey and having the woman chew cloves of garlic. If the physician diagnosed that the uterus was moving upward, he would manually try to push it backwards and keep it down by tying a tight bandage below the rib cage.

The famous Roman physician, Galen, questioned the wandering uterus theory but believed that the uterus was the seat of the problem possibly by releasing vapors that moved upward and affected other organs of the body [27]. The cause was sexual deprivation, and the cure was increased sexual activity. Married women with symptoms of hysteria were told to have sex with their husbands and single women of marriageable age were instructed to marry. For other women (widows, unhappily married, and nuns) treatments included vigorous horseback riding, rocking the pelvis on a swing or chair, and having a physician or midwife massage the vulva with a finger. The notion that hysteria was a uniquely female disorder that could be treated with sexual activity persisted throughout the Middle Ages and well into modern times.

In the early eighteenth century, although traditional gynecological theories on hysteria remained engrained in much of medical practice, physicians in larger metropolitan areas and many of their affluent clientele began to accept a "nervous" origin of the disorder [28]. Terms such as hysteria and vapors in women, and hypochondria and spleen in men were often used to describe a range of nervous disorders. Sir Richard Blackmore, a prominent English physician to royalty and affluent society wrote a book entitled *A Treatise of the Spleen and Vapours: Or Hypochondria and Hysterical Affections* in which he noted that "*Vapours in women and Spleen in men is what neither sex are pleased to own*" [29]. Blackmore recommended against aggressive treatments such as bleeding and purging for these nervous disorders but rather suggested calming medications like bromide and opium that would relax the patient and restore health.

By the mid-eighteenth century, hysteria, still considered a female malady, was thought to be because women were more emotionally labile and had weaker constitutions and more fragile nerves than men. The routine stresses of life made susceptible women vulnerable to hysteria. Stimulants were recommended for treatment and electricity was considered the most powerful stimulant of all. Kratzenstein had suggested that since hysterical women suffered from sluggish blood flow, electrification could thin the blood and improve circulation and should be an excellent remedy for hysteria. Many of the dramatic cures with electrification at the time occurred in women with hysteria.

One of the best documented cures with electricity at the time occurred in America when in 1752 the physician Cadwallader Evans teamed up with Benjamin Franklin to treat a young woman (C.B.) whom Evans had diagnosed with hysterical fits [30]. Evans later published a detailed report of the patient and treatment in *Medical Observations and Inquiries* in 1754. He had first seen the young woman about 10 years earlier when she began having violent fits one after another, 40 within 24 h after the initial attack. The attacks were so violent that three strong people had difficulty holding her down. Evans used the standard treatments of the time including bleeding, blistering, and a variety of *nervous medicines* and there was a gradual improvement so that the fits occurred but once or twice a day.

Over the subsequent 10 years, the woman had a fluctuating course with occasional one-to-two-month remissions but always followed by a relapse. She complained of bouts of muscle cramps involving different parts of the body followed by generalized convulsions. Evans noted that at times she had other typical hysterical symptoms including globus hystericus, a feeling of choking with a lump in the throat. Despite trying a great number of medicines her symptoms persisted and she became frustrated.

In a letter, C.B. later wrote: "*About this time there was great talk of the wonderful power of electricity; and as a person reduc'd to the last extremity, is glad to catch anything; I happened to think it might be useful to me. Altho' I cou'd have no encouragement from any experiment in the like case, I reslov'd to try, let the event be what it might; for death was more desirable than life, on the terms I enjoy'd it. Accordingly, I went to Philadelphia, the beginning of September 1752, and apply'd to B. Franklin, who I thought understood it best of any person here*" [31].

Franklin administered electric shocks four times in the morning and evening and the C. B. reported that initially, the fits were very severe but then they began to abate and eventually completely resolved. She stayed in Philadelphia receiving daily treatment for 2 weeks and when she returned home Franklin provided her with "*a globe and bottle*" to electrify herself for three more months. In this case there was convincing evidence of a permanent cure. Evans noted that he received letters from family members at later dates indicating that C.B. remained in "*perfect health.*" Furthermore, a publication many years later confirmed that C.B. was healthy and lived a vigorous life until the age of 79 years.

Another successful cure of hysteria was reported by the lay electrician Patrick Brydone, in the *Philosophical Transactions of the Royal Society* in London in 1757 [32]. Brydone described a young woman who developed paralysis and loss of sensation along with a variety of other symptoms thought typical of hysteria after her father had died of some type of palsy. Brydone administered a series of electrical treatments over several days and he reported that all her symptoms disappeared, and she remained asymptomatic at short-term follow-up.

In 1759, the English Cleric John Westley (see below) reported curing a woman with hysteria in his pamphlet *Desideratum: Or Electricity made Plain and Useful by a Lover of Mankind and of Common Sense.* Westley described treating a 22-year-old woman with hysterical fits using the *ethereal fire* of electricity. *"On the first Shock her struggling ceased, and she lay still. At the Second her Senses returned. After two or three more, she rose in good Health"* [33].

Although all three of these reported cures of hysteria were by lay electricians without any medical training, the case details are highly consistent with the diagnoses of hysteria. Presumably, they believed the prevailing theories of the day regarding hysteria and felt that electricity was a strong stimulant to the nervous system. Electricity was thought to *tighten* the nerves and increase blood flow in patients with hysteria.

What about the power of suggestion with electricity, the placebo effect? Clearly, Brydone and Westley were zealots who would not have considered that possibility. Ben Franklin, on the other hand, was very skeptical of eighteenth-century doctors and the value of their treatments. In his younger years, he campaigned against quackery and questionable claims of medical cures. In Poor Richard's Almanac Franklin famously wrote: *"He's the best physician that knows the worthlessness of most medicines"* [34]. He understood gullible people were susceptible to the power of suggestion and more likely to respond to quack treatments. Franklin agreed with Abbé Nollet that the miraculous cures attributed to medicated electrical tubes in Italy were almost certainly a form of quackery. In the nineteenth century, often considered the golden age of hysteria, the power of suggestion was shown to be a powerful remedy for hysteria.

Electricity, the Poor Man's Medicine

In London and the English provinces, as public demonstrations of the sensational powers of electricity became fashionable among the upper classes, more and more demonstrators became fascinated with the healing powers of medical electricity. Some branched off to exclusively focus on electrical healing.

Richard Lovett, a lay clerk at Worcester Cathedral, after attending several public electrical demonstrations and reading about electricity in several popular magazines, purchased his own electrical machine [18]. He claimed to have achieved several remarkable cures with *the electrical Virtue*. Lovett was one of the first to suggest that medical electricity was ideal for treating the lower classes. It was cheap, easy to administer, and reasonably safe. He suggested that neighborhoods could share the cost of purchasing electrical instruments and all benefit from the healing powers of electricity. In 1756, Lovett published the first English textbook on medical electricity that rapidly became a must-read for anyone interested in electrical healing.

In Lovett's textbook, treatments were based on the idea that the human body had a natural amount of electricity and that diseases resulted from either too little to too much of this natural electricity. The instruments with sharp points could be used to draw electrical sparks from the patient while the instruments with ball-like endings could provide electrical shocks to the patient. Routine treatments consisted of a combination of sparks and shocks.

A London Gentleman, John Reddall, read Lovett's book and began a series of lectures on medical electricity throughout London. His lectures became so popular that they attracted hundreds of people many from outlying communities. Redall was convinced that electricity would become the dominant force in medical practice and that it was affordable even to the lower classes. He even suggested that every paralytic person should have an electrical machine at home. The only training necessary was familiarity with an electrical machine and reading Lovett's book [35].

Another proponent of Lovett's electrical medicine was the English cleric and evangelist, John Wesley, who was a leader of the revival movement within the Church of England known as Methodism [36]. Lovett's notion of electricity being an effective medical treatment affordable by the lower classes fit well with Wesley's religious beliefs since Methodism was considered the *religion of the poor*. With his evangelical work, Westley traveled throughout England preaching two or three times a day and administering to the poor and sick. He developed a string of medical dispensaries where the sick poor could receive spiritual and medical help. His book on home remedies, Primitive Physick, emphasized moderation in food and drink, temperance of the passions, and prayer to prevent sickness [37].

Westley first introduced electrical treatments in his dispensary at Moorfields in 1756 and it was so popular that within 3 years he offered them at all his other dispensaries. He regularly published the list of names of people who had been healed with the help of the electrical machines. In 1760, he added electricity to the list of home remedies in Primitive Physick.

Medical historian Paola Bertucci wrote: "*If natural sparks were to be seen as signs of God's wrath, and of human imperfection, healing sparks displayed divine benevolence, and could reveal to the faithful the path to salvation*" [38]. Westley preached that electrical machines should be available to everyone There was no need for fancy gadgets. Just line the patients up and provide shocks and sparks, an *electric bath*. For skeptical physicians and apothecaries, Westley had nothing but contempt suggesting that they resented electrical therapy because it was so cheap, easy to administer and effective.

Anton Mesmer and James Graham, Paragons of Quackery

Two of the most famous proponents of medical magnetism and electricity in the late eighteenth century were outright quacks. Though ridiculed by traditional medical practitioners, they established highly lucrative medical practices, Mesmer in Vienna and Paris and Graham in London. At their peak, each became rich treating some of the most powerful and influential people in French and English society. Eventually, both had their downfall slipping into financial ruin and medical obscurity.

Anton Mesmer was born in a small town in Germany in 1734 [39]. After studying theology and philosophy at a Jesuit University in Bavaria, he moved on to the University of Vienna where he first studied law and then medicine. In his medical dissertation entitled *De Planetarum Influxu* (On the Influence of the Planets), Mesmer expanded on the early sixteenth-century teachings of Paracelsus regarding the influence of planets on the functioning of the human body. He proposed a cosmic force, something like gravity, which he called *animal magnetism*, an imperceptible fluid found throughout the cosmos. Through this universal *fluid*, every object could influence every other object including the human body. As an example, he noted how the sun influenced the planets and the moon influenced the earth's tides. In the same fashion heavenly bodies could influence the health of the human body.

In addition to likening his animal magnetism to gravity and magnetism, Mesmer suggested that electricity might also be a related phenomenon. Like many physicians before him, Mesmer believed that all human diseases resulted from obstruction of internal fluids. The physician could remove the obstruction and achieve proper distribution of the fluids by manipulating his newly discovered invisible force, animal magnetism. Although this simplistic idea might seem beyond rational belief, one must keep in mind the mystical qualities of gravity, magnetism and electricity to the average citizen in the eighteenth century.

In 1774, Mesmer began using iron magnets to influence animal magnetism and treat his patients. The first report of a dramatic cure was from a young woman diagnosed with hysteria, Franziska Oesterlin. Franziska (Franzl) suffered from hysterical fits, fainting spells, earaches, and gastrointestinal symptoms. According to Mesmer, she was reduced to skin and bones. During one of her spells, Mesmer placed magnets on her stomach and on both legs. At first, the symptoms were aggravated as she reported feeling currents flowing downwards in all her extremities, but then she felt much better for several hours. Mesmer repeated the therapy over the following days and gradually Franzl recovered.

Mesmer found that he could treat large numbers of patients simultaneously by filling a tub with iron filings from which protruded iron rods that patients could grasp to receive the stored magnetic force, a tub full of animal magnetism. With soft music in the background, the great healer dressed in a lavender silk robe, walked among his thankful patients touching them with a long, magnetized rod. He rapidly became a sensation among the rich and famous of Vienna much to the consternation of the established medical community.

After treating several patients Mesmer began to believe that he did not necessarily need magnets but that he could cure patients based on his unique ability to manipulate animal magnetism. Simply stroking the patient or even moving his hands without touching the patient could often work just as well as a magnet. Furthermore, he had the ability to pass this mysterious power to other objects so that these objects could have the same effect as a magnet. Like gravity and magnetism, Mesmer's mysterious animal magnetism could stream from his fingers through walls and enter the body of a patient even in another room.

In 1777, Mesmer was summoned by Empress Maria Theresa to treat her teenage granddaughter for what was thought to be hysterical blindness. Despite her blindness, the granddaughter had become a successful clavichord player and singer sponsored by the Empress. After a series of treatments, the young woman was cured of her blindness but as often happens with hysteria removing one symptom uncovers other symptoms. She began to experience fainting spells and hysterical fits causing her musical career to rapidly tumble downhill. Concerned that her Royal pension would be lost, her parents dismissed Mesmer and even though her blindness returned, her musical career was salvaged and everyone was happy. With the loss of Royal support and accusations of fraud by the established medical community, Mesmer decided it was time to leave Vienna for Paris.

In Paris, Mesmer teamed up with Charles-Nicolas Deslon who was the personal physician to a brother of King Louis XVI. This provided governmental approval for access to patients that would otherwise be forbidden to foreign physicians. As in Vienna, Mesmer rapidly developed a fashionable medical clinic, set up several miles outside of Paris away from the prying eyes of the Parisian medical academy. His medical facilities included multiple large séance rooms each fitted with magnetic tubs about 4 feet in diameter filled with personally magnetized water, sand, and iron filings. Each tub was covered by lids with holes through which protruded iron bars. Patients were told to touch the iron bars with the affected body part and Mesmer would also touch, stroke, or point to the affected body part with his hand or a wand. In this way, Mesmer could treat large groups of patients at the same time often with patients joining together in rows each touching the neighbors affected body part, passing the invisible magnetic fluid to all (just as electricity could be passed from person to person).

At his peak, Mesmer was treating more than a hundred paying patients a day, patients from all walks of life and levels of society. Although many suffered from hysteria, he treated all medical conditions including neurological disorders such as stroke, epileptic convulsions, and paralysis. As he became wealthy, he sought affirmation from the scientific community for his discovery of *animal magnetism*. He was particularly interested in gaining the favor of Ben Franklin who was in Paris at the time. He considered Franklin to be a great man of science for his work on electricity and considered his own work to be comparable, both having discovered fundamental principles of nature.

Mesmer wrote to Franklin inviting him to visit his clinic to show him his discoveries, but Franklin's friends warned him against such a visit noting that Mesmer's animal magnetism had no scientific basis. Shortly after, the French Royal Academy in Paris commissioned a distinguished commission including Franklin and the well-known French chemist Lavoisier to investigate Mesmer's claims of dramatic cures. The Commission concluded that Mesmer's animal magnetism was a fantasy, a figment of his imagination, and that his claims of cure were simply the result of suggestion. Mesmer decided to leave Paris and after traveling around Europe, he vanished into obscurity although his animal magnetism remained a fascination in parts of affluent Parisian society and would later resurface as hypnotism.

James Graham was born in Edinburgh, Scotland, in 1745, the son of a saddle-maker [40]. He initially decided on a career in medicine but dropped out of medical school, married, and set up an apothecary in Doncaster, Yorkshire. In 1770 he decided to leave England for America where he traveled about the colonies doing eye surgery mainly cataract removal and implanting prosthetic eyes. He finally settled in Philadelphia where he attended lectures at the University of Pennsylvania and developed an interest in medical electricity through interactions with Ebenezer Kinnersley, a friend and collaborator of Ben Franklin. Graham decided to return to England as a war between England and the colonies appeared imminent.

After traveling about Europe in 1776, Graham set up a medical practice in Bath, Somerset, where he advertised cures with *effluvia and vapors* of magnetism and electricity. Controversy with local academic physicians and negative newspaper articles caused Graham to move his practice to London in 1780. His residence that also housed his examination and treatment rooms was located on a terrace overlooking the Thames River not far from the House of Parliament. One of his first celebrity patients was Horace Walpole who sought treatment for gout. He soon attracted many wealthy patrons who heard of his elaborate electrical machines.

His establishment, known as the Temple of Health, was designed to be awe-inspiring. Although Graham catered to the rich and famous, he was not above receiving money from the lower classes, and he even treated a few nonpaying patients. He did his best to separate the rich from the poor, however. The poor waited on wooden benches in a long room lined with artifacts of prior patients who were cured including crutches, canes, ear trumpets, and thick eyeglasses. The rich had a separate entrance into a spacious room lit by stained glass windows, colorful lamps, and candelabras. They first encountered a huge Leyden jar whose outer layer was decorated with figures and ornaments, etched in polished silver and gold foils, just for show since it was not attached to any electrical source.

Nearby was an electric machine consisting of a massive cylinder of white polished glass with a prime conductor more than 10 feet long capped with large metal globes. One end of the prime conductor touched the tail of a 6-foot-long metal dragon mounted on glass pillars. Patients could receive an *electric bath* by sitting next to the machine on an insulated chair. Nearby the prime conductor was a metal shelf lined with glass containers filled with a variety of medicines. Like Pivati's medicated tubes, Graham claimed that electricity passed through the glass causing the medicines to fill the air with healing vapors that could miraculously enter the patients' bodies.

There were several other similar rooms with separate electrical machines some also with large magnets. Graham considered electricity and magnetism to have similar healing properties. In one room electricity was transferred by brass chains and rods to a long bench mounted on glass insulators. The center of the bench contained a powerful magnet so that up to 12 patients at a time could sit on the bench receiving a combination of electrical and magnetic effluvium. Another room had an electrical machine shaped in the form of a cross. A massive room called the Great Apollo Apartment was named in honor of the Greek and Roman gods of healing. Floor to ceiling, stained glass windows highlighted paintings showing themes of fertility and marital bliss. The apartment contained two massive electrical machines. At the center of the apartment was an electrical and magnetic *throne* where patients were bathed in electricity, magnetism, medicinal vapors, and organ music. Graham boasted, "*there is no fever, rheumatism, cramp, spasm or convulsion*" [41] that could not be cured by this treatment.

Of all the many attractions of Graham's Temple of Health, the most talked about was his *Celestial Bed* located in an elegant hall on the second floor. Supported by 28 glass pillars, covered by a richly carved canopy and surrounded by crimson silk curtains, the 12 × 9-foot bed was available for £50 a night. The bed was connected to a large cylindrical electrical machine hidden in a closet providing the occupants with an electrical bath while they conducted their love-making activities. The air was filled with vapors of oriental perfumes and music provided by a small orchestra that could accompany the movements of the copulating couple. Graham guaranteed a cure for sterility and impotence with his celestial bed.

On entering the Temple of Health, patients were greeted by two giant porters, Gog and Magog, named after the Guidhall Giants. Walking among the patrons were thinly clad *Goddesses of Health* selected as models of physical health. The best-known goddess was Lady Emma Hamilton (then known as Emy Lyon) who later became famous as the mistress of Horatio Nelson and a muse for the portrait painter George Romney. The flamboyant Dr. Graham claimed to cure hundreds of patients in a day without the discomfort of bleeding and purgatives used by his more traditional rivals.

Beginning in the mid-1780s Graham began to exhibit strange behavior. He became highly religious, preaching a new theology on the divinity of Jesus Christ. He developed a passion for vegetarianism and experimented with prolonged fasting which he felt would prolong his lifetime. At one point he was confined to his home based on a diagnosis of *lunacy*. He died suddenly at age 49 presumably from an intracranial bleed and was buried in the Greyfriars graveyard near his home in Edinburgh.

Safety of Electricity

Was it safe to shock patients with electricity? As news of the Richmann catastrophic electrocution in St. Petersburg circulated around the world, people began to question the safety of electrifying patients. Richmann made the critical mistake of bending his head near a primary conductor charged by an electric storm.

As with most aspects of electricity at the time, Ben Franklin wrote extensively about his personal experience with electrifying people. In 1755, in a letter to John Liining [42], a physician in South Carolina, Franklin described a dramatic experiment where he lined up six men in a column each holding his hand on the head of the man next to him. The man at the end of the line was connected to the outside of two fully charged Leyden jars. When he brought a discharging rod connected to the inside of the Leyden jars to the head of the man at the beginning of the line, all six suddenly lost consciousness and dropped to the floor. Franklin reported that on recovering consciousness all had retrograde amnesia for the event and could not recall feeling anything or how they fell.

In the same letter, Franklin noted that he personally had suffered an accidental shock to the head that knocked him out but without any residual effect. He also described a young woman whom he was treating who accidentally bent forward so that her head came too close to the prime conductor of his electrical machine. She dropped to the floor briefly losing consciousness but then getting to her feet reporting that she felt fine. Franklin concluded that although electricity to the head could cause a loss of consciousness and retrograde amnesia, there didn't appear to be any long-term consequences. Franklin, himself, lived to age 84 with an active engaged mind right up to his death.

As a rule, the eighteenth-century electrotherapists were cautious and careful not to endanger their patients. With the invention of the battery by Ben Franklin, it became possible to store and discharge a large quantity of electricity, but the average therapist was satisfied with a single Leyden jar and a low-level shock. One area that was poorly understood and largely ignored was the potential risk to the fetus in a pregnant woman. Interestingly, one of the most common uses of electrotherapy in the eighteenth century was for treating an arrest of menses, amenorrhea. Several medical textbooks at the time indicated that electricity was the preferred treatment for lack of menses. Since pregnancy was a leading cause for arrest of menses and since there was no simple test for early pregnancy, it is likely that abortions were either knowingly or unknowingly being induced in women who sought treatment for the arrest of menses. Abortion was universally outlawed at the time, so many felt the subject was best ignored.

In Brief

The early eighteenth century marked the beginning of the age of electricity. Francis Hauksbee built the first efficient static electricity machine and Stephan Gray demonstrated that electricity could be conducted over a distance. Gray's "flying boy" experiments started a trend of electrifying human subjects for public entertainment that ultimately led to early electrotherapy. The Leyden jar provided the first simple way to store electricity for both physical and medical experiments. Ben Franklin was a pivotal character in the history of electricity and neuroscience. His kite experiment proved that lightening was static electricity and his experiments on human subjects showed that electricity applied to the head caused loss of consciousness and retrograde amnesia. Early efforts at electrotherapy were promising although poorly documented and ripe with quackery. The dangers of electricity to the human body were appreciated for the first time.

References

1. Hackmann WD. The relationship between concept and instrument design in eighteenth-century experimental science. Ann Sci. 1979;36:205–24.
2. Schiffer M. Draw the lightning down: Benjamin Franklin and electrical technology in the age of enlightenment. Berkeley: University of California Press; 2003. p. 17–8.
3. Bordeau SP. Volts to hertz—the rise of electricity. Minneapolis: Burgess Publishing Company; 1982. p. 14–6.
4. Priestley J. The history and present state of electricity, with original experiments. 3rd ed. London; 1775. p. 19–31.
5. Gray S. The first Copley Medallist. Nature. 1936;137:299–300.
6. Heilbron JL. Electricity in the 17th and 18th centuries: a study of early modern physics. Berkely: University of California Press; 1979. p. 234–49.
7. Bertucci P. Sparks in the dark: the attraction of electricity in the eighteenth century. Endeavour (Elsevier). 2007;31:88–93.
8. Heilbron J. G M Bose: the prime mover in the invention of the Leyden jar? Isis. 1966;57:264–7.
9. Heilbron JL. Electricity in the 17th and 18th centuries: a study of early modern physics. Berkely: University of California Press; 1979. p. 309–15.
10. Bordeau SP. Volts to hertz—the rise of electricity. Minneapolis: Burgess Publishing Company; 1982. p. 19–20.
11. Franklin B. The kite experiment. The Pennsylvania Gazette, October 19, 1752.
12. Priestley J. The history and present state of electricity, with original experiments. 3rd ed. London; 1775. p. 179–81.
13. Schiffer MB. Draw the lightning down. Benjamin Franklin and electrical technology in the age of enlightenment. Berkley: University of California Press; 2003. p. 163–4.
14. Ashworth WB. Scientist of the day—Georg Wilhelm Richmann. Linda Hall Library, July 22, 2020.
15. Schiffer MB. Draw the lightning down. Benjamin Franklin and electrical technology in the age of enlightenment. Berkley: University of California Press; 2003. p. 165–6.
16. Turner G. North Pole, South Pole. New York: The Experiment; 2010. p. 68–9.
17. Heilbron JL. Nollet, Jean-Antoine. In: Gillespie CC, editor. Dictionary of scientific biography. New York: Charles Scribner's Sons; 1974.
18. Bertucci P. Therapeutic attractions: early applications of electricity to the art of healing. In: Whitaker H, Smith CUM, Finger S, editors. Brain, mind and medicine: essays in eighteenth-century neuroscience. New York: Springer; 2007. p. 271–83.
19. Kaplan PW. The real Dr. Frankenstein: Christian Gottlieb Kratzenstein? J R Soc Med. 2002;95:577–8.
20. Bertucci P. Sparking controversy: Jean Antoine Nollet and medical electricity south of the Alps. Nuncius. 2005;20:153–87.
21. Bertucci P. Sparking controversy: Jean Antoine Nollet and medical electricity south of the Alps. Nuncius. 2005;20:154.
22. Bertucci P. Sparking controversy: Jean Antoine Nollet and medical electricity south of the Alps. Nuncius. 2005;20:155.
23. Bertucci P. Therapeutic attractions: early applications of electricity to the art of healing. In: Whitaker H, Smith CUM, Finger S, editors. Brain, mind and medicine: essays in eighteenth-century neuroscience. New York: Springer; 2007. p. 278.
24. Finger S. Doctor Franklin's medicine. Philadelphia: University of Pennsylvania Press; 2006. p. 80–101.
25. Finger S. Doctor Franklin's medicine. Philadelphia: University of Pennsylvania Press; 2006. p. 97.
26. Finger S. Doctor Franklin's medicine. Philadelphia: University of Pennsylvania Press; 2006. p. 98.
27. Scull A. Hysteria: the disturbing history. New York: Oxford University Press; 2009. p. 11.

28. Faber D. Hysteria in the eighteenth century. In: Whitaker H, Smith CUM, Finger S, editors. Brain, mind and medicine: essays in eighteenth-century neuroscience. New York: Springer; 2007. p. 321–30.
29. Blackmore R. Treatise of the spleen and Vapours: or, Hypochondriacal and hysterical affections. London: Pemberton; 1726. p. 97.
30. Finger S. Doctor Franklin's medicine. Philadelphia: University of Pennsylvania Press; 2006. p. 104–8.
31. Finger S. Doctor Franklin's medicine. Philadelphia: University of Pennsylvania Press; 2006. p. 106.
32. Brydone P. An instance of electrical virtue in the cure of a palsy. Philos Trans R Soc. 1757;50:392–5.
33. Wesley J. The desideratum, or electricity made plain and useful. London: Flexney; 1760.
34. Finger S. Doctor Franklin's medicine. Philadelphia: University of Pennsylvania Press; 2006. p. 35.
35. Lovett R. The subtil medium prov'd, or, that wonderful power of nature: so long ago conjectur'd by the most ancient and remarkable philosophers, which they call'd sometimes aether, but oftener elementary fire, verify'd…. London: Hinton; 1756.
36. Bertucci P. Revealing sparks. John Wesley and the religious utility of electrical healing. Br J Hist Sci. 2006;39:341–62.
37. Wesley J. Primitive Physick: or, an easy and natural method of curing most diseases. 24th ed. London; 1792.
38. Bertucci P. Revealing sparks. John Wesley and the religious utility of electrical healing. Br J Hist Sci. 2006;39:343.
39. Finger S. Doctor Franklin's medicine. Philadelphia: University of Pennsylvania Press; 2006. p. 219–34.
40. Schiffer MB. Draw the lightning down. Benjamin Franklin and electrical technology in the age of enlightenment. Berkley: University of California Press; 2003. p. 148–54.
41. Schiffer MB. Draw the lightning down. Benjamin Franklin and electrical technology in the age of enlightenment. Berkley: University of California Press; 2003. p. 151–2.
42. Finger S. Doctor Franklin's medicine. Philadelphia: University of Pennsylvania Press; 2006. p. 109–10.

Chapter 3
Animal Electricity

> *We believe, therefore, that the electric fluid is prepared by the activity of the cerebrum, that it is extracted in all probability from the blood, and that it enters the nerves.*
>
> Luigi Galvani, 1791

Abstract Two men, Luigi Galvani and Alessandro Volta, dominated the scientific discussion of animal electricity in the latter eighteenth century. Galvani argued that electricity represented the long sought-after "animal spirits" carried in nerves to control body functions while Volta, after initially agreeing with Galvani, changed his mind and became a critic of Galvani's work. In the process Volta discovered the electric pile, a major step in the use of electricity for practical purposes including medicine. The epic confrontation between these two giants changed the history of electricity and neuroscience. Ironically, Galvani's nephew, Giovani Aldini used Volta's electric pile to electrically stimulate the brains of dead animals and humans raising the possibility that electricity could revive the dead. The public became enthralled with the life-giving potential of electricity causing a young Mary Shelly to envision a monster brought to life by electricity.

Keywords Dielectrics · Electrometer · Voltaic pile · Galvanization

In the latter half of the eighteenth century, people began to question whether electricity was part of normal biology and, if so, how. Reports of dramatic cures with electrotherapy suggested an important role for electricity in the normal functioning of the human body but treatments were largely empirical and not based on understanding of the biology or electricity. As suggested in Chap. 2, some electricians and physicians speculated that nerves activated muscles with electricity, but this was purely speculation without objective evidence. Even if nerves carried and released electricity, how would the electricity activate select muscles and not all nearby muscles? After all, when electricity entered the body from an outside source (such as a Leyden jar), it rapidly disseminated throughout the entire body as shown with the "flying boy" demonstrations.

Based on the work of Gray and Franklin, for electricity to flow within a conductor there had to be either a closed circuit such as a connection to the inside and outside of a Leyden jar or a connection to the ground allowing the electricity to

R. W. Baloh, *Brain Electricity*, https://doi.org/10.1007/978-3-031-62994-5_3

dissipate. For electricity to flow within nerves it would have to either complete a circuit back to the source (presumably the brain) or make its way to the ground passing through other body tissues. Neither of these prospects seemed very likely.

Electrical Instruments

With the increased popularity of medical electricity by the mid-eighteenth century, ways of generating and administering electricity were in demand. In England a former patient at one of Wesley's dispensaries, John Read, made the first cheap portable electrical machine [1]. Read started out as a cabinet-maker's apprentice and after attending several public demonstrations of electricity he decided to set up a business of making and selling electrical machines. By the 1750s, small compact medical electricity kits were available throughout England. Read produced a kit that could be taken apart and packed into a small box about a foot square.

The equipment needed to practice electrical healing consisted of an electricity-generating machine, a prime conductor, a Leyden jar, an insulated stool for the patient, and a variety of devices for applying the electricity to different parts of the patient's body. The electricity machines were all variations of the original design of Hauksbee, a large turning wheel driving an axle attached to a glass globe, cylinder, or plate (see Fig. 2.2). A conducting wire or chain attached to the inner and outer metal foils of the Leyden jar was connected to a variety of handheld metal devices for generating sparks and shocks. The devices consisted of metal endings with insulated handles to be held by the practitioner. The metal endings had blunt or pointed tips depending on the desired treatment.

A major problem facing the pioneers in electrotherapy was the lack of a quantitative measurement of the amount of electricity stored in their devices. Recall that Richmann was testing such a measuring device when he was accidentally electrocuted. In 1765, a London apothecary and medical electrician, Timothy Lane invented a discharge electrometer to be attached to a Leyden jar to regulate the intensity of an electric shock [2]. The device consisted of an adjustable gap between two conductors with ball endings. One could make a rough estimate of the magnitude of the electrical discharge by the distance and thickness of sparks jumping across the gap. The instrument became popular with electrotherapy to limit the magnitude of an electric shock. Leyden jars with the attached Lane electrometer became known as "medical bottles."

Gradually, over time electrotherapists observed that the length and thickness of sparks across Lane's electrometer gap varied independently. The spark length depended on the electrical "tension," whereas the spark thickness depended on the "quantity" of electricity. Increasing the charge of a Leyden jar could increase the length of the spark without affecting thickness, whereas joining together multiple Leyden jars in parallel (Franklin's battery) increased the thickness of the spark without affecting the length. In modern terms tension represents electromotive force (EMF) measured in volts and thickness represents current measured in amps but instruments for measuring volts and amps were decades away.

Glass cylinders and plates gradually replaced glass globes in the design of electrical machines in the latter eighteenth century [3]. They were more efficient for generating electricity and easier to make. In 1766, Jesse Ramsden, an instrument maker in London published a pamphlet describing his new electrical machine that used a circular glass plate in place of a glass globe. Ramsden placed two adjustable "rubbers" on both sides of the glass plate at the top and bottom. Charge from the glass plate was drawn to the points of two arms of a Y-shaped prime conductor and passed on to a Leyden jar by a metal chain. Advantages of the plate design included safety (much less likely to crack or explode), efficiency (more electricity produced than with a comparable sized globe), and ease of up-scaling (by increasing the size and number of plates).

Probably the most expensive single piece of scientific equipment made in the eighteenth century was an electric machine built by a Dutch physician, Martinus van Marum (Fig. 3.1). Born in Deft, a town famous for its pottery, Marum oversaw the construction of a massive machine with two large round glass plates measuring 65 inches in diameter, the largest that could be cast at the time. Although starting out as a practicing physician, Marum married well and was later appointed director of the Teyler Museum supported by a local philanthropist. The museum focused on the display and demonstration of scientific instruments. Marum believed that building the world's largest electrical machine could shed light on the nature of electricity and place the Netherlands at the forefront of electricity research.

Contacting the plates were eight rubbers that could be precisely adjusted to obtain maximum charge. The charge was stored on five huge brass prime conductors each supported by five feet glass pillars. Marum typically had two men turning the crank of the machine, but for maximum output, he used four men. In this case, the charge produced could generate visible sparks between conductor knobs about two feet apart. A spark of that length corresponds to about 300–500 thousand volts. To store the electrical charge of the machine, Marum had batteries of Leyden jars, as many as 135 large jars.

Although Marum's massive machine did not provide new insights into the basic mechanisms of electricity, through a series of collaborations with locals and international scientists, the machine did advance the understanding of electrochemistry. As we will see later, large voltages are needed to separate the basic elements of many chemical compounds and electrochemical experiments brought to the Tyler Museum in the Netherlands from England and France contributed to the "new chemistry" being conducted by Davy in London and Lavoisier in Paris.

Alessandro Volta and Electrophorus

The basic unit for the driving force of electricity, the volt, was named after Alessandro Volta, a key contributor to the eighteenth-century electrical revolution. Volta was born in 1745 in the ancient city of Como, then a part of the Austrian Empire, now in northern Italy [4]. As an infant he was slow in learning and his

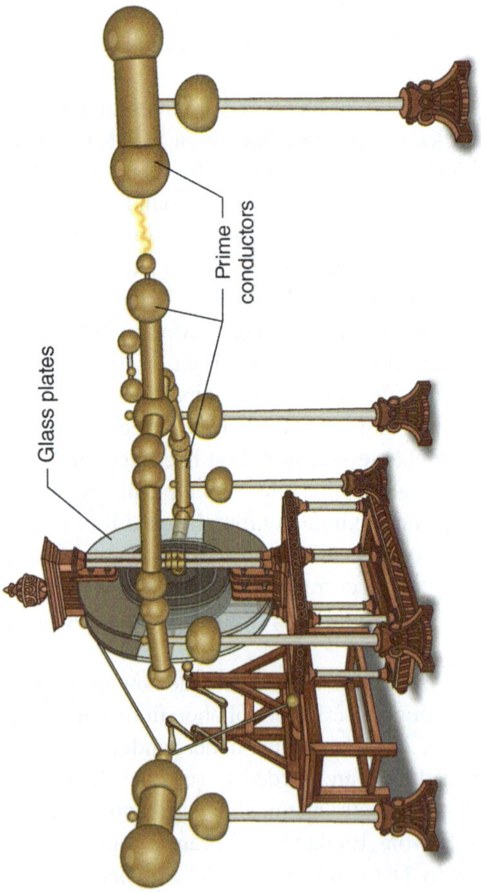

Fig. 3.1 Martinus van Marum's massive static electricity machine. Two 65-inch diameter round glass plates driven by a turning wheel were surrounded by prime conductors for storing the electricity

parents thought he might even be retarded, but by age 7, he turned into a bright and promising student. His father, Filoppo was of noble lineage but not financially savvy, so his children relied on relatives in the church for their education. As was common at the time there was family pressure to enter the clergy, but Volta developed a keen interest in the physical world and decided to become a physicist. He experimented with electricity as a teenager and began corresponding with the famous French electrician Abbe Nollet.

While experimenting with rubbing materials like resin and glass (dielectrics), Volta noted that the discharge of electricity was not instantaneous but consisted of a series of alternations in flow before gradually discharging. It was as though the dielectric material was stubbornly holding on to the electricity. In the process, Volta discovered a way to generate static electricity that he felt was better than the widely used rubbing machines.

Volta described his method called electrophorus in a letter to Joseph Priestly in 1775 [5]. A moveable top plate made of metal with a wooden handle (the conducting plate) sat on top of a plate made of wax, turpentine, and resin (the dielectric plate) (Fig. 3.2). The dielectric plate sat on a fixed grounded metal base. The key feature of Volta's device was a removable grounding chain attached to the top metal plate. He found that after rubbing the resin plate and then replacing the upper metal plate and its grounding chain, the upper plate acquired a charge equal and opposite to that of the resin plate through the process of electrical induction. Of course, without the grounding chain, the metal plate would not hold on to an electrical charge returning to neutral as soon as it was lifted. After removing the grounding chain, Volta lifted the charged metal plate with its insulating handle and passed the charge on to a Leyden jar. The remarkable feature of this device was that he could repeatedly charge the upper plate and pass the charge on to Leyden jars without any

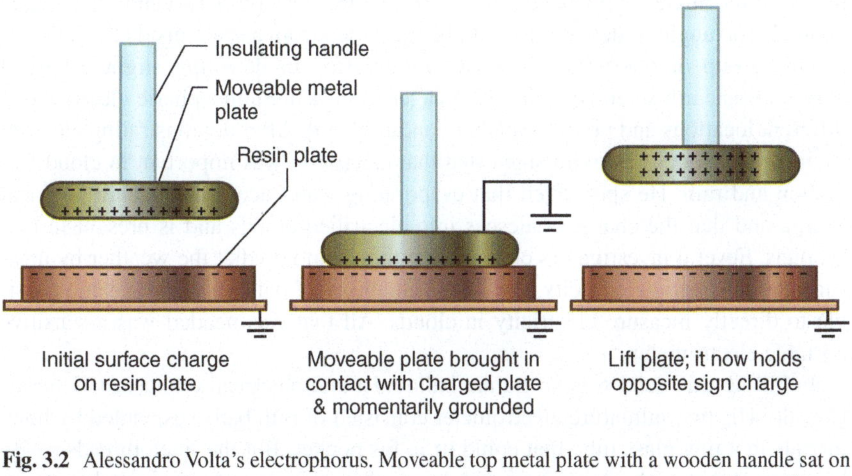

Fig. 3.2 Alessandro Volta's electrophorus. Moveable top metal plate with a wooden handle sat on top of a plate made of wax, turpentine, and resin that in turn sat on a grounded metal base. The top plate had a removable grounding chain (see text for details)

apparent loss of charge in the resin plate. It seemed like a miracle, an endless source of electricity without the need to turn a crank or repeatedly rub the resin.

There was initial great excitement regarding the practical implications of Volta's electrophorus. Instrument makers immediately began to produce devices of variable sizes claiming they could replace electrical machines, particularly for medical applications. In Germany, Georg Lichtenberg made a 6-foot diameter electrophorus with the metal plate raised and lowered by a pulley system. In St. Petersburg, the Russian empress financed a massive electrophorus. It consisted of a 9 × 4.5 feet metal plate on top of 80 pounds of Spanish wax and 180 pounds of pitch. Of course, there was work involved in charging a Leyden jar with electrophorus apparatus—lifting the metal plate and transferring it to the Leyden jar repeatedly.

Early Efforts to Quantify Electricity

It had been well known since the early eighteenth century that like electrical charges repel each other so that one of the simplest electrometers consisted of two gold leaves hinged at the top and attached to a metal bar inside of a glass jar (Fig. 3.3). Gold was known to be a good conductor of electricity. The opposite end of the metal bar extended outside of the jar through an insulating lid to protect the delicate leaves from air currents. When an electric charge was applied to the outside metal bar, such as from a Leyden jar, the electricity was conducted down the bar to the gold leaves causing them to move apart. The distance of the separation of the gold leaves was a rough estimate of the magnitude of the electric charge.

A more quantitative electrometer consisted of a metal bar with a flattened end hinged to a single light gold leaf inside a glass jar. When a charge was applied to the outside end of the bar, the loose end of the gold leaf was repelled from the metal plate and the angle of the deflection of the gold leaf was proportional to the charge applied. The angle of deflection could be measured from a scale inside the jar.

One group of researchers who were interested in detecting slight electrical charges was earth scientists who were eager to measure atmospheric electricity in different locations and predict weather conditions [6]. After demonstrating electricity in storm clouds, Franklin suggested that electricity was important in cloud formation and rain. He speculated that evaporating water acquires a small electrical charge and that the charge coalesces into electrified clouds and is present in rain droplets. Several investigators claimed that they could predict the weather by measuring atmospheric electricity. With the availability of ballooning, it became possible to directly measure electricity in clouds. All that was needed was a sensitive portable electrometer.

Fulfilling this need was an Italian balloonist and electrical enthusiast, Tiberius Cavallo. His first miniature electrometer consisted of pith balls suspended by linen threads in a tiny glass tube that could fit in his pocket. But the linen threads easily became tangled, so he made a second device that consisted of two fine silver wires with small cork balls at the bottom. A brass hemisphere capped the tiny glass tube.

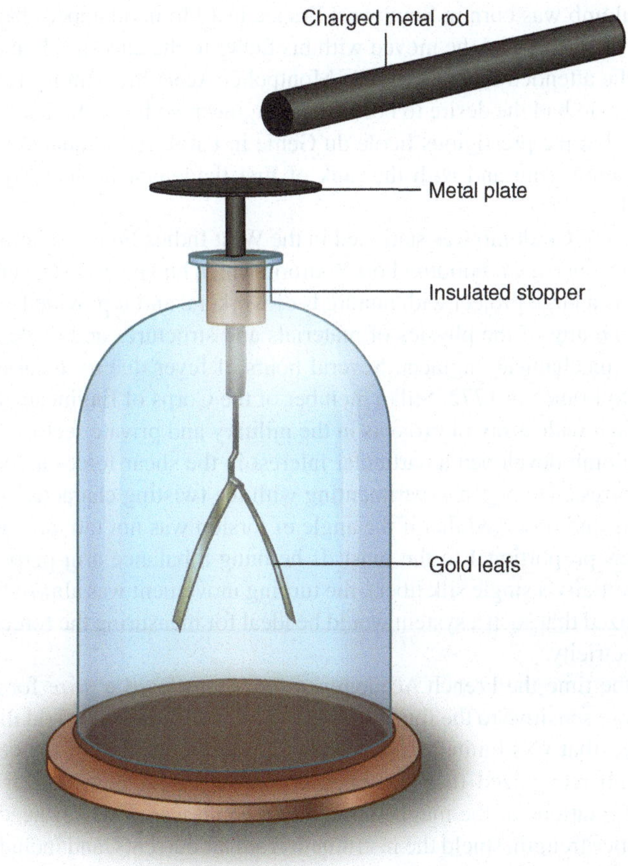

Fig. 3.3 Gold leaf electrometer. When an electric charge was applied to the outside metal plate, it was conducted down the bar to the gold leaves causing them to move apart

With this mobile device, Cavallo was able to measure atmospheric electricity in multiple locations under different weather conditions.

Coulomb's Torsion Electrometer

Charles Coulomb moved the study of electricity and magnetism from a qualitative to quantitative science [7]. He was the first to accurately quantify electrical and magnetic forces and to introduce some basic mathematics to describe these forces.

Coulomb was born in Southwest France in 1736 in an upper-class family. When his parents separated, he moved with his father to the ancestral home in Montpelier where he attended the well-known Montpelier Academy demonstrating a talent for science. He had the desire to become an engineer, so he applied and was eventually accepted at the prestigious Ecole du Genie in Paris. He graduated with a degree in Civil engineering and with the rank of first lieutenant in the Corps of Engineers in 1761.

In 1764, Coulomb was stationed in the West Indies French colony of Martinique where he oversaw rebuilding Fort Bourbon which had been destroyed by the British. This was a large project with hundreds of workers, and it provided Coulomb a good understanding of the physics of materials and structures and made him an accomplished mechanical engineer. Several bouts of fever due to malaria forced him to return to France in 1772. Still a member of the Corps of Engineers, he continued to work on a wide array of projects in the military and private sector.

Coulomb developed a particular interest in the shear forces associated with torsional stress. He began experimenting with the twisting characteristics of a variety of fibers and observed that if the angle of torsion was not too large the force of torsion was proportional to the twist. If he hung a balance arm perpendicular to the fiber (initially a single silk fiber), the turning movement was almost friction free. He recognized that such a system would be ideal for measuring the forces of magnetism and electricity.

At the time the French Academy of Science offered a prize for a compass that was more sensitive to the fluctuations in the Earth's magnetic field than the standard compass that was limited in sensitivity due to the friction of the central pivot [8]. Coulomb recognized that his torsional balance system was ideal for measuring small deviations of the magnetic field. He could use a heavier needle with greater magnetic strength, shield the instrument from air currents, and include a magnifying glass to improve the precision of measurements.

In 1776, Coulomb made his first torsion arm compass and demonstrated that it could reliably measure the daily fluctuations in the earth's magnetic field, and he was awarded the prize by the Academy in 1777. He then went on to test the torsional strength of a variety of metal filaments and noted that for a given metal the torsional strength was proportional to the angle of twist to the fourth power of the diameter and inversely proportional to the length. He could therefore select fine metal threads to produce an instrument with increased sensitivity to measure very small forces.

With his new sensitive compass, Coulomb systematically measured the declination of the earth's magnetic field throughout the day and night. He observed that the declination gradually increased in the morning, reached a peak at about 1 PM, and then gradually decreased throughout the rest of the day. It remained steady overnight. This suggested that the sun was the likely source of the daily variation and Coulomb speculated the sun had a magnetic field just like the earth. The daily fluctuation in declination was due to the interaction of the earth's and sun's magnetic fields.

Coulomb then set about making a highly sensitive electrometer. He used a similar setup to his compass, but instead of the magnetized arm of the compass, the arm of the electrometer consisted of a light insulating bar (initially a piece of straw) with a small pith ball covered with gold foil on one end and a counterweight on the other end (Fig. 3.4). He then inserted a second pith ball covered with gold foil through a hole in the cover of the glass cylinder so that it just touched the movable ball. A knob at the top of the cylinder allowed him to turn the suspended arm and adjust the distance between the two pith balls. The glass cylinder contained a graded scale to measure the separation of the two balls. He could then introduce like or unlike charges to the two balls and accurately measure the attraction or repulsion of the balls when the balls were at different distances apart.

Although others had speculated that the inverse square law was applicable to magnetic and electrical forces just as it was with gravitational force, prior instruments were not accurate enough to confirm the suspicion. With Coulomb's torsion balance instrument, he was able to convincingly show that the inverse square rule did in fact apply to both types of force and therefore Newtonian mathematics was applicable.

Coulomb developed an ingenious method to prove that the attraction and repulsion of two electrified bodies were proportional to the magnitude of the charge on each body. He reduced the charge of a suspended electrified ball next to a charged plate by one-half by touching it with an identically sized uncharged ball. When the ball and plate were like-charged, the distance between them decreased by one-half proving that the force of repulsion was directly proportional to the charge. A century later, the International Electrical Congress meeting in Paris named the unit of static electricity a Coulomb (Q) defined as the quantity of charge when placed 1 cm from an equal charge in space repels it with a force of one dyne.

Electricity to Resuscitate the Apparent Dead

Unlike the organizations today that rescue stray dogs and cats, in the late eighteenth century, humane societies were developed to resuscitate people who were found unconscious from drowning, suffocation, or some medical emergency. Beginning in the Netherlands, these organizations rapidly proliferated in large cities throughout Europe and America [9]. The goal was to train average citizens in resuscitation techniques. For example, for drowning victims several techniques were used to remove water and inflate the lungs including a syringe, bellows, and the recently discovered mouth-to-mouth resuscitation. For someone found unresponsive, if smelling salts didn't work, electricity was considered the most powerful stimulation available. By the 1780s, electric shock treatments were routinely recommended by humane societies for resuscitation of the apparently dead.

In 1796, Dr. Andrew Fothergill provided detailed instructions for the use of electrical shock to the heart in his resuscitation manual for the Royal Humane Society of London. The shock was to be passed through the heart and pulmonary vessels by placing the knob of one discharging rod above the collar bone of the right side and the knob of the other above the short rib of the left. If the shock was not effective,

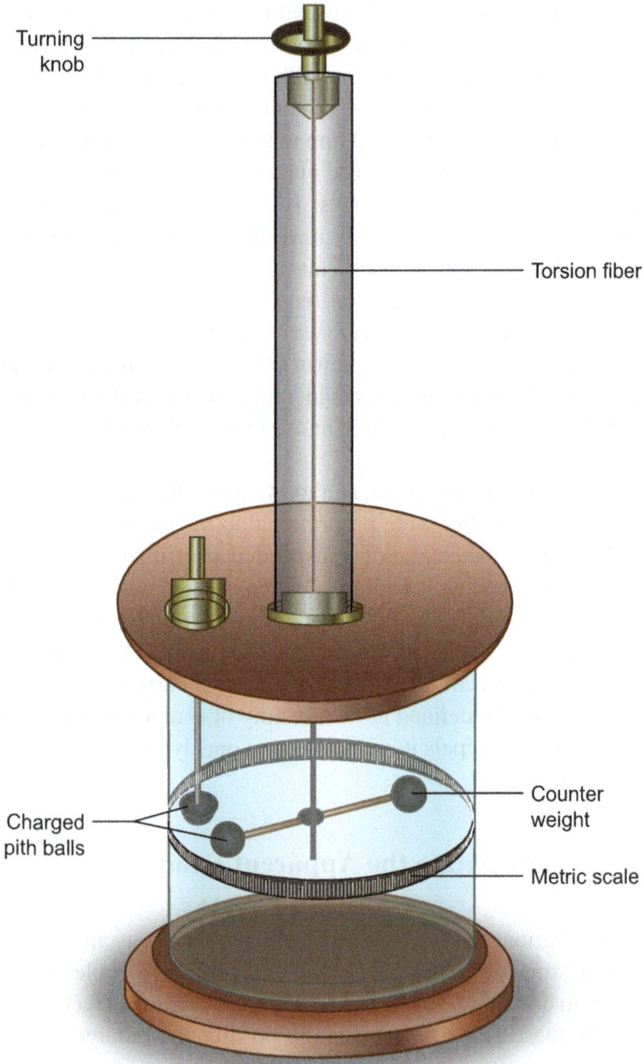

Fig. 3.4 Coulomb's torsion electrometer. Light bar balanced on torsion fiber with pith ball covered with gold foil on one end and counterweight on the other end. A second pith ball covered with gold foil was passed through a hole in the cover plate. Knob at the top of cylinder controlled the distance between the two pith balls

the placements of the knobs could be adjusted, and another shock given. He emphasized the importance of keeping the patient insulated and dry during the procedure. For administering the shock, Fothergill recommended a small electrical machine, a Leyden jar, and two insulated discharging rods.

A surgeon in the Lancashire Humane Society, Mr. Fell, designed an electrical machine specifically for resuscitation. The device was like John Reed's miniature machine also fitting into a box about a cubic foot in size. A glass cylinder, 4 inches in diameter and 6 inches in length, sat on glass supports and turned with a small crank. The charge from the cylinder went directly to a Leyden jar that was connected to glass-handled metal rods via metal chains. A lane electrometer provided a rough estimate of the shock intensity.

Reports of the revival of the apparently dead became common in newspapers and popular magazines. If people could be revived after they stopped breathing and their hearts stopped beating, what was the definition of death? Were people being prematurely buried who could possibly have been resuscitated? In response, some countries passed laws requiring a 3-day waiting period prior to burial. Of course, there was no understanding of the mechanism of the electrical revival. The cardiac pacer and conduction system were not discovered until the early twentieth century.

Electrobotany

Considering that eighteenth-century physicians and electricians were routinely applying electricity to humans, it was not surprising that they would try applying electricity to plants as well [10]. One of the first to assess the effect of electricity on plant growth was the Swiss physician Jean Jallabert mentioned in Chap. 2 for his treatment of patients with electrotherapy. Jallabert began his studies with plant germination. He compared the germination of mustard seeds in electrified water versus nonelectrified water and noted that those in the electrified water began to germinate after just 2 days while those in the nonelectrified water took several days longer. The electrified plants were also the first to develop blossoms and leaves. He repeated the experiments on mustard seeds in pots containing soil and on bulbs from several flowering plants and found the same results. His friend and collaborator in France, Abbe Nollet, conducted similar experiments and reported similar results.

A French priest, Pierre Bertholon, conducted the most extensive work on electrobotany in the eighteenth century. Bertholon was the abbe of St. Lazare and a professor of experimental physics at Montpellier. He left the priesthood late in the century. Bertholon performed a series of experiments like Jallabert and Nollet and convinced himself that electrified plants grew faster and more luxuriantly than control plants. He speculated that plants received electricity in their natural environment from the atmosphere. Rain, snow, and fog carry electricity to the Earth's surface from atmospheric clouds. The electrified water enters the plants through the roots and directly through the pores of the leaves. To test this hypothesis, he compared the growth of identical plants in two pots, one receiving rainwater and the other ordinary garden water. The plant receiving rainwater grew faster.

Bertholon was so impressed with the benefits of electricity on plant growth that he built devices that farmers could use to electrify their crops [11]. As a source of electricity, he used either atmospheric electricity or an electrical machine. To capture the atmospheric electricity he built a collector, called an "electrovegetometer," consisting of a large metal rod with multiple spike endings (like a porcupine) attached to a tall wooden pole. Electricity from the atmosphere or a machine was transferred to a large prime conductor along a metal chain. For a garden, the farmer could stand on an insulated table holding a spray can filled with electrified water or for large fields the farmer could stand on an insulated cart, moving along a track between the rows, spraying from the can. Unfortunately, Bertholon's technology was too advanced for the average farmer, and he found no takers for his electrical watering devices. We now know that electrified water including rainwater contains nitrates that can accelerate plant growth.

Electricity and Fish

As noted in Chap. 1, the fact that certain fish can produce a shock had been known since ancient times. Some ancient physicians even used the shocks from fish to treat patients with a variety of medical conditions. But was the shock due to electricity or some ill-defined mechanical phenomenon? What was the defining feature of an electric shock? Based on Ben Franklin's electrical fluid hypothesis, Joseph Priestley argued that a primary characteristic of electric fluid was its ability to flow through conductors and be blocked by insulators. In 1761, a Dutch researcher Frans van der Lott showed that the "fluid" emitted by the electric eel indeed passed through iron, lead, tin, copper, and silver and was stopped by sealing wax. But Van der Lott's work was not widely disseminated, and skeptics questioned how an electric charge could be stored in an animal surrounded by water. After all water was a good conductor (due to impurities), and if one placed a Leyden jar in a body of water, it would rapidly discharge its electricity.

One unique feature of electricity that was universally accepted was a "spark," with visible light and sound. No one had observed a spark associated with the shock from a fish. This changed when in 1776 the first successful production of a visible spark from an electric eel was described in a scientific journal published in Paris by Abbé François Rozier. John Walsh, an English scientist, and Fellow of the Royal Society conducted the experiment but never published the complete work on his own [12]. He showed that the discharge from a healthy electric eel could jump across a cut in a piece of tinfoil. Not uncommon at the time, Walsh demonstrated the experiment to several groups of scientists including 40 members of the Royal Society to authenticate his findings and wrote to Jean-Baptiste Le Roy, who transcribed lines in the letter in a correspondence that was published in Rozier's journal.

Walsh conducted his research on electric fish from 1772 to 1775 culminating in the spark experiment in 1775. He collaborated with the famous English anatomist John Hunter to study the anatomy and physiology of the electric organs initially of the Torpedo fish and then eels. However, he was not able to demonstrate a visible spark with the Torpedo fish but later was with the eel.

Walsh had read extensively on the question of electricity in fish and after discussions with Ben Franklin, in June of 1772, he traveled across the English Channel to La Rochelle, France where Torpedo fish were available. After a series of experiments on Torpedo fish, he became convinced that their discharge was indeed electricity. He published his experiments in a letter to Franklin in 1773 along with the anatomical studies conducted with John Hunter (Fig. 3.5). This letter and other work of Walsh and Hunter were published in a series of articles in the Philosophical Transactions of the Royal Society between 1773 and 1775 [13, 14].

When Walsh first arrived in La Rochelle, he sat down with local fishermen and was surprised to find that many of them had received a shock from a Torpedo fish. When he gave one a shock from a Leyden jar, the fisherman reported that it felt identical to the shock he had received from a Torpedo fish. A short time later Walsh himself experienced a shock from a Torpedo fish and he also felt sure that it was electrical. Unaware of Van de Lott's earlier work, Walsh performed a series of experiments with his nephew showing that the shock of a Torpedo could be transmitted along a metal conductor and blocked by an insulator such as glass or sealing wax. The shock could also be transmitted around a group of people holding hands in a circle just like the shock from a Leyden jar. Before returning to England, Walsh arranged a series of public exhibitions showing how the electrical properties of the Torpedo were like those of a Leyden jar.

Walsh brought several Torpedos back to England so that John Hunter could perform a detailed analysis of the electric organs. Hunter made beautiful illustrations of the organs that became widely distributed. The organs were made up of hundreds of perpendicular columns each of which consisted of about 150 horizontal discs separated by fluid spaces. The columns ran from the upper to the under surface of the fish body. Several years later, Hunter's detailed description of the electric organ of the Torpedo gave Alessandro Volta the idea to assemble a stack of metal disks to produce the first electrochemical battery.

Hunter noted that the electric organs received a rich nerve supply that upon entering the organ branched in all directions between the columns with a small branch ending upon each partition (Fig. 3.5). In his publication in the *Philosophical Transactions*, Hunter summarized his findings: "*May we not conclude that they* [the nerves] *are subservient to the formation, collection, or management of the electric fluid; especially as it appears evident from Mr. Walsh's experiments, that the will of the animal does absolutely control the electric powers of its body; which must depend on the energy of the nerves*" [15].

On analyzing his research on the Torpedo in France, Walsh wondered why the shock from a Torpedo could be felt and transmitted along a conductor, yet it did not produce a spark across a small gap like a spark from a fully charged Leyden jar. He thought that it must have something to do with the density of electricity stored in the

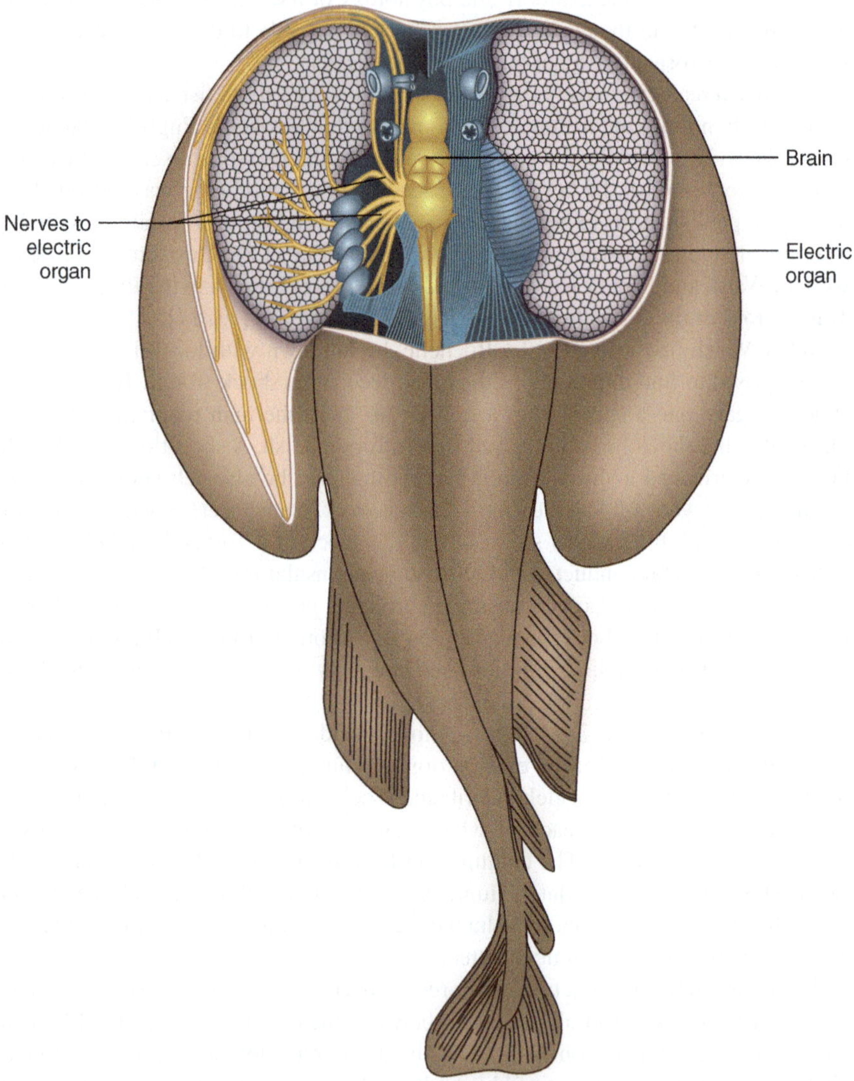

Fig. 3.5 Torpedo fish electric organs. Rich nerve supply entered the organ and branched in all directions between the columns with a small branch ending upon each partition (left side)

fish's electric organ compared to the Leyden jar. He reasoned that the effects of electricity did not simply depend on the quantity of electricity but also on the density or "tension" of the stored electricity. For example, a given quantity of electricity "condensed" in a small Leyden jar might produce a typical spark, whereas the same quantity of electricity in a less condensed state in a large Leyden jar might produce a shock sensation but not a spark.

Around the same time another English scientist, Henry Cavendish was grappling with this very problem [16]. Born into an aristocratic family and trained in physics and chemistry, Cavendish is best known for his discovery of hydrogen but his work on electricity was even more remarkable for the time. Based on a series of quantitative experiments performed in a laboratory in his house, he developed the first comprehensive theory on electricity using mathematics. Cavendish felt that "electrical fluid" exerted a pressure that controlled the flow. This electrical potential that he called the "degree of electrification" was related to the current flow by a simple mathematical relationship that later evolved into Ohm's Law. He also developed a formula for the capacitance in a Leyden jar and introduced an early unit of capacitance.

Cavendish published an early version of his theory in 1771 but most of his electrical experiments did not become known until more than a century later when they were collated and published by James Clerk Maxwell in 1879. By that time others had received credit for his many of his discoveries.

When Cavendish heard of Walsh's experiments on the Torpedo fish, he decided to create an "artificial torpedo" that would mimic a natural torpedo by producing a shock but not a visible spark. He joined together a series of large Leyden jars each with a small amount of electrical charge and demonstrated that the output could produce a strong shock yet not even a small spark. In the description of the artificial Torpedo that he published in 1776, Cavendish concluded that it was the electrical potential (electromotive force later measured in volts) and not the quantity of electricity that determined whether or not a spark occurred [17]. We now know that the electric organ of an eel can generate a much larger potential (around 600 V) than the Torpedo fish (about 50 V) explaining why the electricity from an eel and not a Torpedo was able to generate a visible spark.

Luigi Galvani's Animal Electricity

Luigi Galvani was born in 1737 in Bologna, Italy, part of the Papal States at the time [18, 19]. His mother was the fourth wife of a goldsmith of moderate means able to support his education at the University of Bologna where Luigi studied medicine. He graduated with a degree in medicine and philosophy in 1759 and a few years later became an anatomist and lecturer of surgery at the university. Luigi married Lucia Galeazzi the daughter of one of his professors and he moved into the Galeazzi house where he and Lucia helped her father conduct research experiments.

Galvani took Galeazzi's place as professor and lecturer when Galeazzi died in 1775. He began his academic career studying hearing in animals and humans, but this research was abandoned after a conflict with the famous anatomist, Antonio Scarpa regarding the originality of Galvani's work. He closely collaborated with Lucia throughout much of his career including his experiments on animal electricity but, as was common at the time, Lucia received little credit for her contributions. Later in his career, Galvani teamed up with his nephew Giovanni Aldini who took on the task of defending Galvani's work.

In 1780, Galvani and his wife were dissecting the sciatic nerve (the main nerve to the leg) of a dead frog on a table near an electric machine. When the metal scalpel was in contact with the nerve, it inadvertently picked up a spark from the machine and the frog's leg jumped as if the animal came alive. Every time the machine sparked the frog's leg jumped. The implication of this chance observation came immediately to Galvani. Electricity activated the nerve to the muscle. The long sought after *nervous fluid* was electricity.

Galvani performed a series of experiments to test his hypothesis [20, 21]. He showed that static electricity generated by an electrical machine and stored in a Leyden jar could trigger a muscle contraction whether he stimulated the spinal cord, the sciatic nerve, or the muscle directly. On a stormy evening, he produced muscle contractions by connecting a frog's sciatic nerve to a long wire running upward into the sky showing that atmospheric electricity worked just as well. He then observed that he could trigger muscle contractions without any obvious source of static electricity while manipulating a frog preparation hung on his iron balcony with a brass hook through the spinal cord. He noticed that by touching the hook through the spinal cord against the iron railing in contact with the frog's leg muscle, the muscle contracted. He had produced an electric circuit causing electricity to flow from nerve to muscle. Consistent with this idea, the muscle contractions were stopped by inserting insulating material between the metal conductors, in place of one of the conductors or within the nerve to the leg.

Galvani decided that the animal must generate intrinsic electricity (*animal electricity*) that flowed from nerve to muscle. He speculated that each muscle fiber was like a tiny Leyden jar with an electric charge stored on the inside. Electricity carried in a penetrating nerve could trigger the flow of electric fluid from inside to outside of the muscle fiber causing it to contract like connecting the inside and outside of a Leyden jar. He also speculated that nerve fibers must have some type of surrounding oily insulating material to prevent the electricity from spreading into the surrounding tissue. Galvani published his findings and speculations first in Latin in 1791 and then in an Italian translation in 1792.

Volta and His Bimetal Theory

After his early successes with electrophorus, Alessandro Volta became professor of experimental physics at the University of Pavia in 1779, a position he held for almost 40 years. When he read Galvani's work in March of 1792, Volta initially expressed admiration for Galvani's discovery of *animal electricity* but after conducting a series of his own experiments he became skeptical. He quickly became aware that contacting any two metals forming a bimetallic arc would trigger muscle contraction in a frog, whereas contacting two pieces of the same metal did not cause contraction. For example, in one experiment he attached a silver coin to one thigh of a dead frog and wrapped the other thigh with tin foil. When he connected the two metals with a conducting thread, the frog's legs jerked repeatedly. He concluded that the metals were not triggering the release of intrinsic animal electricity but rather were generating external electricity. In essence the dead frog was acting as an electrometer sensing the electricity produced by the bimetallic arc [22, 23].

It had long been known that applying electricity to the tongue could stimulate the gustatory nerves producing an acidic taste. To further test his bimetallic hypothesis, Volta placed two pieces of different metals on his tongue and after touching the ends of the metals together he experienced an acidic taste. Furthermore, he experienced the same acidic taste if he put the two pieces of different metals in a cup of water and contacted just one with his tongue. He published the first version of the two-metal theory of electricity in late 1792.

Volta had noticed that the vigor of the frog's muscular contractions varied with different metal combinations. To further investigate this phenomenon, he dispensed with the frog preparation altogether and used a more sensitive instrument for detecting small amounts of electricity, which he called a condensing gold leaf electrometer. This device consisted of his electrophorus sitting on top of a gold leaf electrometer. The contact electrical potential between two dissimilar metals was tiny, too small to be accurately detected by an ordinary gold lead electrometer. With the condensing electrometer, these tiny electrical potentials were summated in the dielectric between the upper movable plate and the lower plate attached to the gold leaf electrometer. Volta made a series of upper plates each made of different metals with an insulating handle. He then systematically contacted the different plates together and measured the amount of electric charge on the lower disc using the condensing electrometer.

The fact that different metals produced a different electric charge when in contact not only proved his two-metal theory but also allowed him to rank the electricity-generating potential of different metal combinations. Volta determined whether a charge was positive or negative by rubbing a glass or resin rod and bringing the rod close to the gold leaves to see whether it increased or decreased the separation. For example, he found that with a zinc/copper contact, the zinc became positively charged and the copper negatively charged. Using these techniques, he generated the electrification and polarity relationship of the different metals from positive to negative as follows: positive—zinc, lead, tin, iron, copper, silver, gold, graphite—negative. The further the metals were separated on the list the greater the strength of electrification when they were joined together.

The Galvani/Volta Debate

Based on his findings, Volta concluded that there was no need to postulate "animal electricity" since in all cases the electricity that produced muscle contractions in a frog came from an outside source [23]. For his part Galvani stuck to his animal electricity theory despite Volta's findings. In many ways, they were *comparing apples and oranges*. Both had made important discoveries but neither fully understood the implications.

To counter Volta's argument, Galvani showed that the muscle could be made to contract in several experiments that did not involve metal at all [21]. He first showed that if he immersed the frog leg and the cut sciatic nerve in two separate containers filled with water he could cause the muscle to contract by connecting the water in the two containers with a piece of muscle or moistened paper. He then showed that the leg muscle would contract if he simply touched it with the cut end of the sciatic nerve. Finally, he noted that touching the cut end of the sciatic nerve of one frog nerve—leg preparation to the outside of the nerve of another nerve—leg preparation caused both frog legs to contract.

Although these experiments made Volta reconsider his original criticism of Galvani's animal electricity hypothesis, he still questioned whether there might be some other explanation like a bimetal, possibly contact between two different types of tissue that could generate week electrical effects. One important feature of these experiments that Galvani and Volta both failed to appreciate was the critical importance of contact between a cut and the intact surface of a nerve or muscle to trigger muscle contraction.

Galvani preferred not to become involved in the dispute but his nephew, Giovanni Aldini enthusiastically defended Galvani's position [24]. Although a physicist with no formal training in medicine, Aldini worked in his uncle's laboratory for years and was a strong supporter of animal electricity. Along with Galvani he set out to prove that muscle contraction in the frog could be triggered internally without the need of any external source of electricity including bimetals. They discovered what would later be called an "injury current" which can occur when a nerve is crushed or cut. This was strong support for an internal source of electricity, but it did not refute the bimetal explanation for Galvani's initial experiments.

Gradually Volta moved away from the animal electricity question and focused on the bimetal generation of electricity. He noticed that joining together two metals produced only a transient movement of electricity. However, if the metals were separated by a material soaked with saline or an acidic solution there was a continuous flow of electricity. This was a critical step in his discovery of the electrochemical battery. When Volta heard about Aldini and Galvani's new experiments, he conceded that he probably had gone too far in suggesting the need for an external source of electricity in the frog preparation. But he wasn't ready to accept the notion of animal electricity and stuck by his interpretation of Galvani' original experiments.

For his part, Galvani truly believed that he had discovered the nature of the ancient concept of "animal spirits." *"We believe, therefore, that the electric fluid is prepared by the activity of the cerebrum, that it is extracted in all probability from the blood, and that it enters the nerves.... If this be the case, perhaps at last the nature of animal spirits, which has been hidden and vainly sought after for so long, will at last come out with clarity"* [25]. His idea of an oily insulating material surrounding the nerves explained why the electricity remained in the nerves and did not dissipate into the body fluids. Scientists at the time lined up on each side of the Galvani/Volta debate but some suggested a compromise. Maybe there were different types of electricity? Perhaps Galvani's animal electricity and Volta's bimetallic electricity were just two varieties of a family of different types of electricity.

In the interval, historical political events overshadowed the scientific dispute. Napoleon's army invaded Italy in 1796 forcing Austria to withdraw. Galvani refused to sign allegiance to the new republic and in April 1798 he was discharged from his professorship. He died a few months later. Although not completely in agreement with the new government, Volta made accommodations and continued his research.

Perkins Metallic Tractors

In 1796, an American physician in Connecticut, Elisha Perkins, read about Galvani's experiments and immediately saw a practical application. He invented a new therapeutic instrument that he called the Perkins Metallic tractor [26]. It consisted of two tear-shaped metal rods, one made of iron and the other of brass, each about three inches long. By touching the skin over the affected body part, he could cure a variety of ailments from rheumatism to epilepsy, presumably by releasing animal electricity. The instrument was a huge success marketed to both physicians and laypeople (requiring no special training). George Washington even bought the instrument. It sold for about $25, close to about $1000 in today's currency.

With the initial popularity in America, Perkins decided to branch out to Europe sending his son Benjamin to England to market the instrument. They received an American patent in 1796 and an English patent in 1798. The tractor was widely popular, and the Perkins family became rich. They sponsored the Perkins Medical Institute in London that focused on a school of medical thought called Perkinism. When Elisha died in 1799, Benjamin took over the company and continued to claim miraculous cures. But many scientists and physicians were skeptical of the claims.

In England, several physicians conducted tests using identical instruments made of other materials, including nonmetal materials, and found that the "fake" instruments were equally effective in curing patients if the patients thought that they were being treated with genuine Perkins tractors. Once these experiments were published, Perkins metallic tractors gradually lost their appeal. Cartoonists ridiculed the tractors in magazines and newspapers and physicians stopped using them. Interestingly, these types of controlled experiments became a model for modern placebo-controlled treatment trials. They also demonstrated the importance of the patient's beliefs in medical therapy.

Volta's Electric Pile

Once Volta recognized that he was able to generate a flow of electricity with two metals separated by a liquid conductor, he immediately surmised that he could multiply the electrical output by joining together several bimetal elements. The result was an electric pile or "voltaic pile" consisting of a vertical stack of metal discs, each pair separated by pasteboard discs soaked in salt water (Fig. 3.6). With a stack

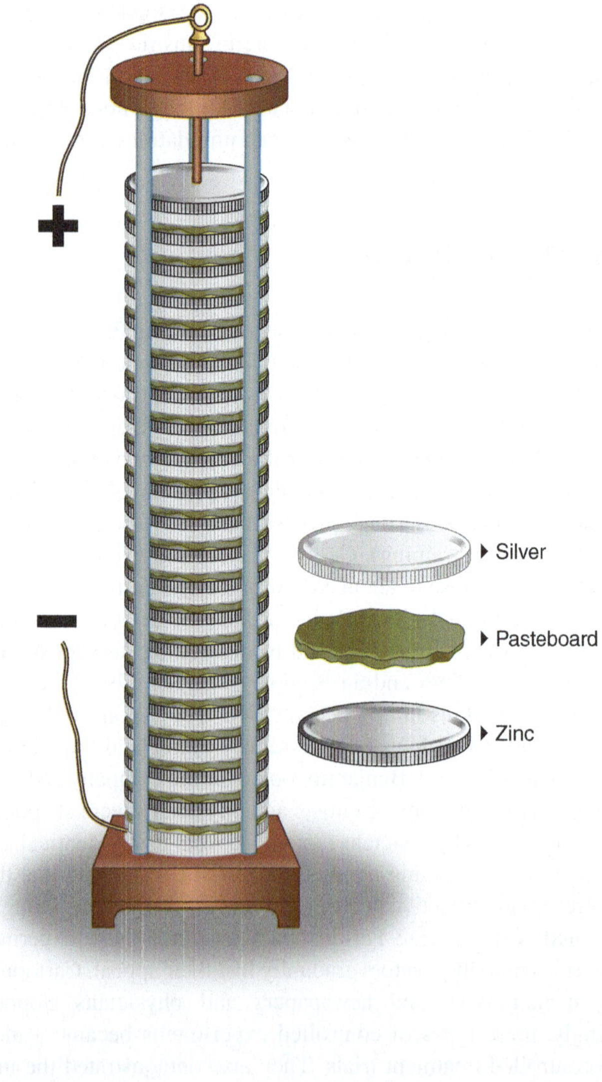

Fig. 3.6 Voltaic pile. Stack of 20 silver-zinc pairs separated by pasteboard discs soaked in salt water

of just 20 silver–zinc pairs, he was able to markedly increase the separation of the gold leaves on his condenser electrometer and produce a palpable shock. Unlike the momentary shock produced by the discharge of a Leyden jar, the electric pile produced a more persistent shock sensation.

Volta described the results of his initial experiments on the electric pile in a letter to Sir Joseph Banks, President of the Royal Society in March 1800. He began the letter with the provocative declaration that, unlike the Leyden jar his invention could provide an "inexhaustible" source of electricity. *"The apparatus to which I will allude, and which will, no doubt, astonish you, is only the assemblage of a number of good conductors of different kinds arranged in a certain manner. Thirty, forty, fifty or more pieces of copper, or rather silver, applied each to a piece of tin or zinc, which is better, and as many strata of water, or any other liquid which may be a better conductor, such as salt water, lye, or pieces of pasteboard, skin, &c well soaked in these liquids, such strata being interposed between every pair or combination of two different metals in an alternate series, and always in the same order of these three kinds of conductors, are all that is necessary for constituting my new instrument…"* [27].

There were technical problems with Volta's electric pile in that the liquid in the soaked pasteboard initially would drip over the pile and the pasteboard discs would easily dry out. To deal with this problem, Volta substituted a series of glass cups partly filled with salt or acid solution in place of the pasteboard. He dipped a plate of copper and zinc into each cup and joined the plates together in a series in the same fashion as with the vertical pile. Volta called this apparatus his "crown of cups." With the crown of cups, he was able to show that the strength of the electrical current depended on the number of cups, the concentration of the liquid in the cups, and the size of the metal plates. For example, he received a greater shock if he placed a finger from each hand into the fluid of cups far apart in the series than if he placed the fingers in cups that were close together.

In his letter to the Royal Society, Volta compared his invention to the electrical organs of the torpedo fish and electrical eels. He wrote: *"To this apparatus, much more similar in its form to the natural electric organ of the torpedo or electric eel than to the Leyden flask, I would wish to give the name of the artificial electric organ; and, indeed, is it not like it, composed entirely of conducting bodies? Is it not also active of itself without any previous charge, without the aid of any electricity excited by any of the means hitherto known? Does it not act incessantly, and without intermission? And in the last place is it not capable of giving shocks which produce the same torpor in the limbs as occasioned by the torpedo?"* [27].

Although he had his condensing electrometer for identifying tiny electrical currents and for quantifying the magnitude of the electricity, for many of his experiments Volta relied on the subjective sensation of shock in his fingers for a quick estimate of the amount of electricity. He was also curious about the effect of electricity on his other senses [22]. As suggested previously, he produced an acid taste by passing an electric current to his tongue. Passing an electric current to his eye on one occasion produced a flash of light and on another occasion produced a quivering of the lips and tongue in addition to a flash of light. His most dramatic self-experiment involved electrifying his ears. He connected two metal rods to an electric pile with about 30 metal pairs and stuck the rods into his ears. He experienced a violent shock in his head along with a sound that was difficult to describe. The shock and concerns about the dangers of electrifying the brain cut short any further such experiments.

One could certainly argue that these experiments supported Galvani's theory of animal electricity. Although Volta eventually accepted that electricity had something to do with nerve transmission in the brain, he felt that it was ordinary electricity, not a separate "animal electricity." He suspected that the brain generated electricity like the mechanism of his electrical pile or the electric organs of fish. Paradoxically, Volta introduced the term "galvanism" to describe the electricity generated by his electrochemical battery and later the term became commonly used to describe the therapeutic use of electricity from batteries.

Volta's Later Years

Volta's letter to the Royal Society was a bombshell to scientists in England and France working on electricity. New improved versions of the Voltaic Pile rapidly proliferated. For example, in 1800 the Scottish professor of chemistry, William Cruickshank, designed the trough battery consisting of a long box with partitioned cells for liquid into which metal plates were inserted. The trough design overcame many of the limitations of Volta's pile or crown of cups design. As to the mechanism of this apparent unlimited source of electricity, Volta simply speculated that contacting two different metals in the presence of a liquid released electricity. The liquid was simply acting as a conductor between the bimetals. He apparently never considered chemical reactions were involved although evidence for a chemical origin of the electricity would soon be uncovered.

Volta suddenly was the most famous scientist in Europe. Voltaic piles rapidly replaced electrical machines and Leyden jars. Instrument makers were soon displaying a wide variety of batteries. Napoleon Bonaparte heard one of Volta's lectures and became a huge fan, showering him with honors. At age 50, Volta made a dramatic change in his lifestyle when he decided to marry and begin a family. He continued to dabble with scientific experiments, but this was no longer his passion. Volta turned to politics becoming a senator and a Count in the Napoleonic empire. When Napoleon was finally defeated, Volta retired to his country estate in Camargo where he died quietly at the age of 82 in 1827.

The same International Electrical Congress in Paris in 1881 that defined the Coulomb (Q) as the unit of electrical charge, defined the Volt as the unit of electromotive force (E) in honor of Volta. Volta was aware of the difference between the tension or electromotive force and the charge or current in an electrical circuit. He pointed out the difference between the high-tension, low-current output of the commonly used electrical machines and the low-tension, high-current output of his electrical pile [28]. The electromotive force implies the ability to do work and the definition of a volt is the potential difference such that one joule (J) of work is done in driving one coulomb (Q) of charge, from one point to another in a circuit.

Giovani Aldini and Early Electrophysiology

By remarkable coincidence, Galvani's experiments on animal electricity led to Volta's discovery of the bimetal electric pile that in turn became a source of electricity (called galvanism by Volta himself) for future studies in electrophysiology. Researchers finally had a tool to study Galvani's theory about an electric nervous fluid (galvanic fluid) that originated in animal bodies and irritated nerves and muscles when flowing through them. First among these researchers was, Galvani's nephew, Giovani Aldini, who took over Galvani's laboratory and performed a series of experiments on the effects of galvanic stimulation of the nervous system in dead animals and humans [29]. Recall that Aldini's training was in physics and not medicine.

According to Aldini, his goal in performing the experiments was to: *"Convey an energetic fluid to the seat of all sensations; distribute its force throughout the different parts of the nervous and muscular systems; produce, reanimate and, so to speak, control the vital forces: this is the object of my research, this is the advantage that I intend to collect from the theory of galvanism. The discovery of the pile of the famous professor Volta served me as a torchlight throughout a long series of experiments and multiplied works that yielded interesting results"* [30].

Despite Aldini's continued debate with Volta regarding animal electricity, he recognized the importance of Volta's discovery of the electric pile.

Aldini began with a series of experiments on the effects of galvanism on the brain in freshly killed animals (birds, sheep, and oxen) [24]. The animals' heads shook from side to side, their eyes turned, and the tongues protruded almost as though they were alive. In one ox brain, he stimulated different regions of the brain to see the relative sensitivity to galvanism. Although movements were obtained from stimulation in many regions the most vigorous movements occurred after the stimulation of the cerebellum and the corpus callosum (the main fiber tract connecting the two cerebral hemispheres).

Even more dramatic reactions occurred after galvanizing fresh human corpses. For example, Aldini connected one terminal of a voltaic pile to a human corpse's hand and the other terminal to an ear and observed a variety of movements of the arms, hands, and fingers at times appearing purposeful. He even speculated that the movements might have reflected the person's premorbid temperament. Since these initial studies were performed on people who had died of diseases and already showed muscle wasting, Aldini was concerned that the brain's vital fluids might be diminished. He therefore asked permission from the government of Bologna to study executed criminals.

In January and February of 1802, with the help of several physicians, Aldini galvanized the bodies and heads of decapitated criminals in a large public square near the Bologna Palace of Justice. He found that the best responses were obtained within about 3 h of death. After filling the ears of a fresh head with salt water, Aldini inserted the terminal wires from a voltaic pile in each ear. All the muscles of the face contracted violently producing a frightful appearance. He noted that the eyelids became particularly animated. Aldini took two heads and after touching them together on a table, he connected the voltaic pile to the right ear of one and the left ear of the other. Even Aldini was startled by the result: *"it was surprising and even frightful to see these two heads making at the same time horrible contortions, as if at each other, so that some of the spectators who were not prepared for such results, were exceedingly terrified"* [31]. One can only imagine the reaction of the casual bystander to these public displays.

Aldini obtained the same muscle contractions throughout the bodies of the beheaded criminals with galvanization of nerves and muscles as he had seen in animals and humans in previous experiments, but he was surprised that the heart was relatively resistant to galvanization. The heart was a mystery. It had a rich supply of nerves, yet electrical stimulation of the nerves did not produce heart muscle contractions. The nerves to the heart appeared to serve a different function than those to skeletal muscles.

With the help of a neuroanatomist, Aldini galvanized various regions of the human brain confirming his findings in the animal experiments. He likened the response to an epileptic seizure. Remarkably he noted that the stimulation of one side of the cerebral cortex caused the movement of the opposite side of the face indicating a crossing of the brain's nervous control of facial movement. It would be many years before the significance of this finding was appreciated.

Late in 1802, Aldini, took a trip to France and England to show the broader scientific community his experiments on animal electricity. He lectured and demonstrated galvanism on animal and human corpses as he had done in Bologna. Public demonstrations were made in large amphitheaters before wide audiences including surgeons, physicians, lay ladies, gentlemen, and often royalty. Local Humane Societies financed many of Aldini's exhibitions.

The beheading of criminals in Italy limited Aldini's experimental options so on arrival in London where criminals were hung and not beheaded, he ordered a freshly hung man for a demonstration at the Royal College of Surgeons in January 1803. Aldini received the body of a 26-year-old criminal who had murdered his wife and child, George Forster. Forster's trial had been well publicized, but Forster experienced more notoriety in death than in life.

With the help of an anatomist and surgeon, Aldini galvanized Forster using a combination of three of Cruickshank's new trough batteries each consisting of forty plates of zinc and copper. As the large crowd watched with nervous anticipation, stimulating rods were applied to Forster's ear and mouth causing his jaw to quiver, his facial muscles to contort into a frightful expression, and his left eye to open. He even appeared to take a breath. Was he coming back to life? The responses were only transient, however, and Aldini later claimed that he had no intention of reanimating Foster. Sensational details of the exhibition were reported in all the London newspapers including *The Times*. Aldini was portrayed as a new Dr. Faust probing the forces that govern human existence.

When Aldini lectured at the Salpêtrière Hospital in Paris, the famous psychiatrist, Philippe Pinel, attended Aldini's demonstrations. Pinel was surprised to find that Aldini had used galvanism to treat a variety of patients with psychiatric illnesses in Bologna. Ben Franklin and others had used a battery of Leyden jars to treat psychiatric patients, but Aldini was the first to use a voltaic pile.

Aldini's best-documented patient was a 27-year-old farmer, Luigi Lanzarini, who was confined to the Santo Orsola Hospital in Bologna in early 1801 for typical symptoms of major depression (melancholy madness). He began Lanzarini's treatment with a small voltaic pile consisting of 15 bimetal discs and increased the size of the pile in later treatments. With the patient sitting next to the voltaic pile, he had Lanzarini place one hand on the bottom of the pile and ran a conductor from the top of the pile to a small moistened, shaved area at the top of Lanzarini's head (over the parietal bone on one side). After 6 weeks of treatment, Lanzarini's mood gradually improved to where Aldini considered him cured. Aldini had Lanzarini stay at his house for a few days to be sure he remained well and then let him rejoin his family. There was no long-term follow-up.

While at the Salpêtrière, Aldini and Pinel tried the voltaic pile treatment on a variety of psychiatric patients but apparently the results were not dramatic. One problem was that the patients were terrified of Aldini's voltaic pile. They had to be restrained to administer the therapy. After Aldini left Paris, Pinel continued with the treatments but never published the results. It is important to note that this was not electroconvulsive therapy. There was no attempt to induce convulsions. The treatment best fits what is now called transcranial electrical stimulation where low voltage electricity is passed through the skull to treat a wide variety of neurological conditions. Along with transcranial magnetic stimulation, these procedures are currently all the rage in modern neurology and both techniques are even being recommended to normal people to improve their brain functions, particularly memory.

On returning to Bologna, Aldini abandoned his work on medicine and biology and focused his scientific career on physics, as chair of physics at the University of Bologna. He worked on several government projects including construction and illumination of lighthouses and using asbestos fibers to protect firefighters. Unlike his uncle, Aldini embraced Napoleon's government and in recognition of his scientific contributions, he was made a knight of the Napoleonic Order of the Iron Crown. In 1807, he left his chair in physics at the University of Bologna and became a councilor of the state of Milan and superintendent of Milan's main library in the state capital of Lombardy. He died at age 72 in Milan and left all his and Galvani's manuscripts to the University of Bologna.

Volta's Electric Pile and Electrotherapy in the Early Nineteenth Century

Toward the end of the eighteenth century, although electrotherapy was still popular among much of the public, a large segment of the medical community began to question its benefit, and many considered it to be outright quackery. Treatments typically consisted of charging the entire body with static electricity derived from an electric machine or a battery of Leyden jars and then drawing off sparks from areas selected for treatment. The problem was that therapists had little control over the quantity of electricity administered and no clear idea of what the electricity was doing to the human body.

With the work of Galvani and Aldini suggesting a role for electricity in nerve and brain function and with Volta's electric pile providing a controllable and continuous source of electricity, electrotherapy had a resurgence in the early nineteenth century. Galvanism with an electric pile was used to treat the same wide variety of medical illnesses as with earlier electrotherapy including rheumatism and gout but there was a new emphasis on treating neurological conditions such as vision and hearing loss, paralysis associated with stroke, spinal cord injuries, and peripheral nerve injuries.

The life and work of the Scottish scientist and inventor, James Dinwiddie, nicely illustrates the transition from electrical machines to voltaic piles in electrotherapy [32]. The son of a rural farmer, Dinwiddie was educated in natural philosophy and mathematics in Edinburgh and after a career as an itinerant lecturer and public demonstrator, he decided to settle in Calcutta, India, where he was offered a teaching position at a newly constructed college. Shortly after arriving in Calcutta, Dinwiddie received word from Europe of Volta's discovery of the electric pile. Although the voltaic pile was relatively easy to assemble it was not at all clear whether the newly described "galvanism" was the same as traditional static electricity generated by electrical machines.

Dinwiddie, who had regularly worked with a traditional static electricity generator, immediately arranged for workmen to prepare metal plates required for a voltaic pile and he ordered sulfuric acid for the separating cloth pads from a local plant. On his 55th birthday in 1801, he produced his first electric pile of 20 pairs of metal plates and shortly after was using it along with a static electricity generator in his student demonstrations. Dinwiddie immediately noted that the voltaic pile was a reliable and powerful substitute for the electrical machine. It was not subject to the variations in humidity so common in India that could disable his static electric machine. Often in the rainy season, static electricity could not be produced. The students were captivated by the "shocks and flashes" induced by his first pile even those who previously showed little interest in his experiments. A minor inconvenience was the need to regularly remove corrosion that built up on the metal plates of the pile.

In early1802, Dinwiddie took a new electric pile consisting of 50 pairs of plates outside the college for demonstrations in the general community. He proudly proclaimed that his pile was as large as those being used in the best European research centers. Dinwiddie's social status improved among the colonial elite, particularly for his role in the use of galvanism for healing. Although Dinwiddie had previously used static electrical machines for healing, he concluded that the voltaic battery had advantages in portability and reliability and was effective in cases that failed traditional static electrical therapy. Although Dinwiddie had no medical training, like Ben Franklin he was often called "Doctor Dinwiddie."

Among his many patients treated with galvanism, Dinwiddie described alleviating the numbness in a man who suffered paralysis of one side of the face (probably Bell's palsy), improving the hearing in a deaf army captain, and improving sensation in a man with paralysis below the elbow of the right arm. He reported treating the eye of a woman who developed blindness after lifting a heavy weight and noted that although he did not restore her vision she did experience a flash of light in the blind eye. Dinwiddie suffered from an array of personal symptoms and illnesses and regularly treated himself with galvanism. He experienced a gradual hearing loss that he thought was improved by galvanism. He described hearing a strange hard to describe sound during the therapy. Over time, he gradually increased the size of his voltaic pile reaching 120 metal plate pairs.

Dinwiddie published a short report in the *Calcutta Gazette* summarizing his experience with galvanism suggesting that it had proven successful in treating rheumatism, headache, paralysis, deafness, and melancholy madness. He followed the reports of successful treatments with galvanism throughout Europe and was particularly fascinated with the work of Giovani Aldini. In June of 1804, Dinwiddie decided to give up galvanic healing after being overwhelmed with potential patients most of whom could not pay. He did continue to carry on lucrative demonstrations using the voltaic pile and started a small business making voltaic piles for medical treatment.

In 1806, having accrued considerable wealth, Dinwiddie decided to return to Britain. He took his voltaic pile with him on his return voyage treating the ailments of his fellow travelers on the way. When he arrived in London, he ordered a new voltaic pile with 120 zinc and copper plates from a Scottish instrument maker and soon began treating himself and selected Scottish friends. He reported curing himself of a large hydrocele in one of his testicles and of ulcers in his legs and successfully treating lumbago and cramps in his friends. He also began to attend the lectures and demonstrations of fellow electricians throughout London. Part of Dinwiddie's appeal in London society was an aura of exoticism associated with his time spent in the Far East.

Dinwiddie was particularly taken by the lectures of Humphrey Davy at the Royal Institution. He regularly attended the lectures and was eventually elected a member himself. Dinwiddie marveled at the size of Davy's audiences, as many as six or seven hundred genteel ladies and gentlemen, but mocked Davy's preening to the women in the audience. Davy had a reputation among the ladies for being a dandy. Dinwiddie boasted with patriotic pride that Davy's voltaic pile consisting of 2000 pairs of plates in a series of wooden troughs was larger than the one that the tyrant, Napoleon had built in Paris. But using the massive battery did have growing pains.

Dinwiddie described an episode in June 1810 that led to a chaotic and embarrassing scene. Apparently, there was faulty wiring that connected the lecture theater with the neighboring room containing the 2000-plate battery causing Davy to repeatedly visit the battery room during the lecture releasing noxious gas into the lecture theater. The audience became overwhelmed with the fumes and made a mad dash for the exits. Davy solved the problem by moving the battery to a basement room below the lecture theater and ran a new set of wires up through the floor. Davy reassured the audience and a few days later he conducted his demonstrations without any further noxious fumes.

As for Dinwiddie, he continued to exhibit galvanism in small gatherings and to use it therapeutically mainly on himself and his friends but he did not have the charisma to attract the large fashionable audiences of Davy and other prominent London lecturers. He thought about returning to his homeland, Scotland but remained in London where he died in 1815.

The Nervous System as Battery and Wires

Aldini's demonstrations that electricity could activate what appeared to be intentional movements in dead bodies was strong evidence for animal electricity independent of Volta's bimetal production of electricity. To many of Aldini's spectators, electricity was the elixir of life. They were convinced that the dead bodies were transiently revived. Electricity was the long-sought nervous fluid that controlled muscular action and perceptions even pleasure and pain. The brain was simply a voltaic pile passing electricity throughout the body via nerves acting as wires.

In Germany, the natural philosopher, Johann Wilhelm Ritter, concluded that animal electricity and bimetal electricity were essentially the same thing [33]. Part of the so-called German romantic natural philosophy movement, Ritter saw animal electricity as an all-pervasive animating fluid for the entire universe. He was convinced that electricity was the driving force of life. Strongly influenced by Ritter, a German physician, Karl August Weinhold set out to demonstrate that the brain worked like a battery with attached wires.

After obtaining his medical degree in Wittenberg in 1805, Weinhold traveled through Europe studying medicine and electrotherapy before holding several medical appointments throughout Germany. He became a professor of medicine and surgery in Halle where he carried out his infamous experiments on kittens described in his 1817 book, *Experiments on Life and its Primary Forces through the Use of Experimental Physiology* [34].

Weinhold was described as an unattractive, abrasive man who cared little about what others thought of him. He was vilified in the press for his crusade to eliminate poverty by preventing poor people from having babies. One of his solutions was to place a ring on men's genitals to prevent ejaculation of sperm. In his book, Weinhold claimed that he could revive kittens after the brain and spinal cord were removed by filling the cavities with bimetal mixtures of zinc and silver or mercury.

Weinhold described one experiment on "a 4-week-old very lively kitten" as follows: *"I removed the cerebrum and cerebellum, as well as the spinal cord, with a small spoon through an opening at the back of the head. After this the animal lost all life, all sensory function, voluntary muscle movement, and eventually its pulse. Afterward, I filled both cavities with the aforementioned amalgam [zinc and silver]. For almost 20 minutes, the animal got into such a life-tension that it raised its head, opened its eyes, stared for a time, tried to get into a crawling position, sank down again several times, yet finally got up with obvious effort, hopped around, and then sank down exhausted. The heartbeat and the pulse, as well as the circulation, were quite active during these observations, and continued after I opened the chest and abdominal cavities 15 minutes later"* [35].

Weinhold believed that he was not simply stimulating nerves with his bimetals but recreating "physical life" in the kittens. In addition to rejuvenating circulation and heart function, he claimed that the treated kittens responded to sensory stimuli such as loud sounds and bright lights.

The notion that nervous fluid and galvanism might be the same thing was taking hold in segments of the scientific community in Europe and England even though there was little agreement as to how the nervous system worked, and how signals were transmitted to and from the brain. There was a developing consensus that electricity existed in the human body but not on what it was doing there. A strong Aldini supporter, the English surgeon and galvanic lecturer, Charles Wilkinson wrote: *"On the supposition that galvanism is the intermediate principle between matter and spirit, I cannot, I must confess conceive the mode in which the agency is effected. To comprehend the essence of our own animation, requires the powers of a principle superior to that we possess"* [36].

The Battle of the Bodies

Although Aldini's flamboyant exhibitions of dead animals and executed convicts were impressive to many in the lay and scientific community, not everyone was ready to accept his claims. Even supporters such as Charles Wilkinson had difficulty reproducing several of Aldini's claims.

At the time in Britain, it was not uncommon for executed criminals to undergo public dissection and galvanization. Two such events, one in 1818 and a second, in response to the first, a year later, nicely illustrate the bizarre nature of these exhibitions (and the sad state of the justice system). The first was conducted at the University of Glasgow, Scotland, by a professor of anatomy, James Jeffrey, and a professor of natural philosophy, Andrew Ure.

Mathew Clydesdale had been convicted of a brutal murder, savagely beating an old man with an axe, and the judge sentenced him to be hanged and delivered to Jeffery for public dissection. Despite failed attempts to escape and to commit suicide by slashing his wrists, Clydesdale was bandaged and taken to the gallows where he was hanged in front of a large crowd. A detachment of troops along with the First Dragoon Guards were there in case the crowd got out of hand. The body was cut down, placed in a coffin, and transported across town in a cart to the university's anatomical theater.

In front of the cart marched town officials, dressed in traditional uniforms of red coats, blue breeches, and white stockings with their battle axes towering above them. The cart was followed by many of the crowd and soldiers from the hanging. The procession moved slowly along High Street to the college gates with people crowding the widows and roofs along the way for a glimpse of the action. A public dissection was to be part of the festivities.

The plan was for Jeffery to dissect and Ure to galvanize the body. Adding to the intrigue, Ure had achieved notoriety earlier in the year when he discovered his wife was pregnant with another man's child and successfully sued for divorce. The other man, a professor of anatomy with a reputation of grave-robbing for cadavers, claimed that Ure planned the whole affair to disgrace his name. The drama, played out in the Glasgow press, was quite a sensation among the highly conservative Presbyterian population.

Ure had acquired a powerful battery with 270 pairs of metal plates in a solution of nitrosulfuric acid. With Jeffery and Ure standing over Clydesdale's body, Jeffery made careful incisions to expose nerves for Ure to galvanize. One wire from the battery was placed in the spinal cord and the other used to stimulate different nerves. When stimulating individual nerves, Ure described powerful and purposeful movements. For example, when the ulnar nerve was stimulated at the elbow, he noted that the fingers *"moved nimbly like those of a violin performer"* and when a nerve in the hand was stimulated Clydesdale *"seemed to point to the different spectators, some of whom thought that he had come to life"* [37].

When Ure connected one end of the battery to the diaphragm and the other to the phrenic nerve in the neck he noted *"laborious breathing instantly commenced. The chest heaved and fell; the belly was protruded. And again collapsed, with the relaxing and retiring diaphragm,"* [37] the battery had caused the corpse to breathe. When he stimulated nerves in the face, he produced facial expressions including *"extraordinary grimaces."* Ure was a showman, and his goal was to convince his audience that electricity brought Clydesdale back to life.

In his report of the Clydesdale galvanization published in the Royal Institution's *Quarterly Journal of Science,* Ure detailed the evidence that Clydesdale returned to life and noted that *"several of the spectators were forced to leave the apartment, and one gentleman fainted"* [38]. In an article in *The Times* entitled *Horrible Phenomena!—Galvanism,* the reporter largely parroted Ure's exaggerated descriptions and noted that many spectators truly believed that Clydesdale had come back to life. Medical historian IR Morus described how the story improved with the telling. Many years later, one of the spectators recalled that Clydesdale "*stared, apparently in astonishment, around him*" and "*made a feeble attempt as if to rise from the chair whereon he was seated*" [38].

In London a year later, academics at Cambridge were highly skeptical of Ure's sensational report. A young professor of Chemistry, James Cummings decided to reproduce Ure's galvanization experiments to see the results for himself. He arranged for the well-known surgeon, Thomas Verney Okes, to perform the dissection and scheduled with local authorities for the delivery of a fresh body.

Tom Weems, an illiterate mill hand, was convicted of killing his young wife, Mary Ann, strangling her with her own garters. As an added nuance, Weems had been forced to marry Mary Ann by Parish officials after she claimed to be pregnant by him but later was found not to be pregnant. Regardless, Weems was sentenced to hanging followed by dissection, the final insult that the law could impose on a convicted murderer.

Within a week of conviction, Weems was hung from the gateway of the Cambridge County goal in front of a large crowd of students, professors, and the lay public. After about an hour, Weems's body was cut down, bundled with blankets, and placed on a cart for transportation to the university. The cart and crowd accompanied by an entourage of sheriff's officers and constables slowly moved down Castle Hill through Cambridge town and to the Chemical Lecture Room. The officers and constables were routinely part of the entourage on such occasions since in the past, friends or family of executed prisoners tried to steal the body to prevent the indignity of public dissection.

James Cummings already had a reputation for experimenting with galvanic stimulation and was determined to follow Ure's experimental protocol as closely as possible so that he could expose the Glasgow exhibit as a flight of fantasy. Cummings used a similar battery and he and his assistants stimulated the same nerves in the same fashion described by Ure. They tried to get Weems to breathe, to move his arms and legs, and to produce facial expressions just as Ure reported. But instead of organized breathing, Cummings produced only slight shuddering and instead of purposeful movements only brief convulsive twitching of muscles.

A report of Cummings' galvanic experiment in the Cambridge Chronicle concluded: *"From the account given by Dr. Ure they had been led to form, perhaps, too sanguine an expectation of the efficacy of galvanism in restoring suspended animation; it is therefore to be regretted, that although undertaken with a battery at least equal powerful, and with every intention to ensure success, these experiments afford no confirmation of the hopes held out by those in Glasgow"* [34].

If Cummings thought that the lack of spectacular results from his galvanizing of Weems body would alter the public's perception of the rejuvenating properties of electricity, he was sorely mistaken. Aldini's and Ure's flamboyant presentations had a much greater effect than the mundane descriptions of Cummings. In the early nineteenth century, electricity had acquired mystical properties that were engrained in the minds of the average person.

Mary Shelly and Frankenstein

It would be impossible to cover this period in the history of electricity and the brain without addressing Mary Shelly's novel, Frankenstein, and its impact on the perception of electricity and galvanism in the public. Although Mary Shelly was just an infant when Aldini made his famous public demonstration on George Forster, the newspaper reports had an enduring effect on almost everyone in London including Shelly's family and friends. Her novel, *Frankenstein, or the Modern Prometheus*, published in 1818 tells the story of a self-trained scientist, Victor Frankenstein and his search for the "*elixir of life*" [39]. After extensive study of traditional subjects such as anatomy, physiology, and chemistry, Frankenstein discovered electricity and galvanism and their potential for reviving the dead. "*I succeeded in discovering the cause of generation and life; nay, more, I became myself capable of bestowing animation upon lifeless matter*" [40]. The novel, often considered the beginnings of science fiction, not only became a classic in the field of gothic horror but in addition was a cautionary tale of the potential risk to society of science pursued unfettered.

Remarkably, Mary Shelly was only 18 years of age when she began writing her novel and 20 years of age when it was published. Her parents were the political anarchist and writer, William Godwin and the feminist activist, Mary Wollstonecraft, who died shortly after giving birth. Her father managed Mary's informal education encouraging a freethinking philosophy. Marry had a tumultuous relationship with her father's second wife and at age 16 she ran away with a friend of her father, the poet Percy Bysshe Shelly, who was married at the time. Mary soon became pregnant, and the couple was ostracized from English Victorian society. After the death of their prematurely born daughter and an attempted suicide by Percy Shelly's first wife, Percy and Mary were finally married in 1816.

The newlywed couple spent the summer of 1816 at Lake Geneva in Switzerland with the poet Lord Byron, when Mary developed the idea for her book. As was common at the time, both Percy Shelly and Lord Byron considered themselves natural philosophers in addition to poets and both were fascinated by recent scientific discoveries. Mary Shelly provided some details on the inspiration for the book in an Introduction for a later edition. "*Many and long were the conversations between Lord Byron and [Percy] Shelley, to which I was a devout but nearly silent listener. During one of these, various philosophical doctrines were discussed, and among others the nature of the principle of life, and whether there was any probability of its ever being discovered and communicated... Perhaps a corpse would be re-animated; galvanism had given token of such things: perhaps the component parts of a creature might be manufactured, brought together, and endued with vital warmth*" [41].

Based on these overheard conversations, Mary Shelly formed the basic premise for Victor Frankenstein. "*I saw the pale student of unhallowed arts kneeling beside the thing he had put together. I saw the hideous phantasm of a man stretched out, and then, on the working of some powerful engine, show signs of life and stir with an uneasy half-vital motion*" [42]. The ghoulish experiments with galvanism on dead bodies initiated by Aldini and later by scientists throughout Europe no doubt influenced Lord Byron and Percy Shelly and through them, Mary Shelly.

Mary Shelly's life was marred by one tragedy after another. Only one of three subsequent births survived and her husband, Percy, drowned when his sailing boat sank in a storm in 1822. She developed a brain tumor in her 40s, and after prolonged suffering, she died at age 53. Shelly wrote several subsequent books, but Frankenstein stood out as her crowning achievement.

In Brief

Two men, Luigi Galvani and Alessandro Volta, dominated the scientific discussion of animal electricity in the latter eighteenth century. Galvani argued that electricity represented the long sought-after "animal spirits" carried in nerves to control body functions while Volta, after initially agreeing with Galvani, changed his mind and became a critic of Galvani's work. In the process Volta discovered the electric pile, a major step in the use of electricity for practical purposes including medicine. The epic confrontation between these two giants changed the history of electricity and neuroscience. Ironically, Galvani's nephew, Giovani Aldini used Volta's electric pile to electrically stimulate the brains of dead animals and humans raising the possibility that electricity could revive the dead. The public became enthralled with the life-giving potential of electricity causing a young Mary Shelly to envision a monster brought to life by electricity.

References

1. Bertucci P. Therapeutic attractions: early applications of electricity to the art of healing. In: Whitaker H, Smith CUM, Finger S, editors. Brain, mind and medicine: essays in eighteenth-century neuroscience. New York: Springer; 2007. p. 276–80.
2. Lane T. Description of an electrometer invented by Mr Lane; with an account of some experiments made by him with it. Philos Trans R Soc. 1767;57:451–60.
3. Schiffer MB. Draw the lightning down. Benjamin Franklin and electrical technology in the age of enlightenment. Berkley: University of California Press; 2003. p. 33–66.
4. Bordeau SP. Volts to Hertz—the rise of electricity. Minneapolis: Burgess Publishing Company; 1982. p. 44–59.
5. Schiffer MB. Draw the lightning down. Benjamin Franklin and electrical technology in the age of enlightenment. Berkley: University of California Press; 2003. p. 56–8.
6. Schiffer MB. Draw the lightning down. Benjamin Franklin and electrical technology in the age of enlightenment. Berkley: University of California Press; 2003. p. 161–83.
7. Bordeau SP. Volts to Hertz—the rise of electricity. Minneapolis: Burgess Publishing Company; 1982. p. 14–31.
8. Turner G. North Pole, South Pole. New York: The Experiment; 2010. p. 70–7.
9. Schiffer MB. Draw the lightning down. Benjamin Franklin and electrical technology in the age of enlightenment. Berkley: University of California Press; 2003. p. 156–9.
10. Schiffer MB. Draw the lightning down. Benjamin Franklin and electrical technology in the age of enlightenment. Berkley: University of California Press; 2003. p. 108–15.
11. Schiffer MB. Draw the lightning down. Benjamin Franklin and electrical technology in the age of enlightenment. Berkley: University of California Press; 2003. p. 1246–9.
12. Piccolino M, Marco B. Drawing a spark from darkness: John Walsh and electric fish. Trends Neurosci. 2002;25:51–7.
13. Walsh J. On the electric properties of the Torpedo. Philos Trans. 1773;63:461–77.
14. Hunter J. Anatomical observation on the Torpedo. Philos Trans. 1773;63:481–9.
15. Hunter J. Anatomical observation on the Torpedo. Philos Trans. 1773;63:486.
16. Schiffer MB. Draw the lightning down. Benjamin Franklin and electrical technology in the age of enlightenment. Berkley: University of California Press; 2003. p. 117–20.

17. Cavendish H. The account of some attempts to imitate the effects of the torpedo by electricity. Philos Trans. 1776;66:196–225.
18. Bordeau SP. Volts to Hertz—the rise of electricity. Minneapolis: Burgess Publishing Company; 1982. p. 46–8.
19. Cohen IB. Introduction to commentary on the effects of electricity on muscular motion by Luigi Galvani, translated by *Margaret* Glover Foleg. Norwalk: Burndy Library; 1954. p. 9–41.
20. Kipnis N. Luigi Galvani and the debate on animal electricity, 1791–1800. Ann Sci. 1987;44:107–42.
21. Piccolino M. Luigi Galvani and animal electricity: two centuries after the foundation of electrophysiology. Trends Neurosci. 1997;20:443–8.
22. Schiffer MB. Draw the lightning down. Benjamin Franklin and electrical technology in the age of enlightenment. Berkley: University of California Press; 2003. p. 125–30.
23. Bordeau SP. Volts to Hertz—the rise of electricity. Minneapolis: Burgess Publishing Company; 1982. p. 51–4.
24. Parent A. Giovanni Aldini: from animal electricity to human brain stimulation. Can J Neurol Sci. 2004;31:576–84.
25. Galvani L. De viribus electricitatis in motu musculari commentarius. Bon Sci Art Inst Acad Commun. 1791;7:363–418. (English Transl. by M Glover Foley, 1953, Burndy)
26. Lanska DJ. Perkins's patent metallic "Tractors": development, adoption, and early dissemination of an eighteenth-century therapeutic fad. J Hist Neurosci. 2019;28:122–46.
27. Volta A. On the electricity excited by the mere contact of conducting substances of different kinds. Philos Trans. 1800;90:403–31.
28. Volta A. Letter of Professor Volta to J.C.Delamethrie on the galvanic phenomena. J Nat Phil Chem Arts. 1802;1:135–42.
29. Scarpelli G. Aldini, Giovanni: a biography. Kos. 1984;2:25–7.
30. Parent A. Giovanni Aldini: from animal electricity to human brain stimulation. Can J Neurol Sci. 2004;31:580.
31. Schiffer MB. Draw the lightning down. Benjamin Franklin and electrical technology in the age of enlightenment. Berkley: University of California Press; 2003. p. 130.
32. Golinski JV. From Calcutta to London: James Dinwiddie's galvanic circuits. In: Lightman B, McOuat G, Stewart L, editors. The Circulation of Knowledge between Britain, India and China: The Early-Modern World to the Twentieth Century. Leiden: BRILL; 2013. p. 75–94.
33. Finger S, Law MB. Karl August Weinhold and his "science"in an era of Mary Shelley's Frankenstein: experiments on electricity and the restoration of life. J Hist Med. 1998;53:161–80.
34. Finger S, Law MB. Karl August Weinhold and his "science"in an era of Mary Shelley's Frankenstein: experiments on electricity and the restoration of life. J Hist Med. 1998;53:166.
35. Finger S, Law MB. Karl August Weinhold and his "science"in an era of Mary Shelley's Frankenstein: experiments on electricity and the restoration of life. J Hist Med. 1998;53:169.
36. Morus IR. Shocking bodies. Life, death and electricity in Victorian England. Gloucestershire: The History Press; 2011. p. 54.
37. Morus IR. Shocking bodies. Life, death and electricity in Victorian England. Gloucestershire: The History Press; 2011. p. 52.
38. Morus IR. Shocking bodies. Life, death and electricity in Victorian England. Gloucestershire: The History Press; 2011. p. 53.
39. Shelley MW. Frankenstein, or the modern prometheus. London: Lackington, Hughes, Harding, Mavor & Jones; 1818.
40. Schiffer MB. Draw the lightning down. Benjamin Franklin and electrical technology in the age of enlightenment. Berkley: University of California Press; 2003. p. 131.
41. Shelley MW. Frankenstein, or the modern prometheus. London: Henry Colburn & Richard Bentley; 1831.
42. Feldman PR, Scott-Kilvert D, editors. The Journals of Mary Shelley 1814–1844. Baltimore: Johns Hopkins University Press; 1987.

Chapter 4
Electricity and Magnetism: Two Sides of the Same Coin

> *The view I am so bold as to put forth considers radiation as a high species of vibrations in the lines of force which are known to connect particles and also masses of matter together ... these vibrations take place along the lines of force.*
>
> *Michael Faraday, 1846*

Abstract Hans Christian Oerstred's simple observation that electric current flowing along a wire generated a surrounding electromagnetic field whose strength was proportional to the amount of current was pivotal in the history of electricity. Based on the discovery Ampere made a quantitative electrometer that he called a galvanometer and Ohm developed his famous eq. I=V/R relating current (I) to the voltage (V) and resistance (R) in an electrical circuit. Wired telegraphy soon revolutionized worldwide communication. Michael Faraday's lines of force theory of electromagnetic waves marked the beginning of modern physics, and his induction coil provided a new and better way to stimulate nerves (faradization). Faraday's experiments on the electric eel inspired Michael Todd's pioneering work on epilepsy and brain electrical stimulation.

Keywords Ion · Cathode · Anode · Galvanometer · Faradization

The invention of the bimetal electrochemical battery by Volta at the turn of the nineteenth century marked a turning point in the science of electricity. For the first time, researchers had a source of electricity with a continuous flow for their experiments. Unlike the electricity generated by electrical machines and stored in Leyden jars that had high tension but relatively low current flow, the electrochemical battery had relatively low tension but high current flow that opened entirely new fields of study. It was soon discovered that high current flow in a conductor generated a magnetic field around the conductor indistinguishable from the magnetism of a lodestone.

One area of debate regarding Volta's battery was the relative role of the metallic plates and the interdigitated cloth soaked with saline or acid solution. Clearly, chemical reactions in the solutions were critical for the electrical flow in the battery.

R. W. Baloh, *Brain Electricity*, https://doi.org/10.1007/978-3-031-62994-5_4

The Birth of Electrochemistry

The earliest experiments in electrochemistry focused on gases, not liquids. With the discovery of hydrogen by Cavendish and oxygen by Priestly, researchers had the components of water, although simply mixing the two gases did nothing [1]. Priestly tried igniting hydrogen and oxygen mixtures with a flame and was surprised to produce an explosion. What was needed was a more controlled explosion in an enclosed space so that the product of the explosion could be collected and analyzed. Several investigators devised ways to use electricity to ignite the gasses in an enclosed vessel and collect the reaction product. Based on his correspondence with Priestly, Volta created an "electric pistol" which consisted of an enclosed metal or glass vessel whose gas contents could be ignited by a spark from two internal metal conductors. The source of the spark could be a Leyden jar or an electrophorus.

With this and similar devices in the early 1780s, Cavendish, Priestly, and Lavoisier each showed that igniting hydrogen and oxygen with an electric spark produced water. Cavendish had priority although he published his results after Priestly and Lavoisier. This was a pivotal point in the history of chemistry because since the time of the great Greek philosophers water was considered a primary element, not a compound.

The reverse process, converting water to hydrogen and oxygen, was delayed a few decades since it required a continuous source of electricity such as that supplied by Volta's battery. Slightly more than a month after Volta reported his discovery to the Royal Society, the English chemists, Sir Anthony Carlisle and William Nicholson, inserted the platinum wire terminals of a Voltaic pile into a jar of slightly acidic water and noted that gas bubbles formed continuously around the terminals. They collected the gases and found them to be hydrogen and oxygen. Mixing the two gases together and igniting them with electricity produced water proving beyond a doubt that water was a compound made of hydrogen and oxygen. This was a dramatic moment in the history of chemistry illustrating the important connection between electricity and chemistry.

Into this new era stepped an apothecary's apprentice who set his sights on a career in chemistry, Humphrey Davy [2, 3]. Davy, who was born in Cornwall, England, in 1778, was mostly self-taught reading the chemistry textbooks of Nicholson and Lavoisier. He gained some early recognition with his experiments inhaling the newly discovered gas, nitrous oxide. Davy described bizarre cognitive effects including hallucinations and even speculated on its possible use as an anesthetic for surgery. It would be almost another century before an American dentist, Horace Wells, would first use the gas as an anesthetic for tooth extraction. Known as "laughing gas," it became a public sensation and sniffing the gas a popular English parlor amusement. Based on this and other early promising work in chemistry, Davy was offered a position at the newly formed Royal Institution in 1801 where he set up a laboratory and provided public lectures in chemistry.

Nicholson and Carlisle's work decomposing water into hydrogen and oxygen with a Voltaic pile had a profound influence on Davy. Within a few years, he had constructed a huge voltaic pile composed of 500 bimetal pairs in his laboratory at the Royal Institution. One of his first experiments with the new battery was to place pieces of carbon on each terminal producing an extremely bright light between the pieces of carbon, the first arc light. Davy compared the brightness of the light to that of the sun. It was so hot that it melted and fused metals.

With this powerful source of electricity, Davy began to isolate the metallic elements into a variety of mineral salts. He started with solutions of potash and soda salts. With his Voltaic pile, he was able to decompose the salts into metallic elements that he called potassium and sodium. Sodium was a remarkable silvery-white metal that was so soft that it could be cut with a knife. The process that used electricity to tear molecules apart into basic elements became known as electrolysis. With this process, Davy went on to identify several new metal elements including calcium, barium, magnesium, and strontium. He also discovered a new pungent greenish gas when electrifying muriatic acid that he called chlorine.

How was electricity able to tear molecules apart? Some suggested that electricity in a solution was like an acid that could cause chemical decomposition as acids had previously been shown to do. In a lecture before the Royal Society in 1806 titled *On Some Chemical Agencies of Electricity*, Davy speculated that electricity separated elements of a compound in solution by attracting them to either the positive or negatively charged terminal, where the elements were released or deposited. In the process, he was anticipating the concept of an "ion," an atom or molecule in which the number of electrons is different from the number of protons. The positively or negatively charged ions are attracted to the oppositely charged electrical terminal.

Davy's electrochemical experiments raised questions regarding Volta's bimetal explanation for his electric pile. Davy noted that iron was positively charged in an iron–copper pile with water, but it was negatively charged when the water was replaced by a solution of potassium sulfate. This suggested that the liquid in the electric pile played an active role in generating electricity rather than the passive role suggested by Volta. Davy also showed that an electric pile could be made with a single metal and cells with two different liquids, one alkaline and one acid, connected with strips of paper moistened with water. The amount of electricity generated with a single metal and two different liquids was less than the amount generated by two different metals and a single liquid, so Davy acknowledged that bimetal contact was also important in the electric pile. But with just bimetal contact there would be only a brief discharge of electricity. The persistent electrical discharge of the bimetal pile was dependent on chemical reactions in the liquid that replenished the charge in the metals.

Hans Christian Oersted Discovers Electromagnetism

As indicated in Chap. 1, electricity and magnetism were originally thought to be different varieties of the same phenomena. It wasn't until the sixteenth century that Gilbert pointed out the major differences between electricity and magnetism and suggested that they were two unrelated phenomena. Beginning in the mid-eighteenth century hints of an association between electricity and magnetism were again observed but it would not be until a chance discovery during a physics classroom demonstration in Copenhagen, Denmark, that the full significance of the association was appreciated.

Ben Franklin reported that he was able to magnetize a steel needle with the discharge of a Leyden jar and the Italian professor of physics, Giovanni Beccaria, reported that when he discharged a Leyden jar into a watch spring it produced a magnet with poles at each end of the spring. Shortly after Volta described his electric pile, several observers noted that if two close parallel wires were connected to a pile they would cling together if brought in contact. We now know they were held together by magnetism. Finally, in 1802, the Italian physicist Gian Romagnosi noted that he could cause a compass needle to deviate from its baseline position by touching the needle with a silver knob connected to an electric pile. Any one of these observations could have led to a breakthrough in understanding electromagnetism but the observers were not primed to recognize the significance of the findings.

In the spring of 1820, Hans Christian Oersted, a professor of physics at the University of Copenhagen, began a routine demonstration for his students on the ability of electric current to generate heat and light in a metal wire attached to a voltaic pile [4, 5]. He had the students stand close to the laboratory table so they could see the glow of a red hot wire as his assistant connected the wire to the pile terminals. By chance, there was a compass sitting on the table near the wire and Oersted was surprised to see the needle suddenly swing from its initial position and point at the wire. Everyone was focusing on the wire and did not notice the compass needle move, but Oersted was watching the compass and when he asked his assistant to disconnect the wire from the battery, the compass needle immediately returned to its original north–south position. When the assistant reconnected the wire to the battery terminals, the compass needle again turned toward the wire. This simple observation along with the recognition of its significance turned out to be one of the most important discoveries in the history of science.

Oersted went on to assess the nature of the magnetic force surrounding a wire by moving the compass 360° around the wire and noting the changing position of the compass needle (Fig. 4.1). The compass needle maintained a perpendicular position relative to the wire, so that when above the wire, it pointed in the opposite direction as when it was below the wire. Just as heat and light radiated from a wire carrying an electric current, Oersted wondered whether magnetism might also radiate from such a wire. He had a longstanding interest in the relationship between electricity and magnetism noting the similarities of the attractions and repulsions and even suggesting in a book written in 1813, that an attempt should be made to see if electricity had any action on a magnet.

In July 1820, Oersted privately printed a 4-page pamphlet describing his findings entitled "*Experiments on the Effect of an Electric Current on the Magnetic Needle*" in Latin and simultaneously sent the pamphlet to scientific societies and prominent scientists throughout Europe. Within a few weeks the pamphlet was translated into English, German, French, Italian, and Dutch and Oersted's further experiments

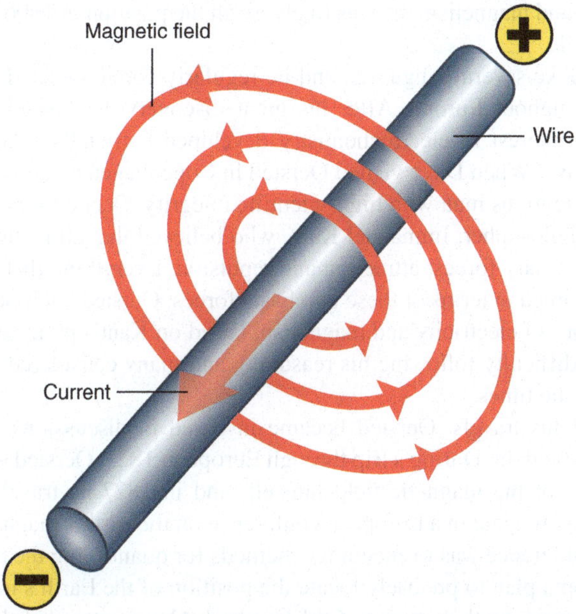

Fig. 4.1 Magnetic field surrounding a wire with a moving electrical current. The force is perpendicular to the wire so that it is in opposite directions above and below the wire

were published in *Schweigger's Journal of Chemistry and Physics*. Oersted described how the magnetic force surrounding a conducting wire assumed transverse circles, the planes of which were at right angles to the wire. The magnitude of the force depended on the distance from the wire, the power of the battery, and the quality of the metal wire. A circular closed circuit carrying an electric current attracted or repelled the pole of a magnet, could be turned by a magnet and had a north and south pole just like a magnet. Furthermore, the effect of a closed circuit on a magnet depended on the quantity of electricity in the circuit and not upon the tension of the electricity.

Hans Christian Oersted was born in 1777 in a small Danish town, the eldest son of a village apothecary. There were no local schools, so Hans relied on the help of friendly villagers. He began working in his father's apothecary as a boy and rapidly developed an interest in science and medicine. He entered the University of Copenhagen at age 16, supporting himself with tutoring and odd jobs. Hans received a degree in pharmacy in 1797 and a Doctor in Philosophy in 1799. After graduation, Oersted took the customary post-graduate tour throughout Europe meeting prominent scientists in Germany, France, and the Netherlands. He was particularly excited about Volta's discovery of the electric pile and the electrochemical studies that soon followed. He became a strong proponent of the chemical origin of Voltaic current. On returning to Copenhagen in 1803, Oersted was initially turned down for an appointment as Professor of Physics but ultimately after several years of lecturing on electricity and magnetism, he was finely given the position in 1806, a position he held until 1851.

Oersted spoke several languages and he regularly communicated with eminent scientists throughout Europe. After his great discovery, he visited England and based on his interest in electrochemistry developed a friendly relationship with Humphrey Davy. When Davy visited Oersted in Copenhagen he described him as a simple, unpretentious man with high scientific integrity. Oersted was an admirer of the German philosopher, Immanuel Kant, who believed that all matter was dependent on two primary forces: attraction and repulsion. Everything that one observed could be explained in terms of these two basic forces. Oersted attempted to develop a unified theory of electricity and magnetism based on Kant's premise but others in the field had difficulty following his reasoning and many considered it to be hopelessly behind the times.

As part of his travels, Oersted became involved in discussions regarding the earth's magnetic field. During a trip through Europe in 1823, Oersted made multiple measurements of the magnetic field himself, and in 1827 he traveled to Altona, Germany, to participate in a European conference on the earth's magnetic field. The goal of the conference was to encourage methods for quantifying the magnetic field and to develop a plan to precisely locate the position of the Earth's magnetic north pole. The German mathematician, Carl Friedrich Gauss, attended the conference and would soon measure the field's absolute intensity. Four years after the conference, the British Royal Navy officer, James Clark Ross, discovered the magnetic north pole which at that time was at the northern tip of mainland Canada. He used a magnetic needle hung by a silk thread to identify a 90-degree inclination at the pole. The pole has been slowly moving since that time first south and then north and recently moving into northern Russia. The Altona meeting led to the development of the Magnetic Union of Gottingen, the first large-scale international scientific collaboration, and the forerunner of CERN, the European Organization for Nuclear Research, one of the world's largest and most respected centers for scientific research.

Based on Oersted's experiments, for the first time, it was possible to reliably identify and measure electric current in a conducting wire. One immediate practical application was to transmit signals over a distance, the telegraph. All that was needed were transmitting wires connecting two locations, an electrochemical battery that could be switched on and off, and a magnetic needle at the receiving terminal that would respond each time the battery was connected at the sending terminal. A variety of techniques were developed to transmit signals from those using a single wire to those using a bundle of wires. The French mathematician and physicist Andre Marie Ampere (see next section) developed an elaborate scheme that used a bundle of wires, one for each letter in the alphabet. When activated by its key at the sending end, the current in the wire at the receiving end would turn a magnetic dial and uncover the appropriate letter at the receiving end. With time, a single wire carrying a coded signal (Morse code) won out mainly because of the simplicity and reliability. By the mid-nineteenth century, the telegraph was the most important electrical technology in use driving the development of new electrical equipment including batteries, long-distance transmission wires and switching devices.

Ampere and the Quantification of Electromagnetism

Andre Marie Ampere was born in Lyon, France, in 1775, the son of an affluent merchant, self-taught scholar, and justice of the peace [6]. His father took responsibility for Andre's early education encouraging him to read extensively, particularly in history and science. Andre displayed a remarkable aptitude for mathematics as a child, and at age 12, he taught himself to read Latin since the most famous works in mathematics at the time were written in Latin. Andre's mother was a devout Catholic who insisted on his being brought up in the Catholic tradition.

Ampere's adult life was mired by personal tragedies starting with the witnessing of his father's execution on the guillotine at age 18 during the French Revolution. He married at age 24 followed shortly by the birth of his son, but his happiness was short-lived when his young wife died a few years later. He moved from Lyon to Paris, remarried, and had a daughter but was divorced shortly after when he found the relationship unbearable. His mother and aunt moved to Paris to manage the household, but there were constant problems with both children. and eventually his relationship with them completely deteriorated. Based on his mother's influence, he often turned to religion for comfort, but he also turned to mathematics and his scientific experiments where there was less ambiguity and more certainty.

Oersted's 1820 paper had a dramatic impact on Ampere. He recognized that electricity in motion, electric current in a circuit supplied by a voltaic pile, was different from static electricity at rest as produced by electrostatic machines. He considered electricity to be like a fluid, static electricity a fluid at rest, and electric current a fluid in motion. Electromagnetism was a feature of electric current depending on the amount and direction of the current flow. Ampere felt that electric current was in fact behind all magnetism. Magnets must have electric current in motion within them. Even the earth must have electric currents circulating within it to explain its magnetic field. With his strong background in mathematics, Ampere rapidly developed a series of equations to describe the magnetic field surrounding a wire carrying an electric current. These equations served as a foundation for later understanding of electromagnetism.

Ampere conceived of electric current as "*a double current, one of positive electricity, the other one of negative electricity, starting out in opposite senses from points where the electromotive action arises, and going out to reunite in the parts of the circuits remote from these points. The currents of which I am speaking are then accelerated until the inertia of the electric fluids and resistance in which they encounter because of imperfection of even the best conductors make equilibrium with the electromotive force, after which they continue indefinitely with constant velocity so long as this force has the same intensity, but they always cease on the instant that the circuit is broken...*" [7]. The basic unit of electrical current would later be named the *ampere.*

For the first time, there was a simple means for quantifying electrical current. Standard gold leaf electrometers measured electrical tension, not electrical current. Since Oersted showed that the magnetic field surrounding a wire carrying an electric current was proportional to the magnitude and direction of the current, Ampere saw that a compass needle could be used to quantify the magnitude and direction of electric current in a conductor. By wrapping several turns of the wire around a compass, the sensitivity of the device could be improved. Ampere wrote: "*I think that to distinguish this instrument from the ordinary electrometer we should give it the name galvanometer and that it should be used in all experiments on electric currents, as we are accustomed to use an electrometer on electric machines, so as to see every instant if a current exists and what is its energy*" [8].

But what was occurring at the molecular level to explain electric current and magnetism? Ampere hypothesized that each molecule in a conductor was associated with an electric current whirling about it in a random fashion. When connected to an electromotive force such as a voltaic pile, there was a mass movement of these tiny electric currents in one direction along the conductor. Furthermore, each of the molecules of the conductor was a tiny magnet due to the whirling electrical current but the polarities were randomly oriented so there was no net magnetic field. An electrical current or an external fixed magnet could align the polarities of the individual molecules producing a fixed magnetic field. The whirling electrical currents of individual molecules would later be attributed to the spin and orbital motion of tiny particles called electrons some of which freely moved about the molecule.

Ohm's Law

Ampere made a clear distinction between the electric tension and the electric current in a closed electrical circuit. He also introduced a new property of the conductor within the circuit that he called resistance. He attributed the resistance to imperfections in the conductor that limited current flow, but he made no attempt to define the properties of a conductor that determined resistance or to quantitatively relate the magnitude of current flow to electric tension or resistance. This distinction goes to a relatively unknown German mathematician, George Simon Ohm. His simple eq. $i = v/r$ indicates that the current (i) in an electrical circuit is directly proportional to the tension (or voltage, v) and inversely proportional to the resistance (r) in the circuit. It is also known as Ohm's law, one of the most famous equations in the history of science, simple yet revolutionary. It ranks alongside Newton's $f = ma$ (force equals mass times acceleration) and Einstein's $e = mc^2$ (energy equals mass times the speed of light squared).

George Ohm was born in 1789 in Erlangen, Germany, the eldest of two sons of a locksmith [9]. Although financially limited, his father had an interest in science and took a direct role in the education of his sons. He arranged for tutoring in mathematics from a university student in exchange for free rent in his home. The boys received 5 years of basic classical training at a local gymnasium in preparation for university and both were eventually enrolled at the University of Erlangen. Both brothers were uniquely talented in mathematics and even compared to the famous eighteenth-century Bernouilli brothers. George received a Doctor of Philosophy degree in 1811 and this was followed by several modest positions teaching mathematics ultimately at a Jesuit Gymnasium in Cologne, Germany, where he spent almost 10 years.

It was during his stay in Cologne that Ohm heard about and became captivated with the published work of Oersted, Ampere, and Davy. Although his teaching load was heavy, he decided that he must begin his own program of experiments on electricity and magnetism hopefully leading to publication and eventually promotion, freeing up more time for personal pursuits. He decided to focus on the conductance (the inverse of resistance) of the wires in an electrical circuit.

Ohm published his first scientific paper in 1825 investigating the relationship between the conductance of electrical current in a closed circuit and the characteristics of the wire in the circuit. He used a simple setup consisting of (1) a voltaic battery, (2) a fixed short segment of thick wire over which hung a compass needle suspended from a torsion head (like Coulomb's torsion compass), and (3) a test wire that he varied in length. The torsion of the compass needle measured by returning the deflected needle to zero by twisting the torsion head provided a measure of the electrical current in the circuit with increasing wire length. He found that the conductance was inversely proportional to the length and directly proportional to the diameter or cross-section of a test wire so that a wire with a diameter of 2 cm and a length of 10 cm had the same conductance (resistance) as a wire with a diameter of 1 cm and a length of 5 cm.

With the same apparatus, Ohm went on to test the conductivity of nine different metals in the test wire. He used copper as the reference metal and arbitrarily set the conductance reading at 1000. He then made other wires all with the same diameter but varied the length of the wire to give the same reading as the standard copper wire. With this technique, he observed the following ranking for conductivity: copper—1000, gold—574, silver—356, zinc—333, brass—280, iron—174, platinum—171, tin—168, and lead—97. Ohm later discovered that he had made an error in the measurement of the silver wire and that silver was a slightly better conductor than copper, but otherwise, the measurements have stood the test of time.

Having identified the factors that determined conductivity, Ohm moved on to address a question that had long intrigued him, the relationship between the three fundamental elements in an electrical circuit, (1) the electromotive force or tension, (2) the electrical current, and (3) the resistance to current flow. Around this time, he read the book *The Analytic Theory of Heat* by the famous French mathematician Joseph Fourier in which Fourier provided a mathematical description of the conduction transfer of heat in a material from an area of higher temperature to one of lower temperature. The quantity of heat in a given time flowing in a given direction was the product of the conducting area normal to the path of flow, the temperature gradient along the path and the thermal conductivity of the material. Ohm immediately saw an analogy of heat flow in a metal rod to electrical flow in a conductor. If so, the electrical current was proportional to the electromotive force and the conductance (or inversely proportional to the resistance).

Ohm surmised that the current in a conductor resulted from a transfer of electric fluid, particle to particle, within a conductor. With this transfer, there was a steady drop in tension, particle by particle, as the electric fluid moved against the resistance of the conductor. The electrical current was proportional to the sum of all the tensions and inversely proportional to the resistance in the conductor. In two papers published in early 1826, Ohm described a series of experiments regarding electrical tension and the quantity of electricity in a closed circuit and for the first time presented his famous equation, I=V/R. A year later, he published a book summarizing his work entitled *The Galvanic Circuit Mathematically Investigated.* Remarkably, Ohm's work initially was not well received. Some felt that his findings were too simple for such a complex subject. Others were outright hostile suggesting that Ohm's theories were a "web of naked fancies" [10]. The Minister of Education in Berlin stated, "*a professor who preached such heresies was unworthy to teach science!*" It was more than a decade before the broad scientific community became aware of the importance and implications of Ohm's law.

A problem facing Ohm and his contemporaries was the lack of information about the particulate structure of matter. It wasn't until the early twentieth century that scientists theorized the existence of atoms composed of electric charges. The atomic nucleus was composed of positively charged protons balanced by surrounding negatively charged electrons. Within a metal conductor with a crystalline structure some of the electrons were free to move about no longer localized to an individual atom but belonging to the crystal. The free electrons left a residual lattice of positive ions in the metal, within which the free electrons wandered about.

Atoms including the electrons are in constant motion, the higher the temperature, the greater the motion. If an electrical potential is applied to a metal conductor, the net effect acting on the free electrons is to superimpose a *drift velocity* onto their normal random motion. The electrons randomly collide with the ionic lattice so that the net drift velocity is only a tiny fraction of the random velocity of free electrons. Yet, the total number of electrons moving at the drift velocity can be enormous. For example, assuming one free electron per atom, a cubic centimeter of copper would have about 8.5×10^{23} free electrons to constitute an electric current. We will return to this topic in Chap. 12 after addressing quantum theory and atomic structure.

Michael Faraday, the Early Years

One of the most inspiring stories in the history of science is how the son of an impoverished blacksmith became the founder of modern physics [11, 12]. With no formal training in physics or mathematics, Michael Faraday conducted a series of ingenious experiments that led to the development of electric motors, generators, and transformers and to the field-of-force theory that is the foundation for our modern understanding of electromagnetism. In addition, Faraday had excellent communication skills presenting hundreds of lectures and demonstrations with the goal of explaining his science to the lay public. He also had a good sense of humor. When the British Prime Minister asked Faraday what use could be made of his discoveries on electromagnetism, Faraday responded, "*some day it might be possible to tax them*" [13].

Michael Faraday was born in Surrey on the outskirts of London in September 1791. His father who worked as a blacksmith fell into hard times and as his health began to deteriorate the family was forced to move into cramped quarters above a coach house in Manchester Square, London. Despite having to live in meager circumstances the family stayed together, and Michael attended a common day school for a few years where he learned to read and write. When he was age 14, Faraday became bound for 7 years as an apprentice to a bookmaker, Mr. G. Riebau, a position that turned out to be ideal for the inquisitive and industrious novice. With the many volumes of books on the shelves of the store, Faraday had access to a new and intriguing world of reading and learning. Fortunately, Riebau and his wife liked Michael and were supportive of his intellectual pursuits. Early on, Faraday developed the habit of taking copious notes from his reading material, a habit that persisted throughout his scientific career.

Faraday's introduction to scientific methodology came in 1810 when he began attending lectures at the "City Philosophical Society." A London scientific enthusiast, John Tatum, provided his house and scientific equipment for weekly evening lectures on different scientific topics. Tatum lectured bi-weekly and other members lectured in between. At age 19, Faraday gave his first lecture at the Society championing the two-fluid theory of electricity. He used the traditional argument of two different types of static electricity, resinous and vitreous and their attractive and repulsive powers. At the Society lectures, Faraday was exposed to a wide range of scientific topics from electricity to mechanics, optics, chemistry, and astronomy. They were an ideal training ground for the young Faraday to develop writing and speaking talents and for acquiring the self-confidence necessary for a future career in science.

In 1812, when Faraday completed his bookbinding apprenticeship, he was sure he wanted a career in science, but the possibilities seemed bleak. He had no formal training so traditional university positions were not an option. With the support of a friend from the Society, Faraday had the opportunity to attend a series of four lectures on chemistry given by Humphrey Davy who had just been knighted and was the new president of the Royal Institution. Davy was at the peak of his career having identified a series of elements by using a voltaic pile to separate the elements in solution. Faraday was so impressed with Davy that he was determined to obtain a position in his laboratory. As was his practice, Faraday took copious notes at Davy's lectures, and he used his bookmaking skills to bind the notes and send them to Davy as an introduction. Davy wrote back to Faraday noting that he was impressed with the notes and when an opening became available in the spring of 1813, Faraday was hired as an assistant in the laboratory at a wage of 2.5 shillings a week along with 2 small rooms at the top of the Institute for housing.

Faraday immediately made himself indispensable to Davy, helping prepare the chemicals for experiments and working on the organization of lectures and presentations. When Davy decided to take a tour of Europe to give a series of lectures and meet with prominent scientists, he asked Faraday to accompany him and his new wife as a secretary and experimental assistant. Despite the raging Napoleonic War, Davy, who had previously been given a Prize by Napoleon was given special clearance by the emperor to travel in Europe. In Paris, Davy presented to the French Academy of Sciences and introduced Faraday to Ampere. In Italy, they had long discussions with Volta in Milan and after touring Switzerland and Germany they returned to England in 1815. Faraday, who had not previously been outside the London area, had a remarkable opportunity to meet the most famous scientists in Europe and be introduced to new cultures and languages. When he returned to London, he could speak French and Italian and was a confident scientist. Davy promoted him to Superintendent of Apparatus at the Royal Institute along with increased salary.

Although Faraday's greatest accomplishments came in the area of electromagnetism, prior to Oersted's discovery in 1820, Faraday made significant contributions to the fields of chemistry, mechanics, and metallurgy. Working with Davy he synthesized several compounds of chlorine with carbon. His studies on coal tar led to the discovery of isobutylene, used in the production of rubber and benzene. His expertise in chemistry led to demands for his services as a consultant to pharmaceutical companies and as an expert witness in technical trials. Faraday's pioneering work in the field of metallurgy led to the production of the first stainless steel and the alloying of steel with other metals. In 1815, along with Davy, he developed the widely popular miner's safety lamp.

In the early years of his scientific career, Faraday was focused on his work and showed little interest in social activities or the pursuit of the fairer sex. In a short poem jotted in the margins of a notebook, he berated love, "*What is the pest and plague of human life? And what is the curse that often brings a wife?*" [14] But in the winter of 1820, he met and rapidly fell in love with Sarah Bernard, the nineteen-year-old sister of a friend at the City Philosophical Society. He began a love letter to Sarah "*You know my former prejudices and my present thoughts – you know my weaknesses, my vanity, my whole mind; you have converted me from one erroneous way, let me hope you will attempt to correct what others are wrong…*" [14] Despite initial efforts by the girl's father to prevent the relationship, Faraday persisted and even left work and followed Sarah to her aunt's home on the coast of Kent where she had been sequestered. He made slow progress but then came a breakthrough when Faraday and Sarah took a tour of the cliffs of Dover.

In his journal, Faraday wrote, "*I can never forget this day. Though I had ventured to plan it, I had little hope of succeeding. But, when the day came, from the first waking moment in it to the last it was full of interest to me; every circumstance bore so strongly on my hopes and fears that I seemed to live with thrice the energy I had ever done before*" [15]. Sarah had accepted his proposal and they were married in June 1821. Although they never had children, they were the favorite of their nephews and nieces and Faraday enjoyed having the children sit and watch him work in his laboratory often making special displays for their entertainment.

When the exciting news of Oersted's discovery reached the Royal Institution in the fall of 1820, Davy set aside his other projects and tried to verify Oersted's findings. Although he saw the compass needle move toward a current-carrying wire, Davy apparently misunderstood Oersted's description of the tangential nature of magnetic force assuming it was a straight-line attraction between the compass needle and wire. At the same time, an English physicist, William Hyde Wollaston, read Oersted's report and immediately saw the implication of the proposed tangential magnetic force surrounding a current-carrying wire. He suggested to Davy that such a wire should rotate on its axis if it was brought next to a magnet. Davy attempted Wollaston's experiment but failed to see the expected rotation of the wire although the wire did move toward the magnet. While this was going on Faraday was working on other projects and just beginning his courtship of Sarah Bernard so he was not directly involved in Davy's experiment. He did see Davy's demonstration, however, and this would later be the source of embarrassment.

The First Electric Motor

After returning to the laboratory after his marriage, Faraday became immersed in electromagnetism when he was asked to write a historical review of the subject for the *Annals of Philosophy*. He was immediately struck by the observation that magnetism was transverse to the direction of current flow and not the expected push–pull effect. He decided to revisit the Wollaston experiment that Davy had tried and failed.

With Sarah's 14-year-old brother, George, watching, Faraday developed a clever apparatus in which he fixed a bar magnet in wax upright at the center of a cup filled with mercury (salt water would also work) with the upper pole of the magnet exposed (Fig. 4.2). Directly above the exposed pole of the magnet, he hung a copper wire with its lower end dipping into the mercury next to the magnet. When he connected the upper end of the hanging wire to one terminal of a voltaic pile and a wire contacting the mercury to the other end of the pile creating a circuit, the hanging wire began to rapidly rotate around the exposed magnetic pole. The rotation resulted from an interaction between the magnetic field around the wire due to current flow and the field of the magnet. Electric energy from the voltaic pile was being converted to mechanical energy. Faraday had made the world's first electromagnetic motor. The motion continued until the apparatus was disconnected from the battery. In his journal Faraday wrote, "Very satisfactory, but make a more sensible apparatus" [16].

Excited with his finding, Faraday rushed to publish the result in the next edition of the *Quarterly Journal of Science*. The article came out within a month. It was an instant sensation in the European scientific community but there were problems at home. Faraday had failed to give the customary recognition to his mentor, Davy and made no mention of Wollaston and his prior suggestion for such an experiment. Some even accused Faraday of plagiarism. Faraday apologized to Wollaston and the matter was largely forgotten, although Davy's relationship with Faraday was never quite the same.

In the following years as Davy's health deteriorated Faraday took over the duties of running the laboratory at the Royal Institution. Davy suffered his first stroke in 1826 and finally died after a stroke in 1829. During this time, Faraday was busy with

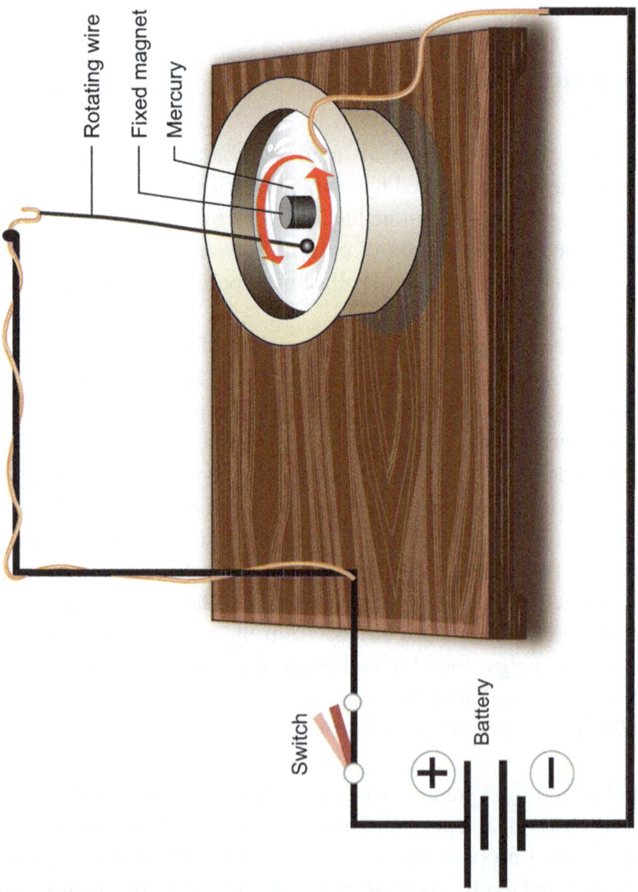

Fig. 4.2 Diagram of Faraday's first electromagnetic motor. The hanging wire rotated around the fixed magnet when the switch was closed due to interaction of the magnetic fields of the wire and the magnet

the development of an optical glass for improved lenses used by the Admiralty. He put aside his research in electromagnetism in part due to the Wollaston controversy. Even though his work on optical glass was highly successful he continued to think about new experiments and in 1831 he decided to return his focus to electromagnetism.

The Induction Coil

Faraday reasoned that since electric current generated magnetism, magnetism should produce electricity. He had previously experimented with wire coils next to magnets but saw no evidence of an electric current. Something else must be involved. It occurred to him that since it was the flow of electric current that produced magnetism maybe a moving magnet would produce electricity.

By 1831, it was well known that one could produce a powerful magnet by wrapping an iron rod with an insulated coil of wire and then connecting the wire terminals to a voltaic pile. The more turns in the wire coil, the greater the magnet strength. The iron rod with north and south poles remained magnetized even after the wire coil was removed.

Faraday decided to see what would happen if he linked two electric circuits together magnetically. Would electric current in one circuit induce electric current in the other circuit? To achieve this linkage, he had a wrought-iron ring six inches in diameter forged. He wrapped a wire coil around each side of the circular iron ring being careful to insulate the wire from the ring with cloth and from neighboring wire with string (Fig. 4.3). At the time insulated wires were not available. One coil of wire was then connected to a voltaic battery and the other to a galvanometer.

To his surprise, when he attached the coil to the battery, the needle on the galvanometer spun several times and then stopped even though the current in the first coil remained constant. The galvanometer needle moved the same amount but in the opposite direction when the first circuit was disconnected from the battery. Faraday realized it was the changing magnetic field that induced the electric current in the second coil not the fixed magnetic field. Faraday called the process *volta-electric induction,* the basic mechanism of modern transformers where electric current is transferred from one coil to another to increase or decrease the voltage depending on the number of turns in each coil. The apparatus called an *induction coil* also provided a means to convert a constant electrical current in the first coil into a pulsed alternating current in the second coil.

The First Electric Generator

Faraday then decided to try to produce an electric current by moving a fixed magnet. He made a hollow wire coil with the terminals attached to a galvanometer and plunged a magnetized iron bar in and out of the coil. The galvanometer needle

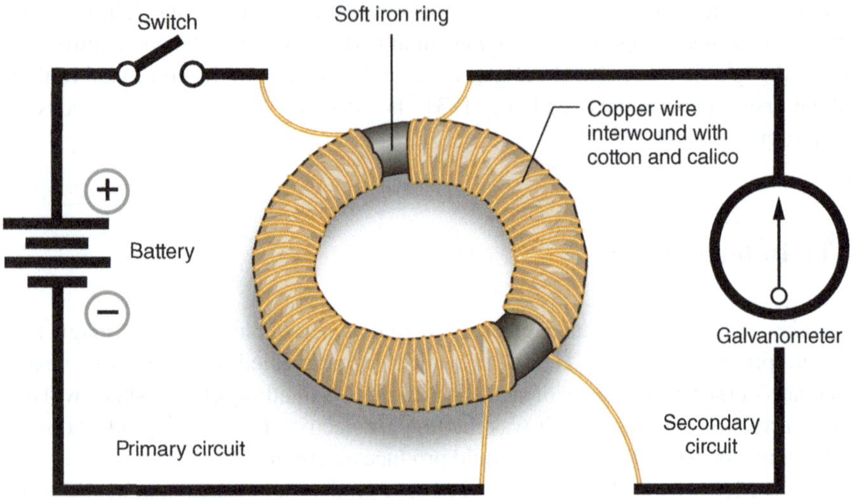

Fig. 4.3 Diagram of Faraday's first induction coil (*volta-electric induction*). Opening and closing the switch in the primary circuit produced an alternating current in the secondary circuit measured with the galvanometer

reacted to the movement of the magnet, swinging in opposite directions when the bar moved in and out of the coil. He called this apparatus a solenoid. Faraday had discovered the principle used by modern electric power generators. All that was needed was a way to generate a continuous source of electricity rather than the intermittent current made by making and breaking the circuit.

The final breakthrough experiment that Faraday performed in 1831 was to develop an apparatus for continuous generation of electric current. To do this he used a rotating copper disk in a fixed magnetic field. He manufactured a thin circular copper plate a foot in diameter fixed on a brass axel mounted on a frame so that the outer edge of the copper disk moved between the two poles of a fixed magnet (Fig. 4.4). He attached wires to sliding contacts on the outer edge of the disk and on the axel and attached the end of the wires to a galvanometer producing a circuit to measure electric current.

Faraday wrote in his notebook, "*the instant the disk moved the galvanometer was influenced, and by revolving the disk quickly the needle could be deflected 90° or more*" [17]. When Faraday spun the disk, the needle stayed in its new position, and there was a steady current produced. If he spun the disk in the opposite direction, the needle moved in the reverse direction and held the new position. The copper disk was acting like an array of radial wires which were cutting across the magnetic field as the disk turned causing electric current to be continuously produced. This simple experiment provided underpinning for Faraday's fields-of-force electromagnetic theory which states that the induced electromotive force in an electrical circuit is proportional to the rate of change of magnetic flux lines in the circuit. The key

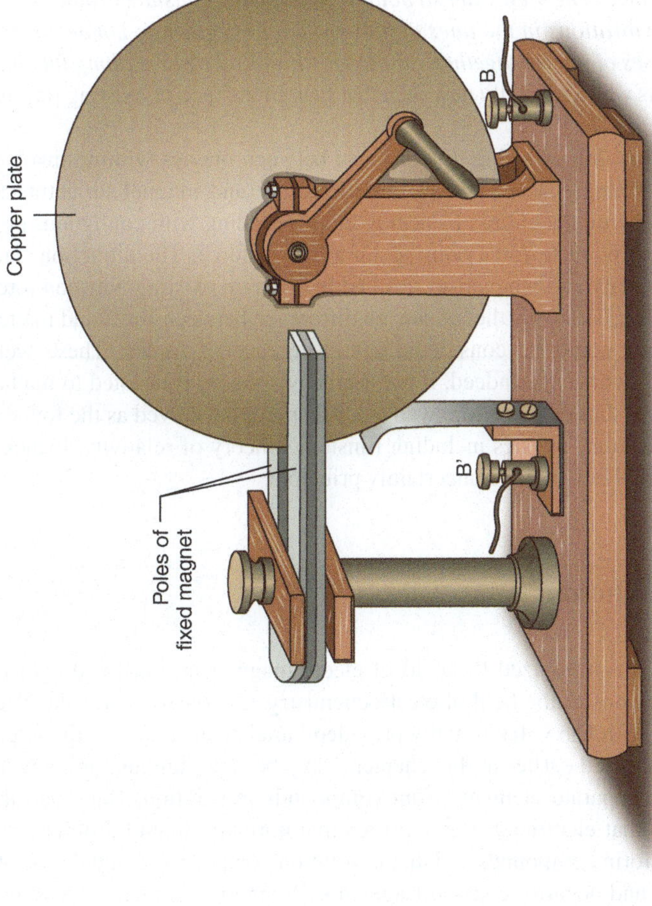

Fig. 4.4 Faraday's first electric generator. A circular copper plate was rotated between the two poles of a fixed magnet. An electric current, generated between wires attached to sliding contacts on the axel (B) and the outer edge of the disc (B'), was measured with a galvanometer

element is the motion of a conductor relative to a magnet so that the conductor intersects or cuts through the force lines of the magnetic field.

But what did Faraday mean by lines of force? He knew little about mathematical equations, so he had to use physical models to explain his ideas. No doubt he was heavily influenced by how iron filings arranged themselves in a pattern of lines when he sprinkled them on paper held over the poles of a magnet. Faraday felt that the lines were real representing the lines of an invisible field of force existing in space. Furthermore, vibrations in these lines of force could radiate outward into space. He later wrote, "*The view I am so bold as to put forth considers radiation as a high species of vibrations in the lines of force which are known to connect particles and also masses of matter together…these vibrations take place along the lines of force*" [18]. This was a radically new idea that flew in the face of existing physics at the time.

Isaac Newton saw forces acting at a distance between masses with no involvement of the intervening space. Faraday saw electricity and magnetism acting as lines of force occupying the entire physical world proceeding with contiguous progression from force point to force point along a line of force. The attractions and repulsions of electricity and magnetism resulted from strain existing between force centers. At its basics, Faraday's theory saw no difference between force and matter. Physical substances could be considered a convergence of forces. These were remarkably modern concepts. Indeed, it was Faraday's theory translated to mathematical equations by James Clerk Maxwell (see Chap. 6) that served as the foundation of twentieth-century physics including Einstein's theory of relativity, Planck's quantum theory and Heisenberg's uncertainty principle.

Electricity in Solutions

Faraday not only revolutionized the field of electromagnetism, but he also made important contributions to the field of electrochemistry. Of course, it was the electrochemical work with Davy that initially provided Faraday an entry into the scientific world. As discussed earlier in this chapter, Davy became famous for his work with electricity to separate elements from compounds in solution. Davy had the insight to suggest that electricity was the force that normally bound different elements together to form compounds and that in solution compounds could dissociate into their positive and negative charged fragments. When the terminals of a voltaic pile were placed in a solution, the positively charged fragments migrated to the negative terminal and the negatively charged fragments to the positive terminal where the charge was taken up. As a result, the remaining uncharged elements accumulated at the terminals.

Faraday agreed with Davy and went further noting that the chemical separating power was in direct proportion to the absolute quantity of electricity passing through the solution. "*The equivalent weights of bodies are simply those quantities of them which contain equal quantities of electricity, or have natural equal electric power, it being the electricity which determines the combining force. Or, if we adopt the atomic theory or phraseology, the atoms of bodies which are equivalent to each other in their ordinary chemical action have equal quantities of electricity naturally associated with them*" [19]. He made this statement more than a half century before the concept of atoms with protons and electrons was formulated.

Faraday showed that the same fixed quantity of electricity freed one gram equivalent of hydrogen, ½ gram equivalent of oxygen, and ½ gram equivalent of copper. We now know that the chemical equivalent of an element is the reciprocal of the valence of the element in a compound which is determined by the number of electrons that are added, lost, or shared.

To describe these new concepts, Faraday needed new words, so he turned to a close friend and writer on the history of science, Dr. William Whewell at the Trinity College, Cambridge, for help. Together they called the migrating fragments in the solution, *ions,* from the Greek word for "wanderer." The overall fluid electrochemical process they called *electrolysis* and the conducting solution, the *electrolyte*. The terminal (previously called the pole) through which the electrical current entered the electrolyte was called the *anode* and the terminal from which it left was called the *cathode*. The negatively charged ions of the electrolyte that migrated to the anode were called the *anions* and the positively charged ions that moved to the cathode, the *cations*.

Faraday and the Electric Eel

Although mainly focusing on the nonorganic aspects of electricity, Faraday followed the exciting developments in animal electricity and maintained correspondence with many of the leading researchers in the field. In 1835, he presented a series of 14 lectures on electricity to the medical students at St George's Hospital. He was particularly interested in the ongoing work on electric fish and made inquiries about obtaining electric fish for study. He sought the advice of a contemporary who had been actively involved in such research, the German natural philosopher, Baron Alexander von Humboldt.

Humboldt, who by then was considered the "grand old man" of German science, had worked extensively on animal electricity throughout his career and was well known for his excursion to South America at the turn of the nineteenth century to study the electric eel [20, 21]. Humboldt had studied animal electricity in a wide variety of animals including many invertebrates and felt that the electric organs in fish could provide a model for understanding how electricity was produced in the nervous system of animals including humans.

Humboldt was famous for his self-experimentation, not uncommon among electricity researchers at the time, although Humboldt took it to another level. He frequently used his body as an electrometer. Several of his experiments led to severe pain even bordering on self-mutilation. For example, he applied chemical irritants such as potassium carbonate and dilute hydrochloric acid to various parts of his body to make the injured region more sensitive in his galvanic stimulation experiments. He conducted experiments on pain sensation after an accidental wound on his wrist and in his mouth after a tooth extraction.

In the early 1800s, Humboldt visited the Spanish Crown Colony of New Granada (now Venezuela) to study gymnotes (electric eels). After several frustrating months searching for the eels in oppressing heat and mosquito-infested swamps Humboldt and a botanist colleague were directed by natives to a riverbed where the eels congregated in the shallow muddy water during the dry season. Humboldt's detailed description of what he witnessed when he arrived at the riverbed was so dramatic that it immediately caught the imagination of the European public and subsequently was illustrated and reprinted in numerous books and articles. The local natives drove a herd of horses and pack mules into the water to lure the eels out of the mud producing a shocking frenzy that resulted in the deaths of several of the animals. The natives called the process *embarbascar con caballos* (stupefying with horses). Injured by the violent shocks, several of the animals sank beneath the water and drowned.

According to Humboldt, "*The eels, stunned by the noise, defend themselves by repeated discharge of their electric batteries*" [22]. Having used the animals as bait, the natives stood on the shore or sat on overhanging trees and used harpoons attached to dry cords to capture the eels without wounding them too severely. Some were as large as 5 feet in length.

Humboldt conducted dissections and physiological experiments on captured eels often subjecting himself to painful shocks. The eel's electric organ was huge, more than two-thirds of its body size. Humboldt received shocks by accident and in planned experiments: "*If by chance a stroke be received before the fish is wounded or wearied by long pursuit, the pain and numbness are so violent that it is impossible to describe the nature of the feeling they excite. I do not remember having ever received from the discharge of a large Leyden jar a more dreadful shock than that which I experienced by imprudently placing my feet on a gymnotus just taken out of the water. I was affected during the rest of the day with a violent pain in the knees, and in almost every joint…*" [23].

A key question that Humboldt grappled with was whether the animal electricity produced by the eel's electric organ was the same as electricity produced by a traditional electrical machine He couldn't measure the discharge from an eel with his gold leaf electrometer and despite the prior report of Walsh, he could not produce a visible spark with the discharge. Further, he noted that the discharge of an eel would not pass through a small piece of bone or coal, whereas "genuine electricity" easily passed through these materials. Maybe organic galvanic electricity was different from inorganic electricity. On the other hand, the shock of an eel felt like the shock from a Leyden jar or a voltaic pile. "*The sensation caused by the feeble shocks of an electric eel appeared to me analogous to that painful twitching with which I have been seized at each contact of two heterogeneous metals applied to wounds that I had made on my back by means of cantharides*" [23].

When Faraday reached out to Humboldt for advice regarding obtaining and handling an electric eel for study, he received a detailed reply: "*I would advise that they be transported from Surinam (from Essequibo, Demerara, Cayenne) in summer, for the Gymnotus in its native country lives in water of 25° centigrade (or 77° Fahrenheit). Some are five feet in length, but I would advise that such as are about twenty-seven or twenty-eight inches in length be chosen. Their power varies with their food, and their state of rest... I retained them in a box or trough about four feet long, and sixteen inches wide and deep. The water must be fresh, and be changed every 3 or 4 days: the fish must not be prevented from coming to the surface, for they like to swallow air. A net should be put over and round the trough, for the Gymnotus often springs out of the water. These are all the directions that I can give you. It is, however,* ***important*** *that the animal should not be tormented or fatigued, for it becomes exhausted by frequent electric explosions. Several Gymnoti may be retained in the same trough*" [24].

Subsequently, Faraday found out that an electric eel had just been brought to London by the Gallery in Adelaide Street and they were willing to lend him the fish for experimentation in the name of science. The Adelaide Gallery, founded by the American inventor and entrepreneur Jacob Perkins, was one of several lay scientific institutions in London at the time whose goal was to bring "*practical science to the public in spectacular fashion*" [25]. The Gallery had a range of electrical devices including a collection of voltaic batteries, a ferromagnetic globe to illustrate the earth's magnetic field, and a large electromagnet that produced electric sparks. One of the main attractions of the Gallery was feeding time at the pool containing the exotic electric eels.

Faraday accepted the offer and the fish that was 40 inches long was brought to his laboratory where his assistant worked to get it gradually to eat small fish. Faraday performed a series of four experiments on the fish beginning in early September 1838 with weeks in between to allow the animal to rest and recover. The goal of the experiments was to see if the electricity discharged by the fish had the same properties as inorganic electricity.

Like others before him, Faraday first tested the subjective sensation of shock produced by the fish's electric organ directly with his hands and with a metal probe. He described the shock as very powerful and noted that the strongest shock occurred when he touched the animal with one hand near the tail and the other hand near the head. The nearer the hands were together on the body of the fish, the less powerful the shock.

In order to study the electrical discharge in detail he needed a way to collect the discharge from the fish without it being dissipated into the surrounding water [24]. To do this he constructed two insulated metal "jackets" that fit over the back and sides of the fish made of a thin plate of copper eight inches long and two and a half inches wide bent into a saddle shape, one near the tail and the other near the head. He attached a thick copper wire to each jacket to convey the electricity to the test apparatus. The outside of each jacket and wire were insulated with natural rubber (caoutchouc) that extended over the edges of the jacket and touched a glass plate at the bottom of the water. Faraday noted that the Gymnotus was insulated almost as well as if it had been in the air.

When wires from the metal jackets were connected to a galvanometer, there was a strong 30–40° deflection of the needle when the fish produced a discharge. The direction of the electric current through the galvanometer was always from the head to tail direction so that the jacket near the head was positive and the jacket near the tail, negative (recall that electricity was thought to flow from the positive to negative). When the discharge of the Gymnotus was passed through an induction coil, it produced a magnet just like a voltaic pile with polarity of the magnet consistent with electric current moving from the head to tail direction of the fish. Finally, Faraday found that electricity from the eel could decompose potassium iodine in solution and produce a spark as previously reported by Walsh. In other words, the electricity generated by the eel was no different from electricity produced by an electrical machine or a voltaic battery.

It was clear that the fish had voluntary control over the discharge of the electric organ. Humboldt had previously shown that removing the brain prevented the organ from discharging. Faraday felt that "nervous fluid" was somehow related to electricity. He argued that electricity could be converted to magnetism and heat and magnetism and heat into electricity so possibly nervous fluid could be converted to electricity in the electric organ and electricity from the organ could be converted to nervous fluid. "*This may seem to some a very wild notion, as assuming that the nervous power is in some degree analogous to such powers as heat, electricity, and magnetism. I am only assuming it, however, as a reason for making certain experiments, which, according as they give positive or negative results, will regulate further expectation. And with respect to the nature of nervous power, that exertion of it which is conveyed along the nerves to the various organs which they excite into action, is not the direct principle of* ***life****; and therefore I see no natural reason why we should not be allowed in certain cases to* ***determine*** *as well as observe its course*" [26]. Faraday tried to obtain additional Gymnotus for study, but without success, and he never conducted further experiments to test his hypothesis.

Robert Todd, Electricity and the Brain

Present during all four of Faraday's experiments on the electric eel was a young physician, Robert Todd, who had just been appointed Chair of Physiology and Morbid Anatomy at the newly established King's College, London [27, 28]. Todd, who was born in Ireland, trained with Robert Graves (of Grave's Disease fame) at the Richmond Hospital in Dublin before moving to Oxford where he obtained his medical degree in 1833. Todd had a keen interest in animal electricity, and he jumped at the chance when colleagues at King's College offered to introduce him to Faraday.

At the time, little was known about the gray and white matter of the brain. As microscopes became available, it was possible to study the microstructure of the brain for the first time. The German zoologist, Theodor Schwann had just formulated the so-called cell theory, that all living things were composed of individual cells, the basic unit of life. Schwann noted that the covering of nerve fibers that gave them the glistening white appearance was generated by a cell separate from the nerve cell, now called a Schwann cell. Schwann thought that the cell secreted the sheath, called myelin, but modern electron microscopy has shown that the multilayered sheath is composed of a folding of the Schwann cell membrane wrapped around the nerve fiber.

With his microscope, Todd focused on the gray matter of the cerebral cortex, subcortical nuclei, and the spinal cord. He identified "nerve vesicles" (nerve cells) that had one or more nerve fibers attached to them. He surmised that the fibers in the white matter were all connected to vesicles in the gray matter and that the vesicles were the "point of departure" of one or more fibers. Furthermore, he emphasized that the nerve vesicle, the fiber emanating from it, and the surrounding blood vessels formed "a distinct apparatus for the development of nervous polarity" [28]. In the process, he anticipated the modern concept of a neurovascular unit.

Based on his conversations with Faraday, Todd applied the concept of polarity to the nervous function. He compared the brain to a huge battery made up of many smaller batteries (the vesicles). "*Adopting the language of the illustrious Faraday which expresses with clearness and precision the fundamental phenomena of the electric force, we may call the nervous power a polar force, generated in the centres and propagated by the rapid polarization of the neighbouring particles in various directions*" [29].

Although Todd and Faraday saw an analogy of nerve cells to small batteries, they had doubts that the nervous force was the same as electricity. After all, nerves were not very good conductors of electricity, particularly compared to metal conductors. Others have shown that tying a ligature around a nerve stopped the transmission of nerve signals but not electricity. Also, placing a piece of metal between two ends of a cut nerve allowed the transmission of electricity but not nerve signals. Since electricity could be converted into heat and magnetism, Faraday and Todd viewed nerve transmission as another force associated with electricity, one that could be converted to and from electricity just as the electricity in the eel's electric organ could be converted to and from nervous force.

Todd was the first to use the terms afferent and efferent to describe nerves carrying signals to and from the central nervous system, respectively. Although changes in nervous polarity in the vesicles in the gray matter explained the propagation of efferent nerve signals, the propagation of afferent nerve signals was more difficult to explain. Todd speculated that some mechanical, chemical, or even galvanic stimulus at the sensory nerve terminal could alter the state of polarity and trigger the production of an afferent nerve signal. The change in polarity was rapidly transmitted throughout the length of the nerve likely by some type of "molecular change." These ideas are remarkably modern considering the complete absence of information regarding nerve structure and function at the time.

Electricity and Epilepsy

Being a physician, Todd immediately began to consider the implications of the "nerve force theory" on the health and disease of the brain. He surmised that a healthy nutrition was critical for maintaining normal polarity for motor, sensory, and mental activity in the brain. "*The rapidity with which a nerve loses its power after it has been separated from the nervous centres clearly denotes that connection with the centre is a necessary condition for nutritive activity of nerves*" [28]. We now know that in addition to triggering nerve signals, nerve cells are critical for providing nutrients and growth signals to peripheral nerves. Todd speculated on how drugs including tobacco and alcohol could alter the function of the brain by altering nervous polarity. Although Todd is best known for his description of paralysis after a focal epileptic seizure, called Todd's paralysis, he was the first to propose an electrical theory to explain epilepsy.

People with epilepsy have long suffered from social stigmata and ostracism. Even though Hippocrates suggested that epilepsy was a brain disease as far back as the fifth century BCE, many people in the nineteenth century still believed that epilepsy was caused by demons and treatments were aimed at driving out the evil spitits. Remarkably, one of the main medical treatments for epilepsy in the early nineteenth century was electrotherapy, first with static electricity and then voltaic batteries when they became available. The rationale for using electricity was poorly defined other than the general principle that all diseases resulted from an imbalance in body electricity.

Todd, relying heavily on the electrical concepts of Faraday, concluded that epileptic convulsions resulted from an abnormal buildup of nervous force in the brain, "*these periodical evolutions of the nervous force may be compared to the electrical phenomena described by Faraday under the name of 'disruptive discharge'*" [27]. Faraday had shown that when the buildup in electrical tension reached a threshold there was a sudden discharge in polar tension resulting in a spark. By analogy, Todd saw the polar tensions in the gray matter of the brain reaching a threshold and then discharging "*exciting the other parts of the brain and spinal cord with all the violence of the discharge from a highly charged Leyden jar*."

Todd described his clinical and experimental work on Epilepsy in a series of three Lumleian lectures at the Royal College of Physicians and in four articles published in the London Medical Gazette in 1849 [30]. He noted the remarkable variability of clinical features from tonic-clonic convulsions with loss of consciousness to brief absence spells that could occur at any time of life with a wide range of severities. He attributed the loss of consciousness to the involvement of the highest cortical vesicular matter and noted that motor convulsions in the limbs could occur on the opposite side of localized cortical damage. He emphasized that the extremities could remain paralyzed after a focal motor seizure, the sign that still bears his name—Todd's paralysis. Although he recognized that a wide variety of brain abnormalities were associated with epilepsy, he was also aware that many patients with epilepsy had no evidence of brain damage. Finally, he noted that heredity was important in some types of epilepsy and that repeated seizures could lead to loss of memory and overall mental function.

Todd conducted a series of experiments with electrical stimulation of the brain in rabbits to test his electrical theory of epilepsy. Todd first stimulated the gray matter of the brain stem and spinal cord of a rabbit and noted continuous muscle contractions which he described as being like "tetanus or laryngismus" unassociated with any change in consciousness. On the other hand, when he stimulated the cortical and subcortical gray matter, the animal developed tonic-clonic muscular movements with loss of consciousness very similar to an epileptic seizure. "*Having passed fine bradawls into the cranium in such a direction as I had previously satisfied myself would lead to this organ [the brain], I subjected it to the influence of the machine; general convulsions were produced of a character essentially different from those which resulted from stimulating the spinal cord or the medulla oblongata [brain stem]. They were combined movements of alternate contraction and relaxation, flexion and extension, affecting the muscles of all the limbs, of the trunk and of the eyes, which rolled about just as in epilepsy*" [31].

Instead of a constant current voltaic battery as used by Aldini, Todd used a new magnetoelectric machine designed by Faraday that produced a high voltage pulsed current that had recently been shown to be more effective for stimulating nerves than a low voltage direct current battery (discussed in Chap. 5). The machine was based on Faraday's induction coil with an inner coil consisting of a small number of turns of thick wire and an outer coil with a much larger number of turns of thin wire. The connection of the inner wire to a battery was made intermittent by turning a spoked rotating disk. As we will see, electrical stimulation using faradic current, called faradization, rapidly became popular for electrotherapy in the latter half of the nineteenth century.

A current-day epileptologist at King's College in London, Edward Reynolds, questioned why much of Todd's pioneering work on epilepsy was ignored by the next generation of English epileptologists, in particular, David Ferrier and Hughlings Jackson [27]. Although their work will be discussed in detail in Chap. 7, it is useful to briefly mention it here. Jackson, a clinician with little basic science background, is often credited with being the first to suggest that epilepsy was due to a "nervous discharge." He is best known for his description of the slow progression of a focal motor seizure. Ferrier, on the other hand, was a neurophysiologist interested in the basic mechanisms of nerve function.

Jackson was aware of Todd's work on epilepsy and quoted Todd in many of his early publications including Todd's suggestion of an electrical discharge. In fact, it was Jackson's description of post-seizure paralysis in an article 27 years later and the mention of Todd's idea that it was due to nervous exhaustion that solidified the term "Todd's paralysis." However, in Jackson's famous 1890 Lemleian Lectures on epilepsy, he did not mention Todd and his electrical discharge theory. He described the nervous discharge as the liberation of energy by nervous elements. He likened it to gunpowder in a cannon in which the stored energy is released when the gun is fired. He did not mention batteries or electricity but rather suggested a chemical process. Presumably, he felt that his theory on epilepsy was completely different from Todd's theory.

In the case of Ferrier, the reasons for not mentioning Todd's work may be more sinister. Ferrier, a Scottish physician and physiologist working at King's College, used faradic electrical stimulation on the brains of dogs and later monkeys to develop a specific map of motor cortical areas. Like Todd's work on the rabbit, he noted that high-intensity faradic stimulation of the motor cortical regions of dogs caused repetitive movements of the face and extremities reminiscent of epileptic seizures in humans. He also observed a spread of the motor activity along the motor cortex consistent with the Jacksonian "march" in a focal motor seizure.

Reynolds pointed out that Ferrier had to have been aware of Todd and his findings since there was a large statue of Todd at the entrance to the hospital that Ferrier passed by daily [27]. Furthermore, Ferrier failed to reference earlier similar work on the motor cortex by the German neurophysiologists, Fritsch and Hitzig. My UCLA colleague and neuroscience historian Mary Brazier concluded that the oversight was likely intentional.

The New Breed of Medical Instrument Makers

Faraday had a major influence on the medical uses of electricity in the latter half of the nineteenth century, even though he himself had little interest in electrotherapy. He was suspicious of the claims of the electrical instrument makers that proliferated in London. He considered them vultures stealing the ideas and discoveries of true scientists for monetary purposes.

Faraday maintained a long-standing feud with the London instrument maker, William Sturgeon, who was the first to make a functioning electromagnet [32]. Sturgeon wrapped insulated copper wire around a small horseshoe-shaped bar of soft iron and connected the ends of the wire to a voltaic battery. He submitted his electromagnet along with several other tabletop electrical devices that he used in his displays to the Royal Society of Arts in 1824 and was awarded a silver medal. Joseph Henry, a young American teacher at the Albany Institute in New York, soon found a way to tightly wind a horseshoe magnet that could lift the unbelievable weight of 750 pounds.

Like Faraday Sturgeon was born into a poor London family and apprenticed to a local shoemaker at age 13 years. Unlike Faraday, he was treated cruelly in his apprenticeship and at the age of 19 left to join the military. He developed an interest in electricity in the Army and taught himself the basics of science from books borrowed from his sergeant. After leaving the army in 1820 he set up as a shoemaker but soon began to supplement his income by making and selling electrical instruments and providing occasional lectures.

Sturgeon became an accomplished showman appealing to the common man. His view of experimentation and electricity was the exact opposite of Faraday's view. He was interested in the spectacular with an emphasis on the size and appearance of the equipment and not the science behind the experiment. For example, Sturgeon constructed a globe wrapped with silver and platinum wires that rotated around a central magnet. When he heated the globe electricity was produced at the junction of the wires by the process of thermoelectricity. He felt that electricity "flowing in the earth" could explain global weather patterns. The increased heat from the sun near the equator led to more thermoelectricity producing more atmospheric electricity and increased thunderstorms.

In 1837, Sturgeon got together with a group of other instrument makers and a variety of other electrical enthusiasts to form the London Electrical Society that met regularly at the shop of the instrument maker, E.M. Clarke. The society was open to all comers regardless of background or training. Since a large part of the instrument makers' clientele were physicians, unsurprisingly, many of the lectures and displays addressed the medical uses of electricity. Clarke demonstrated his new magneto-electric machine and its effect on nerves and muscles. He bragged that "*no person out of hundreds who have tried it could possibly endure the intense agony it is capable of producing*" [33]. The American inventor Thomas Davenport sent a team of promoters to attend a Society meeting and demonstrate his "electromagnetic engine," the first practical electric motor.

At an early Society meeting, a letter was read describing the bizarre claim of the generation of insect life from electricity by Society member, Andrew Crosse. Crosse was a squire in rural Somerset, England, with a keen interest in electrical experiments. The trees of his estate were covered with wires which he used to collect electricity for his experiments. He was known locally as the "thunder and lightning man." Crosse was keen to produce artificial crystals with electricity. He used a large bank of voltaic batteries to pass current through various chemical solutions for months without success.

In one of his setups, Crosse steadily dripped a solution of silicate of potash and muriatic acid onto a piece of volcanic rock that was electrified with batteries. He was trying to produce crystals of silica but to his surprise: "*At the end of fourteen days, two or three very minute specks or nipples were visible on the surface of the stone, between the two wires, by means of a lens. On the eighteenth day, these nipples elongated and were covered with fine filaments. On the twenty-second day their size and elongation increased, and on the twenty-sixth day each figure assumed the form of a perfect insect, standing on a few bristles which formed its tail*" [34]. Several hundred of these insects appeared and began to move about over the next few weeks. Crosse claimed that he was able to obtain the same result when repeating the experiment.

As news of Crosse's discovery leaked out to the media, public reactions varied from amazement to incredulity. Some even considered his research blasphemy. The religious community condemned Crosse for interfering with God's laws and there was a rumor that a local vicar attempted to perform an exorcism at Crosse's laboratory. Many in the Society took Crosse's report seriously and one, W.H. Weekes reported at a subsequent meeting that he was able to repeat the experiment and produce insects from electricity. Crosse sent samples to Michael Faraday, but Faraday did not enter the fray ignoring the ludicrous claims.

In his mind, Faraday considered himself to be above the public show presented at the Electrical Society lectures. He was a true "scientist" whose goal was to protect the public from false profits and show them what was "true" science. He rehearsed and meticulously prepared for his famous Friday evening lectures regularly attended by hundreds of people from all walks of life. He would not present any experiment unless he was confident of the result. Furthermore, he carefully screened guest lecturers to be sure they would not present any material that might be outside the realm of "science" or might be offensive to the audience.

Like his predecessor Sir Humphrey Davy, Faraday was a favorite of the ladies. A member of the audience nicely summarized Faraday's charisma: "*There was a gleaming in his eyes which no painter could copy, and which no poet could describe*" [35]. Cornelia Crosse, wife of the "insect maker" Andrew Crosse observed: "*No attentive listener ever came away from one of Faraday's lectures without having the limits of his spiritual vision enlarged, or without feeling that his imagination has been stimulated to something beyond the mere exposition of facts*" [36]. There can be little doubt that Faraday worked very hard to cultivate this sentiment in his audience and provide a romantic, highly idealized version of science.

But to William Sturgeon and many other members of the Electrical Society, Faraday was pompous and his theories about electricity were esoteric and of little practical use. They were all about creating bigger and better electrical equipment. Members of the Electricity Society had little interest in the difference between science and invention that Faraday insisted upon. But the Society soon began to suffer from internal rivalries many instigated by the combative Sturgeon. Several members complained that Sturgeon was claiming discoveries for himself that rightfully belonged to the overall Society. Sturgeon became frustrated and left for Manchester in 1840 where he supervised the Royal Victoria Gallery for Practical Science. The London Electrical Society lasted a few more years before it became financially insolvent.

Faraday's Induction Coil and Nerve Stimulation

One finding in his frog experiments that perplexed Galvani was that the type of the electrical stimulus seemed to be more important than the quantity of the electrical stimulus [37]. For example, he noted that the frog leg twitched with each spark of the nearby electrical machine yet connecting the electrical machine directly to the nerve via a conducting wire produced only a single muscle twitch. Similarly, a Leyden jar that had previously been discharged on multiple occasions could be as effective as a fully charged Leyden jar for producing a muscle twitch. These observations suggested to Galvani that the electrical stimulus may be acting as a trigger for some internal force, an electrical force generated intrinsically within the nerve and muscle tissue, rather than being the primary force for the muscle twitch. This intrinsic electricity represented Galvani's original idea of *animal electricity*.

As Volta's electric pile began replacing electrical machines for nerve stimulation, it became apparent that muscle twitches were reliably induced only when the current from the battery was turned on and off by connecting and disconnecting the circuit. A brief pulse of electrical current, positive or negative, was a better trigger for inducing a muscle twitch than a continuous current. With Faraday's invention of the induction coil, researchers and clinicians had a new tool for electrical stimulation. Not only did the induction coil produce sharp pulses of current, but the voltage of the pulsed current also varied with the number of turns of the thin wire of the outer coil. The frequency of the electrical pulses was controlled by the frequency of making and breaking the circuit of the battery with the inner coil.

This new type of electrical stimulation became known as faradization, and it gradually replaced galvanic stimulation for both experimental and clinical uses of electricity in the latter nineteenth century. Both faradic and galvanic electrotherapy used pulsed direct current, but the frequency of faradic current (50—100 Hz) was greater than that of galvanic current (about 30 Hz) and the duration of faradic pulses (<1 ms) was shorter than the duration of galvanic pulses (>1 ms). The brief sharp pulses of faradic current proved ideal for stimulating peripheral nerves and the brain.

In Brief

Hans Christian Oerstred's simple observation that electric current flowing along a wire generated a surrounding electromagnetic field whose strength was proportional to the amount of current was pivotal in the history of electricity. Based on the discovery Ampere made a quantitative electrometer that he called a galvanometer and Ohm developed his famous eq. I=V/R relating current (I) to the voltage (V) and resistance (R) in an electrical circuit. Wired telegraphy soon revolutionized worldwide communication. Michael Faraday's lines of force theory of electromagnetic waves marked the beginning of modern physics, and his induction coil provided a new and better way to stimulate nerves (faradization). Faraday's experiments on the electric eel inspired Michael Todd's pioneering work on epilepsy and brain electrical stimulation.

References

1. Schiffer MB. Draw the lightning down. Benjamin Franklin and electrical technology in the age of enlightenment. Berkley: University of California Press; 2003. p. 206–25.
2. Bordeau SP. Volts to hertz—the rise of electricity. Minneapolis: Burgess Publishing Company; 1982. p. 112–5.
3. Forbes N, Mahon B. Faraday, Maxwell and the electromagnetic field. Guilford: Prometheus Books; 2019. p. 31–40.
4. Bordeau SP. Volts to hertz—the rise of electricity. Minneapolis: Burgess Publishing Company; 1982. p. 60–9.

5. Mitchel A. The spinning magnet. London: Oneworld Publications; 2019. p. 117–31.
6. Bordeau SP. Volts to hertz—the rise of electricity. Minneapolis: Burgess Publishing Company; 1982. p. 70–85.
7. Bordeau SP. Volts to hertz—the rise of electricity. Minneapolis: Burgess Publishing Company; 1982. p. 75.
8. Bordeau SP. Volts to hertz—the rise of electricity. Minneapolis: Burgess Publishing Company; 1982. p. 76.
9. Bordeau SP. Volts to hertz—the rise of electricity. Minneapolis: Burgess Publishing Company; 1982. p. 86–109.
10. Bordeau SP. Volts to hertz—the rise of electricity. Minneapolis: Burgess Publishing Company; 1982. p. 99.
11. Forbes N, Mahon B. Faraday, Maxwell and the electromagnetic field. Guilford: Prometheus Books; 2019.
12. Morus IR. Michael Faraday and the electrical century. London: Icon Books Ltd.; 2004.
13. Bordeau SP. Volts to hertz—the rise of electricity. Minneapolis: Burgess Publishing Company; 1982. p. 123.
14. Forbes N, Mahon B. Faraday, Maxwell and the electromagnetic field. Guilford: Prometheus Books; 2019. p. 55.
15. Forbes N, Mahon B. Faraday, Maxwell and the electromagnetic field. Guilford: Prometheus Books; 2019. p. 56.
16. Forbes N, Mahon B. Faraday, Maxwell and the electromagnetic field. Guilford: Prometheus Books; 2019. p. 59.
17. Bordeau SP. Volts to hertz—the rise of electricity. Minneapolis: Burgess Publishing Company; 1982. p. 122.
18. Bordeau SP. Volts to hertz—the rise of electricity. Minneapolis, MI: Burgess Publishing Company; 1982. p. 127.
19. Bordeau SP. Volts to hertz—the rise of electricity. Minneapolis: Burgess Publishing Company; 1982. p. 129.
20. Finger S, Piccolino M, Stahnisch FW. Alexander von Humboldt: galvanism, animal electricity, and self-experimentation part 1: formative years, *Naturphilosophie*, and galvanism. J Hist Neurosci. 2013;22:225–60.
21. Finger S, Piccolino M, Stahnisch FW. Alexander von Humboldt: galvanism, animal electricity, and self-experimentation part 2: the electric eel, animal electricity, and later years. J Hist Neurosci. 2013;22:327–52.
22. Finger S, Piccolino M, Stahnisch FW. Alexander von Humboldt: galvanism, animal electricity, and self-experimentation part 2: the electric eel, animal electricity, and later years. J Hist Neurosci. 2013;22:330.
23. Finger S, Piccolino M, Stahnisch FW. Alexander von Humboldt: galvanism, animal electricity, and self-experimentation part 2: the electric eel, animal electricity, and later years. J Hist Neurosci. 2013;22:334.
24. Faraday M. Experimental researches in electricity-fifteenth series. Notice of the character and direction of the electric force of the *Gymnotus*. Phil Trans R Soc. 1839;129:1–12.
25. Morus IR. Michael Faraday and the electrical century. London: Icon Books Ltd.; 2004. p. 144.
26. Faraday M. Experimental researches in electricity-fifteenth series. Notice of the character and direction of the electric force of the *Gymnotus*. Phil Trans R Soc. 1839;129:11.
27. Reynolds EH. Todd, Hughlings Jackson, and the electrical basis of epilepsy. Lancet. 2001;358:575–7.
28. Reynolds EH. Todd, Faraday, and the electrical basis of brain activity. Lancet Neurol. 2004;3:557–63.
29. Todd RB. On the pathology and treatment of convulsive diseases. London Medical Gazette. 1849;8:726.
30. Todd RB. On the pathology and treatment of convulsive diseases. London Medical Gazette. 1849;8:661–71; 724–29, 766–72, 815–22, 837–46

31. Todd RB. On the pathology and treatment of convulsive diseases. London Medical Gazette. 1849;8:821.
32. Morus IR. Michael Faraday and the electrical century. London: Icon Books Ltd.; 2004. p. 134–43.
33. Morus IR. Michael Faraday and the electrical century. London: Icon Books Ltd.; 2004. p. 140.
34. Morus IR. Michael Faraday and the electrical century. London: Icon Books Ltd.; 2004. p. 77–9.
35. Morus IR. Michael Faraday and the electrical century. London: Icon Books Ltd.; 2004. p. 119.
36. Morus IR. Michael Faraday and the electrical century. London: Icon Books Ltd.; 2004. p. 120.
37. Piccolino M. Animal electricity and the birth of electrophysiology: the legacy of Luigi Galvani. Brain Res Bul. 1998;46:381–407.

Chapter 5
Electricity and the Nervous System

We shall probably never attain the power of measuring the velocity of nervous action; for we have not the opportunity of comparing its propagation through immense space, as we have in the case of light.

Johannes Müller, 1848

Abstract In the tradition of Galvani and Aldini, Carlo Matteucci used a newly developed galvanometer to record a current flow between the cut and intact surface of a frog nerve or muscle that he called the demarcation current. The cut surface had a negative potential relative to the intact surface. Emil du Bois-Reymond identified the electrical event associated with nerve excitation of muscle that he called the negative variation, later known as the action potential. Herman Helmholtz measured the speed of the action potential as it moved down the nerve to the muscle and it was surprisingly slow (about 30 m/s) compared to the speed of electricity moving along a wire. His student Julius Bernstein showed that the action potential lasted about 1 ms and its time course was independent of the stimulus strength. The debate over localized versus generalized function of the cerebral cortex was convincingly resolved by clinical–pathological studies in patients and electrical stimulation of the brain in primates. Functions were clearly localized within the cerebral cortex. Finally, Charles Sherrington used electrical stimulation of the spinal cord to study the integrative function of the nervous system. He introduced the concept of a synapse and showed how complex behaviors such as walking and scratching could still occur in a rudimentary form after the spinal cord had been resected from the brain.

Keywords Electrocytes · Biological pile · Negative variation · Action potential · Synapse

Recall that Galvani and Volta had a heated debate regarding the interpretation of Galvani's frog leg experiments. Galvani was convinced that his experiments proved the existence of *animal electricity,* whereas Volta felt that electricity was generated by the bimetal that Galvani used to stimulate the nerve and that the nerve-frog leg preparation was simply acting as an organic electrometer [1, 2].

R. W. Baloh, *Brain Electricity*, https://doi.org/10.1007/978-3-031-62994-5_5

A major problem facing these pioneer researchers was the lack of instruments for accurately measuring electrical current and an inability to control the tension and amount of electrical current used to stimulate the nerve-muscle preparation. With the discovery of electromagnetism, two new devices became available—the galvanometer and the induction coil—both key instruments for solving the problem. As the Nobel Prize–winning twentieth- century British electrophysiologist, Edgar Adrian wrote: "*The history of electrophysiology has been decided by the history of electrical recording instruments*" [3].

Nobili's Galvanometer

With the discovery of the galvanometer by Oersted and Ampere, researchers immediately tried to measure animal electricity in nerves using a galvanometer. But initial efforts failed. Even though the sensitivity of the galvanometer could be improved by increasing the number of turns in the coil surrounding the magnetized needle, technical problems arose. A major problem was extraneous electromagnetic fields that influenced the magnetized needle independent of a weak magnetic field associated with the current in the surrounding coil. Although the earth's magnetic field is fixed, it interacts with the magnetic field produced by electrical current and slight changes in the position of the recording apparatus altered the measurements.

To deal with this problem, an Italian physicist, Leopoldo Nobili, invented the astatic galvanometer in 1825. He mounted a pair of magnetic needles vertically apart on a silk thread, parallel to each other but with the poles reversed. The combination of needles had a net magnetic dipole movement of zero with no preferred direction relative to the earth's magnetic field. The lower magnetic needle was located inside a horizontal coil that carried the test current and was sensitive to the resulting magnetic field. The angle of deflection of the lower needle provided a quantitative measure of the current in the coil.

With this new device, Nobili set out to measure the electrical current in the nerve-frog leg preparation used by Galvani. The preparation consisted of the skinned muscles of a frog leg and the nerve to the leg cut near the spinal cord. He placed the intact muscles into one container and the cut nerve into another container, both filled with salt water. As Galvani had observed connecting the two containers with moist cotton caused the frog leg to twitch. When Nobili connected the two containers via the platinum electrodes attached to his astatic galvanometer, he observed a clear deflection of the lower needle with the muscle contraction indicating a current flow. No needle deflection was observed between the two containers without frog tissue. He was therefore the first scientist to observe what he called "frog current" by a galvanometer.

Nobili published his findings in 1828 coming to the erroneous conclusion that the current resulted from a thermoelectric effect produced by differential cooling of the nerve and muscles due to evaporation [1]. As we will see later in this chapter, he was recording what would become known as the *demarcation current* due to the electrical potential difference between the intact and cut surface of a nerve or muscle. In other words, the difference in electrical potential between the inside and outside of the cells.

Carlo Matteucci and the Demarcation Potential

An Italian physicist, Carlo Matteucci, following the tradition of Galvani and Aldini, used the new instrumentation available in the mid-nineteenth century to make several key discoveries regarding "*animal electricity.*" Matteucci was born in Forlí, Italy, in 1811 and graduated in physics from the University of Bologna at the young age of 16 years. After studying in Paris, he returned home to begin research in electrophysiology. He started by repeating several of Galvani's experiments on the nerve-frog leg preparation and performing a series of studies on the nervous control of the electric organ of the torpedo fish [4].

Since the studies of Walsh and Hunter in the late eighteenth century, it was clear that the electric organ of the torpedo fish was under nervous control (see Chap. 3). Galvani, himself, had shown that severing the nerve to the electric organ prevented the voluntary discharge of the organ. Matteucci decided to investigate the brain regions responsible for the discharge of the organ. The torpedo fish brain has four lobes or ganglia and Matteucci found that only the fourth lobe, which he called the electric lobe, was important for the discharge of the electric organ. If he touched or pricked the fourth lobe, he triggered a discharge of the electric organ and if he lesioned the fourth lobe, the organ no longer discharged. He could discharge the electric organ by applying the electric current to the fourth lobe or the nerves exiting the fourth lobe.

Matteucci was struck by the analogy of the nervous control of the electric organ and the nervous control of striated muscle. "*The link existing between the fourth lobe and the nerves emerging from it, on one side, and the substance of the organ, on the other side, is precisely the same which exists between any ordinary nerve and the muscles into which it ramifies...The action of the electric current is identical, and in this regard we must note that the electric current may still evoke a discharge when all other stimulating actions are no longer capable of doing so. This is the greatest possible analogy between the unknown force of nerves and electricity*" [5].

We now know that the electric organs of fish evolved from modified muscle and nerve cells called electrocytes. Electrocytes arise from transformed skeletal muscles or specialized axons originating in the spinal cord, and the mechanism of generating electricity is like that of normal muscles and nerves. The electric field measured near the fish is the result of a synchronized activation of thousands of electrocytes [6].

In his initial lesion experiments, Matteucci noted that cutting the animals' spinal cord did not prevent discharge of the electric organ after stimulating the fourth lobe, so the nerves from the fourth lobe ran directly to the electric organs, not through the spinal cord. However, in later experiments, he found that cutting the spinal cord prevented reflex discharge of the organ if the animal was touched below the level of the spinal cord section. Touching the animal above the level of the spinal cord section caused the electric organ to discharge. He correctly surmised that sensory nerves from the skin ran to the fourth lobe via the spinal cord and relexify triggered discharge of the electric organ via nerves running outside the spinal cord.

Matteucci then moved on to study the frog nerve-muscle preparation using Nobili's astatic galvanometer [7, 8]. In the process, he made a key discovery that explained many prior confusing findings in experiments using this preparation, what he called the *demarcation current*. He reliably recorded a current flow between the cut and intact surface of a nerve or muscle. Furthermore, the direction of the current flow was always the same, the cut surface had a negative potential compared to the intact surface. This explained why touching a muscle with the cut end of its nerve caused contraction (Galvani's experiment) and touching an intact nerve with the cut and intact surface of a piece of muscle caused contraction (von Humbolt's experiment).

To prove that the measured current was biological and not due to contact between the muscle or nerve tissue and the metal electrodes of the galvanometer, Matteucci came up with the ingenious idea of creating a battery of cut frog thigh muscles arranged in a series so that the intact surface of one thigh muscle was always in contact with the cut surface of the next thigh muscle (Fig. 5.1). With each addition of a cut thigh muscle, the galvanometer needle deflected the same amount so that the total deflection was proportional to the number of muscle thighs in his "biological pile." Matteucci had produced a structure akin to the torpedo electric organ.

Matteucci made sure that his findings were not unique to the frog. "*I have also obtained a well evident current, sometimes of 20 to 30 degrees, by wounding the breast muscle or the thigh of a living animal (pigeon, rabbit, ewe) and then plunging one of the plates within the interior of the wound and placing the other on the bare surface of the wounded muscle. The current was constantly directed, in the animal, from the inside of the wound to the external surface of the muscle*" [9].

Another important observation made by Matteucci was that the negative demarcation measured in a cut muscle disappeared during tetanic contraction of the muscle whether the tetany occurred spontaneously or was induced by a drug such as strychnine. The demarcation potential came back after the tetanic contraction stopped. Even though he did not recognize it, Matteucci had identified the principle behind what would become known as the *action potential,* an impulse of electrical depolarization moving along the membrane of a nerve or muscle. The subsequent debate over the interpretation of Matteucci's finding led to a bitter dispute with the German physiologist, Emil du Bois-Reymond, and had a profound impact on the careers of both these pioneering electrophysiologists [10].

Johannes Müller and Nineteenth-Century German Neurophysiology

Before addressing the work of Müller's famous pupils Emil du Bois-Reymond and Hermann von Helmholtz on the action potential, we will briefly consider the work of their teacher [11]. Müller was born in Koblenz at the turn of the nineteenth century, the son of a shoemaker. He would go on to be the most influential physiologist

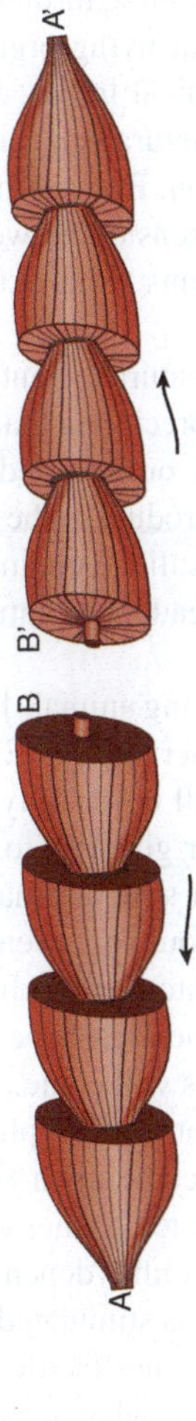

Fig. 5.1 Matteucci's frog thigh-muscle "pile." The electric current produced was proportional to the number of half-thighs in the pile. (B) cut surface, (A) intact surface. Arrows indicate direction of current flow

of his time, straddling the fence between the old and the new physiology, observation vs. experimentation, vital forces vs. natural forces. Although he recognized the value of carefully planned experiments, he preferred the simplicity of observation and even though he strongly supported the naturalistic views of his students, they considered him a vitalist.

Vitalism, the notion of supranatural vital force driving all living creatures, dominated medical teaching for millennia. In the mid-nineteenth century, a new doctrine, naturalism (or materialism), had a significant impact on the teaching of physiology beginning in Germany and rapidly spreading throughout the world. In direct conflict with vitalism, naturalism assumed that living organisms were driven by no other forces than natural physical and chemical forces. All natural phenomena could be explained on physical grounds. With naturalism came experimental physiology and an emphasis on animal experimentation. Because many vitalists were also antivivisectionists, this further increased the tension between the proponents of these two opposing views. Müller and his students were at the center of this clash between cultures and generations.

The vitalists argued that the new natural scientists only believed in their numbers, experiments and analysis not appreciating that there are supernatural phenomena that cannot be measured, weighed, or dissected. At the center of the debate was the meaning of the soul. Is the soul a product of the brain working according to pure physical laws or does the brain just facilitate communication of an immaterial soul that remains even after the brain is dead? The naturalists were clear on this issue; when the brain dies, the soul dies.

Vitalists felt that experiments on living animals had not contributed very much to our understanding of the function of nerve cells. Even if we knew the function and connections of each nerve cell, we will never fully understand the basis of reasoning. How can the uniform gray matter give rise to such a diversity of mental life? How can people with the same brain structure have so many different ideas and opinions? The naturalists of the time could not even begin to answer these questions since so little was known about the anatomy and physiology of the brain. Of course, they argued that as the physical and chemical processes of brain function are better understood, answers to these questions will come.

Müller's most important contribution to neurophysiology was his "*law of specific nerve energies*" [12]. He apparently came upon the idea when he pressed on his eyeball while sitting in the dark and noted flashes of light. Sensation did not necessarily adhere to the outside world but rather depended on the nature of the sensory organs. No matter how the optic nerve is stimulated, whether by light, electricity, or pressure, one has a visual experience, not tactile, auditory or gustatory. In other words, the sensory receptors determine what sensory stimulus is experienced not the nerves supplying the receptors. The sensory receptors in the eye are sensitive to light, whereas the sensory receptors in the ear are sensitive to sound. All sensory nerves work the same, carrying the sensory signals to the brain where they are interpreted. This may seem a trivial conclusion but at the time it was monumental. Müller's pupil Herman Helmholtz ranked the law of specific nerve energies alongside Newton's law of gravity.

An important implication of the law of specific nerve energies is that sensory perceptions rely on the sensitivity and specificity of the sensory receptors. External objects produce sensory stimuli such as light and sound and the sensory receptors do their best to recreate the stimuli in the form of nerve impulses. But what if they aren't very efficient? How much can we really know about the outside world? This notion that one's perception of the external world is limited by the number and sensitivity of sensory organs was initially proposed by the English physician, Charles Bell, but Müller is credited with expanding and popularizing the idea in his many publications.

Regarding animal electricity, Müller had misgivings about Galvani's notion that nerve transmission was electrical. He had no doubt that nerves could be activated by electrical stimulation since he had conducted such experiments himself but was not willing to dismiss the possibility of "*an imponderable*" vital force. Müller had tried to detect electrical current in a frog nerve-muscle preparation with a Nobili galvanometer next to the nerve but saw no deflection of the needle when the nerve was stimulated. Furthermore, Müller questioned whether the electrical force in nerves moved too fast to be measured? If electricity in nerves moved near the speed of light, there were no instruments available that could measure it. Müller wrote: "*We shall probably never attain the power of measuring the velocity of nervous action; for we have not the opportunity of comparing its propagation through immense space, as we have in the case of light*" [13].

Emil du Bois-Reymond and the Negative Variation (Action Potential)

Emil du Bois-Reymond's biographer, Gabriel Finkelstein, considered him "*the most important forgotten intellectual of the nineteenth century*" [14]. At the time, du Bois-Reymond was an international celebrity, considered one of the greatest scientists Germany ever produced. He served as professor, dean, and rector at the University of Berlin and was well known for his public lectures. "*His picture could be seen hanging for sale in German shop windows alongside those of the Prussian royal family*" [14]. Yet, he has been largely forgotten by contemporary neuroscientists. Du Bois-Reymond main scientific discovery, the action potential, was mired in the controversy with Carlo Matteucci and was overshadowed by the measurement of the speed of nerve conduction by Helmholtz.

Du Bois-Reymond was born to a middle-class family in Berlin, Germany, in 1818, and with the exception of a few years in Bonn, he spent his entire working career in Berlin. Emil enrolled in the University of Berlin in 1838 beginning in physics before selecting medicine. It was natural that he would gravitate to the famous professor of anatomy and physiology, Johannes Müller since he was considered the best teacher in the medical school.

In the spring of 1841, as Emil's formal studies were completed, Johannes Müller handed his young pupil a copy of Carlo Matteucci's most recent publication, *On the electrical phenomena of animals*. Emil had been searching for a research area to focus on for more than a year and Müller thought that Matteucci's area of work was ideal for Emil based on his interests and talents. Müller suggested that Emil first verify Matteucci's findings and then expand upon them as he saw fit. Du Bois-Reymond would spend the rest of his scientific career working on this project.

Müller had judged well in selecting animal electricity for du Bois-Reymond to study. There was a rich tradition of German researchers in the field including Alexander von Humboldt who founded the University of Berlin. Du Bois-Reymond's background in physics seemed ideal for designing experiments driven by hypothesis, in this case, the hypothesis that electricity governed the action of nerves. Du Bois-Reymond, for his part, saw research on animal electricity as the road to making his name in scientific research.

Du Bois-Reymond began planning his research expecting to be up and running by mid-summer of 1841. His first step was to purchase a galvanometer from the instrument maker Hirschmann. This was no ordinary galvanometer but the largest one ever built, consisting of wire more than a half mile long wrapped in 4100 turns. To his dismay, it took more than two months to be delivered and when he finally received the instrument, he was disappointed to find that it was not ready for use. Not only did he need to calibrate the galvanometer, but he had to "*take an unforeseeable series of precautions against the recurrent nightmare of errors induced by foreign sources of electricity*" [15]. To add insult to injury he found that the instrument had a defect in the wire, so he had to unwind, repair the wire, and then rewind it himself.

Du Bois-Reymond carried his research materials to and from the laboratory each day and was at the mercy of the laboratory assistants. He noted that often an assistant would not be available and "*then you've walked through the city in the heat for half an hour with a sack of frogs and ice for nothing*" [16]. He finally decided to do his experiments at home and to make his own galvanometer. He had Hirschmann make the frame and he completed the device including winding the wire. Although frustrating at the time, the hands-on experience with the instrument he gained proved to be one of the keys to his later success.

Come September 1841 with still no progress, du Bois-Reymond had to put aside his research to serve as a "*voluntary compulsory surgeon in the service of the Fatherland*" [17] so it wasn't until the spring of 1842 that he was able to begin his experiments. One of the problems facing researchers trying to record animal electricity with a galvanometer was that touching the animal tissue with the electrodes could generate electricity as Volta had shown.

After trying a variety of techniques, du Bois-Reymond came up with the idea of placing blotting paper soaked with salt solution between the electrodes and the tissue. He dropped the electrodes into two varnished glass containers filled with saturated salt water and hung several layers of blotting paper over the edge of the containers next to each other allowing the salt water to soak through the paper. He could then complete the circuit with the galvanometer by laying the tissue being examined across the gap between the two soaked paper pads. As a control he could place a soaked piece of blotting paper across the gap.

With this setup, du Bois-Reymond confirmed Matteucci's finding of a demarcation potential in frog muscle arguing that his apparatus was much more sensitive than Matteucci's apparatus. Matteucci had used cotton soaked in a weak salt solution between the electrodes and the tissue which according to du Bois-Reymond caused him to lose control over his setup. Du Bois-Reymond tested a wide variety of muscle tissues from different animals from the whole muscle down to a single muscle fiber, and while he placed the cut section on one paper cushion and the intact section on the other paper cushion, he recorded a deflection of the galvanometer needle indicating current flow. Based on his findings he hypothesized that each muscle fiber was like a small battery with a positive exterior and a negative interior.

Even more important, the increased sensitivity of his apparatus allowed du Bois Reymond to identify the electrical events associated with nerve excitation of muscle that he called the negative Schwankung (the negative variation), which represented the first instrumental recording of what would become known as the action potential. To measure the negative variation, du Bois-Reymond took advantage of the unique properties associated with muscle tetany. A muscle could be sent into prolonged contraction (tetany) by a variety of methods including trauma, drugs, and repetitive electrical shocks. Du Bois-Reymond proved that muscle tetany represented a series of short muscle contractions that summed to a continuous contraction. He found that he could best produce muscle tetany with brief pulses of electrical current generated by an induction coil similar to the one used by Todd. The alternating current of the induction coil did not damage his preparation and prevented the galvanometer electrodes from becoming polarized.

To measure the electrical events associated with muscle tetany, du Bois-Reymond placed the nerve-muscle preparation onto the salt water-soaked pads connected to the galvanometer and cranked the wheel of the induction coil to produce tetany in the muscle. Simultaneously, the muscle contracted and the galvanometer needle passed straight into the negative quadrant of the dial and oscillated around the position of equilibrium. He was recording the action current or action potential associated with each brief muscle contraction during tetany.

Dubois-Reymond then tackled the much more difficult task of recording the action potential of a nerve. This would take several years and require improvements in his instruments. He first tried to tetanize a nerve and record the negative variation as he had done with muscle, but no matter how he produced the tetany (trauma, drugs, electricity), he could not identify negative oscillation. He realized that he would need better control of the stimulating electrical current and improved the sensitivity of his galvanometer. He designed an induction coil that included an automatic switching device to repeatedly break the battery circuit (discussed later in the chapter) and a sliding outer coil to control the voltage of the electrical impulses. His new galvanometer consisted of more than three miles of wire with 24,160 turns, four times larger than his prior instrument. Calibration of the galvanometer was painstaking since the device was enclosed in a glass bell to protect the needle from air currents and he had to lift the bell off for each adjustment. After repeated attempts with questionable findings, du Bois-Reymond was finally able to measure negative variation in a frog nerve in late 1847.

The Matteucci—du Bois-Reymond Controversy

When Emil du Bois-Reymond first read Matteucci's paper given to him by Müller, he was favorably impressed with Matteucci's findings and there can be no question that Matteucci's work provided a framework for du Bois-Reymond's early experiments. However, a key difference between these two pioneer investigators was trust and reliance on instrumentation in their research [10].

Matteucci tried a version of Nobili's galvanometer, but the device was not very sensitive or reliable and Matteucci preferred what he called a "rheoscopic frog," a frog leg muscle with the exposed nerve attached, to detect electricity in the animal tissue. Touching the nerve of the preparation with even the slightest electrical charge triggered a muscle contraction. One of Matteucci's greatest discoveries was the "induced twitch" that led to him being awarded the Copley Medal by the Royal Society of London in 1844. Matteucci prepared two frog-nerve preparations and placed the nerve of preparation one on the muscle of preparation two. When he triggered a contraction of the muscle in preparation two (electrically or mechanically), he elicited a contraction of the muscle in preparation one. Matteucci provided a confusing explanation for the finding: "*the induced twitch is the first expression of a distant action, or more appropriately, of an induction which is exerted by a contracting muscle on a nerve*" [18].

By contrast, du Bois-Reymond was enamored with instrumentation and considered improvements in his instruments as the key to success in his research. He spent months designing and building a sensitive galvanometer. Du Bois-Reymond argued that Matteucci's rheoscopic frog only responded to electrical charge, whereas his galvanometer could measure the strength and direction of an electric current. Furthermore, he wanted to show that animal electricity was no different than any other type of electricity and closed the door on vitalist interpretations of the data.

There are conflicting reports of how the conflict began between these two pioneers in animal electricity, but du Bois-Reymond was much the aggressor from the start. He was young and eager to make a name for himself and he felt that he was not getting proper respect from colleagues, particularly in France and England. He was frustrated that Matteucci had received the Copely Medal, and he did not despite the efforts of his friends in England. Du Bois-Reymond complained that Matteucci used "*the unique tactic of self-plagiarism*" publishing multiple versions of the same article and citing all of them [19]. Making matters worse Matteucci frequently changed his opinions, so it was difficult to know exactly what he thought. For his part Matteucci referred to du Bois-Reymond as "*the young physiologist from Berlin*" and complained that du Bois-Reymond was trying to destroy his scientific reputation.

Du Bois-Reymond interpreted Matteucci's "*induced twitch*" experiment to indicate that the contraction of muscle one produced an electrical impulse that activated the nerve to the second muscle which in turn caused muscle two to twitch. In response, Matteucci wrote: "*It has been impossible for me to imagine the physical meaning of these statements*" [18]. Du Bois-Reymond's sarcastic response was typical of the interaction between the two men: "*With his observations, Matteucci has done me, without knowing, a real friend's service*" [20].

The Italian neurophysiologist, Giuseppe Moruzzi, summarized his view of the relative contributions of Matteucci and du Bois-Reymond in 1964. "*History must attribute to Matteucci the discovery of the action current which was demonstrated for the first time with the experiment of the induced twitch (1842). The impartial historian must also recognize that he was the first to observe the effects of the depolarization associated with muscle contraction by using the galvanometer, in 1838 and in 1842. It is certain, nevertheless, that modern electrophysiology begins with du Bois-Reymond's discovery (1848–49) of the 'negative Schwankung' which Matteucci had accidentally observed in 1838 but was unable to interpret correctly*" [20].

Helmholtz and the Speed of Nerve Conduction

Of his many outstanding pupils, Müller's most famous pupil was probably Hermann von Helmholtz [21]. His close friend, du Bois-Reymond, was in awe of Helmholtz. Helmholtz had an extensive understanding of physics and mathematics and he was comfortable working on a wide variety of scientific topics. To those in the medical field, Helmholtz is best known for his discovery of the ophthalmoscope. He made basic discoveries in how the eye perceives colors and the ear perceives sound frequencies. In neurophysiology he is best known for being the first to measure the speed of nerve conduction. Considering the instruments available at that time, this was a remarkable achievement, pivotal in the modern understanding of the nerve action potential.

Helmholtz was born in Potsdam, Prussia (now Germany) in 1821, the eldest son of a teacher of philology and classical literature at a local Gymnasium. At the time Potsdam was a garrison city of the Prussian Army and his mother was the daughter of an artillery officer. Hermann was baptized in the Lutheran church but was not religious. As a child he was sickly spending much of his time indoors reading having learned to read at a young age. He was considered gifted in mathematics in elementary school and at age 11 he enrolled at the Gymnasium where his father taught. Hermann was interested in a career in mathematics and physics but considering the financial circumstances of the family, his father convinced him to study medicine instead. It was the only area where scholarships for higher education were available.

Helmholtz received a government grant to study medicine at the Royal Friedrich-Wilhelm Institute of Medicine and Surgery in Berlin by promising to work as a doctor in the Prussian army, two years for each year of financial support. In 1837, at age 16 years, he left for Berlin where in addition to studying medicine he attended lectures at the University of Berlin in chemistry and physiology. He did not have formal training in mathematics but rather was self-taught reading works of the great masters including Laplace and Bernoulli. For recreation he played Mozart and Beethoven on a piano and read literature voraciously.

In 1841, Helmholtz began to work on his medical dissertation concerning the relationship between nerve cell bodies and nerve fibers under the supervision of Professor Müller. Müller had a longstanding interest in the problem and previously with another student, Robert Remak, had concluded that the two had to be related even though there was no definitive evidence using primitive microscopes. Recall that Robert Todd in London came to a similar conclusion around the same time (see Chap. 4). The Czech physiologist, Jan Purkinje, probably was the first to clearly show that nerve cell bodies gave rise to the nerve fibers in the brain and spinal cord in 1837.

Helmholtz decided to attack the problem by studying the nerve ganglia of invertebrates. He used wine to fix the specimens and after saving money for a new microscope, he was able to demonstrate that nerve axons are connected to the nerve cell bodies in the ganglia. Furthermore, he showed that each axon was connected to only one cell body. In his dissertation, he described his findings in a wide variety of invertebrates including leeches, crabs, and crayfish. His dissertation, entitled "The Structure of the Nervous System in Invertebrates," was dedicated to Müller and successfully defended in 1842.

After graduation, Helmholtz did a year of internship at the Charité Hospital in Berlin where despite his heavy clinical work he found time to continue experiments in Müller's laboratory. As with several of Müller's other pupils, Helmholtz became interested in the question of vital force. He felt that there was "*something contrary to nature*" in the concept and was determined to prove it wrong. At that time, many believed that the process of fermentation was an example of a vital force, so he decided to investigate the process. With his microscope, he was able to demonstrate that fermentation depended on tiny living organisms and that the process could be explained by organic chemistry without the need for a vital force.

After completing his internship, Helmholtz returned to Potsdam to fulfill his military commitment serving as an assistant surgeon in the army. Since it was peacetime, his work schedule was slow, so he built a small laboratory in his barracks and continued his attack on the vital force theory. He studied the chemical changes associated with muscle contraction in the frog and concluded that the animal's body heat could be fully explained based on the food it ingested and the kinetic activity associated with muscle contraction without the need to postulate a vital force. This work led to an essay "On the Conservation of Energy" in which he used mathematical principles to argue that energy-like matter could change its state, but the quantity always remained the same [11]. For example, a chemical reaction could produce energy in the form of heat, light, or even electricity, but the total amount of energy did not change. He first presented the essay to the Physical Society of Berlin in 1847 and later in the year published it as a small monograph.

With Conservation of Energy, Helmholtz had made a name for himself in German science. He was introduced to the Society by fellow Müller pupil, du Bois-Reymond who wrote: "*He* [Helmholtz] *is a fellow who has devoured chemistry, physics and mathematics with a spoon, who completely holds our worldview, and who is rich in thoughts and new ways of looking at things*" [22]. Even more important for his future, Helmholtz caught the attention of the great Alexander von Humboldt, a regular dinner companion of King Friedrich Wilhelm IV. Humboldt arranged for a ministerial release for Helmholtz from his remaining years of military service so that he could return to Berlin and rejoin Müller's laboratory. A year later, at the young age of 28 years, Helmholtz was appointed professor of Physiology in the historical capital of East Prussia, Königsberg (now Kaliningrad, Russia).

On accepting the position at Königsberg, Helmholtz married Olga von Velten, the daughter of a military physician whom he met during his military service. He finally had the financial resources to start a family. Königsberg was in a rural area outside of the scientific mainstream but Helmholtz wrote to his father that it was an ideal place for his research career since there wasn't much else to do. In late 1849, with his newly pregnant wife serving as an assistant, Helmholtz began his work on the speed of nerve conduction. He, of course, was aware of Müller's speculation that it might be impossible to measure the speed of nerve conduction, but he was also aware of du Bois-Reymond's recording of the negative variation or action current in a frog nerve.

At the time, there was a race by scientists to construct an electronic device that could measure very brief intervals of time to accurately measure the velocity of projectiles such as bullets or cannon balls. In 1840, the English physicist and inventor, Charles Wheatstone made an electromagnetic device that could accurately measure 1/500 of a second and others rapidly improved the sensitivity. Helmholtz could not afford to purchase one of these expensive pieces of equipment so he made his own device consisting of a sensitive galvanometer that could be turned on by an electrical stimulus to a nerve and turned off by contraction of the muscle it supplied. The device was difficult to calibrate but Helmholtz persevered and was able to get accurate measurements of very short time intervals.

For the nerve conduction velocity experiments, Helmholtz made a frog nerve-muscle preparation consisting of the femoral nerve cut just outside the spinal cord and the gastrocnemius muscle that the nerve supplied. A small hook in the muscle was attached to a sensitive lever that broke the electrical circuit when the muscle contracted. By measuring the brief time interval between the electrical stimulus to the nerve and the muscle contraction and knowing the length of the nerve, he could estimate the velocity of nerve conduction. His initial measurements were much slower than he or anyone else expected, in the range of 25–38 m/s.

Helmholtz then made measurements by stimulating the nerve at different distances from the muscle. By subtracting the measurement with electrical stimulation at a point close to the muscle from the measurement with stimulation at a point farthest from the muscle, he had a measure of nerve conduction velocity between the two points. With this method, he obtained a mean speed of nerve conduction of approximately 27 m/s. Finally, he repeated the experiment using an entirely different system to measure the time intervals, a myograph (a smoked drum) that he designed to record the stimulus artifacts and the muscle contraction. His measurements of nerve conduction velocity with the myograph were about the same as those obtained with his galvanometer. Helmholtz was convinced of his findings and published a short report detailing the findings in January, 1850.

The speed of nerve conduction was remarkably slow ruling out the possibility that nerves were simply transmitting electricity. Of course, if the speed of nerve conduction was even close to the speed of light, Helmholtz could not have measured it. Helmholtz suspected that the nerve transmission was a chemical process possibly involving the movement of ions. Consistent with a chemical reaction, he observed that the speed of nerve conduction slowed as the temperature in the testing room cooled. Helmholtz's finding was a nail in the coffin for the notion espoused by many that nerve conduction represented an immaterial vital force that could not be measured. The nerve impulse was a physicochemical phenomenon with defined features that could be measured.

Having devised a reliable method for measuring nerve conduction velocity he moved on to measuring the speed of conduction of sensory nerves in human subjects. In this case he measured the reaction time between an applied weak shock to different locations on the body and a motor response where the test subject moved a finger or clenched the teeth as fast as possible. He used both the galvanometer and myograph to measure the time interval between the sensory stimulus and the motor response. For example, he noted that a shock of the big toe arrived at the brain about one-thirteenth of a second later than a shock to the face. Based on numerous measurements, he concluded that the average speed of human sensory nerves was about 30 m/s not much different than the frog motor nerves. Helmholtz later wrote: "*Happily, the distances our sense-perceptions have to travel before they reach the brain are short, otherwise our consciousness would always lag behind the present*" [23].

Helmholtz tried to measure the motor nerve conduction velocity in human subjects but ran into technical problems and put the project aside. It was almost 20 years later that he along with Russian student working in his laboratory devised a method to place the forearm in a special cast so that thumb movement could be reliably recorded by the myograph. They stimulated the median nerve at the wrist and elbow and calculated a mean human motor nerve conduction velocity of about 33 m/s. By this time a student of du Bois-Reymond, Julius Bernstein, had developed instruments to measure movement of the negative variation (action potential) as it moved along a motor nerve (see below).

Du Bois-Reymond had mixed feelings about Helmholtz's report on nerve conduction velocity. He was excited because it supported his interpretation of the negative variation that he had recorded. But he was frustrated that he had been scooped by his friend on a problem that he had been working on for several years. He had tried to measure the negative variation as it moved along a nerve, but he just didn't have equipment sensitive enough to make the measurement. Helmholtz had solved the problem using a different, relatively simple technique. Why didn't he think of it? As for their teacher, Müller was thrilled with Helmholtz's findings even though they were contrary to his expectations. He wrote to Helmholtz congratulating him on his work and even suggested sensory nerve studies that Helmholtz later performed.

Julius Bernstein and His Differential Rheotome

With the discovery of the *demarcation potential* by Matteucci and the *negative variation* by du Bois-Reymond, it was apparent that nerve fibers maintained a resting electrical potential with the inside of the nerve fiber negative relative to the outside. When excitation (action potential) moved along the nerve fiber, the outside of the membrane briefly became negative relative to a distant inactive region. Helmholtz measured the speed of nerve conduction based on muscle contractions and found it to be several orders of magnitude slower than the speed of electricity moving along a wire. This raised the critical question: did the negative variation correspond to the actual nerve signal? A student of both du Bois-Reymond and Helmholtz, Julius Bernstein showed this was the case by measuring the speed of the negative variation (action potential) as it moved along a motor nerve with his *differential rheotome*.

Bernstein was born in Berlin in 1839 [24]. His father was a progressive writer, journalist, and publisher active in the reform Judaism movement. Even as a young gymnasium student, Julius attended lectures and visited laboratories at the Physiological Institute at the University of Berlin where he met du Bois-Reymond. He attended medical school at the University of Breslau after which he returned to Berlin to obtain a doctorate in medicine working on muscle physiology with du Bois-Reymond. He then spent seven years working with Helmholtz. At age 34 he was appointed Chair of Physiology and Director of the Physiological Institute at the University of Halle where he remained until his retirement in 1911.

It was in Heidelberg, four years after he began working with Helmholtz, that Bernstein published his famous paper describing the mechanical instrument that allowed him to resolve the time course of the electrical activity in a nerve [25]. The device consisted of a high-speed rotating wheel with a small cam that opened and closed a stimulation and recording electrical circuit for a fraction of a millisecond. After delivering a brief electrical stimulus to the nerve, the rotating cam opened a sampling window for a galvanometer to record the response. Using repeated stimulations at high frequency, he was able to record the electrical event in the nerve within a definite time interval after the stimulation at a distant site. It was then simply a matter of varying the interval between the stimulus and the sampling window and averaging the responses to determine the time course of the electrical process.

Bernstein measured the nerve conduction velocity by changing the stimulation site and observing the change in the delay of the electrical event. He obtained an average rate of propagation of the negative variation (action potential) of 29 m/s. which was very close to Helmholtz's measurement of the velocity of the signal from nerve to muscle indicating that they were indeed the same thing. Even more important, Bernstein provided the first recording of the time course of the action potential, showing that it was a transient phenomenon lasting about 1 ms and its time course was independent of stimulus strength (Fig. 5.2). Surprisingly, the amplitude of the current during the negative variation exceeded the level of the resting current measured between an intact and an injured segment of the nerve (Matteucci's demarcation potential). This overshoot suggested that the negative variation was more than simply a breakdown in the resting state.

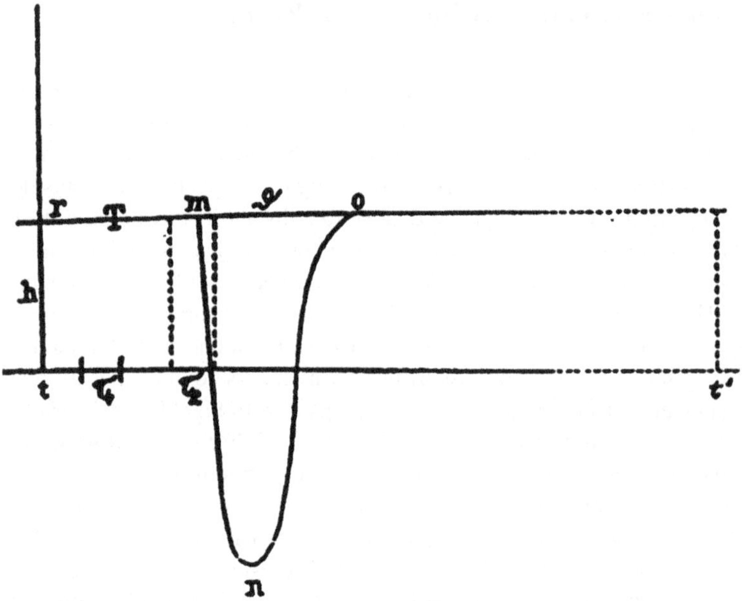

Fig. 5.2 First published recording of the negative variation (action potential) measured with Bernstein's differential rheotome. The downward deflection (peak current, *n*) overshoots the baseline resting current (*h*) by a large amount. The duration of the action potential (m–o) was about 1 ms. (From Bernstein J. Ueber den zeitlichen Verlauf der negative Schwankung des Nervenstroms. *Pfluger's Arch Ges Physiol* 1868;1:173–207)

As we will see in Chap. 10, over the next several decades Bernstein went on to develop a general theory of membrane excitation based on membrane permeability that was published in 1902. Curiously, he did not trust his finding of an overshoot of the baseline during the action potential and his theory did not predict an overshoot. It would be up to others to recognize the critical importance of the baseline overshoot and explain its origin.

Brain Control of Nerve Activity

By the mid-nineteenth century, there was a growing consensus that the brain was the source of animal electricity. It was common to compare the nervous system to a telegraph system. The brain contained the battery to generate the electrical signals and the nerves were telegraph cables carrying signals to and from the brain.

But the notion of the brain as a battery was an obvious over-simplification considering the complex anatomical structure of the brain. Indeed, relatively little was known about the microanatomy of the brain. If it was the source of electrical fluid for nerves and muscles, how was electricity generated and distributed? As noted in Chap. 4, Robert Todd and others had observed nerve vesicles (cells) in the cerebral cortex which Todd assumed were connected to the underlying nerve fibers. Each vesicle might act as a small battery sending electricity through the nerves to other nerves and muscles. But how were the vesicles, nerves, and muscles connected and how was the electricity produced and transmitted between them?

For centuries the cerebral cortex was dismissed as an unimportant "rind" (Latin word for cortex) surrounding the nervous fluid in the ventricles [26]. Aristotle viewed the brain as a radiator to cool the blood passing through while Hippocrates considered it the source of phlegm. With the invention of the microscope in the seventeenth century, the Italian anatomist Marcello Malpighi saw little glands with attached tiny ducts, so he considered the brain a glandular organ. Malpighi's globules turned out to be artifacts from boiling the tissue to harden it prior to viewing. Even with better microscopes and fixatives in the mid-nineteenth century, investigators like Todd had difficulty identifying the relationship between nerve cells and the underlying nerve fibers. Todd thought that multiple nerve fibers (axons) could originate from a single nerve cell.

The galvanic stimulation studies of Aldini and others on animal and human brains suggested there might be localization of function in different regions of the cerebral cortex, but this was not systematically evaluated, and it was unclear whether they were stimulating part or all the cerebral cortex or the underlying nerve fibers. At the time these studies were being conducted, there was already an ongoing battle between the so-called "localizers" and "generalizers" of cerebral cortical function. The localizers argued that different areas of the cerebral cortex had unique functions, whereas the generalizers argued that the cerebral cortex functioned without specialized areas.

Localizers Versus Generalizers

A strong proponent of localization of functions to different regions of the cerebral cortex was Franz Joseph Gall [27]. Gall was born in Germany, the son of a merchant of Italian descent and he attended medical school at the University of Vienna. He soon established himself as one of the most prominent physicians in Vienna and was even offered the position of physician to Holy Roman Emperor Francis II. He turned down the position to continue work on phrenology.

Gall postulated that the brain was composed of many organs since there were so many faculties and feelings. He went on to identify 35 brain centers (organs) with specific functions, including centers for speech, vanity, talent, kindness, and even religion. Since he felt that the skull was "molded" to the brain, it was possible to examine the size and shape of the different brain organs by palpating the skull. With his collaborator JC Spurzheim, Gall set out to prove his theory by palpating the skulls of a wide variety of people including lunatics, criminals, aristocrats, and scientists.

After concerns were raised about possible materialistic and antireligious aspects of phrenology by the Emperor, Gall moved to Paris in 1807. Skull palpating rapidly became popular in progressive Parisian society and Gall's medical practice flourished. But before long similar concerns were raised regarding phrenology in Paris and Napoleon asked the Paris Academy of Science to investigate Gall's theory. In 1822 the Academy commissioned the well-known French physician and physiologist, Marie Jean Pierre Flourens to investigate Gall's claims.

Flourens was a child prodigy, having enrolled at the famous Montpellier medical school in southern France at age 15 years, receiving his medical degree at age 19 years. He studied brain function in Paris under Georges Cuvier and then conducted a series of key studies on the vestibular organs of the inner ear. Flourens devised a simple scheme to test Gall's theory. He systematically removed anatomically defined regions of the brain in frogs, ducks, pigeons, and rabbits and observed the animals' behavior after recovery.

When Flourens removed the cerebellum, the animal could no longer coordinate movements and if he damaged the medulla of the brainstem the animal lost vital functions and stopped breathing. But when he removed the different parts of the cerebral cortex, he found little evidence of localization and concluded that the cortex mainly functioned as a whole. "*A large section of the cerebral lobes can be removed without loss of function. As more is removed, all functions weaken and gradually disappear. Thus the cerebral lobes operate in unison for full exercise of their function*" [28].

Flourens did conclude that the cerebral cortex was important for willing, remembering, and perceiving just not the punctate localization suggested by Gall. No doubt if he had studied other animals, particularly primates, he would have identified more evidence for localization of function. But even in primates, he would have had a hard time finding 35 brain organs proposed by Gall. As it turned out, Gall had some visionary ideas, but he was led astray by poor methodology and biased experiments. The most obvious flaw was the notion that he could examine the cerebral cortex by palpating the skull.

For centuries physicians had observed that patients with major trauma to one side of the head exhibited paralysis of the face and extremities on the opposite side of the body. They also noted that when the right side was paralyzed, it was common for patients to also lose the ability to speak. In 1825, a French physician Jean-Baptiste Bouillard published a book summarizing his extensive experience with patients who lost the ability to communicate with speech due to brain damage [26]. Several had postmortem examinations of the brain and Bouillard concluded that speech function was localized to the frontal region (he did not comment on the side even though in three out of four of his patients the damage was on the left side). Interestingly, Bouillard was a proponent of phrenology and Gall had also localized speech to the front of the brain. On the other hand, in 1840, another French physician, Gabriel Andral, reported 14 cases with loss of speech but without evidence of frontal brain damage.

Brain comparisons across species suggested that an enlarging brain and cerebral cortex was associated with increasing intellect. However, in February 1861, the anatomist Pierre Gratiolet made a startling presentation before the Anthropological Society in Paris. He studied the skull of an "Indian" brought back from Mexico by occupying forces and noted the disturbing finding that the brain volume was larger than that of most contemporary Frenchmen. How could a "savage" have a larger brain than highly intelligent Frenchmen?

At a meeting of the Anthropological Society in early April 1861, a young Parisian physician, Simon Aubertin who was married to Bouillard's daughter, described a patient he examined after the man had shot himself in the head exposing the front of his brain. The man appeared to have normal speech and intellect. Aubertin told the audience that when he applied light pressure to the frontal cortex the man's speech was interrupted even in the middle of a word and immediately returned to normal once the pressure was released. Aubertin concluded his presentation by making a strong argument for cerebral localization only to have a generalizer in the audience stand up and counter: "*I do not hesitate to conclude that all attempts at localization, which up to know have been tried, lack any foundation. They are no doubt great efforts! But when one attempts to grasp the truth at the height of these babbles, the edifice crumbles*" [29].

Into this heated debate stepped Paul Broca, a French surgeon with a reputation of being cautious and fair-minded [30]. Broca had followed the debate between the localizers and generalizers and leaned toward the localizers position primarily based on the clinical data of Bouillard and Aubertin. As a founding member of the Anthropological Society, Broca attended Gratiolet's and Aubertin's presentations and made his own presentation to the Society in March 1861. In addition to presenting clinical data, he argued in favor of cerebral localization based on his microscopic studies that showed distinct cellular variation in different regions of the cerebral cortex.

On April 12, 1861, a 51-year-old man named Louis Victor Leborgne presented to Broca's hospital surgical service for the treatment of leg ulcers. This landmark case provided strong support for the localizers. "Tan," the nickname given to Leborgne based on the sound he made when asked his name, was first hospitalized at age 31 with a slow loss of the ability to speak. He had previously suffered from epilepsy and after losing speech he gradually became paralyzed on the right side. Broca noted gangrene in his paralyzed leg and knew that Tan did not have long to live.

Broca suspected that Tan had damage to the left frontal lobe and that he would be able to confirm his suspicion with post-mortem examination. First, Broca wanted to document Tan's language and cognitive function and he invited Simon Aubertin to assist with the examination. Even though Tan appeared to be alert and understand verbal and written questions, he responded to all queries with the same two syllables—"Tan, Tan," along with expressive gestures with his face and left hand. He could not write due to the paralysis of the right hand. Despite these limitations, Broca and Aubertin were able to show that Tan understood spoken and written language and was oriented to time, person, and place based on responses with his left hand.

Leborgne died a week after being hospitalized and when Broca examined the brain at autopsy, he identified a localized "softening" in the left frontal cerebral

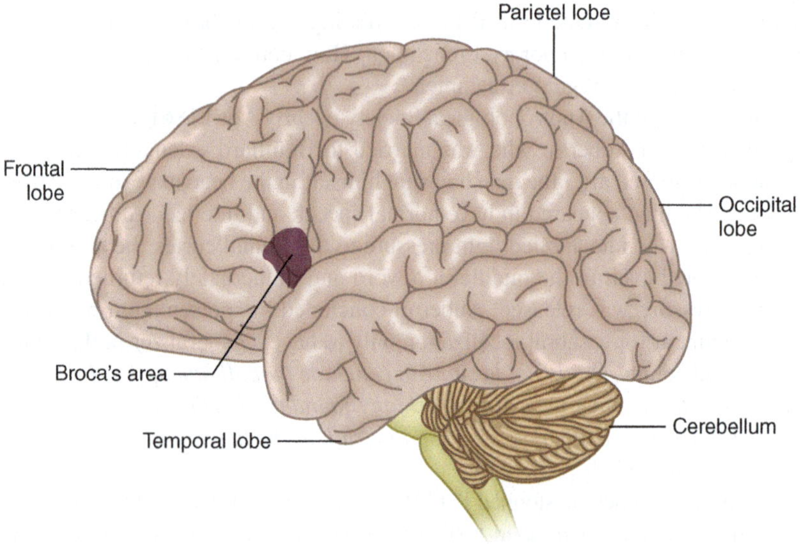

Fig. 5.3 Area of localized "softening" of the cerebral cortex in Broca's patient Leborgne with aphasia. Now known as Broca's area

cortex, with the most severe damage to the third gyrus roughly corresponding to current Brodmann's areas 44 and 45 (Fig. 5.3). The cause of the cerebral softening was central nervous system syphilis. Broca called Leborgne's deficit aphemia (without speech), later called aphasia (Broca's aphasia or motor aphasia). Broca distinguished this from the inability to hear or understand speech which is now known as Wernicke's aphasia or receptive aphasia. This more devastating type of aphasia results from focal damage to the superior temporal gyrus.

Broca presented his clinical and autopsy findings in the Leborgne case at a meeting of the Anthropological Society four months later. Broca was by then fully committed to the localizers and he complimented Bouillard and Aubertin for their prior work localizing speech to the frontal cortex. Broca suggested that the frontal lobes might also be important for executive functions such as judgement and intellect since Tan was losing these functions as the disease process spread late in his life. Perhaps intellect was not related to brain size but rather to the size of the frontal lobes.

Over the next few years, Broca was able to identify eight additional cases of motor aphasia and to his surprise, all eight showed damage to the left frontal lobe [31]. He was initially cautious about generalizing but he finally introduced the term "cerebral dominance" in an article published in the Bulletin of the Anthropological Society in 1865. He speculated that even though the two hemispheres looked alike and probably had a similar innate capacity one hemisphere dominated the other in the development of speech.

Across the channel in England, John Hughlings Jackson was following Broca's findings with great interest and though initially skeptical, after examining 70 of his own cases with aphasia and one-sided paralysis, he noted that in all but one case the paralysis was on the right side indicating left hemisphere damage [32]. There could be no doubt that the left cerebral hemisphere was nearly always dominant for language. But what about the right hemisphere? On examining his patients with left-sided weakness (indicating right hemisphere damage) he noted that many had problems with spatial awareness. They had difficulty identifying places and things. By contrast his patients with aphasia and left hemisphere damage had no problems with spatial awareness. Interestingly, in 1872 he reported a man with left-sided paralysis who could not recognize people including his own wife even though his vision was normal. More than a century later, the well-known English American neurologist and writer, Oliver Sacks, published a best-selling book entitled *The Man Who Mistook His Wife for a Hat* describing a variety of unusual patients with damage to the right nondominant hemisphere [33].

Electrical Stimulation of the Frontal Motor Cortex

Despite the clinical work of Broca and Jackson, the question of localization within the cerebral cortex was still in dispute. After all, clinical observations were notoriously unreliable compared to controlled animal experiments and prior electrical stimulation studies in animals had been inconclusive. Possibly the most compelling clinical evidence for cortical localization was Hughlings Jackson's observation that focal motor seizures, including those in his own wife, had a consistent pattern in the spread in muscle contractions from one body part to the next. For example, a seizure could begin with twitching in a finger and gradually move into the hand, the forearm, the upper arm, the face, and then the entire side of the body.

By 1870, Jackson surmised that there must be a somatotopic display of the motor control of the entire body laid out over the convolutions of the frontal cortex. During a focal seizure the electrical activity began at one point and radiated out from that point. In the same year, two German physicians, Gustav Fritsch, and Eduard Hitzig, showed that cortical localization of movement control was not unique to humans but was also present in dogs and that electrical stimulation was a powerful tool for studying cortical localization [34]. Each had clinical experiences that spurred an interest in motor control by the cerebral cortex. While serving as a battlefield surgeon in the Franco-Prussian war, Fritsch noted that when cleaning head wounds, if he accidentally irritated the exposed brain, muscles would twitch on the opposite side of the body. While conducting electrotherapy in his patients, Hitzig noticed that electrical stimulation at the back of the head triggered eye movements.

Fritsch and Hitzig were aware that these observations did not necessarily indicate that the cerebral cortex initiated the movements since they could have activated subcortical structures such as the corpus striatum. Hitzig, who was apparently unaware of Robert Todd's work, conducted preliminary electrical studies of the cerebral cortex in rabbits that were promising but not conclusive. Fritsch and Hitzig met in Berlin and decided to join forces and conduct a series of experiments stimulating the cerebral cortex in dogs.

Neither had access to a research laboratory to conduct the studies so they decided to perform the studies in Fritsch's home on Frau Fritsch's dressing table. The procedure was simple—surgically remove the skull and expose the cerebral cortex and then stimulate different cortical regions with *pulses* of electricity from a battery. They were appropriately concerned about the spread of the electrical current, so they used the lowest levels possible. They tested the electrical stimulus on their tongues before touching the cortex with the electrode. The brief weak current pulses produced localized muscle contractions, whereas stronger current pulses activated multiple muscles, even an entire side of the body.

After perfecting their technique and exposing larger areas of the cortex, they were able to find a region of the frontal cortex where electrical stimulation reliably triggered the movement of the forepaw on the opposite side. Stimulating nearby regions produced movements of the hind paw, the neck, and the face. Importantly, the same low-level stimulation to other areas of the cortex produced no such movement. As Jackson had suggested for humans, the excitable sites seemed to form a topographic map of body movements laid out on the surface of the cerebral cortex.

Fritsch and Hirtzig then decided to see what would happen if they surgically removed the area of the cortex that controlled the forepaw. After two dogs recovered from the surgery, the paw was not paralyzed but the dogs did have difficulty controlling the paw and the forelimb frequently slipped out from under them as they attempted to run. In their report of the experiments published in 1870, Frisch and Hirtzig concluded that there must also be other motor centers (probably subcortical) involved in the motor control of the limb and that the role of the cortex was primarily for the willful or voluntary control of movement.

These two German physicians made quite a team [35]. Fritsch had a longstanding interest in anthropology and later went on to search for body features that distinguished the superiority of the white race. Hitzig, who was from an affluent Jewish family, was described as "*a stern and forbidding character of incorrigible conceit and vanity complicated by Prussianism*" [34]. They operated on their first few dogs including the family pet without using anesthesia. The American zoologist and founder of the *Journal of Comparative Neurology*, Clarence Herrick described them as "*splendid specimens of physical development and German culture at its best*" [36].

David Ferrier and the West Riding Lunatic Asylum

Two Scots who met in medical school, David Ferrier and James Crichton-Browne, joined forces and expanded the work started by Fritsch and Hirtzig [34, 35]. Ferrier received top honors at a medical school in Scotland and then left for England. After a brief try at private practice, he accepted joint appointments at King's College Hospital and the National Hospital for the Paralyzed and Epileptic at Queen Square. It was at the National Hospital, devoted exclusively to people with diseases of the Nervous System, that Ferrier became friends with Hughlings Jackson. Ferrier was a stickler for medical tradition and was one of the last physicians who regularly conducted his rounds in a top hat and black tailcoat.

Ferrier's friend, Crichton-Browne, was appointed head of the West Riding Lunatic Asylum in West Riding, one of the three ridings (sections) of Yorkshire, shortly after graduating from medical school. His father was a pioneer in reforming mental hospitals and Crichton-Brown was determined to make the West Riding Lunatic Asylum a major brain research center. In the process he founded the first English journal exclusively devoted to brain research, West Riding Lunatic Asylum Medical Reports. This was a forerunner of the journal *Brain* which he founded along with Ferrier and Hughlings Jackson seven years later.

Shortly after Fritsch and Hitzig published their work, Ferrier visited his friend Crichton-Browne in West Riding, and among many topics, they discussed the Fritsch and Hitzig experiments. From prior discussions with Hughlings Jackson, Ferrier had developed a keen interest in cerebral localization, and Crichton-Browne immediately saw this as an opportunity to improve the scientific reputation of his West Riding Institute. He enticed Ferrier to come to West Riding and conduct his own electrical stimulation studies in animals offering him a large fully equipped laboratory and all the animals he would need.

Ferrier began his electrical stimulation studies at West Riding on the brains of rabbits, cats, and dogs in 1873. Instead of using galvanic stimulation like Fritsch and Hitzig, Ferrier used faradic stimulation with an induction coil like the one developed by Du Bois-Reymond. The high-frequency pulsed current produced by the induction coil proved to be a much better stimulus for eliciting movements than the low-frequency pulsed current used by Fritsch and Hitzig. With induction coil stimulation (faradization), Ferrier found many more excitable areas of the cortex than the Germans reported. He found that he could evoke movements of different body parts such as the hand, foot, face, mouth, and tongue with stimulation of different regions of the frontal cortex and many of the movements seemed natural such as grimacing, blinking, scratching, and flexing the digits. Furthermore, by increasing the duration of stimulation to more than 5 seconds, he could trigger a focal motor seizure that would move over the motor cortex just as Jackson predicted. With even stronger stimulation he could produce a generalized seizure.

On comparing the responses in rabbits, cats, and dogs, Ferrier noted that the overall organization of movement control was similar, but some parts of the body had larger areas of representation in the cortex in each species. He suggested that these differences depended "*on the habits of the animals*" [37]. In other words, the organization of movement control in the brain was related to behavioral specialization in each animal. Like Fritsch and Hitzig, Ferrier then went on to damage localized areas of the motor cortex to see how it impaired the animal's movements. He found that animals higher on the phylogenetic scale showed greater impairment than those lower on the scale. This explained why Flourens and other earlier investigators who studied frogs, birds, and rabbits had difficulty finding movement impairment after damaging the cerebral cortex.

Soon after starting his experiments at West Riding, Ferrier decided that he would need to study primates if he wanted to understand the organization of the human cerebral cortex. He applied to the Royal Society for a grant to study monkeys and the grant was approved by the end of 1873. In early 1874 he moved back to King's College Hospital to begin an extended period of study of the cerebral cortex of monkeys. He used the same faradic electrical stimulation and identified 19 centers in the frontal cortex related to different complex movements such as arm retraction, opening the mouth and protruding the tongue, and a sneering facial expression. Ferrier emphasized the importance of maintaining the electrical stimulation for longer durations to generate more complete movements rather than muscle twitches produced by very brief stimulations. "*A slight stimulus of short duration causes only a part of a complex action, which is manifested in its completeness when the stimulus is of somewhat greater intensity and duration*" [38].

Ferrier extended his electrical stimulation outside of the motor cortex located on the side of the head near the junction of the frontal and parietal lobes and noted that he could elicit movements that were like the reactions of normal monkeys to different sensory stimuli. For example, electrical stimulation of the top of the temporal lobe caused the animal to perk up the opposite ear and turn the head as if reacting to a sound. Stimulation of the bottom of the temporal lobe caused them to close a nostril, raise a lip, and grimace as if they were sniffing a disagreeable odor. He correctly localized the sensations of hearing and smell to these regions of the cerebral cortex. Based on inducing eye movements, he incorrectly localized vision to the back of the parietal lobe. The front part of the frontal lobes behind the orbits appeared to be insensitive to electrical stimulation.

When Ferrier damaged the frontal motor cortex in monkeys and allowed them to recover, the animals exhibited much more dramatic motor impairment than anyone had previously seen. The monkeys were paralyzed on the opposite side and could not move their extremities. The animals looked just like a patient with a stroke. Damage to the upper and lower parts of the temporal lobes confirmed their role in hearing and smell, respectively. Damage to both frontal lobes in front of the motor cortex caused the monkeys to become inattentive to everything around them. They were apathetic, dull, and listless and would aimlessly wander about their cage.

Ferrier published his work on the motor cortex in rabbits, cats, and dogs performed in Yorkshire in *The West Riding Lunatic Asylum Medical Journal* in late 1873 and submitted his work on all animals including monkeys to The *Philosophical Transactions of the Royal Society* in 1874. In the introduction, he emphasized the clinical contributions of his friend Hughlings Jackson but did not mention Robert Todd and only briefly mentioned the work of Fritsch and Hitzig. The reviewers felt that he had not given Fritsch and Hitzig proper credit for their work and sent it back for revisions. Ferrier made a few minor revisions, but the reviewers were still not satisfied, and the paper was rejected. Ferrier did present his work at a meeting of the Royal Society and a short abstract was published in the proceedings with no mention of Fritsch and Hitzig. The next year he decided to separately write up his work on monkeys which he saw as a way around the priority issue with Fritsch and Hitzig. This paper was accepted for publication in the *Philosophical Proceedings*.

Ferrier continued his electrical stimulation work in monkeys with particular attention on sensory areas of the cortex. In 1876, he published a book summarizing his work on all animals, *The Functions of the Brain* [39]. The book contained much of the work included in the previously rejected paper and although he still did not mention Todd, he did give Fritsch and Hitzig more credit for their work. In his book, Ferrier was so confident that the cortical localization findings in the monkey were applicable to humans that he included a figure of a human brain with a functional map of the monkey cortex superimposed on the human cortex (Fig. 5.4). He even suggested that his topographical maps could be used to localize brain lesions in patients prior to surgery. Although there were critics who felt that Ferrier had gone too far in his conclusions without mentioning contradictory data, overall, the book received laudatory reviews. A second edition in 1886 provided much more experimental data including expanded sections on the sensory cortex.

Ferrier Versus Goltz, the Last Great Battle of the Localizers

In 1881, more than 120,000 scientists, physicians, and royalty from all parts of Europe congregated in London for the Seventh International Medical Congress [34]. Three great men of medicine were honored Paget from England, Charcot from France, and Langenbeck from Germany. All participants received a medallion with Queen Victoria on one side and Hippocrates on the other. Of the more than 100 sections, the section on physiology included a session on the contrasting views about cortical localization of David Ferrier and Fredrich Goltz. The session was highly anticipated and turned out to be the most talked about session of the Congress.

Goltz, a physiologist from Strasburg, Germany, had performed numerous experiments removing large parts of the cerebral cortex in dogs. He was a flamboyant speaker with a flair for the dramatic. Goltz stated that despite removing large parts of the cerebral cortex in many dogs, he had never produced paralysis, deafness, or blindness. He even brought a dog along with him who had much of his cerebral cortex removed yet showed no localized deficits.

Ferrier then stepped up to the podium to make his case for cortical localization. There was a stark contrast in the appearance of the two protagonists. Goltz was heavy-set with an unruly Walrus-like mustache, whereas Ferrier was thin with a neat handlebar mustache. Ferrier began by politely pointing out that dogs have a relatively small cerebral cortex compared to primates and that Goltz probably did not destroy as much of the cortex as he thought. He then described his findings in two monkeys, one that was paralyzed on the right side after having the left motor cortex removed and another that was deaf after the upper temporal cortex was removed on both sides.

After the presentations ended, the chairman of the session asked the men if they would agree to examination of their animals, and both agreed. Later in the day a

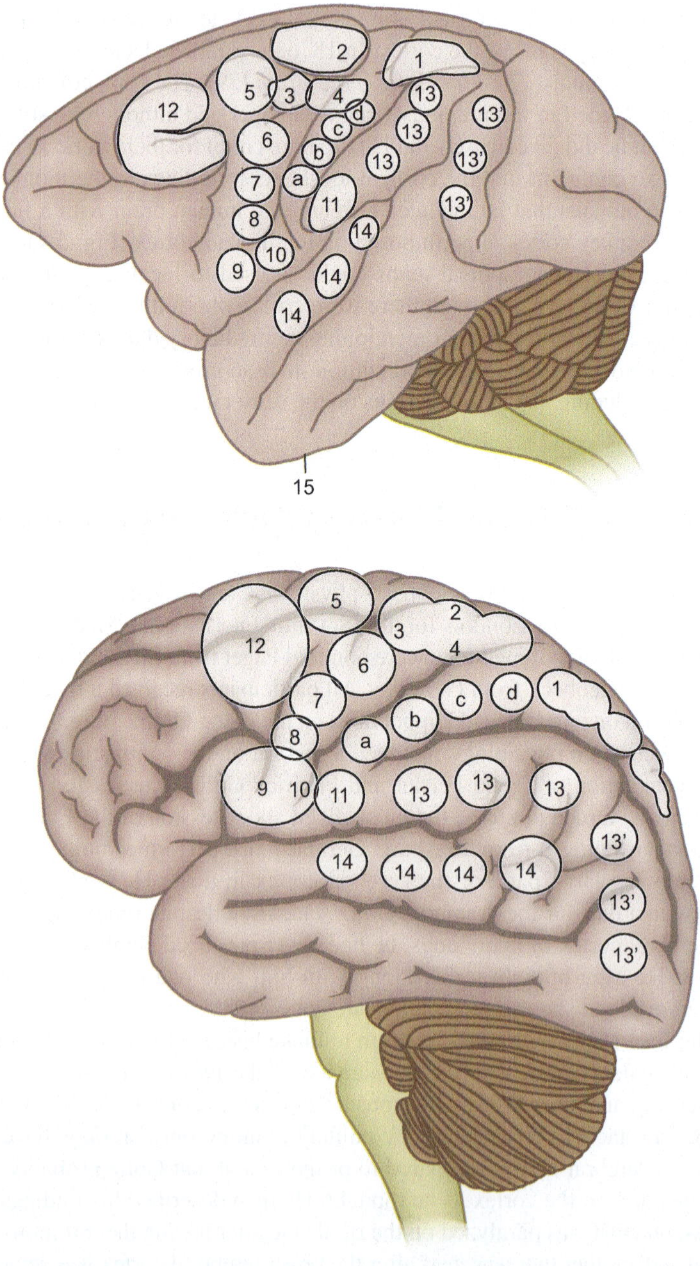

Fig. 5.4 David Ferrier's electrical stimulation cortical localization map of a monkey brain (above) superimposed on a human brain (below). (From Ferrier [39])

large contingent of the morning audience along with the presenters walked over to King's College Hospital to examine the animals. As described, Goltz's dog ran about the room and responded to sensory stimuli like a normal dog. Also as presented, Ferrier's monkey with damage to the left motor cortex could not lift his right arm or use his right hand and hobbled about the room and the monkey with damage to both temporal lobes did not respond to a percussion cap exploded next to his ear but otherwise appeared normal.

The chairman then asked the men whether they were willing to have a postmortem examination of the dog and one of the monkeys to see if the lesions were as presented and both agreed. After the animals were sacrificed the brains were examined by a panel of experts. In the case of Ferrier's paralyzed monkey, the damage was localized to the left frontoparietal region involving the motor cortex as predicted but in the case of Goltz's dog much more motor and sensory cortex remained intact than he thought. The localizers claimed victory and Ferrier was widely acclaimed.

An unexpected consequence of the publicity associated with the debate on localization at the International Medical Congress was that Ferrier was subsequently prosecuted under the Cruelty to Animals Act of 1876. In the early 1870s, there was strong sentiment in Britain to ban experimentation on live animals spearheaded by Frances Power Cobbe and her well-known Victoria Street Society. The Society had Royal influence and a Royal Commission was set up to investigate the matter. The result was a compromise Cruelty to Animals Act of 1876 that required yearly licensing for experimental procedures at specific sites. The experiments had to minimize pain and be justified based on their medical importance. The Victoria Street Society checked and found out that Ferrier was not licensed for the experiments he described at the Congress and they had him prosecuted under the act.

Ferrier was given the choice of a trial by jury or by a judge's ruling and he chose the latter. Ferrier's defense attorney argued that he was not liable because his surgical colleague at King's College Hospital, George Yeo, performed the surgery and Dr. Yeo had a proper license. Yeo also cared for the animals in accordance with the act. The prosecution was taken by surprise by this disclosure, but they argued that keeping the animals alive for months after the anesthesia wore off was painful and cruel and was a clear violation of the Cruelty to Animals Act. Ferrier was acquitted and although some considered it a defeat for the anti-vivisectionists the trial had a chilling effect on future animal experimentation in Britain and Europe. Experiments such as those carried out by Fritsch and Hitzig would no longer be tolerated.

The Human Cerebral Cortex

The first well-documented case of electrical stimulation of the brain in a conscious human patient was performed by an American physician, Roberts Bartholow, in 1874 [40]. A 30-year-old domestic worker, Mary Rafferty, was hospitalized for treatment of a purulent ulcer on her scalp that had eroded the underlying bone. The cause of the ulcer was felt to be due to pressure from a piece of whalebone used to attach her wig. When Bartholow examined her scalp, he noted a large, approximately two inches in diameter, hole at the top of the skull with the underlying pulsating brain clearly visible. Bartholow was an experienced electrotherapist with access to an electric room at the hospital and he received Mary's permission to conduct a few tests on the exposed brain. The question of informed consent was later raised since he described Mary as somewhat feeble minded.

Bartholow took the patient to the electrical room where he began by pricking the exposed brain with a needle electrode. He noted that Mary experienced no pain and had no movements or abnormal sensations of the extremities. He next applied pulses of galvanic current to the right side of the exposed area of the brain and noted muscle contractions of the right arm and leg. "*The arm was thrown out, the fingers extended, and the leg was projected forward. The muscles of the neck were thrown into action, and the head was strongly deflected to the right*" [41]. Bartholow observed similar reactions of the left extremities after stimulating the right side of the exposed area. He then decided to try increasing the strength of the electrical current and ran into trouble. Mary became agitated, began to cry and lost consciousness. She then began to convulse on the left side lasting for 5 min followed by coma lasting 20 min. When she regained consciousness, she complained of weakness and dizziness but otherwise appeared back to baseline.

Bartholow planned a further series of experiments but Mary's condition rapidly deteriorated and she died before any further tests could be performed. At autopsy the area of the brain exposed was said to be over the parietal lobes. Bartholow's publication of Mary's case was controversial and led to international criticism [42]. Bartholow was forced to leave Cincinnati where he was on the faculty of the Medical College of Ohio. He defended himself as best he could and eventually obtained a position at Jefferson Medical College in Philadelphia.

As noted earlier in the chapter, Ferrier suggested that his transposed functional map of the human cerebral cortex could be useful for localizing brain lesions prior to surgical removal. The first surgeon to follow Ferrier's suggestion was a Scot, William Macewen, who was also a follower of Joseph Lister and surgical antisepsis. Macewen's first case was a teenage girl with focal seizures of the right arm and face suggesting a tumor involving the motor cortex on the left side [43]. He successfully removed a meningeal tumor located exactly where he predicted, and the girl survived. Macewen went on to perform a series of successful surgeries based on Ferrier's functional map removing tumors, abscesses, and blood clots in the late 1870s.

One of the most famous surgical cases of the era, described in detail in the journal *Lancet,* was performed by a London surgeon Rickman Godlee on a patient referred to him by Hughes Bennett [44]. The young surgeon, Victor Horsley, watched the operation with great interest. A 20-year-old Scottish farmer had experienced focal motor seizures for three years and had residual weakness on the left side indicating a probable tumor in the right motor cortex. The operation was the first at the National Hospital for Epilepsy and Paralysis and since there was no operating room a make-shift room was prepared, and instruments were brought over from King's College Hospital including a carbolic spraying machine for disinfecting the room. After anesthesia with chloroform, the suspected area of the brain was exposed but the surface of the brain appeared normal. Godlee was not deterred and probed deeper where to everyone's relief, he identified a walnut sized tumor that he successfully removed. Unfortunately, despite his efforts to prevent infection, the patient developed postoperative meningitis and eventually died.

Victor Horsley would go on to be considered the father of functional neurosurgery and was the first to use intraoperative electrical stimulation of the cerebral cortex for the localization of epileptic foci in humans [45]. Horsley was born in London and received his medical training at University College London. After working with the German physiologist, Edward Schäfer, on lesions of the cerebral cortex of monkeys in Berlin, Horsley retuned to London in 1881 and became a house surgeon at the University College Hospital. He developed a reputation for his speed in operating, a skill that proved useful at a time when anesthesia was in its infancy and intraoperative infections were common.

In 1886, Horsley was appointed Assistant Professor of Surgery at the National Hospital for Paralysis and Epilepsy where he teamed up with Hughlings Jackson and performed surgery on Jackson's patients with focal epilepsy. Horsley used faradic electrical stimulation to identify the exact localization of the epileptic focus and then surgically removed the focus [46]. As Ferrier did in the monkey brain, Horsley used faradic stimulation to develop functional maps of the human brain and found a remarkable overlap with Ferrier's maps. Horsley's epileptic surgeries were performed decades before the discovery of electroencephalography (EEG) (see Chap. 10).

Sherrington and the Integrative Action of the Nervous System

Charles Sherrington's long life spanned a critical period of rapid advances in electrophysiology from the late nineteenth century through the early twentieth century and his ideas on reflex behavior marked the beginning of modern understanding of brain function. His decision to focus on the spinal cord and the relatively simple reflexes rather than the more complex cerebral cortex was key to his early success. In the process, he integrated knowledge of anatomy, electrophysiology, and behavior to provide a better understanding of how the nervous system works.

Charles Sherrington was born in London in 1857, the oldest of three brothers [47]. His father died when he was a child and his mother remarried Caleb Rose, a prominent physician in Ipswich. Rose, a Renaissance Man with interests in the classics, geology, and archeology, was a strong influence on Charles' decision to study medicine. When just 18 years of age, he passed the examination given by the Royal College of Surgeons and he began clinical training at St. Thomas's Hospital. After studying medicine in London and Edinburgh, Sherrington made his way to Cambridge where he was officially admitted in 1880. Based on reading the work of David Ferrier, he decided to choose the Department of Physiology where he was offered a position in the laboratory of John Langley later known for his work on the autonomic nervous system (see Chap. 10).

Langley had been asked to study the brain and spinal cord of Goltz's dog that he brought to the 1881 International Medical Congress, and he asked Sherrington to help him with the tedious process. Sherrington's first publication was a description of the neuronal degeneration in the dog's brain with Langley in 1884. Shortly after Sherrington received a scholarship to travel to Germany to study spinal reflex actions with the electrophysiologist Eduard Pflüger in Bonn and study neuronal degeneration with Goltz in Strasburg. At the time there was a consensus that nerve fibers formed a net interconnected with all other nerve fibers. Reports of separate individual nerve cells (the neuron doctrine) by the Spanish anatomist, Santiago Ramón y Cajal were just beginning to be published. Sherrington was impressed that the pattern of neuronal degeneration after focal cortical lesions was highly limited and not diffuse as would be expected with the nerve net doctrine. He was convinced that neurons must be separate entities and that there must be some type of junction between neurons. This would also explain why electrical signals only traveled in one direction, away from the nerve cell body. He called the junction a *synapsis,* from the Greek word meaning "to clasp," which would later evolve into *synapse*.

The problem was that the resolution of microscopes of the time was not good enough to see the junctions between neurons. As he began to study nerve reflexes in the spinal cord, it became apparent that the delay times in the spinal cord were longer than expected based on Helmholtz's measure of the speed of nerve conduction. A delay in the signal as it crossed the gaps between neurons in the reflex could explain the lost time. This would mark the beginning of the war between the *sparkers* and the *soupers.* The *sparkers* believed that an electric charge jumped across the synaptic gap, like a spark plug, while the *soupers* believed that when the electric charge reached the nerve terminal it triggered the release of chemicals that crossed the synapse and activated the neuron on the other side of the gap.

Sherrington began his studies with the simple knee reflex [48]. The reflex is now part of the routine physical examination. The physician taps on the tendon just below the kneecap causing the lower leg to jerk forward. Although at the time many felt that the muscle contraction was simply due to a mechanical twitching of the muscle from the tap, Sherrington performed detailed studies on a variety of animals including monkeys and clearly demonstrated that it was a reflex. He identified the sensory and motor nerves in the reflex and showed the sensory nerve originated in the muscle tendon and the motor nerve supplied the quadriceps muscle. One of the key observations in the study was that there were specialized sensory receptors in the muscles and tendons that sensed stretch and contraction providing proprioceptive feedback to the nervous system, a type of muscular sense. This so-called *sixth sense* provided key information for the control of fine motor activity.

Surprisingly, Sherrington found that up to 50% of the fibers in a muscle nerve remained intact after he cut the motor nerves to the muscle at the spinal cord. This indicated that sensory feedback was vital to normal muscle function. He also identified the rodlike spindles in the muscles and the Golgi organs in the tendons that served as the tension recorders for the reflex. The knee reflex turned out to be a simple two neuronal reflex with a single synapse between the sensory and motor nerves located in the spinal cord.

From the simple knee reflex, Sherrington went on to study more complex reflexes such as walking and scratching. He studied how muscles influence each other in the process of *reciprocal innervation*. For example, when walking some leg muscles extend while others flex and head and arm movements are synchronized to maintain balance. All these movements were reflexively controlled most occurring at the level of the spinal cord. To study these complex reflexes, Sherrington sectioned the brain stem of anesthetized animals to remove the lower nervous system from the higher cortical voluntary centers. Although the decerebrate animals were stiff (decerebrate rigidity) they could still carry out complex behaviors such as walking and scratching [49]. For example, in a decerebrate cat if the right foreleg was stimulated it moved forward and the right hind leg moved backward. At the same time, due to reciprocal innervation, the legs on the opposite side of the body moved in opposite directions creating a normal walking pattern (Fig. 5.5).

Even more impressive, such a "brainless" cat could vigorously scratch an area of skin stimulated by a flea as it moved across the body. Sherrington studied the nerves and muscles involved in the complicated scratch reflex. To simulate a flea bite he used a thin pin attached to an electric stimulator and multiple pins to simulate the

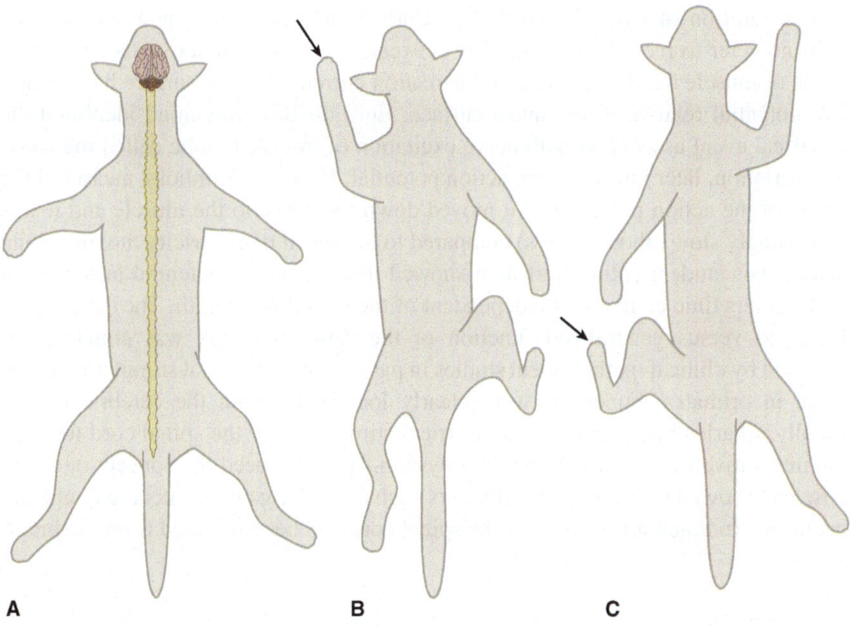

Fig. 5.5 Sherrington's decerebrate cat illustrating reciprocal innervation in the spinal cord. Stimulating the paw (arrows) triggered a walking pattern. (From Sherrington [49])

flea jumping across the skin. When the skin was stimulated, the appropriate leg extended the digits to form a claw and then went through an alternating series of knee, ankle, and hip movements to generate repetitive scratching movements. Reciprocal innervation activated other reflexes to maintain balance and prevent falling while the scratching was ongoing.

Sherrington was aware that even these complex reflexes were at the lowest level of the integrative action occurring in the central nervous system. Although the integrative process began at the level of the spinal cord, there was a longitudinal integration reaching its highest level at the cerebral cortex. He believed that the synapses were the key to this integrative process. The relatively simple reflexes taking place in the spinal cord were tied together by excitation and inhibition at the synapses in the brain to produce complicated behavioral responses. Just after the turn of the twentieth century, Sherrington was invited to Yale University to summarize his findings on central integration in a series of lectures. Sherrington emphasized the complexity of integration in the brain which could initiate motor acts without any peripheral sensory input. He concluded the lectures with a challenge to his colleagues in the new century, "It is then around the cerebrum, its physiological and psychological attributes, that the main interest of biology must turn" [50].

In Brief

In the tradition of Galvani and Aldini, Carlo Matteucci used a newly developed galvanometer to record a current flow between the cut and intact surface of a frog nerve or muscle that he called the demarcation current. The cut surface had a negative potential relative to the intact surface. Emil du Bois-Reymond identified the electrical event associated with nerve excitation of muscle that he called the negative variation, later known as the action potential. Herman Helmholtz measured the speed of the action potential as it moved down the nerve to the muscle and it was surprisingly slow (about 30 m/s) compared to the speed of electricity moving along a wire. His student Julius Bernstein showed that the action potential lasted about 1 ms and its time course was independent of the stimulus strength. The debate over localized versus generalized function of the cerebral cortex was convincingly resolved by clinical–pathological studies in patients and electrical stimulation of the brain in primates. Functions were clearly localized within the cerebral cortex. Finally, Charles Sherrington used electrical stimulation of the spinal cord to study the integrative function of the nervous system. He introduced the concept of a synapse and showed how complex behaviors such as walking and scratching could still occur in a rudimentary form after the spinal cord had been resected from the brain.

References

1. Piccolino M. Animal electricity and the birth of electrophysiology: the legacy of Luigi Galvani. Brain Res Bul. 1998;46:381–407.
2. Kipnis N. Luigi Galvani and the debate on animal electricity, 1791–1800. Ann Sci. 1987;44:107–42.
3. Adrian ED. The mechanism of nervous action. Electrical studies of the neurone. Philadelphia: University of Pennsylvania Press; 1932.
4. Moruzzi G. The electrophysiological work of Carl Matteucci. Brain Res Bul. 1996;40:69–91.
5. Moruzzi G. The electrophysiological work of Carl Matteucci. Brain Res Bul. 1996;40:73.
6. Nelson ME. Electric fish. Curr Biol. 2011;21:R528–9.
7. Matteucci C. Electrophysiological researches. Third memoir. On induced contractions. Philos Trans R Soc. 1845:303–17.
8. Matteucci C. Electrophysiological researches. Fourth memoir. The physiological action of electric-current. 1846;483–499.
9. Moruzzi G. The electrophysiological work of Carl Matteucci. Brain Res Bul. 1996;40:80.
10. Finkelstein G. Matteucci and du Bois-Reymond: a bitter rivalry. Arch Ital Biol. 2011;149(Suppl):29–37.
11. Finger S, Wade NJ. The neuroscience of Helmholtz and the theories of Johannes Müller Part 1: nerve cell structure, vitalism and the nerve impulse. J Hist Neurosci. 2002;11:136–55.
12. Finger S, Wade NJ. The neuroscience of Helmholtz and the theories of Johannes Müller. Part 2. J Hist Neurosci. 2002;11:234–54.
13. Finger S, Wade NJ. The neuroscience of Helmholtz and the theories of Johannes Müller Part 1: nerve cell structure, vitalism and the nerve impulse. J Hist Neurosci. 2002;11:150.
14. Finkelstein G. Emil du Bois-Reymond: neuroscience, self, and society in nineteenth-century Germany. Cambridge, MA/London: The MIT Press; 2013. p. XV.
15. Finkelstein G. Emil du Bois-Reymond: neuroscience, self, and society in nineteenth-century Germany. Cambridge, MA/London: The MIT Press; 2013. p. 60.
16. Finkelstein G. Emil du Bois-Reymond: neuroscience, self, and society in nineteenth-century Germany. Cambridge, MA/London: The MIT Press; 2013. p. 61.
17. Finkelstein G. Emil du Bois-Reymond: neuroscience, self, and society in nineteenth-century Germany. Cambridge, MA/London: The MIT Press; 2013. p. 63.
18. Moruzzi G. The electrophysiological work of Carl Matteucci. Brain Res Bul. 1996;40:83.
19. Finkelstein G. Emil du Bois-Reymond: neuroscience, self, and society in nineteenth-century Germany. Cambridge, MA/London: The MIT Press; 2013. p. 82.
20. Moruzzi G. The electrophysiological work of Carl Matteucci. Brain Res Bul. 1996;40:84.
21. Cahan D, editor. Hermann von Helmholtz and the Foundation of Nineteenth century science. Berkeley: University of California Press; 1993.
22. Cahan D, editor. Hermann von Helmholtz and the Foundation of Nineteenth century science. Berkeley: University of California Press; 1993. p. 27.
23. Finger S, Wade NJ. The neuroscience of Helmholtz and the theories of Johannes Müller Part 1: nerve cell structure, vitalism and the nerve impulse. J Hist Neurosci. 2002;11:152.
24. Seyfarth E-A. Julius Bernstein (1839–1917): pioneer neurobiologist and biophysicist. Biol Cybern. 2006;94:2–8.
25. De Palma A, Pareti G. Bernstein's long path to membrane theory: radical change and conservation in nineteenth-century German electrophysiology. J Hist Neurosci. 2011;20:306–37.
26. Gross CG. The discovery of the motor cortex and its background. J Hist Neurosci. 2007;16:320–31.
27. Finger S. Minds behind the brain. A history of the pioneers and their discoveries. New York: Oxford University Press; 2000. p. 119–39.
28. Baloh RW. Vertigo. Five physician scientists and the quest for a cure. New York: Oxford University Press; 2017. p. 11.

29. Finger S. Minds behind the brain. A history of the pioneers and their discoveries. New York: Oxford University Press; 2000. p. 140.
30. Schiller F. Paul Broca: founder of French anthropology, explorer of the brain. New York: Oxford University Press. p. 174–99.
31. Finger S. Minds behind the brain. A history of the pioneers and their discoveries. New York: Oxford University Press; 2000. p. 145.
32. Finger S. Minds behind the brain. A history of the pioneers and their discoveries. New York: Oxford University Press; 2000. p. 149–50.
33. Sacks O. The man who mistook his wife for a hat and other clinical notes. New York: Summit Books; 1970.
34. Finger S. Minds behind the brain. A history of the pioneers and their discoveries. New York: Oxford University Press; 2000. p. 155–75.
35. Taylor CSR, Gross CG. Twitches versus movements: a story of motor cortex. The Neuroscientist. 2003;9:332–42.
36. Taylor CSR, Gross CG. Twitches versus movements: a story of motor cortex. The Neuroscientist. 2003;9:335.
37. Taylor CSR, Gross CG. Twitches versus movements: a story of motor cortex. The Neuroscientist. 2003;9:336.
38. Taylor CSR, Gross CG. Twitches versus movements: a story of motor cortex. The Neuroscientist. 2003;9:339.
39. Ferrier D. The functions of the brain. London: Smith, Elder and Company; 1876.
40. Bartholow R. Experimental investigations into the function of the human brain. Am J Med Sci. 1874;134:305–13.
41. Bartholow R. Experimental investigations into the function of the human brain. Am J Med Sci. 1874;134:310.
42. Thomas RK, Young CD. A note of the early history of electrical stimulation of the brain. J Gen Psych. 2010;120:73–81.
43. Macewen W. Tumour of the dura mater removed during life in a person afflicted with epilepsy. Glasgow Med J. 1879;122:210–3.
44. Bennett A, Goodlee R. Excision of a tumour from the brain. Lancet. 1884;2:1090–1.
45. Vilensky JA, Gilman S. Horsley was the first to use electrical stimulation of the human cerebral cortex intraoperatively. Surg Neurol. 2002:425–6.
46. Isitan C, Yan Q, Spencer DD, Alkawadri R. Brief history of electrical cortical stimulation: a journey in time from Volta to Penfield. Epilepsy Res. 2020;166:363–9.
47. Eccles JC, Gibson WC. Sherrington: his life and thought. New York: Springer International; 1979.
48. Sherrington CS. Note on the knee jerk. St. Thomas's Hosp Rep. 1891;21:145–7.
49. Sherrington CS. Decerebrate rigidity, and reflex co-ordination of movements. J Phys. 1898;12:319–32.
50. Sherrington CS. The integrative nature of the nervous system. New Haven: Yale University Press; 1906. p. 390.

Chapter 6
Electrotherapy and Neurology

> *I am quite satisfied that some affections of the nervous system can be cured by electricity alone, but in the majority of cases simultaneous internal treatment should not be neglected if we wish to increase our chances of success.*
>
> *Julius Althaus, 1859*

Abstract Electricity played a prominent role in the development of medical neurology in the nineteenth century. Nearly all the pioneer neurologists in Europe and America were electrotherapists. Since electricity stimulated nerves, nerve cells, and muscle cells, it seemed logical that electrotherapy was most useful for treating neurological diseases. Hospitals set up special areas to house the electronic equipment along with electricians to maintain the equipment. In France, Duchenne demonstrated how electricity could distinguish between nerve and muscle diseases. Neurologists treated a wide range of neurological conditions with electrotherapy from paralysis to dystonia and epilepsy. The best results were obtained with the common diagnoses of hysteria and neurasthenia. These disorders were greatly dependent on suggestion and electrotherapy was a powerful placebo. Electricians and neurologists developed a tenuous relationship as each tried to establish their role in the potentially lucrative field of electrotherapy.

Keywords Electric bath · Duchenne smile · Neurosis · Neurasthenia · Electrocution

The nineteenth century marked the beginning of a new medical subspecialty, *neurology*—the study of diseases of the nervous system. The term *neurologie* was first used in the seventeenth century by the English physician Thomas Willis in the English edition of his famous book *Cerebri Anatome* [1]. Willis used the term to refer to the study of spinal, cranial, and autonomic nerves but gradually over time the use of the term expanded to signify the study of the health and disease of the entire nervous system including the brain. As we will see, there was a close interrelationship between the developing specialty of neurology and electrotherapy. Most of the founding fathers in neurology in the Western World were practicing electrotherapists—Todd, Bird, and Althaus in England, Duchenne and Charcot in France, Erb and Möbius in Germany, and Beard and Mitchell in America.

R. W. Baloh, *Brain Electricity*, https://doi.org/10.1007/978-3-031-62994-5_6

Golding Bird and the New Electrotherapy

An early example of the new approach to electrotherapy can be seen with the opening of the "electrifying room" at Guy's Hospital in London in 1836 [2]. Electricity had been used in hospitals since the mid-eighteenth century, but it was entirely under the control of the treating physicians who were responsible for the equipment used. The electrifying room at Guy's Hospital was sanctioned by the hospital administration who were responsible for maintaining the equipment and appointing a director. Their goal was to evaluate the therapeutic effects of electricity with standardized equipment and establish its role in treating specific diseases.

To run the electrifying room, the hospital administrators selected a young medical student, Golding Bird, who had been at Guy's hospital since 1832. Bird had been an apprentice to a London apothecary prior to entering medical school, and at Guy's, he caught the attention of Thomas Addison, of Addison's Disease fame, then an assistant physician at Guy's Hospital. Bird had already published several papers on chemical analysis of blood and urine. He was appointed a lecturer at Guy's Hospital in 1836 and received his MD degree in 1838. Bird was an electricity enthusiast and joined the London Electricity Society soon after its foundation in 1837.

Bird along with a student acting as a Clinical Clerk treated outpatients and inpatients with electricity daily at three o'clock keeping detailed notes regarding the patient's symptoms and response to electrical therapy. The electricity room was equipped with all the usual sources of electricity including traditional electricity machines and Leyden jars along with batteries and induction coils. Static electricity for an "electric bath" and the drawing of sparks was gradually replaced by prescribed doses of galvanic and faradic electricity applied to specific locations.

Like most London hospitals at the time, patients in Guy's Hospital were mostly working-class people drawn from the poorer sections of the city. They were expected to accept treatments, many of which were considered experimental, without complaint. It was an ideal setting for a young physician such as Golding Bird to practice his new electrical treatments and establish a name for himself in electrotherapy. He had a seemingly endless supply of "willing" patients to test the therapies and technical support to maintain the electrical equipment.

A unique feature of Bird's electrotherapy at Guy's Hospital was that he restricted its use to "nervous disorders," particularly chorea and hysterical paralysis. He argued that since electricity clearly stimulated nerves, it should be effective for treating nervous disorders. Since many of the women suffering from these nervous disorders also had amenorrhea, he also treated amenorrhea with galvanic stimulation. In his first publication in *Guy's Hospital Reports,* Bird reported that 35 of 36 patients with chorea and 20 of 24 patients with amenorrhea benefited from electrotherapy [3]. Bird's mentor Thomas Addison was impressed noting: "*although I have often ordered it myself, and have more frequently witnessed its employment by others, I had never for a moment entertained the belief that it possessed the power over the disorders alluded to, which I am now inclined to concede*" [4].

It is likely that most if not all the patients treated for involuntary movements (chorea) and paralysis fit a diagnosis of hysteria. For example, a young girl, Sarah Wheeler, admitted on November 5, 1939, with a diagnosis of chorea, had involuntary movements of the right arm and shoulder which she attributed to fear caused by threats from her schoolmaster. A series of electrotherapies resulted in cessation of the movements and resumption of her menses. As indicated in Chap. 3, hysteria was a poorly understood nervous condition that was highly susceptible to suggestion, and in retrospect it is very likely that Bird was taking advantage of the dramatic placebo effects of electrotherapy. There is no other reasonable explanation for the rapid reversal of chorea and paralysis with electrotherapy.

One of Bird's most lasting contributions to electrotherapy was the development of an induction coil that could produce rapid repetitive impulses of electricity for stimulating nerves. Recall that the induction coil that Faraday made for Robert Todd (see Chap. 4) had a wheel that was manually turned to repetitively make and break the contacts between the primary wires from the battery to the induction coil. Bird developed a method to automatically make and break the primary circuit using a spring-loaded switch that was magnetically activated by the current from the battery (Fig. 6.1) [5]. When the battery wire was connected, it produced a magnetic field around the metal rod inside the inner coil activating the switch to break the circuit. This shut off the magnet and the spring closed the circuit producing a continuous cycle of making and breaking the circuit.

Bird was unique among the practitioners of electrotherapy in the mid-nineteenth century in that he could cross disciplines and was accepted in both the electrical and medical communities. As an early member of the London Electrical Society, he maintained close connections with London electricians and instrument makers and

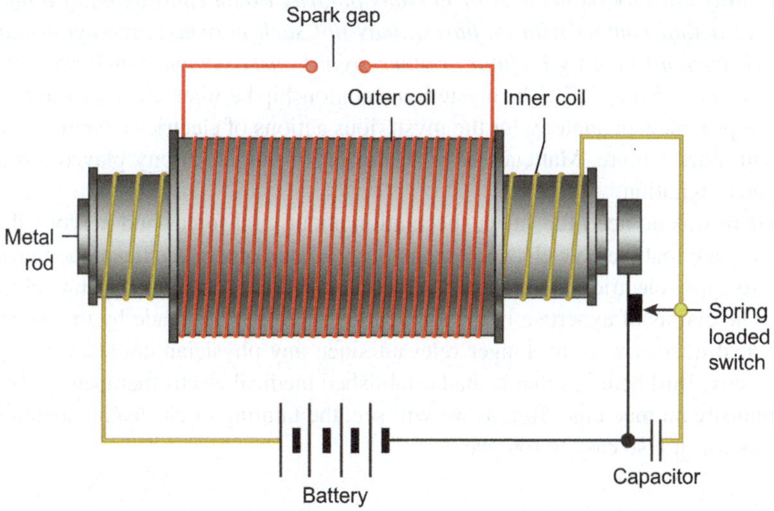

Fig. 6.1 Schematic drawing of Bird's induction coil for generating continuous alternating current (see text for details)

with his position at Guy's Hospital, he was accepted by London's most eminent physicians.

At the time, there was still a debate among electricians and physicians as to who was best suited to perform electrotherapy. As far as most electricians were concerned, medical training was not necessary since electricity was the vital force of nervous fluid, the force that guided the universe. Understanding electrical instrumentation was more important than understanding the human body. The electrician Andrew Crosse, mentioned earlier for his claim to have produced insects from electricity, felt that animal electricity was compatible with his strong religious beliefs: "*It has long been my belief, that the Electric influence is the great principle by which the Almighty puts together, and separates; and that it might be called, metaphorically speaking, the right arm of God*" [6]. Golding Bird was one of several investigators who conducted experiments showing that Crosse's claim of spontaneous generation of life with electricity was bogus.

Golding Bird worked hard to dissociate himself from the quacks who used electrotherapy to enrich themselves. He considered his electrotherapy to be based on a rational understanding of the action of electricity on the nervous system. Beginning in March of 1847, he gave a series of lectures before the prestigious Royal College of Physicians outlining what he considered a scientific underpinning for the medical uses of electricity. He quoted the work of Michael Faraday and Carlo Matteucci to provide a rationale for the use of electricity for treating nervous disorders. He was particularly enamored by Faraday's experiments showing that electricity could produce magnetism, and magnetism could generate electricity.

"*May not one of the uses of the electricity so freely developed in the body, especially that existing in muscles, be to excite in the nervous chords the vis nervosa, just as currents, if passing near a bar of iron at right angles to its axis, excite magnetism? May not this vis nervosa or nervous polarity excite contraction of a muscle without actual contact with its fibres...may not such nervous force again induce electric currents in any glandular or other organs, just as magnetism in motion will re-excite electricity?*" [7]. The mysterious relationship between electricity and magnetism provided an analogy for the mysterious actions of electricity on the nervous system. Furthermore, Matteucci's work was proof that electricity played a role in nervous excitation.

There was no need for vague speculation about vital forces and nervous fluids. Bird argued that the new electrical technology he and others had developed made it easy to apply electrical treatments without the need for specialized technical training. The claims of expertise in the application of electricity made by the so-called electrical quacks were no longer relevant since any physician could safely apply electricity. Bird believed that he had established medical electrotherapeutics as part of mainstream medicine. But, as we will see, the tainting of electrotherapeutics by quacks was not so easy to remove.

Harry Lobb and the London Galvanic Hospital

Despite Bird's claim for the benefits of electrotherapy for curing nervous disorders, relatively few reputable medical institutions in London followed his lead. When Harry Lobb founded the London Galvanic Hospital in 1861, the prestigious medical journal *Lancet* ran an editorial that facetiously questioned what next, "*a Quinine Hospital, an Hospital for treatment by Cod-liver Oil, by the Hypophosphites, or by the Excrement of Boa-Constrictors?*" [8]. Lobb responded by pointing out that he had no choice but to start his own hospital since his alma mater St Bartholomew's Hospital did not allow him to practice electrotherapy on its wards [9]. Like Bird, he argued that all he wanted to do was to rescue electrotherapy from the "quacks."

Harry Lobb studied medicine at St. Bartholomew's Hospital and attended lectures at Guy's Hospital where he developed an interest in electrotherapy [10]. William Lobb, Harry's father and a prominent London physician with a lucrative practice, was not sure Harry had what it took to be successful in practice. In response, Harry decided to spend two years in the Far East as a ship's surgeon with the East India Company. When he returned to London, he set up his own practice at his home in the newly fashionable Bayswater district. There, he had the good fortune to become friends with Charles Locock, Queen Victoria's physician. Locock began to regularly refer wealthy patients with neurological symptoms to Lobb for electrotherapy. By the end of the 1850s, Lobb published 2 books on the use of galvanism for a wide range of diseases of the nervous system dedicating the second book to Locock for "*the valuable assistance afforded by him in the establishment of galvanism as a recognized agent in the treatment of disease*" [11].

With the income from Locock's wealthy patients, Lobb opened the London Galvanic Hospital in 1861 noting a long list of aristocratic patrons. He dedicated the hospital to "*treatment, with the aid of Electricity, of all forms of Nervous and Muscular Diseases, for which the force is particularly adapted*" [12]. The hospital would provide hope for the "*helpless, abandoned, and so-called incurable cases*" that other hospitals could not treat. Lobb also saw the hospital as a nidus for a school for electrotherapy and electrophysiology.

An important supporter of the Galvanic Hospital was the telegraph engineer and flamboyant editor of the journal *Electrician*, Desmond FitzGerald. He prophesized that the hospital would be "*the means of affording an asylum for this science, a home from whence all our English discoveries will emanate*" [12]. Fitzgerald published a series of Lobb's lectures in the *Electrician* with glowing editorial comments. He argued that Lobb's electrotherapy could "*cure all manner of diseases, which had baffled all remedies, of whatever nature they might be*" [12].

Unfortunately, Lobb's grand ambitions did not come to fruition and after only a few years of operation, the hospital shut down. A major factor in the closure was Locock's retirement and loss of his wealthy patient referrals. To salvage his electrical practice, Lobb made the fatal mistake of running a series of advertisements in *The Times* promoting his books and lauding the benefits of electrotherapy. This was an activity that could not be tolerated by the medical establishment of Victorian

England. Such advertisements might be acceptable for a tradesman but not for a gentleman and above all, a physician must be gentlemen. Lobb was threatened with loss of membership in the Royal College of Surgeons, and he was shunned by many colleagues.

Electrotherapy Devices

In the mid to latter parts of the nineteenth century, newspapers and magazines were full of advertisements lauding the benefits of a wide variety of electrical devices for maintaining health and fitness and for treating all types of diseases. The medical electrical business was highly lucrative. Devices that received the greatest notoriety were Pulvermacher chains and electrical belts. Although very popular and widely used by both respected physicians and quacks, there is little evidence that these devices provided any therapeutic benefit.

Isaac Lewis Pulvermacher was a Viennese telegraph engineer and inventor who published the details of his electric chain in Germany in 1850 [13]. He immediately traveled to London to demonstrate the device to physicians who practiced electrotherapy. The chain was a series of identical small batteries that could be joined together to generate electricity for electrotherapy (Fig. 6.2). Each link in the chain consisted of a small wooden rod wound with thin zinc and copper wires intertwined. Each end had a positive and negative pole with loops of wire to join the links together. The device was activated by soaking the links in a solution of vinegar and water. The amount of electrical current generated depended on the number of links in the chain.

Shortly after arriving in London, Pulvermacher gave one of his battery chains to Golding Bird to try in his practice [13]. Bird found the device to be a useful portable source of electricity that patients with neurological symptoms could use in their own homes. His initial impression was so positive that he provided a testimonial for the company to use when contacting other physicians and wrote a supportive article that was published in the journal *Lancet*. But it was not long before Bird began to question his endorsement of Pulvermacher's chain.

Bird prided himself as a staunch opponent of quack practitioners who claimed immediate miraculous cures with electrotherapy. Yet this was the main sector of the medical community that the Pulvermacher Company was seeking with its advertisements. Furthermore, they were using Bird's endorsement in the advertising. Bird tried to make it clear that he only recommended the Pulvermacher chain as a ready source of electricity and did not support the Company's claims of miraculous cures. He even suggested that the electrical current was often too low to be of therapeutic value and that wrapping the chain around an affected extremity would be of little use since the skin would prevent electricity from entering the body. Despite having their advertisements banned from several medical journals and threats of legal action from Bird, the Pulvermacher Company continued to advertise with notable names including Bird and continued to make claims of miraculous cures.

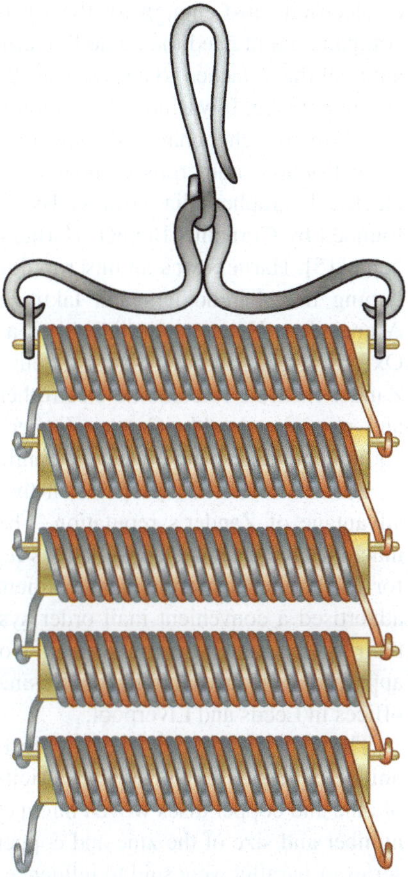

Fig. 6.2 Pulvermacher chain. Each link was made of a wooden rod wound with zinc and copper wires intertwined. The chain was soaked in vinegar water

Harry Lobb was another early advocate of Pulvermacher's chains and his books contained detailed descriptions of the chains including some that he had specially made [14]. He recommended the chains for treating neuralgia (nerve pain) and paralysis. For example, to treat neuralgia he used a chain with 40–60 links running from the nape to the painful area. He published a series of cases in the *Lancet* that were successfully cured with Pulvermacher's chains. When he opened his Galvanic Hospital, patients were discharged with instructions for continuing electrical treatments on their own at home with Pulvermacher's chains.

Pulvermacher patented his chain battery in the United States in 1853 and soon after he patented a wearable chain battery belt, the so-called electric belt. A unique feature of the electric belt was that it did not require soaking in vinegar but could be activated by body sweat. The electric belt became so popular in the United States that Pulvermacher eventually moved the company headquarters from London to Cincinnati in 1880 where belts were manufactured for all parts of the body including the waist, arms, legs, chest, and neck. It was particularly popular for treating lumbago, low back pain. Activated by body sweat, these belts produced tiny amounts of electricity just enough for the wearer to feel and know it was "working." The company maintained the name Pulvermacher of London for prestige. It even made a model that attached to a sack over the male genitalia to treat erectile dysfunction and impotence. It was not uncommon for people to be seen in public wearing multiple Pulvermacher chains like jewelry. It became a fashion statement.

In London, numerous companies were competing in the burgeoning medical electrical paraphernalia business. By the mid-1880s, the Medical Battery Company founded by Cornelius Bennett Harness dominated the field particularly in electric belts [15]. Harness was an unscrupulous businessman who began his company by duping the shareholders and taking over the assets of the Pall Mall Electric Association. His company inhabited a large building with an imposing facade on Oxford Street. Harness was all about image and he acquired the nearby London Zander Institute to combine electrotherapy with physical exercise for maintaining and regaining health. Gustav Zander was a highly respected Swedish exercise equipment maker who owned the English patents on several gymnastic appliances.

Harness renamed his operation the Electropathic and Zander Institute to take advantage of Zander's reputation. The Institute provided patients with Zander's machines for exercise, consultations regarding which electrical appliance was best for them, and electrotherapy treatments by trained associates. The Institute also advertised a convenient mail order system. People could send a letter describing their symptoms and Harness' staff would make a diagnosis and recommend the appropriate electrical appliance. Business was so good that Harness opened branch offices in Leeds and Liverpool.

The electropathic belts were not just any electric belts. Harness claimed he had a unique understanding of how electricity worked on the body. His belts were made of zinc and copper discs woven into a canvas webbing and connected by wires. The number and size of the zinc and copper discs and whether they were connected in series or parallel were said to influence the therapeutic benefits. The belts had to be worn next to the skin so that they could be activated by sweat. The price of the belts ranged from 5 to 12 guineas with the more expensive belts recommended for the most serious illnesses.

Ironically, it was the sale of a 5 guinea electropathic belt to a middle-aged bank clerk, a Mr. D. Jeffery, that led to the downfall of Harness and his electrotherapy empire [16]. In 1892, Jeffery walked into Harness's Institute and asked for a consultation regarding discomfort which he thought might be due to a "rupture." A Mr. Simmons examined Jeffery and told him he did not have a rupture but that he would benefit from an electropathic belt. Jeffery agreed to buy a belt paying 2 guineas out of pocket and signing an IOU for an additional 3 guineas and 5 shillings.

Despite faithfully wearing the belt, his symptoms persisted and Jeffery decided to consult a physician. The doctor told Jeffrey that he had a hernia and that the belt was probably making his symptoms worse rather than better. Jeffery was naturally upset, and he refused to pay the IOU. After giving him time to reconsider without success, the Medical Battery Company decided to take Jeffrey to court to recover the unpaid part of his bill. Why Harness would take such a risk for such a small sum of money is difficult to phantom. He presumably thought that Jeffery would be easily intimidated but was he ever wrong.

Jeffrey counter sued the company for misrepresentation. His lawyer claimed that the Medical Battery Company made false claims and misrepresented its products. When Jeffrey entered the building, he expected to see a qualified man. On cross-examination, Mr. Simmons admitted that he had no medical or electrical training and that he previously worked as a furniture salesman. When asked if the belt could cure a hernia, Simmons replied: "I cannot say, but the belt would give support" [17]. Jeffrey's doctor testified that the belt aggravated Jeffrey's complaint. The defense also called an electrical expert who testified that if the belt was used as recommended. it produced no electricity and even if soaked in salt water. the tiny amount of electricity generated would pass through the belt rather than through the body.

The judge ruled in favor of Jeffrey and against the Medical Battery Company. He indicated that the ruling was primarily based on misrepresentation and that he was not qualified to make a judgement regarding the efficacy of the electric belt. The negative publicity associate with the trial was the beginning of the end for the Medical Battery Company. Other trials followed and both the medical and electrical community turned against Harness and his electrical appliances. Within a year of the Jeffrey trial, the Medical Battery Company was in receivership and Harness was exposed as a disreputable quack.

Julius Althaus and Electrotherapy for Epilepsy

In 1866, a German immigrant, Julius Althaus, founded the London Infirmary for Epilepsy and Paralysis that would later become the Hospital for Diseases of the Nervous System [18]. Althaus obtained his degree in Berlin, studied with Charcot in Paris, and worked as an assistant of Robert Todd learning electrotherapeutics at King's college. The Infirmary was located near Harley Street, already known for its upscale medical practices. It provided a large space for electrical instruments and therapy.

Althaus rapidly became a leading proponent of electrotherapy in London and three editions of his textbook *Treatise in Medical Electricity* were published between 1859 and 1873. His book became the "bible" for electrotherapy providing both a detailed description of the electrical equipment required and application for different diseases. Althaus was well aware of the skepticism of electrotherapy by many physicians and in the Preface of the first edition of his book, he noted: "*electricity is as little infallible as any other remedy we possess*" and "*I am quite satisfied that some affections of the nervous system can be cured by electricity alone, but in the majority of cases simultaneous internal treatment should not be neglected if we wish to increase our chances of success*" [19]. Althaus even provided a list of trusted instrument makers and emphasized the need for physicians to work with a skilled electrician. He emphasized that the dose of electricity needed to be tailored to the patient's height, weight, sex, and age.

Modern physicians might have a difficult time understanding how Althaus could recommend treating epilepsy with electrotherapy. After all, epileptic seizures result from an abnormal electrical discharge from the brain and can be triggered by electrical stimulation of the brain. But at the time Althaus wrote his book, the role of the brain in epileptic seizures was not well understood.

Certainly, Althaus was aware of Robert Todd's description of epileptic-like seizures in rabbits after electrical stimulation of the brain and by the time of his later editions of the work of Fritsch, Hitzig, and Ferrier (see Chap. 5). It was also known that strokes and brain tumors were commonly associated with seizures but as Althaus pointed out, in most patients with epileptic seizures, no pathology could be identified in the brain even with the best microscopes available at the time. These cases were thought to result from irritation outside of the central nervous system, involving peripheral nerves or other organs. Epilepsy was considered an essential condition involving other organ systems and not necessarily associated with focal brain damage.

According to Althaus: "*Some observers have gone so far as to deny that epilepsy is a real disease, and consider it only a symptom, the epileptic attack being, in their opinion, invariably owing to coarse structural disease affecting the nervous system or its remote parts. Such an opinion can only be attributed to insufficient observation of the complaint and will most certainly never be endorsed by those who have studied the symptoms and habits of epileptics more particularly; for it is then seen that epilepsy is a true general neurosis*" [18].

Althaus felt that the vascular system played a key role in the cause of epilepsy. This idea was based on the observation that people who blacked out from a drop in blood flow to the brain (often from a common faint) could suffer a seizure particularly if the blackout occurred while sitting with the head upright. Placing the patient supine with the head below the feet could stop the seizure. Dating back to the work of Nollet and Kratzenstein discussed in Chap. 2, electrotherapy was thought to improve circulation to the brain.

Althaus recommended different forms of electricity for different diseases of the nervous system. Direct current from a battery had a strong effect on the overall electricity of the nervous system and was particularly useful for treating "functional nervous affections" such as hysteria. On the other hand, faradic current could restore the molecular equilibrium of motor nerves and muscles and was particularly beneficial for neuromuscular disorders. He saw the grey matter of the brain as a galvanic battery that generated electrical current and the white matter as telegraph wires that directed electricity to where it was needed.

Duchenne and Localized Nerve Stimulation

Guillaume Benjamin Amand Duchenne was born in Boulogne on the northern coast of France in 1806 [20, 21]. His father was a naval commander from a family with a long line of seafarers and fishermen. Duchenne broke with the family tradition and went to Paris to study medicine. After graduating in 1831, he returned to Boulogne to practice medicine.

Duchenne soon became intrigued with electricity and in the mid-1830s, he used electrical stimulation to treat a patient suffering from neuralgia after a nerve injury. He was so impressed with the results that he developed his own machine for electrical stimulation after Faraday's induction coil with the unique feature of surface electrodes he designed for localized nerve stimulation. He first tried a subcutaneous needle for stimulation but found that surface electrodes worked better. His metallic electrodes had a cloth cover that when moistened provided good electrical transmission. He realized that he could initiate the movements of individual muscles by varying the placement of the electrodes. In addition to treating nerve and muscle diseases, he began to use electrical stimulation to diagnose different neuromuscular disorders.

When Duchenne's young wife died in childbirth, he went into a severe depression. He eventually recovered and remarried and in 1842, he decided to leave Boulogne and set up practice in Paris focusing on neurology. As an outsider, he was unable to obtain a faculty or hospital position, so he visited hospital wards across Paris searching for unusual patients with neuromuscular diseases that he could study. At the time, understanding of neuromuscular diseases was primitive at best with many different disease conditions having similar clinical appearance, muscle wasting.

With his electrical stimulation technique, Duchenne had a simple method to determine whether neuromuscular disorders were primarily due to nerve or muscle damage. If a muscle did not contract with stimulation of its nerve supply but did contract with direct stimulation, it was a nerve disorder whereas if it contracted equally with both nerve and muscle stimulation, it was a muscle disorder.

With this technique, Duchenne was the first to describe two similar but clinically different motor nerve disorders, poliomyelitis and progressive muscular atrophy (amyotrophic lateral sclerosis (ALS) and progressive bulbar palsy). Both conditions had electrical features of a motor nerve disorder but markedly different clinical profiles. Duchenne also noted that he could reliably identify hysterical paralysis since despite the apparent muscle weakness the electrical stimulation studies were normal.

Duchenne described a variety of primary muscle disorders although interestingly he was not the first to describe the muscle disease named after him, Duchenne muscular dystrophy (pseudohypertrophic muscular dystrophy of childhood). That distinction goes to the English physician, Edward Meryon [22]. The condition is characterized by childhood onset of enlarged calf muscles along with progressive muscle weakness often leading to death in the late teens. It is inherited as a sex-linked trait meaning it only affects boys in a family.

When challenged that he was not the first to describe the disorder, Duchenne fought back claiming that he was the first to recognize the disease and that the naming was a matter for the honor of France. Duchenne's initial paper on childhood muscular dystrophies published several years after Meryon's paper, included a variety of early onset muscle disorders some with features of pseudohypertrophic dystrophy but others without such features. One of the boys described as having pseudohypertrophic dystrophy had a sister with the same disorder. Duchenne was the first to describe the pathology of pseudohypertrophic dystrophy.

In these early years of the discipline of neurology, it was common to name diseases after the physician who first described them. Battles over priority for naming a disease were common. After all, having a disease named after oneself often meant academic honors and a guaranteed place in neurological history. In many instances, the name that eventually won out was not the first to describe the condition but the one who was best known or generated the most publicity. Sometimes the disease was named after a famous person with the disease. In America, amyotrophic lateral sclerosis (ALS) was called Lou Gehrig disease after the famous baseball player who died of the disease. This practice persisted well into the twentieth century before a consensus developed to stop naming diseases after people and to name diseases based on characteristic clinical features, for example, motor neuron disease and sex-linked pseudohypertrophic muscular dystrophy.

When Duchenne arrived in Paris with his provincial mannerisms, his peers found him argumentative and overly assertive, and he had difficulty making friends. He had the good fortune to meet and become friends with a young physician, Jean-Martin Charcot, who was completing an internship with the well-known French physician M. Rayer. Charcot would go on to be the Director of the medical service at the Salpêtriére, a public hospital for destitute women on the outskirts of Paris. Often considered the father of modern neurology, Charcot applied the revolutionary new approach of clinical–pathological correlation to study neurological diseases. Charcot and his students meticulously documented symptoms and signs of patients over decades and since most of his patients were poor and friendless, postmortem examinations were assured. Like Duchenne, he had several neurological diseases named after him.

Charcot would become a valuable ally to Duchenne's career in neurology. He was impressed with Duchenne's electrodiagnostic techniques and invited him to examine patients with neuromuscular disorders at the Salpêtriére. Charcot taught Duchenne to use a microscope to study neuromuscular histopathology and Duchenne developed a harpoon-like biopsy needle, called the Duchenne trocar, that allowed him to obtain muscle tissue samples without the need for anesthesia [23]. The combination of electrodiagnosis and muscle pathology were powerful tools for improved classification of neuromuscular diseases.

Duchenne and the Facial Muscles

After spending years electrically stimulating the individual muscles of the arms and legs, in the 1850s, Duchenne turned his attention to the facial muscles [24]. Unlike the muscles of the body and extremities, the muscles of the face played an important role in expression of emotions, providing a window to the soul. Using his induction coil stimulator, he produced "localized faradization" to the skin over individual nerves and muscles employing the new technology of photography to document the results. Duchenne observed that most natural movements were produced by groups of muscles rather than a single muscle. He made a series of photographs showing how the different muscles of the face were used to produce different facial expressions such as a smile or a frown. He was one of the first medical researchers to use photography to document his research.

One of Duchenne's most famous observations became known as a "*Duchenne smile,*" a smile that resulted from true happiness, involving the muscles around the eyes as well as the mouth. With a forced smile, such as a smile that one produces for a photograph or being introduced to a stranger, only the muscles about the mouth contract with no movement about the eyes. It was impossible to produce a Duchenne smile on command.

Duchenne had a difficult time finding normal subjects and patients willing to undergo electrical stimulation of the face, especially for the long periods of time needed to take photographs in the mid-nineteenth century. Although he paid a few actors and poor Parisians, he obtained most patients from the Salpêtriére and other indigent hospitals. His most photographed patient was a man with a facial palsy described by Duchenne as "*an old toothless man, with a thin face, whose features, without being absolutely ugly, approached ordinary triviality, and whose facial expression was in perfect agreement with his inoffensive character and his restricted intelligence*" [25]. A picture of Duchenne stimulating the facial muscles of the "old man" to produce a forced smile in his book *The Mechanism of Human Facial Expression* published in 1862 is a classic in the history of medical photography (Fig. 6.3).

Duchenne was intrigued with how people used facial expressions to communicate with body language. He went on to catalog which muscles were used with different expressions and suggested that one could guess what a person was thinking based on the pattern of their facial muscle contraction. A decade later, Charles Darwin referred to Duchenne's photographs to support his premise that facial expressions are universal among humans and animals in his famous book *The Expression of the Emotions in Man and Animals* [26].

Duchenne used electrical stimulation in patients with facial palsy both for prognosis and treatment. If stimulating the facial nerve at the point where it exits its bony canal and enters the face resulted in muscle contraction on that side of the face, the prognosis for recovery was good. If the nerve was not responsive to stimulation, electrical stimulation of the individual muscles on that side of the face prevented muscle atrophy until nerve function might recover. The most common cause of one-sided facial paralysis is Bell's palsy, a condition resulting from a viral inflammation of the sensory ganglia of the facial nerve. The facial nerve is damaged by compression of the swollen nerve within its bony canal. In most cases (>95%), the nerve recovers over several months and facial muscle function returns to normal. Neurologists today still use electrical stimulation for prognosis and treatment of patients with Bell's palsy.

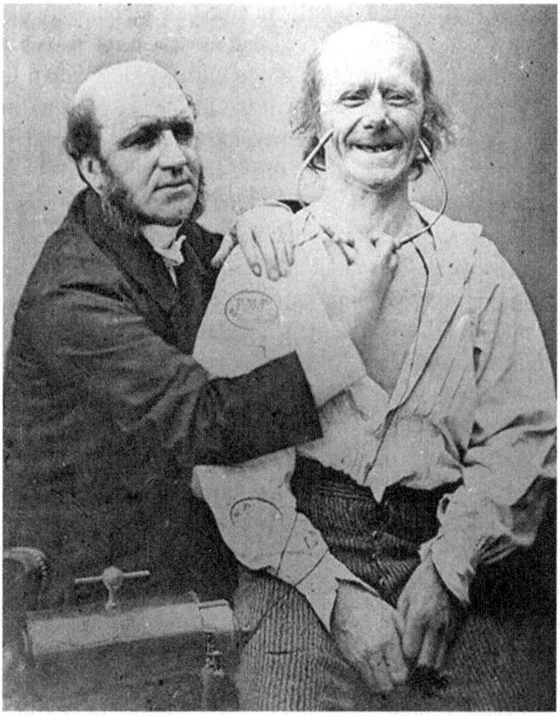

Fig. 6.3 Photograph of Duchenne stimulating the facial muscles to produce a "forced smile." See text for details. (From Duchenne de Boulogne [25])

Erb–Duchenne Brachial Plexus Palsy

Paralysis of an arm due to injury to the brachial plexus, known as Erb–Duchenne brachial plexus palsy, is a common obstetrical complication occurring in about 2 in every 1000 live births. Risk factors include large babies, prolonged labor, and a difficult instrument delivery. The brachial plexus is a network of nerves that innervate the shoulder, arm, and hand formed from the lower four cervical and the first thoracic nerve roots. The injury occurs when the nerves are stretched, pulled apart, or even torn away from the spinal cord. The paralysis can lead to a lifetime of disability. The history of how the hyphenated name came about provides insight into the naming process and the cause of the nerve injury [27].

Although descriptions of shoulder injuries at birth resulting in arm paralysis date back to the mid-eighteenth century, Duchenne provided the first detailed description of a case in 1868 suggesting that the paralysis resulted from nerve injury during delivery. He coined the term obstetrical palsy. A few years later, he described 4 cases in which he used electrical stimulation to confirm that the main muscles involved were the deltoid, infraspinatus, biceps, and brachialis consistent with an injury to the upper part of the brachial plexus. "*In this kind of paralysis of the upper limb from obstetrical manipulations, the arm falls motionless along the side of the body and is rotated inwards; the forearm remains extended but the movements of the hand are preserved*" [28].

Duchenne speculated that the injury was not due to the use of forceps but rather due to the physician placing his fingers under the armpits to pull the fetus from the birth canal. Although at the time the injury was considered untreatable, Duchenne recommended the use of electrical stimulation to determine the extent of the nerve damage and for treatment although it is unclear what he expected to achieve with electrotherapy other than maintaining muscle bulk should there be nerve recovery. He left it to others to explain how pressure in the armpit would lead to selective damage to the upper brachial plexus.

Wilhelm Heinrich Erb was a German physician born in 1840 in a small village in Bavaria. His father was a woodcutter in the Black Forest. He obtained his doctorate degree from Munich in 1864 and then assumed academic positions first in Heidelberg and then Leipzig. He is probably best known for his efforts to establish Neurology as a separate discipline from psychiatry. He fought for a separate rotation for neurology in the medical school curriculum.

Like Duchenne, Erb's research focused on neuromuscular disorders, and he had a neuromuscular disease named after him, Erb dystrophy, a progressive muscular dystrophy beginning in late childhood. He was the first to describe the knee jerk, the biceps femoris reflex. According to legend, Erb noticed that when a friend accidently hit himself below the patella with a heavy key, his knee jerked. Erb was a prolific writer with several books including one on electrotherapy and another on peripheral nerve diseases.

In 1877, he published a paper describing 4 patients with brachial plexus palsy involving the same four muscles described by Duchenne. He concluded that the injury could not have been produced by pressure in the armpit as proposed by Duchenne since the nerves to the affected muscles originated from the upper brachial plexus and would not be damaged by pressure in the armpit. Using electrical stimulation, he determined that the site of the nerve damage was in the low neck where the cervical roots 5 and 6 originated. The location in the neck 2–3 cm above the clavicle in front of the transverse process of the sixth cervical vertebra became known as Erb's point.

Erb concluded that the most likely cause of the obstetrical nerve injury was the physician grabbing the infant's neck and forcibly extracting the child from the birth canal, the so-called Prague maneuver. Electrical stimulation at Erb's point is still used for diagnosis of obstetrical brachial nerve palsy. Although electrotherapy is of little use for treatment, modern surgical techniques including nerve graphs, elbow and shoulder reconstructions, and muscle tendon transfers have improved the prognosis.

Charcot and Electrotherapy for Hysteria

As discussed in Chap. 2, electrotherapy was a popular treatment for hysteria beginning in the mid-eighteenth century, a time when hysteria was considered a unique female disorder due to dysfunction of the reproductive organs. But by the mid-nineteenth century, physicians in France concluded that hysteria was a brain disorder that could affect men and women. Duchenne and Charcot were strong advocates for the use of electrotherapy for this new version of hysteria.

In 1859, the French physician Paul Briquet described the clinical and epidemiological features of patients with hysteria seen at the La Charité hospital in Paris over a 10-year period [29]. Briquet concluded that hysteria was a "neurosis of the brain" that could affect children and men as well as women. He was bothered by the misogynistic connotations of the word hysteria, but he argued that hysteria had been generally accepted since the time of Hippocrates and that hopefully with time it would lose its pejorative implications.

Despite providing several fully documented cases of hysteria in men, Briquet emphasized that the disorder was much more common in women. On reviewing the hospital records, he found that 1000 women and only 50 men were diagnosed with hysteria; women were 20 times more likely to be diagnosed with hysteria than men. When he surveyed the women patients in the hospital at any given time, he found that about 25% had a diagnosis of hysteria. Of course, this marked predominance of women was biased by the still general conception that hysteria was a female disease.

Conflicting with the traditional view of hysteria as a disease of middle-aged women, Briquet found that 20% of patients with hysteria had onset before puberty and more than half had onset before the age of 20. This observation becomes even more important when one considers the overall poor prognosis in these patients; Briquet noted that "*one quarter never recover or have the illness their entire life... Some young girls who became hysteric before the age of 12 or 13 years are condemned to a lifetime of suffering, malaise, and sometimes serious illnesses. They are always sick, abort easily, or if they go to term, give birth to more hysterics*" [30].

Briquet reviewed the large variety of treatments that were commonly used for hysteria including bloodletting, cupping glasses or leeches placed on the vulva or the anus, exciting drugs, lotions, enemas, and surgical procedures including cauterization of the genitals and concluded that none were effective. The only treatment he found effective was electrical stimulation with faradization. He noted that he had referred several patients with hysteria to Duchenne for electrotherapy all with good outcomes.

Charcot was impressed with Briquet's findings and decided to study patients with hysteria at the Salpêtriére. He became convinced that hysteria was an organic disease rooted in the brain even though he was unable to find brain abnormalities on postmortem examinations. This was not that surprising since at the time he and others could not find a specific brain pathology for many neurologic diseases, particularly epilepsy. Presumably, there were diseases of the nervous system that left no material trace in the postmortem brain. Of course, his primitive microscopes limited the ability to identify subtle traces.

Charcot felt that the characteristic symptoms and signs of hysteria developed in someone with an underlying inherited susceptibility, "an inherent weakness" [31]. The idea of a familial disposition in patients with hysteria was not new but the suggestion of a genetic susceptibility was a new wrinkle. Charcot believed that the inherent weakness predisposed the patient to triggers such as physical or emotional trauma or even the presence of another person with hysteria. He had previously observed that when hysterical patients were housed on the same ward with epileptic patients, the hysterical patients often developed epileptic-like fits.

The Salpêtriére was a women's hospital and Charcot felt that hysteria was most common in women, but he did see men with identical symptoms. In women, he noted that ovarian tenderness was a common physical finding and applying pressure on the ovaries could trigger or modify their symptoms. In men, squeezing the testicles could have the same effect [32]. This observation seems a bit odd considering his premise that hysteria was a brain disease, but he considered some type of interaction between the genitals and the brain, either neural or chemical that could trigger hysteria.

Charcot moved from the field of clinical science to the theatrical with his weekly hysteric "*circus.*" Initially these were simply medical rounds attended by fellow physicians and students but as Charcot's fame and fortune spiraled upward, his rounds became well known throughout affluent Parisian society. Writers, artists, academics, businessmen, and royalty alike attended the rounds during which Charcot displayed lightly clad young women with hysteria performing bizarre contortions and behaviors.

Charcot paraded these young women before the lay audience like animals in a circus tent, apparently oblivious to any pain and anguish they may have suffered. Many patients returned on a regular basis for Charcot's rounds some becoming stars in Parisian society. Marie Wittman (Blanche), known as Charcot's pet hysteric and the queen of hysterics, was made famous in an 1887 painting by the French academic history painter, André Brouillet (Fig. 6.4). In this painting that captures Charcot presenting Blanche to his neurological colleagues, Blanche, with her breasts barely covered, faints into the outstretched arms of Charcot's assistant Joseph Babinski as she turns her head to the side with her face contorted.

In Brouillet's painting, there is a small table next to Charcot containing the electrical stimulation equipment he commonly used. It was generally assumed that the electrical stimulator was the one developed by Duchenne but the stimulator was in fact the medical induction coil developed by Emil du Bois-Reymond [33]. Recall that du Bois-Reymond's devise consisted of a primary coil and a secondary, movable induced coil that slid on rails so that the induced current could be modified by changing the position between the two coils (see Chap. 5). The devise powered by a Grenet battery (bichromate cell), transmitted the induced electrical current to the patient through electrodes with large flat bottoms covered by chamois leather. To improve conduction through the skin, the electrodes were moistened with salt water. The glass of water on the corner of the table, previously thought to be for the patient, was to keep the electrodes moistened.

In 1875, Charcot asked a local physician interested in electrotherapy, Marie Gabrie Romain Vigouroux, to set up an electrotherapy department at the Salpêtriére [34]. Vigouroux used the induction devise shown in the Brouillet painting primarily for faradization of patients with a diagnosis of hysteria. Like Briquet, Charcot became an enthusiast for the use of electrotherapy in treating hysteria. Charcot noted that the induction device could provide high voltage in large quantity.

Fig. 6.4 Photo of 1887 group tableau portrait painting by André Brouillet of Charcot presenting his famous patient Blanche Wittman to his neurological colleagues as she experiences a hysterical seizure

Vigouroux is best known for the discovery of what is now known as the electrodermal response. Human skin involuntarily changes resistance to applied electricity based on external stimuli. This would later lead to the development of the psychogalvanometer, a devise that measures changes in a person's skin resistance in different psychological states, a component of modern-day polygraph machines.

Like Duchenne, Charcot became interested in electrical stimulation of the face to study different facial expressions. This led to the bizarre discovery that if he delivered electrical stimulation to the face of a hysterical patient during the catalepsy phase of hypnosis, the expression became fixed and remained even though electrical stimulation was stopped. "*Once produced, the movement imprinted on the features of the face cannot be erased, even after removing the electrodes that had generated it. The facial expression remains immobile during catalepsy, as well as the accompanying attitude and gesture. The subject is thus transformed into a sort of an expressive statue, an immobile model representing with striking truth the most diverse expressions*" [35]. Likely, the contorted facial expression of Blanche Wittman was produced by electrical stimulation of her face and Brouillet took advantage of the "imprinting" to accurately portray the expression in his painting.

At the time, photographs required a prolonged exposure due to low quality of the lenses and film emulsions. Charcot pointed out that the prolonged facial expressions induced in hysterical patients were useful for obtaining good photographs of otherwise fleeting expressions. Blanche Wittman later appeared in a series of photographs showing persistent frontalis muscle contraction after faradization during a cataleptic episode.

Neurology and Electrotherapy in the Late Nineteenth-Century America

In nineteenth-century America, the new subspecialty of neurology was intertwined with the burgeoning field of electrotherapy. The brain was perceived as an "electric organ" that used electricity to generate thoughts and communicate with the body.

As in Britain and Europe, by the late nineteenth century, electrotherapy became an accepted form of treatment by the American medical establishment as evidenced by the fact that the 1890 United States census included a special survey on the uses of electrotherapy in the state of New York [36]. Five medical schools had courses in electrotherapeutics, fifty-one hospitals made 17,071 electrical treatments and sixty-five physicians treated 16,072 patients with electricity in that year. Electrical treatments by medical professionals were only the tip of the iceberg. Newspapers and popular magazines advertised electric belts, corsets, and medallions claiming miracle cures. Electric devices including medical batteries were sold in drug stores and mail-order catalogues from Sears-Roebuck.

Batteries were a fundamental part of the electrotherapy equipment, but wet cell technology of the early nineteenth century was fraught with difficulties. Wet cell batteries lost strength over time and either the electrolyte solution or the whole battery had to be replaced. Connections became loose and metallic parts rusted. Furthermore, wet cell batteries contained a caustic fluid that could be dangerous if spilled during a carriage ride over a rough road.

Manufacturers were in competition to develop dry cell batteries and one of the first to be successful was the Chloride of Silver Dry Cell Company of Baltimore Maryland. Their catalogue boasted, "*The cells being hermetically sealed there can be no leakage, and should one of these batteries be tipped over, an accident liable to happen to every instrument of the kind, it can remain in a reversed position for an indefinite period of time without being injured in the slightest*" [37]. They also claimed that each cell produced exactly the same amount of current so that physicians always knew how much current they were applying. Even though dry cells were more expensive than wet cells by the end of the nineteenth century, they dominated the medical equipment market.

The medical profession was adamant that cures did not simply result from the application of electricity but from the skill and knowledge of the physician performing the treatment. Furthermore, electrotherapists needed to be grounded in the fundamental principles governing the action of electricity. Pioneering neurologists argued, in addition, knowledge of the anatomy and physiology of nerves and the brain were important for successful application of electricity. Electrotherapy was considered the treatment of choice for a variety of neurological disorders including paralysis, neuralgia, and hysteria.

Tainted with a misogynistic history and a changing symptom complex, the diagnosis of hysteria began to lose its luster in the late nineteenth century. Many physicians felt uneasy making a diagnosis of hysteria in their male patients. Diagnoses such as spleen and hypochondriasis had been used for men with hysterical symptoms, but these terms were rooted in archaic concepts that were no longer relevant. In Germany, the term "nervosity" (nervosität) and in England and America the terms "nerves" and "neurasthenia" became popular accounting for a wide range of physical and mental symptoms seen in men and women felt to be of "nervous" origin.

The term neurasthenia, a general physical and mental exhaustion, had been used in England since the mid-nineteenth century but it was redefined in 1869, by an American neurologist and electrotherapist, George Beard [38]. Symptoms ranged from fatigue, headache, body pain, and insomnia to numbness and paralysis. He concluded that the symptoms were due to lack of "nervous force" resulting from subtle changes in the brain. Beard felt that electricity could restore the nervous force. Neurasthenia rapidly took hold in America and Europe where it became a fashionable diagnosis of the rapidly expanding affluent middle classes.

Beard, the son of a Connecticut preacher, graduated from Yale College and entered Yale Medical School in 1862. In his first year of medical school, he developed typical symptoms of what he would later call neurasthenia. He was frustrated when doctors could not make a diagnosis, so he treated himself with electrotherapy and the symptoms gradually resolved. After serving as an assistant surgeon in the Navy during the Civil War, he returned to medical school at the College of Physicians and Surgeons in New York graduating in 1866. He then began a practice in New York with A.D. Rockwell specializing in neurology and electrotherapy.

In 1872, Beard and Rockwell published what was considered the definitive textbook on electrotherapy in America, *The Medical and Surgical Uses of Electricity*. Like most textbooks on the subject, they began with a section on the physics and technology of electricity. They also emphasized the importance of understanding electrophysiology, "*Any old country granny, the stupidest of nurses, an infant even, can hold two sponges on the surface of the body, and let the current run. Those who aim no higher than this – the indiscriminate holding of electrodes on patients – need give no thought to electro-physiology*" [39].

Beard saw neurasthenia from a unique American perspective. He felt that the American drive to be successful and the pace of modern life provoked nervous disorders. Beard noted that scientific discoveries such as the telegraph and steam engine increased the pace of daily life and this along with the constant stride to get ahead overtaxed the nervous system leading to a breakdown. Furthermore, Beard pointed out that neurasthenia was a disease of high achievers, those who were most exposed to the stresses of modern living. Thus, neurasthenia became a status diagnosis often affecting the rich and famous.

Despite his strong belief in electrotherapy, Beard made a chance observation late in his career that forced him to consider psychogenic factors in the cause and treatment of neurasthenia. He was about to treat a patient with electrotherapy when he noticed that his battery was dead [40]. On a whim, Beard decided to go ahead with the treatment and to his surprise the patient had a good response. Could a patient's expectation influence the therapeutic outcome? It seemed so. Beard even went so far as to test the effect of telling patients that a given symptom would disappear at a certain time and day and dramatically the symptom disappeared. Beard presented his observations at a meeting of the newly formed American Neurological Association, but the other twelve members were not impressed, some even criticizing his experiment as immoral.

One of those present at the meeting, Silas Weir Mitchell (he preferred Weir), shared Beard's interest in electrotherapy and neurasthenia. Elected the first president of the Association, Mitchel was a prolific writer who in addition to more than a hundred scientific articles and books published more than fifty literary titles including novels, essays, and poems. In his later years, he published at least one novel a year and at the turn of the twentieth century, he was the leading-selling American author of fiction.

Mitchell was born in 1829, the son of a prominent Philadelphia doctor [41]. He graduated from the Jefferson Medical College at age 21 in 1850 and traveled through Europe for a year spending some time in Paris with the famous physiologist, Claude Bernard. With the onset of the American Civil War, Mitchell accepted a position as a contract surgeon for the Union Army. Although he arranged a commission for his brother, Mitchell did not join the Union Army himself. His father became ill and had to give up his practice in 1855. After his father's death and Mitchell's marriage in 1859, Mitchell was the sole provider for an extended family, and he chose to pay someone else to take his place in the army. This decision haunted Mitchell throughout his life and he often wondered what it would have been like to have fought in a military battle.

Mitchell first worked at the old armory building on 16th street in Philadelphia that was serving as a Union Hospital. It was there that Mitchell first became interested in nerve disorders and he convinced Surgeon General Hammond to devote a large ward solely to the treatment of such disorders [42]. As the number of wounded soldiers exponentially increased, Mitchell was moved to a much larger hospital at Turner's Lane, called the "Stump Hospital" by the soldiers. Here, Mitchell along with colleagues George Reed Morehouse and William Williams Keen, focused on nerve injuries due to gun shots.

Mitchell, who coined the term phantom limb, noted that 95% of men experienced some feeling of the limb still being present after amputation. Most reported that the phantom limb felt shorter than the real limb and most complained of pain in the phantom limb. Mitchell along with Morehouse and Keen published a classic textbook on pain and wartime injuries in 1864 [43]. The book provided detailed descriptions of the wide range of painful conditions associated with nerve injuries. Like most of his colleagues, he recommended electrotherapy for treating paralysis and chronic pain associated with nerve injury, called neuralgia. The evidence that electrotherapy could help paralysis was meager but there really were not any other treatment options and it did provide some hope to the devastated soldiers.

On the other hand, there was evidence that electrotherapy was helpful for treating neuralgia. The most dramatic pain syndrome described in the book would later become known as causalgia, combining the Greek words for heat and pain. The soldiers described the pain as "burning," "mustard red hot," or like a "red hot file rasping the skin." It typically affected the palms of the hands or the back of the feet. Mitchell and colleagues noted that the pain made the sufferer anxious and irritable and markedly interfered with sleep. Some of the soldiers with causalgia responded to electrotherapy while others were resistant to all therapies including morphine.

After the war, Mitchell returned to a lucrative neurological private practice in Philadelphia that was largely funded by affluent women suffering from hysteria and neurasthenia. Based on his experience with soldiers during the Civil War and patients in his private practice, he developed the so-called "rest cure" that received worldwide notoriety. In a conversation with his close friend and fellow physician William Osler, Mitchell suggested that a woman patient provided the initial inspiration for the rest cure [44].

The highly intelligent woman completed her degree at a prominent Boston college in three years and then married and had four children in rapid sequence. This was followed by a complete mental breakdown with extreme fatigue and a diagnosis of nervous exhaustion or neurasthenia. Doctors recommended exercise but she refused to get out of bed. When Mitchell saw her, she was pale and thin and was eating relatively little. He decided to let her stay in bed, increase her food intake, and in the place of exercise, receive regular sessions of electrical stimulation of muscles and massages, so-called "passive exercise." Remarkably she had a complete recovery returning to her wifely duties including having several more children.

In his best-selling book on nervous disorders, *Wear and Tear, or Hints for the overworked,* Mitchell summarized his misogynistic views on women's mental health. Young women whose nervous systems were overtaxed with misguided mental pursuits were destined to "*the shawl and the sofa, neuralgia, weak backs, and the varied forms of hysteria, that domestic demon which has produced untold discomfort in many households…only the doctor knows what one of these self-made invalids can do to make a household wretched*" [45]. Mitchell believed that both men and women were susceptible to hysteria and neurasthenia, but women were particularly susceptible because of their weak constitution.

Mitchell outlined his rest cure for nervous disorders in another medical advice book *Fat and Blood: An Essay on the Treatment of Current Forms of Neurasthenia and Hysteria* [46]. Mitchell noted that men with neurasthenia did not do so well with the rest cure so for them, which included Walt Whitman and Teddy Roosevelt, he recommended the "West cure," sending them to visit the West to engage in vigorous physical activity and to write about their experiences.

Although there were exceptions particularly among educated women, many saw Mitchell's rest cure (with electrotherapy) as a compassionate, effective treatment for hysteria and neurasthenia. At the time, there were not a lot of other treatments available. Regardless of one's opinion of the effectiveness of Mitchell's rest cure, it did make him a wealthy man. The demand for the treatment was so great that Mitchell developed a series of satellite treatment centers in and around Philadelphia run by women nurses he personally trained. At its peak, Mitchell's medical practice was one of the largest in the United States with a yearly income of more than $100,000 or $2,500,000 in today's currency [47].

Electrotherapy and Suggestion

In the late nineteenth century, electrotherapy practices and clinics flourished throughout Central and Western Europe. Spa and seaside resorts of any repute advertised electrotherapy facilities. Both lay and medical publications lauded the medical benefits of electrotherapy.

With the advances in understanding of animal electricity, European neurologists saw the nervous system as a sophisticated electrical circuit. A sick brain was deficient of electricity and electrotherapy could remedy the problem. According to Duchenne, "*No sincere neurologist could practice without the use of electrotherapy*" [48]. In Germany, Wilhelm Erb's *Handbook of Electrotherapy* was the most widely read manual for therapy. In 1882, Erb wrote, "*It is no exaggeration to say that the majority at least of German neuropathologists began in electrotherapy. Full mastery in neuropathology is impossible without a full understanding of the electrodiagnostic and electrotherapeutic methods*" [49]. At the time in Germany, the term "neuropathologist" was used for a practitioner of the newly developing subspecialty of neurology.

Into this exuberant atmosphere stepped Paul Julius Möbius, a practicing electrotherapist and neurologist who gradually became skeptical of the hype associated with electrotherapy [50]. Möbius was born in Leipzig in 1853, the grandson of the famous German mathematician and astronomer, August Ferdinand Möbius. After obtaining his medical degree from the University of Leipzig in 1876, Möbius opened a private practice with the doorplate reading "Doctor for the nervously ill and electrotherapist." He also served as an assistant to Wilhelm Erb and Adolph Strümp at the Leipzig University Medical Clinic. Möbius specialized in using electrotherapy for treating neurasthenics and for pain relief in patients with neuralgias.

Möbius lacked the credentials for an academic career, but he became well known and respected for a series of six extended reviews of the German medical literature on electrotherapy and diagnostics published in the journal *Schmidt's Jahrbücher* between 1882 and 1893. With these reviews he was able to enter the scientific debate on electrotherapy despite his lack of academic credentials and according to medical historian Holger Steinberg, he became one of the most respected authorities in medical criticism. His reviews made him adored by some and feared by others.

By 1891, Möbius was convinced that in four out of five cases, the positive results of electrotherapy were suggestive rather than physical in nature. He wrote, "*So far nothing has proved that electricity has a curing effect in paralysis. On the contrary, paralysis following the destruction of central nervous cells cannot heal and paralysis following destruction of peripheral nerves or muscle fibres heals, if curable, by itself…There is no doubt that electricity is effective with certain kinds of pain, paraesthesia, motor irritations or certain organic irregularities (e.g. vasomotor disturbances, underactivity of the bowels, menstruation abnormalities, insomnia). The same disturbances are subject to suggestive influence. Hence the effects of electricity may be suggestive in nature. The assumption is supported by a variety of reasons, primarily by the irregularity of the curative effects. The latter could easily be explained if one were to accept that the effects were mediated psychologically but would remain inexplicable on the assumption that there was a physical relation*" [51].

In reaction, German supporters of electrotherapy organized a conference in September 1891 with the goal of responding to the "inflaming subversive ideas" of Möbius. The Frankfurt Council, as Möbius later called it, included 35 electrotherapists meeting in Frankfurt on Main to present and publish papers defending electrotherapy. In the opening presentation, co-organizer Leopold Laquer stated that he considered the curative effects of electrotherapy for a wide variety of neurological diseases to be an "empirical fact." Although he admitted that some "nervously ill patients" might be susceptible to suggestion, he argued that most strongly resisted electrotherapy and even feared application of electric current. Therefore, it was unlikely that they would expect a cure from electrotherapy.

Although the organizing committee selected speakers who were mostly enthusiastic about electrotherapy, some presented data that was at least partially supportive of Möbius' view. Professor Ottomar Rosenbach from Breslau, Germany (now Wroclaw, Poland) summarized his extensive experience treating organic brain diseases with electrotherapy at the All-Saints Hospital. He noted that many acute brain diseases spontaneously heal with or without treatment while others particularly chronic diseases affecting the spinal cord were resistant to all forms of treatment and he doubted that electrotherapy was helpful in either case. Similarly, Ludwig Bruns had "serious doubts" whether traumatic or neuritic paralysis could be cured with electrotherapy.

Regarding the "suggestive" nature of electrotherapy, Ewald Hecker who operated a neuropsychiatric spa in Johannisberg (Rhine District) and then Wiesbaden, noted that concerns regarding the therapeutic benefits of electricity were even more justified for functional diseases of the nervous system. He felt that the large number of symptoms that occurred with functional diseases were due to autosuggestion and could be cured by reassurance from a caring physician without the need for more aggressive intervention. Electrotherapy was simply acting as a "carrier of suggestion" or as German neuropsychiatrist, Emil Kraepelin, later called "hypnotizing patients through galvanization of their head."

Möbius agreed with Hecker but disagreed with Laquer's use of the term suggestion. Möbius considered suggestion an unconscious mental process, that could cause or eliminate symptoms both with organic and functional diseases. Suggestibility was a basic human attribute based on beliefs. Hypnosis was a rare specific case of it. Mobius felt that it was impossible to exclude suggestion as part of the efficacy of any type of medical treatment. In many ways, Möbius was anticipating Kraepelin and Freud and a more modern biopsychosocial model of medicine.

At the time, it was common practice for physicians to base their evidence for therapeutic efficacy on individual case reports. In fact, most of the evidence for the effectiveness of electrotherapy presented at the Frankfurt conference was in the form of testimonials and individual case studies. Such evidence can be very misleading. What was needed were large comparative studies of patients with specific diseases and matched controls or controlled studies in animals. Unfortunately, such studies were not done with electrotherapy.

Neurologists and Electricians

In Britain in the late nineteenth century, a tenuous relationship existed between the electrical and medical professions. Each profession had its own opinions and goals related to the use of electricity for treating patients and each had suspicions about the other's opinions and goals.

As noted earlier in the chapter, Julius Althaus emphasized that physicians needed to work with skilled electricians in his popular textbook *Treatise in Medical Electricity*. In 1886, a colleague of Althaus, Herbert Tibbits founded the West-end School of Massage and Electricity in conjunction with the West-end Hospital for Diseases of the Nervous System, Paralysis and Epilepsy [52]. Like Mitchell in America, Tibbits believed in the joint application of electricity and massage for his patients. With Tibbits' system, the masseuses themselves were electrified so that they could direct the electricity to a particular part of the body during the massage. The goal of the school was to train nurses to apply electrical massages under medical supervision.

Like Althaus, Tibbits preached the importance of proper training in electricity and the use of electrical equipment. He developed his own battery manufactured by Mr. Hawksley of 300 Oxford Street and recommended the use of telegraph wire to connect the battery to the electrodes. He strongly recommended that the electricity be applied by a physician or a skilled operator under the supervision of a physician. "*The experience of the National Hospital for the Paralyzed and Epileptic, and my own experience in private practice, show conclusively that when patients apply electricity, the result has been most frequently unsatisfactory*" [53].

Tibbits' reputation solidified and he found the electrical treatment business to be very lucrative. But problems arose when he spent a large sum of money on a fund-raising event for his hospital that was a complete failure [54]. The hospital management committee refused to reimburse him and soon after he declared bankruptcy. This occurred shortly after the Jeffery versus Harness Medical Battery Company trial and Harness was looking for a chance to rehabilitate his reputation. Harness approached Tibbits and asked him to conduct a series of tests on his electropathic belt to prove its effectiveness for a fee. Tibbits needed the money, so he agreed.

In a series of experiments that Tibbits described in a published pamphlet, he claimed that when the electropathic belt was worn as directed, it did produce electricity and that the electricity entered and circulated in the body. A mirror galvanometer showed that the belt generated a small amount of electricity when activated by slight moistening or by perspiration. Tibbets then inserted a platinum needle through the skin insulated to within one-sixteenth of an inch of the tip and connected one end of the galvanometer to the belt and the other to the needle. Again, the galvanometer indicated passage of a small amount of current. In his pamphlet, Tibbits wrote: "*these experiments prove conclusively that a current of electricity penetrated the skin and influenced the subjacent tissue*" [55]. Harness advertised that Tibbits' experiments vindicated his Company.

The editors of the *Electrical Review* were not impressed and responded with a scathing editorial. They noted that Tibbits exhibited "*a most incredible ignorance of electrical laws*" and that he was unfit to speak as an authority on electricity. They even suggested that he should be censored by the British Medical Association. Tibbets was irate and sued the *Electrical review* for libel.

In the jury trial that took place before the Queen's Bench Division of the High Court in February 1893, the reputations of both Tibbets and Harness were on the line. Tibbits lawyer, Sir Richard Webster, argued that Tibbits experiments were carried out on Harness' belts to criticize material that they found to be factually incorrect. They called as witnesses many of the leading electrical scientists of the era.

There was general agreement that the current that Tibbets recorded with the transcutaneous needle was due to a reaction between the belt and the platinum needle and not from electricity penetrating the body. A professor of Experimental Physics from University College Bristol, Silvanus P. Thompson, said: "*This dear old doctor does not know what he is talking about*" [56]. Others questioned Tibbets' electrical qualifications to be conducting such experiments. Probably the most respected physicist of the time, William Thomson (Lord Kelvin), testified that Harness' electropathic belt could produce "*no more than an infinitesimal quantity of electricity*" [56]. The jury took only five minutes to rule in favor of the *Electrical Review.* The downfall of Tibbits and Harness was complete.

Electricians and electrical engineers saw the financial possibilities of electrotherapy and felt that they should get a piece of the action [52]. In 1888, the Institute of Medical Electricity was organized as a joint-stock company by a consortium of electricians and electrical engineers. The managing director was Henry Newman Lawrence, a prominent member of the Institute of Electrical Engineers, and the resident physician was Arthur Harries. The shareholders included aristocrats and professors of physics and physiology.

An example of the Institute's advanced technology was the "Electrified Room System" consisting of a large room with an insulated metal floor and metal ceiling, each separately connected to the outputs of a powerful induction coil. This produced an electric field between the ceiling and floor so that anyone in the room was exposed to the pulsed electrical current as well as a static charge. When standing, sitting, or lying in this electrified room, the patient received the benefits of electricity without experiencing an electric shock. As part of the treatment, the patient could undergo a massage from a grounded masseuse so that the electricity would flow to the earth through the hands of the masseuse. The cost to the patient for a half-hour electrical massage session was 10 shillings.

The Institute's listed aims were to: (1) provide electrical treatments under the direction of qualified medical men at moderate fees, (2) encourage scientific investigation in the use of electricity in medicine, and (3) provide medical men with reliable electrical apparatus and instruments for diagnosis and treatment of patients [57].

These seem laudatory aims shared by their medical colleagues but when the Institute's board of directors attempted to appoint two members of the Royal College of Physicians as consultants, the College refused to grant permission. The College felt that it was unprofessional for any of its members to be associated with a commercial company whose goal was to treat medical diseases. Not without a modicum of hypocrisy, they concluded that administering electricity to the body for commercial gain was considered inappropriate for a medical gentleman.

To many electricians and electrical engineers, medical electrotherapists were naïve, simply applying an electrical current and observing its effect. Terms such as galvanization and faradization were frowned upon as nonspecific and nonquantitative. Rather, it was important to control the strength, direction, and duration of any electrical current applied to the body. With the general acceptance of Ohm's law and the development of quantitative electrical equipment, it was no longer acceptable to apply electricity to the body without precise knowledge of the quantity.

Electrical engineers believed that the knowledge they acquired running telegraph offices was applicable to operating electrotherapy clinics. Up-to-date telegraph equipment including batteries, galvanometers, induction coils, and rheostats were all applicable to medical electrotherapy. The importance of precise measurements was central to the culture of electrical engineers of the time. If electrotherapy was to be scientifically acceptable, it had to become a quantitative discipline.

In the field of telegraphy, it was critical to know the exact characteristics of the different components of the telegraphic network for precise operation. Understanding the resistance of the conducting wires proved particularly important. Ohm's law of resistance in conducting wires was critical for pinpointing damage in telegraph lines. Many engineers and some electrotherapists compared the human body to a telegraph line with the skin as the outer insulation and the central fluid with electrolytes as a mass conductor. In his *Handbook of electro-therapeutics* translated into English in 1883, Wilhelm Erb stated, "*the human body is nothing more than a large conducting mass of definite resistance; and the laws controlling the distribution of large conducting masses therefor apply to it without limitation*" [58]. He went on to emphasize the importance of understanding Ohm's laws to the practicing electrotherapist.

But measuring the resistance of the human body was no easy task. The resistance was unexpectedly high and varied greatly with the dryness of the skin, the temperature of the body and the type of current (direct vs. alternating). There was marked variability in the strength of electrical current that the human body could tolerate without danger or even death. There was also debate regarding the best instruments to measure the resistance of the human body. The well-known electrical engineer Latimer Clark suggested that the human body's resistance acted like a simple galvanometer and resistance coil. He compared the electrical conductivity of the human body to the duplexed Atlantic submarine telegraph cable, a major achievement of late nineteenth century electrical theory.

By the end of the nineteenth century, most large metropolitan hospitals in Britain had an electrical department with a doctor in charge and a hired electrician handling the day-to-day activities. The electricians kept the equipment in working order and were in charge when the medical director was away. They were notified about the patients requiring electrical treatments and were responsible for supplying the necessary batteries and induction coils used in the treatments. Both inpatients and outpatients were treated although the largest number were outpatients. For example, St Bartholomew's Hospital had one of the largest electrical departments and by far the most common diagnosis in outpatients receiving electrotherapy was infantile paralysis.

The physicians in charge of these large electrical departments considered themselves at the cutting edge of medical science and prided themselves in understanding the latest electrical physics. One of these "new age" electrotherapists, a former army surgeon and director of the electrical department at the London Hospital, William Snowden Hedley boasted that the advanced treatments offered at his hospital were not like those "*of the comparatively recent past, when an unmeasured current was 'dabbed' on to the body at the end of a couple of sponges, and it was orthodox practice to administer faradic torture by means of a couple of metal cylinders grasped by the patient*" [59]. As a sign of the acceptance and prestige of these hospital electrical departments at the turn of the twentieth century, the British Medical Association formed a subsection for medical electricity and the British Electrotherapeutics Society was established.

Electrical Deaths

As we will see in the coming chapters as electricity became more widely used by the public in the late nineteenth century, the risks of injury and death from electrocution became a great concern. Naturally, the new bread of hospital electrotherapists was called upon for their expertise in treating such patients. Articles on the mechanism of tissue damage and cause of death from electric shock were common in medical journals in the 1890s. In his short article *The Pathology and Treatment of Electrical Accidents* published in the *Lancet* in 1894, William Hedley summarized the medical understanding of electrocution at the time [60].

Hedley suggested that death from electrocution resulted from either tissue destruction due to the heat generated by the electricity or from arrest of respiration, asphyxia. If the tissue was heated by the passage of electric current, more than 45 °C irreparable tissue damage occurred and nothing could be done. In the case of asphyxia, the patient could sometimes be saved with artificial respiration.

Hedley went on to provide an anecdotal report of a man who was successfully resuscitated after 750 mA of alternating current at 4500 V was accidentally short-circuited through his body. The current entered the body through his hand and exited from the buttock. The Laborde procedure (grabbing the tongue between the index finger and thumb and rhythmically pulling and relaxing at a rate of 15–20 times per minute) was performed about a half hour after he was presumed dead. The man began to breathe and within an hour he was conversing. The only residual was the burn marks where the current entered and exited. How did the electricity lead to asphyxia? Hedley suggested a range of possibilities including direct or indirect effects on the lungs, heart, and nervous system. Stimulation of afferent nerves such as the vagus nerve could inhibit brainstem respiratory centers or have direct effects on the lung and heart muscle. He noted that induction currents "*applied to the heart of a dog produced those irregular, arhythmic, fibrillary contractions spoken of as 'delirium cordis*" [60].

The tissue damage from electrocution was obviously related to the magnitude and duration of the current flow but the current pathway was also critical in determining the type and degree of injury. Current passing through the head or thorax commonly led to a fatal outcome. Current through the thorax could produce cardiac and/or respiratory arrest whereas current through the brain could cause seizures and paralysis in addition to respiratory arrest. Hedley dramatically summarized the effect of electricity on the brain in an article entitled *The Painlessness of Electric Death* published in the *Electrical Review* in 1897: "*Entering the skull, a current diffuses itself through the brain, sweeping away the faculties of sensation, intellect, and will; concentrating itself, it makes for the foramen magnum, smiting in its path the vast collection of conducting fibres and nerve centers – the medulla oblongata. The delicate mechanism by which vital functions (respiratory, vaso motor, cardio-inhibigtory &c.) are governed is thus destroyed, and it is on the balanced relationship of such functions that life depends*" [61].

Cure for Aging

We end this chapter with a brief return to John Althaus, one of the founding fathers of neurology in Britain, at the turn of the century when he published an article in the *Lancet* describing his experience with electrotherapy for reversing the effects of aging on the brain. This article epitomizes the grandiose expectations for electrotherapy entering the twentieth century.

In his article entitled, *Old age and rejuvenescence,* Althaus recognized two types of aging, premature and normal [62]. With premature aging, the brain showed degenerative changes due to a variety of diseases whereas with normal aging, the brain remained grossly normal with the potential of survival to about 100 years. He argued that galvanic electrical therapy was useful for both types of aging but had the greatest potential for maintaining longevity, a real fountain of youth.

Regarding premature aging, Althaus noted that patients showed improved body strength and mental energy after galvanic stimulation even though the underlying disease process often continued unchecked. He admitted that there was controversy among neurologists regarding the true benefit of electrotherapy for degenerative brain diseases but emphasized that based on his many years of experience electrotherapy represented the only hope for such patients. Constant current galvanism improved the nutrition of neurons and over time could reverse the neurodegeneration. He dismissed newer electrotherapy techniques using alternating current as being tedious and inconvenient.

Althaus saved his greatest praise for electrotherapy in normal aging. "*The old man takes fresh interest in the affairs of daily life, he resumes his work with some amount of vigour, he has a more erect attitude, he walks and stands better, and he has a quicker digestion and a healthier sleep. His general aspect is so much changed that he looks five or ten years younger than before the treatment was commenced. His peevish and querulous temper is no longer a source of trouble to his friends*" [62]. Althaus claimed that the results of electrotherapy were very good in about 40%, fair in 30%, and insignificant in about 30% but never completely useless. As was common in medical reports at the time, Althaus went on to provide testimonials from a few grateful patients. After several treatments, a veteran Shakespearian actor noted "*I no longer crawl but walk*" and an aged composer boasted "*I no longer walk but fly; and young fellows have now difficulty in keeping up with me*" [62].

The electrotherapy should be started in the 60s or "*at any time when age has begun decidedly to tell upon them*" [62], in daily or every other day sessions with regular repeat sessions to maintain the benefit. Althaus warned that the therapist's experience and technique were critical for successful treatment. "*It is notorious that electricity is frequently applied in the most careless and haphazard manner, and this explains why so manty persons are sceptical concerning the curative effects* [62]. *An application of constant current to the brain during which not the slightest fault has been committed is in a certain sense an artistic performance which requires not only knowledge but also much practice and some talent like a good musical performance*" [62]. Of course the same could be said for most medical therapies—the art (or placebo effect) of medicine.

In Brief

Electricity played a prominent role in the development of medical neurology in the nineteenth century. Nearly all the pioneer neurologists in Europe and America were electrotherapists. Since electricity stimulated nerves, nerve cells, and muscle cells, it seemed logical that electrotherapy was most useful for treating neurological diseases. Hospitals set up special areas to house the electronic equipment along with electricians to maintain the equipment. In France, Duchenne demonstrated how electricity could distinguish between nerve and muscle diseases. Neurologists treated a wide range of neurological conditions with electrotherapy from paralysis to dystonia and epilepsy. The best results were obtained with the common diagnoses of hysteria and neurasthenia. These disorders were greatly dependent on suggestion and electrotherapy was a powerful placebo. Electricians and neurologists developed a tenuous relationship as each tried to establish their role in the potentially lucrative field of electrotherapy.

References

1. Finger S. Minds behind the brain. A history of the pioneers and their discoveries. New York: Oxford University Press; 2000. p. 85–99.
2. Morus IR. Marketing the machine: the construction of electrotherapeutics as viable medicine in early Victorian England. Medical Hist. 1992;36:34–52.
3. Bird G. Report on the value of electricity as a remedial agent in the treatment of diseases. Guy's Hosp Reps. 1841;6:81–120.
4. Addison T. On the influence of electricity, as a remedy in certain convulsive and spasmodic diseases. Guy's Hosp Reps. 1837;2:493–4.
5. Bird G. Observations on induced electric currents, with a description of a magnetic contact-breaker. Philos Mag, 3rd Series. 1838;12:18–22.
6. Morus IR. Marketing the machine: the construction of electrotherapeutics as viable medicine in early Victorian England. Medical Hist. 1992;36:43.
7. Morus IR. Marketing the machine: the construction of electrotherapeutics as viable medicine in early Victorian England. Medical Hist. 1992;36:47.
8. Editorial. The march of specialism. Lancet. 1863;i:183.
9. Lobb H. Special hospitals – The London Galvanic Hospital. Lancet. 1863;i:219.
10. Morus IR. Shocking bodies. Life death and electricity in Victorian England. Stroud: The History Press; 2011. p. 102–10.
11. Morus IR. Shocking bodies. Life death and electricity in Victorian England. Stroud: The History Press; 2011. p. 105.
12. Morus IR. Shocking bodies. Life death and electricity in Victorian England. Stroud: The History Press; 2011. p. 106.
13. Coley NG. The collateral sciences in the work of Golding Bird (1814–1854). Med Hist. 1969;13:369–70.
14. Morus IR. Shocking bodies. Life death and electricity in Victorian England. Stroud: The History Press; 2011. p. 108–9.
15. Morus IR. Shocking bodies. Life death and electricity in Victorian England. Stroud: The History Press; 2011. p. 137–44.
16. Morus IR. Shocking bodies. Life death and electricity in Victorian England. Stroud: The History Press; 2011. p. 172–82.
17. Morus IR. Shocking bodies. Life death and electricity in Victorian England. Stroud: The History Press; 2011. p. 174.
18. Bladin PF. Julius Althaus (1833–1900). Neurologist and cultural polymath; founder of Maida Vale Hospital. J Clin Neurosci. 2008;15:495–501.
19. Althaus J. A treatise on medical electricity, theoretical and practical; and its use in the treatment of paralysis, neuralgia, and other diseases. London; 1859.
20. Goldblatt D. A gentleman of Bologna: Duchenne, Duchesne, etc. Semin Neurol. 1988;8:115–6.
21. Campbell ER. The Achievement of Duchenne. President's address. Proc R Soc Med. 1972;66:18–22.
22. Bach JR. The Duchenne de Boulogne – Meryon controversy and pseudohypertrophic muscular dystrophy. J Hist Med Allied Sci. 2000;55:158–78.
23. Nelson KR, Genain C. Duchenne de Boulogne and the muscle biopsy. J Child Neurol. 1989;4:315.

24. Duchenne de Boulogne GB. The mechanism of human facial expression. edited and translated by Cuthbertson RA. Cambridge: Cambridge University Press; 1990.
25. Duchenne de Boulogne GB. The mechanism of human facial expression. edited and translated by Cuthbertson RA. Cambridge: Cambridge University Press; 1990. p. 42.
26. Darwin C, Ekman P, Prodger P. The expression of the emotions in man and animals. 3rd ed. London: HarperCollins; 1998.
27. Schmitt C, Mehiman CT, Meiss AL. Hyphenated history: Erb-Duchenne Brachial Plexus Palsy. Am J Orthop. 2008;37:356–8.
28. Schmitt C, Mehiman CT, Meiss AL. Hyphenated history: Erb-Duchenne Brachial Plexus Palsy. Am J Orthop. 2008;37:357.
29. Briquet P. Traité de l'Hystérie. Paris: JB Bailiére et Fils; 1859.
30. Briquet P. Traité de l'Hystérie. Paris: JB Bailiére et Fils; 1859. p. 584.
31. Shorter E. From paralysis to fatigue: A history of psychosomatic illness in the modern era. New York: Free Press; 1992. p. 166–200.
32. Scull A. Hysteria: the disturbing history. New York: Oxford University Press; 2009. p. 111.
33. Brigo F, Balasse A, Nardone R, Walusinski O. Jean-Martin Charcot's medical instruments: electrotherapeutic devices in La Leçon Clinique à la Salpêtrière. J Hist Neurosci. 2021;30:94–101.
34. Walusinski O. The cruelty and failings of therapies for neurological diseases in French literature. Prog Brain Res. 2013;206:171–99.
35. Brigo F, Balasse A, Nardone R, Walusinski O. Jean-Martin Charcot's medical instruments: electrotherapeutic devices in La Leçon Clinique à la Salpêtrière. J Hist Neurosci. 2021;30:99–100.
36. Rosner L. The professional context of electrotherapeutics. J Hist Med Allied Sci. 1988;43:64–82.
37. Rosner L. The professional context of electrotherapeutics. J Hist Med Allied Sci. 1988;43:77.
38. Ellerbroek WC. Dr. George M. Beard pioneer in psychosomatic medicine. Psychosomatics. 1972;13:57–60.
39. Beard GM, Rockwell AD. A practical treatise on the medical and surgical uses of electricity. 2nd ed. New York: William Wood and Co.; 1878. p. 99.
40. Beard GM. The influence of mind in the causation and cure of disease – the potency of definite expectations. J Nerv Mental Dis. 1876;3:429–34.
41. Earnest ES. A. Weir Mitchell, Novelist, and Physician. Philadelphia: University of Pennsylvania Press; 1950.
42. Mitchell SW. Some personal recollections of the civil war. Philadelphia: Transactions of the College of Physicians of Philadelphia; 1905.
43. Mitchell SW, Morehouse GR, Keen WW. Gunshot wounds and other injuries of nerves. Philadelphia: Lippincott and Co.; 1864.
44. Earnest ES. A. Weir Mitchell, Novelist, and Physician. Philadelphia: University of Pennsylvania Press; 1950. p. 81.
45. Mitchell SW. Wear and tear for the over worked. 5th ed. Philadelphia: Lippincott; 1891. p. 32.
46. Mitchell SW. Fat and blood: an essay on the treatments of certain forms of neurasthenia and hysteria. Philadelphia: Lippincott; 1889.
47. Tucker BR. S. Weir Mitchell: a brief sketch of his life with personal recollections. Boston: Richard G. Badger; 1924. p. 15.
48. Harm E. The origin and early history of electrotherapy and electroshock. Am J Psychiatry. 1955;111:933.
49. Steinberg H. Electrotherapeutic disputes: the 'Frankfurt Council' of 1891. Brain. 2011;134:1233.
50. Steinberg H. Electrotherapeutic disputes: the 'Frankfurt Council' of 1891. Brain. 2011;134:1229–43.
51. Steinberg H. Electrotherapeutic disputes: the 'Frankfurt Council' of 1891. Brain. 2011;134:1232.
52. Morus IR. The measure of man: technologizing the Victorian body. Hist Sci. 1999;37:249–82.
53. Morus IR. The measure of man: technologizing the Victorian body. Hist Sci. 1999;37:256.

54. Morus IR. Shocking bodies. Life death and electricity in Victorian England. Stroud: The History Press; 2011. p. 176–82.
55. Morus IR. Shocking bodies. Life death and electricity in Victorian England. Stroud: The History Press; 2011. p. 177.
56. Morus IR. Shocking bodies. Life death and electricity in Victorian England. Stroud: The History Press; 2011. p. 178.
57. Morus IR. The measure of man: technologizing the Victorian body. Hist Sci. 1999;37:259.
58. Morus IR. The measure of man: technologizing the Victorian body. Hist Sci. 1999;37:262.
59. Hedley WS. The scope and value of electricity in medicine. Lancet. 1896;I:1279–81.
60. Hedley WS. The pathology and treatment of electrical accidents. Lancet. 1894;144:437.
61. Hedley WS. The painlessness of electric death. Electr Rev. 1897;41:205–6.
62. Althaus J. Old age and rejuvenescence. Lancet. 1899;I:149–52.

Chapter 7
Electromagnetic Waves

The theory I propose may be called a theory of electromagnetic field because it has to do with space in the neighborhood of the electric and magnetic bodies, and it may be called a dynamical theory because it assumes that in space there is matter in motion, by which the observed electromagnetic phenomena are produced.

James Clerk Maxwell, 1864

Abstract James Clerk Maxwell converted Faraday's lines of force theory into mathematical equations that provided the foundation of modern physics. The equations showed how electromagnetic waves pass through space as vibrations in an all-pervasive ether.

Maxwell began with a mechanical model of spinning spheres and vortices and ended with a model that linked quantities varying in space and time represented by abstract mathematical symbols. One of the few contemporary scientists that appreciated the significance of Maxwell's equations, Heinrich Hertz discovered a way to detect the ubiquitous electromagnetic waves. He showed that the waves passed through a thick door and were deflected and refracted by a metal surface. His work led to the development of wireless telegraphy by Marconi and voice radio by Fessenden.

Keywords Photoelasticity · Diamagnetic · Paramagnetic · Displacement current · Spark gap

When Michael Faraday envisioned magnetic lines of force, he was thinking of how iron filings arranged themselves in a pattern of lines when sprinkled onto paper held over the poles of a magnet. He felt that the lines represented invisible forces existing in space. He further speculated that vibrations in these lines of force could produce electromagnetic waves that radiated outward. Attractions and repulsions of electricity and magnetism resulted from strain existing between force centers. Lines of force could curve and fill the space between the centers of force.

Lacking a background in mathematics, Faraday relied on his remarkable intuition to describe the lines of force based on visual imagery. What was needed was someone to translate Faraday's visualizations into mathematical equations that could be tested and revised based on experimental results. In stepped James Clerk Maxwell, whose mathematical description of electromagnetic waves marked the beginning of modern physics.

R. W. Baloh, *Brain Electricity*, https://doi.org/10.1007/978-3-031-62994-5_7

Gauss and Weber Measure Electromagnetism in Absolute Values

Before addressing Maxwell's mathematical description of electromagnetic waves, it is useful to consider the contributions of the German scientists Carl Friedrich Gauss and Wilhelm Eduard Weber that provided a foundation for Maxwell's work [1]. Gauss, a mathematician, and Weber, a physicist, worked together to develop a rational system for measuring the absolute values of magnetic lines of force and later Weber used the system to measure the absolute values of voltage, current, and resistance in an electrical circuit. They used the fundamental units of length, mass, and time of the metric system to describe a nonmechanical force for the first time.

Carl Friedrich Gauss was born into a poor family in Brunswick, Germany in 1777. His father was an uneducated labor foreman at various trades and his mother was semi-literate. Gauss learned to read and count with little help and at age 7 years he was enrolled in an elementary school. His mathematical skills at such a young age immediately impressed his teachers.

In his book on the history of electricity, Stanford Bordeau illustrated these skills with a story describing an encounter between Gauss and a demanding school master known for caning his students for the slightest failure [2]. The schoolmaster gave the class a problem: *"Write down all the numbers from 1 to 100, add them, and give me the sum."* Knowing the answer, the schoolmaster anticipated a leisurely hour as the class worked on the task of summing the numbers.

The students were told to come forward and place their slate with the answer upside down on the schoolmaster's desk as soon as they finished the calculation. The first to finish would be at the bottom and the last to finish at the top of the pile. The schoolmaster had barely finished his instructions when Gauss arose from his seat and walked up and placed his slate on the schoolmaster's table. The rest of the class spent the hour toiling over their calculations.

When the schoolmaster examined the slates, most of the results were wrong, and the culprits accepted a rap from the cane. When he finally got to Gauss's slate it contained a single number—5050, the correct answer. The schoolmaster was astonished and asked Gauss how he came to the correct answer so fast. Gauss responded that he had used the symmetry of pairing numbers: 100 + 1 = 101, 99 + 2 = 101, 98 + 3 = 101 … to 50 + 51 = 101, a total of 50 pairs each adding to 101, or 50 × 101 = 5050.

With the help of supportive teachers, Gauss moved on to a local Gymnasium despite initial resistance from his father who expected Carl to join him as a laborer to help support the family. As at the elementary school, teachers at the Gymnasium could not challenge him in mathematics so he focused on the study of languages another area in which he excelled. At age 15, Gauss matriculated at the local Caroline College with financial support of the Duke of Brunswick, Carl Wilhelm Ferdinand. At the time it was common for local aristocrats to support needy and worthy students who showed great academic promise.

Remarkably, during his 3 years at the college Gauss read all the important works in mathematics and began his own independent research including work on prime numbers and statistics [3]. His innovative work on statistics was not only crucial to Maxwell and other nineteenth-century researchers but is still routinely used by current-day investigators. He derived a bell-shaped curve to describe the distribution of a random variable about the mean value and he discovered the *least squa*res method to best fit a curve to a set of points obtained experimentally. The so-called normal distribution or *Gaussian distribution* is a probability distribution symmetrical about the mean so that values near the mean are more frequent in occurrence than values far from the mean.

In 1795, still under the patronage of the Duke of Brunswick, Gauss was admitted to the prestigious University of Gottingen to study mathematics. Again, Gauss demonstrated his creative abilities in developing a new method for expressing equations that contained imaginary numbers using Euclidian geometry. The mathematical idea of an imaginary number $i = \sqrt{-1}$ arose from the need to solve the equation $x^2 + 1 = 0$. It followed that numbers could be expressed in the form, $a + ib$, where a is the real part, and b with the operator i the imaginary part. Gauss suggested that such numbers be called complex numbers and that they could be represented in a two-dimensional plane with a and b coordinates and the rotator i.

The system, which became known as a Gaussian plane, was useful for solving equations involving quantities having both magnitude and direction (vectors) such as electromagnetic waves. Their vector components could be solved by adding or subtracting their complex algebraic expressions since the sum or difference of two waves is the sum or difference of their vector representations.

After completing his studies at the University of Gottingen, Gauss went on to have a distinguished career in mathematics receiving a Doctor of Philosophy degree from the University of Helmstadt in 1799. He applied his mathematical skills to the study of planetary movements devising completely new techniques for astronomical calculations described in his 1809 book, *Theory of the Motion of Heavenly Bodies Revolving about the sun in Conic Sections* (written in Latin). After reading the book, the French mathematician, Pierre-Simon Laplace called Gauss the world's greatest mathematician [4].

Gauss had become a celebrity and began receiving offers of appointments from several European universities. Baron von Humboldt along with his influential friends convinced Gauss to stay in Germany and accept the position as Director of the University of Gottingen observatory. Part of his duties was to design and construct a new observatory that was finally completed in 1821. Gauss and his colleagues and students spent the next 10 years calculating and observing the positions and movements of heavenly bodies.

At the age of 53 in 1830, Gauss decided on a major change of his research focus to the field of physics and terrestrial magnetism. He had developed a theory of geomagnetism in his early twenties but circumstances at the time were not right for pursuing the subject. Again, Baron von Humboldt provided the impetus for this career change. Humboldt had been working in Berlin for many years attempting to understand and chart variations in the earth's magnetic field. As noted in Chap. 2, it was known that the magnetic north pole of the earth did not correspond to the geographic north pole and the magnetic pole was slowly drifting westward. In addition, there were daily variations in the magnetic intensity associated with magnetic storms from the sun.

For several years, Humboldt had been unsuccessfully trying to entice Gauss to come to Berlin to use his mathematical skills to help him develop better measurements of the earth's magnetic field. Finally, he persuaded Gauss to attend a scientific convention in Berlin and stay at his home for 3 weeks. Humboldt showed Gauss his plans for a joint European investigation of the earth's magnetic field, and he convinced Gauss to make a career change and join his study of geomagnetism. Humboldt had been instrumental in getting other European governments to establish stations for charting the strength and direction of the magnetic field, and he had already collected extensive data on the declination and inclination of the compass needle at different locations around the world.

One of the incentives that Humboldt offered Gauss was an assistant of his choice to help with the work. A young assistant professor from the University of Halle, Wilhelm Eduard Weber presented a paper at the Berlin convention that impressed Humboldt and Gauss and Weber was chosen as Gauss's assistant. Fortunately, Weber turned out to be an excellent experimental physicist and over the next 6 years working in Gottingen the two became close friends and highly successful collaborators.

Unlike Gauss, Weber grew up in an academic household, one of 12 children of a professor of theology at the University of Wittenberg, then in the Kingdom of Prussia [1, 5]. Two of his brothers became famous professors of anatomy. Weber began experimenting with science as a teenager and matriculated at the University of Halle at age 16. He was drawn to physics and mathematics and developed an interest in acoustics and eventually wrote his doctoral thesis on organ pipes. After serving as a lecturer, he was appointed assistant professor at the University of Halle. Weber was offered a professorship of physics at the University of Gottingen in 1831 so he could work with Gauss.

Working together at Gottingen, Gauss and Weber constructed a magnetic observatory completely free of iron that might introduce irregularities in the magnetic field. They developed a variety of instruments to make precise measurements of the magnetic field. Up to that time magnetic measurements were simply comparisons of arbitrary quantities without universal comparative standards. Gauss recognized the need for a standard of comparison and decided to represent magnetic measurements in the metric system of units, something that had never been tried before for a non-mechanical force.

Gauss and Weber constructed an apparatus with a bar magnet placed at a right angle and fixed distance from the needle of a magnetometer, a suspended magnetized needle with a mirror. Deflection of the magnetometer due to the bar magnet and the oscillation period of the magnet in the earth's magnetic field provided two equations each with an unknown variable that when solved yielded the earth's magnetic strength in metric units of length, mass, and time. Using the measurements collected from the worldwide network of observation stations, Gauss was able to quantitatively express the earth's magnetism at any point.

Gauss summarized his work on the earth's magnetic field in his book *The General Theory of Magnetism* published in 1838. He argued that the earth's magnetic field originated in its interior, the result of electrical currents generated by rotation of the hot metallic central core. The differential rotative forces between the core and the outer mantle caused local shifts in the current circulation explaining the variability of magnetic field intensity around the globe. A small part of the variation in magnetic intensity came from outside the earth by electrical currents moving high in the atmosphere, the result of magnetic storms on the sun. There was only one true north and south pole on the earth's surface.

Gauss's collaboration with Weber ended in 1837 when Queen Victoria came to the throne in England and her uncle, Ernst August became king of Hannover. At the time, Germany was a group of kingdoms governed by individual rulers. King August immediately disbanded the liberal constitution of Hannover and required that all public servants swear allegiance to the new king. Weber and several other Gottingen professors protested and were fired from their positions. The more conservative Gauss did not agree with the opposition and remained in his mathematical position, but he did no further work with physical experiments.

Weber went on to have his own brilliant career in experimental physics. After traveling to London and Paris, he was appointed a professor in physics at the University of Leipzig in 1843. When in 1848 King August's power was diminished, Weber was able to return to his old position in physics at the University of Gottingen.

Based on his work with Gauss, Weber concluded that electric current could be measured in absolute terms since it was accompanied by a magnetic field that was proportional to the current and the magnetic field could be measured in absolute terms. He developed a tangent galvanometer that compared the magnetic field produced by a given current to that of the earth's magnetic field. The instrument consisted of two circular coils and a suspended compass needle such that the coils and needle were lined up with the earth's north–south magnetic meridian. When current was passed through the coils, the needle deflected to line up with the coil's magnetic flux. The deflection was opposed by the earth's magnetic field, which was known in absolute terms, so the intensity of the current could be calculated in absolute terms.

Working with Rudolph Kohlrausch, a professor of mathematics and physics at the University of Marburg, Weber estimated the velocity of electricity moving along a conducting wire. They obtained a value of 310,740,000 m/s. The velocity of light had been measured a few years earlier at 314,000,000 m/s leading to the conclusion that electricity was propagated along a wire near the speed of light. These measurements of the velocity of light and electricity became a key piece of the theory of the propagation of electromagnetic waves by Maxwell a few decades later.

In his later years, Weber became fascinated with the electrical properties of matter that allowed electricity to flow along a wire. He envisioned atomic structures with negative charges spinning in orbits around positive charges. The positively charged particles were fixed in the lattice of the conducting material while the negatively charged particles could move about the atomic lattice. A directional migration of the negatively charged particles explained the flow of electricity along a conductor.

Based on the work of Gauss and Weber, the MKS (meter-kilogram-second) system became the main system used for describing the properties of electromagnetic waves in technical journals. Both Gauss and Weber have units of magnetic flux in the MKS system named after them, 1 weber/m^2 is equal to 10^4 Gauss (a unit of flux intensity). Weber's name was originally proposed for the unit of electric current at the First International Electrical Congress in Paris in 1881, but political forces were aligned against Weber and the unit was named for Ampere (amp). Gauss's name lives on in the many concepts and methods he devised in mathematics and electromagnetism (e.g., gaussian plane, gaussian distribution, gaussian integers, gaussian noise, and gaussian su*rface*). In World War II, the term *degaussing* became widely known as the process of neutralizing the external magnetic field of a ship so that it would not be blown up by a magnetic mine.

James Clerk Maxwell, the Early Years

Building on the work of Faraday, Gauss, and Weber, James Clerk Maxwell created eight famous mathematical equations that became the basis of modern electromagnetic theory. These equations predicted that energy could be transmitted as electromagnetic waves moving through space at the speed of light. His work was not only the foundation of all modern communications but also the underpinning of Einstein's theory of relativity and the beginning of modern physics.

James Clerk Maxwell was born in 1831 in the rolling hills of Galloway in southwest Scotland [6, 7]. His father had inherited an estate called Glenlair consisting of 15,000 acres along the river Urr. Despite Maxwell's remarkable achievements and subsequent fame, he was at heart a country lad who returned to his home in Glenlair whenever he could for inspiration and solace.

James was born into a well-known Scottish family, the Clerks of Penicuik, near Edinburgh. James' father, John Clerk inherited Glenlair, which had descended to the Clerks through a previous family marriage to the Maxwell's. With the marriage came the stipulation that whoever inherited the estate would add Maxwell to his name. The Clerks had their own estate at Penicuik inherited by John's older brother, sir George Clerk.

In marked contrast to Michael Faraday, James Maxwell grew up in an affluent household with parents who were devoted to his education. James was an only child; an older sister died shortly after birth. His mother, who supervised his early education, described him as a happy child who was curious about everything, particularly mechanical devices. His favorite phrase was "What's the go o' that" and when the answer was not satisfactory "but what's the *particular* go of it" [8]. James was fascinated by nature and spent much of his childhood exploring the countryside along with the other estate children. He had a distinct Galloway accent that remained with him throughout life.

His carefree childhood came to an abrupt ending when at age eight James' mother was diagnosed with stomach cancer. After 2 years of excruciating pain and suffering, she died at age 47 and James and his father were desolate. His father initially tried a local tutor, but James did not respond to the mechanical recitation and physical discipline, so he arranged for James to attend the Edinburgh Academy, one of the most prestigious schools in Scotland. Although it meant leaving his father and Glenlair, two aunts lived near the school to provide emotional support.

Initially, he had trouble fitting in at the Academy with his strange country clothes and thick accent leading to the nickname "Dafty," but gradually, his intelligence and resourcefulness won the respect of his fellow students. His interest in mathematics was kindled, and at age 14, he wrote his first scientific paper describing a method for tracing perfect oval curves with pins and string. How to draw an eclipse with pins and string was well known but James used complex looping to produce a whole family of curves. His proud father showed the paper to his friend J.D. Forbes, professor of natural philosophy at Edinburgh University, who felt it was worthy of presentation at the Royal Society of Edinburgh. James was too young to appear before the Society, so Forbes presented the paper for him.

James' father was a lawyer and he had always assumed his son would follow in his footsteps even though his aptitude seemed more appropriate for a career in science. At the time, a career in science was thought to be fine for an independently wealthy gentleman but a poor career choice for a gentleman who had to earn a living. As for James' opinion, although he was drawn to mathematics and science, he had a broad interest in literature and philosophy and felt that law might also be interesting if he got to know it. At age 16, he matriculated at Edinburgh University to complete his general education before studying law.

At Edinburgh University, James was inspired by two professors that had a major influence on his future career, his father's friend Professor Forbes, chairman of natural philosophy and Sir William Hamilton, chairman of logic and metaphysics. At the time, natural philosophy was the same as what we now would call science. Forbes was an experimental physicist who let James use his equipment and learn how to conduct scientific experiments. Hamilton, on the other hand, taught James to think logically and to be willing to speculate on experimental results traits that were natural to him.

At the young age of 19, James had two additional papers read to the Royal Society, *The Theory of Rolling Curves* and *On the Equilibrium of Elastic Solids.* He spent most of his time focusing on mathematics and physics experimenting in optics, electromagnetism, and chemistry. During one of his many visits to Edinburgh when James was at the Edinburgh Academy, his father took him to an exhibition of "electromagnetic machines," illustrating the remarkable discoveries of Michael Faraday. This was James' first exposure to the concept of forces acting in space, a concept that shaped the focus of his education and later experimental work.

Polarized Light and Lines of Stress

Some of James' best work during his time at Edinburgh University was conducted on holiday at a makeshift laboratory he set up in a room over a washhouse at Glenlair. As it turned out, his most important piece of equipment was a broken window glass. He had read that if polarized light was shown through glass under strain, it would produce beautiful colored patterns. With polarized light, the transverse wave vibrations are all lined up in a particular plane rather than being randomly oriented as in ordinary sunlight. To investigate the phenomenon, he heated pieces of glass until they were red hot and then rapidly cooled them causing the outer surface to cool faster than the inner surface. This produced permanent strains within the glass. He made a polarizing prism with a thin elastic plate of mica and two pieces of glass cut to shape and held at the correct angle by sealing wax. When sunlight passed through the polarizing lens and struck the glass under strain, he observed vivid colored lines connecting points of equal strain within the glass.

To document the results of his experiment, he placed a thin piece of paper over the glass and used watercolors to copy the image. Maxwell sent his findings to a famous English optician, William Nicol who was so impressed with the images that he sent Maxwell a pair of his prized polarizing lenses for his future experiments. These were the same polarizing lens that Michael Faraday had used in his experiments on the influence of magnetism on light rays.

With his new source of polarized light, Maxwell expanded his research to include gelatin from the Glenlair kitchen that he twisted to produce different patterns of torsional strain. The colored lines that he copied with watercolors and crayons showed the lines of strain. He had invented the process of photoelasticity, a process that would be widely used by structural engineers. By making a scale model of a structure such as a bridge out of transparent material and shining polarized light on it, the engineer could identify areas of strain that might need strengthening. He described his findings in *On the Equilibrium of Elastic Solids* also read for him before the Royal Society because he was still considered too young to read the paper himself. The paper included a full mathematical theory of photoelasticity based on strain functions. His work marked the beginning of a lifelong interest in optics and the perception of light.

With completion of his work at Edinburgh University, J.D. Forbes tried to convince James' father that his son's talents were in mathematics and science and that he should consider Cambridge University since it was the best place for young aspiring scientists. His father was initially reluctant even though he agreed with Forbes' assessment since his son would be far from home and potentially under the influence of rich English aristocrats. But Forbes persisted and James obviously needed a more challenging curriculum, so his father gave in and in the autumn of 1850, James packed his experimental equipment including his Nicol prisms in a trunk and headed for Cambridge.

Cambridge was *the* University for mathematics dating back to the time that Isaac Newton held the Lucasian Chair of Mathematics. By contrast, Oxford was known for the humanities. It typically took 4 years for a bachelor's degree at Cambridge, and in the final year, all students had to pass a difficult mathematics exam before they could receive a degree of any kind.

The exam was given in January at the Senate House, a large building shaped like a Greek Temple without any source of heat. All students had to take the first 3 days of exams solving math problems against the clock for five and a half hours each day in freezing conditions. Students who wanted honors, regardless of degree, continued for four more days with more difficult problems. Those who achieved the highest scores gained first-class honors and the title of wrangler. Being the top wrangler was almost like winning an Olympic gold medal. Not only did it provide a boost in his future career regardless of field, but the winner also became an instant celebrity in the local press. It was not unusual for large sums of money to be bet on who would be the top wrangler. The top wranglers would later compete in an even harder examination for the Smith's Prize.

Maxwell originally enrolled at St. Peter's College but transferred to Trinity College after the first semester where he fit in much better. He became a private pupil of William Hopkins, one of the most distinguished mathematicians at Cambridge. Hopkins had tutored hundreds of the top mathematicians, and he described Maxwell as *"unquestionably the most extraordinary man I have ever met within the whole of my experience"* [9].

Like Michael Faraday, Maxwell liked to solve problems by visualizing the solution. On one occasion, Hopkins covered the entire blackboard with equations to solve a problem that Maxwell solved with a simple diagram. Maxwell had interests outside of his required courses participating in social activities and delving into literature and philosophy. He was selected as an associate of the "Apostles," a group of 12 gifted students involved in discussions and speculative essays on subjects outside the regular courses of the university.

In January 1854, Maxwell completed his undergraduate training with distinction. Unlike many of his fellow students who stayed up nights studying for the examinations, Maxwell did not believe in cramming and continued his daily routine including regular physical exercise. On the math exams, he was up against E.J. Routh, who would later become a celebrated mathematician and math tutor at Trinity. Maxwell came in second to Routh in the wrangler competition but tied Routh for the Smith Prize.

After discussions with his father, James decided to stay on at Trinity as a Fellow for a few years until he could decide on his future career. At the time, Fellows at Trinity had to remain unmarried and be ordained into the Church of England within 7 years of becoming a Fellow and Maxwell had no intention of making either of those commitments. In addition to tutoring and lecturing, he was able to return to the experimental work which he enjoyed most.

Color Vision and the Light Spectrum

Maxwell had a longstanding interest in colors and prisms since childhood and wondered how the eye perceives colors. In 1849, he began some experimental work with Forbes in his laboratory in Edinburgh, but this was put aside when he enrolled at Cambridge. They experimented with spinning disks with different colored sectors so that when spinning fast, the colors on the disk merged into a single color. They tried mixing a variety of colors on the spinning disk to see what color would appear. The results were confusing, and they could not get white no matter what colors they mixed.

At the turn of the nineteenth century, the English physician and natural philosopher, Thomas Young, proposed that the eye had three different types of sensory receptors each sensitive to a particular color and that the brain combined the signals from the three receptors to form a single perceived color. Young came to this conclusion since he felt it was inconceivable that the eye could have receptors that responded to every wavelength of light. Furthermore, it was widely known that artists could produce any color by mixing different proportions of just three basic colors, red, yellow, and blue. Young even suggested using multicolored spinning disks to study what was happening in the human visual system.

Since the work of Isaac Newton, it was well known that prisms separated sunlight into light rays of different wave lengths, the shortest wavelengths perceived as blue or violet, the middle wavelengths as green, and the longest wavelengths as red. Maxwell realized that mixing colors of light that reach the eye is quite different from an artist mixing colored pigments to generate different colors. He felt that the three basic colors were red, green, and blue and not the red, yellow, and blue used by artists that he and Forbes had initially tried. To test this hypothesis, Maxwell ordered sheets of different colored paper and made a mixing disk that he called a "color top." The top was 6 inches in diameter with percentage markers around the rim and included a handle and a shank for winding a pull string. With this device, he found that he could produce a perception of white by mixing equal percentages of red, green, and blue and produce any other color by simply changing the percentage of red, green, and blue on the disk.

Maxwell then tested a large group of colleagues of different ages with his color top and found a remarkable uniformity in normal color vision. He then recruited people who were color-blind and found that the great majority lacked functioning red-sensitive receptors, so they were unable to distinguish between red and green. Maxwell then went on to develop a "color box" that more accurately reflected the nature of color perception than his color top. With the color box, he used a prism to produce a color spectrum from sunlight and with adjustable slits in the box he could select different colors in the spectrum that were then recombined optically to produce new colors. With this device that he kept improving over the years, he clearly showed how all colors could be created from combing the three basic colors.

At the same time Maxwell was developing his theory of color vision, in Germany Hermann Helmholtz was conducting his own experiments on color mixing using

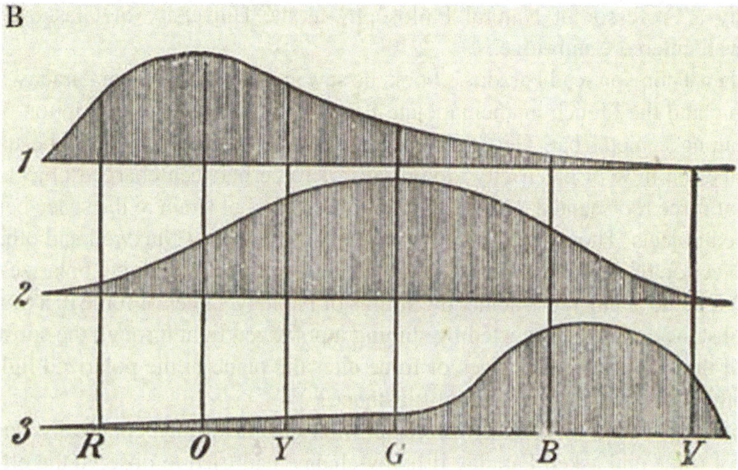

Fig. 7.1 Helmholtz's drawing of the response curves to different wavelengths of light of three receptor types in the retina. Type 1 had a peak response between red and yellow, type 2 between yellow and blue, and type 3 between blue and violet. R – red, O – orange, Y – yellow, G – green, B – blue, V – violet. (From Helmholtz H. Ueber die Theorie der zusammengesetzten Farben. *Poggendorffs Annalen der Physik und Chemie* 1852;87:45–66)

wheels and prisms and came to similar conclusions [10]. Maxwell and Helmholtz corresponded keeping each other informed of their progress. Helmholtz speculated that the retina had three distinct sets of "nervous fibers" each responding to light of all wavelengths. The response curves of the three receptor types overlapped each with a different peak response, one with the peak between red and orange, one with a peak near green, and one with the peak between blue and violet (Fig. 7.1). Like Maxwell, Helmholtz found that color-blind people were missing one of the three retinal "nervous fibers" usually the one with the red–orange peak.

William Thomson and Faraday's Lines of Force

When Faraday's book *Experimental Researches in Electricity* was published, most contemporary physicists and mathematicians had difficulty understanding what exactly he meant by lines of force or what the concept implied. One exception, a young Scottish mathematician, William Thomson saw the implications of the concept and attempted to describe it mathematically. Thomson's own work and his communications with Faraday and Maxwell were influential in defining the concept of lines of force [11].

William Thomson, who later became Lord Kelvin, was born in Belfast, Northern Ireland but moved to Glasgow, Scotland as a child and spent nearly all his academic career at the University of Glasgow. Like Maxwell, he was a mathematical prodigy and attended Cambridge University. Thomson was 7 years older than Maxwell and already a Professor of Natural Philosophy at the University of Glasgow when Maxwell entered Cambridge.

When Thomson read Faraday's book, he saw an analogy between Faraday's lines of force and the French mathematician, Fourier's mathematical description of heat flow along a metal bar. He suggested that the same equations could be used to describe the flow of electricity along lines of force between charged objects. The lines of force represented a true physical force, a kind of strain in the space between charged objects. Thomson was aware of the earlier work of Maxwell and others on the detection of internal strains in a transparent substance by shining polarized light through it. He wandered whether the strains of Faraday's lines of force in a transparent substance could be detected by shining a polarized light through the substance. Would the electromagnetic lines of force alter the plane of the polarized light just like other strains in the transparent substance?

Thomson wrote to Faraday describing his efforts to mathematically describe lines of force and asked Faraday if he by chance had tried to observe the effect of lines force on polarized light. Faraday was impressed with Thomson's work and responded that he had indeed tried the experiment but without success. But Faraday vowed to try again based on Thomson's new interest. He used an electrostatic generator to electrify a variety of transparent substances including fluids and shone the polarized light parallel and across the direction of the electrical lines of force, but he did not observe any effect on the polarized light.

Faraday then decided to try magnetic lines of force since they could be made stronger than electrical ones. He initially began with electromagnets he had in the laboratory but also ordered a more powerful electromagnet from the Royal Military Academy in Woolwich. After numerous attempts with a variety of transparent substances, he was able to see what appeared to be a faint response using a piece of heavy glass, made of borosilicate of lead, that he had left over from another project. When he received and tried his new larger magnet, the effect on the polarized light was much clearer. The plane of the polarized light rotated as it passed through the magnetized heavy glass, and he was able to show that the angle of twist was proportional to the strength of the magnet (by changing the number of turns in the wire coil). He then retested other transparent substances including liquids with the new magnet, and they all showed the same effect.

These results supported Faraday's lines of force theory by showing how magnetic lines of force altered light rays moving through them. He further showed that a variety of transparent substances that were previously thought to be nonmagnetic were also influenced by magnetism. An obvious question was: If substances previously thought to be nonmagnetic were placed between the poles of a strong enough magnet, would they move like a compass needle? To answer this question, Faraday made an even larger electromagnet consisting of a large piece of iron bar from a ship's anchor wound with 522 feet of wire all weighing 238 pounds. With this powerful electromagnet, he found that his piece of heavy glass suspended by a thread aligned itself at a right angle to the magnet's lines of force. The same occurred with just about all substances including pieces of wood and fruit. These substances that aligned themselves across the magnetic lines of force he called *diamagnetic*, whereas substances that aligned themselves parallel to the lines of force, such as iron, nickel, and cobalt, he called *paramagnetic*. Magnetism appeared to be a universal feature of matter.

Around this time, Faraday first used the term *magnetic field* to describe the lines of force surrounding a magnet or a moving electrical charge.

James Maxwell's Mathematical Models of Electromagnetism

While working with color disks and color boxes, Maxwell continued with his duties of a Fellow supervising classes and proctoring examinations at Cambridge. At the same time, he continued to read and think about electricity and magnetism, topics that had interested him since childhood when he wandered about how an electric bell worked. Maxwell decided it was time to get directly involved in research into this fascinating subject. It was only natural that he should turn for advice to, William Thomson, who he had met several years earlier through a family member [12]. Thomson was more than pleased to serve as a mentor to Maxwell and his first suggestion for Maxwell was to read Faraday's book *Experimental Researches in Electricity* in great detail. As noted earlier, Thomson was one of only a few scientists who had taken Faraday's line of force theory seriously and Thomson's analogy between electric lines of force and heat flow immediately caught Maxwell's attention.

Maxwell decided to expand on Thomson's analogy with the goal of expressing Faraday's entire force field theory in mathematical language. He began with the general analogy of the flow of an imaginary fluid through a porous substance. The fluid was weightless, friction free, and incompressible. With these assumptions, the speed of a unit of fluid flowing outward from a point source was inversely proportional to the square of the distance from the source and could be represented in simple mathematical terms. It was well known that electric and magnetic forces followed a similar inverse square law. The direction and speed of flow of fluid at any point in space represented the direction and strength of the electrical or magnetic force, the faster the flow, the stronger the force.

Maxwell introduced the term *flux* to describe the amount of force acting through any cross-section in space. A high density of flux corresponded to a high concentration of Faraday's lines of force, the higher the density of flux the stronger the electric or magnetic force. To manage the fluid's flow through space, Maxwell imagined tubes for the fluid to flow in, all fitting together leaving no gaps between. The fluid moved fast where the diameter of the tubes narrowed and slow where they widened. By analogy, the electric and magnetic lines of force were strong where the tubes were narrow and weak where the tubes were wide.

Maxwell did not suggest that the mechanics of fluid flow represented a physical model of electricity and magnetism. Rather he was trying to "*shew how, by strict application of the ideas and methods of Faraday, the connection of the very different order of phenomena which he has discovered may be clearly placed before the mathematical mind*" [13]. He simply used the fluid flow model and its mathematical properties as an analogy to create a mathematical representation of electrical and magnetic force without commitment to any specific physical model. Remarkably, the fluid flow model reproduced the known properties of electric and magnetic fields very well. This early model was simplistic compared to some of his later models but represented the same general principle that a model must lead to an accurate mathematical description of the physical properties of electrical and magnetic forces. Maxwell was the first to show that the laws of electricity and magnetism could be modeled using Newtonian mathematics.

At the time Faraday read Maxwell's initial report, he was 60 years old nearing the end of his research career while Maxwell was 26 years old and just beginning his career. In correspondence with Maxwell, Faraday indicated how impressed he was that Maxwell's equations predicted so many of his findings but queried Maxwell "*...would it not be a good thing if mathematicians working on these subjects, were to give us the results in the popular, useful, working state, as well as in that which is their own and proper to them?*" [14].

It would be 6 years later after moving to King's College in London that Maxwell returned to the mathematical description of Faraday's lines of force [15]. In his new model, he tackled the difficult problem of explaining the relationship between a moving electric current and the surrounding magnetic field and electromagnetic induction. He came up with an ingenious thought model consisting of rows of tiny rotating spheres ("particles of electricity") surrounded by large circular vortices representing the magnetic field (Fig. 7.2). When revolving in place, the tiny spheres were just stationary electric charges but under the influence of an electromotive force, they would translate along the rows causing the surrounding vortices to rotate producing magnetic flux. To explain induction, he showed how the rotating vortices (magnetic field) could transiently set in countermotion neighboring rows of electrical particles and the on through the next layer of vortices to other more distant rows of electrical particles.

Maxwell's imaginary spheres and vortices were everywhere, in space and in any matter that occupied space. In conducting materials, like a copper wire, the spheres were relatively free to move along their row whereas in insulating materials the spheres were relatively fixed. With an applied electromotive force, electrical current rapidly moved along a conductor but in an insulator (dielectric), there was little moving electrical current generated, what Maxwell called a *displacement current*. Maxwell gave the spheres elastic walls so that even though they did not move, they could distort in one direction, the beginning of a current. With the removal of the electromotive force, the distorted spheres would snap back producing a transient current in the opposite direction. Furthermore, if the spheres were set into elastic motion by an oscillating electric field, there would be a wave-like displacement of current along with transverse waves of magnetism.

The process would not be instantaneous since the motion had to overcome the inertia of the system. Waves of energy were sent out through space, electromagnetic waves. Just as Faraday had speculated, oscillation in the electric and magnetic lines of force would propagate as waves. Maxwell's equations predicted that the

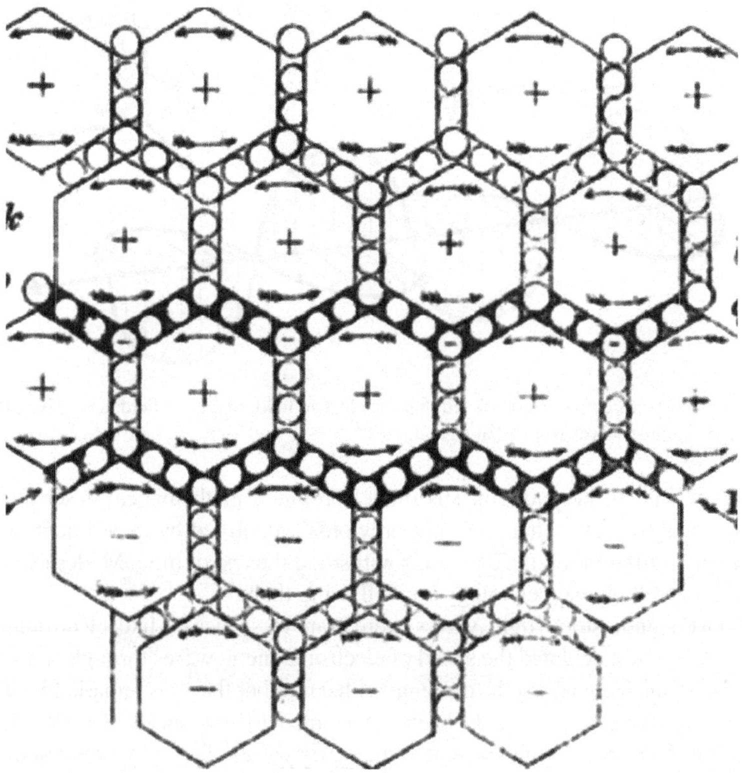

Fig. 7.2 Maxwell's mechanical model of electromagnetism. Under the influence of an electromotive force the tiny spheres representing electric charge translated along the rows causing the surrounding vortices to rotate producing the magnetic flux. (From Maxwell J. On physical lines of force. *Philosophical Magazine* 1861;90:11–23)

electromagnetic waves were double transverse, meaning that the electric and magnetic fields were at right angles, and the wave direction was at right angles to both (Fig. 7.3).

In Maxwell's model, the imaginary spheres and vortices were present in space without matter. He speculated that the mass density of the spheres and vortices could be extremely low so that they provided no significant obstruction to ordinary matter and could not be detected by any known type of instrument. Again, he emphasized that this was only a conceptual model, a way of thinking about electromagnetism.

Scientists in the late nineteenth century were still locked into a Newtonian universe where all physical phenomena had a mechanical explanation, and it was just a matter of finding the right mechanical model. Not surprisingly, many scientists had trouble with Maxwell's model even though he repeatedly emphasized that he did not think it necessarily represented nature's actual mechanism. He even agreed that it was somewhat awkward. His model was just a means to get at the relevant mathematical equations to describe the relationship between electricity and magnetism.

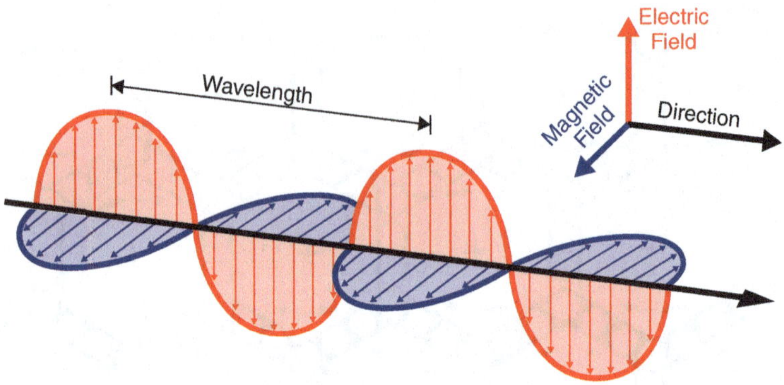

Fig. 7.3 Maxwell's equations predicted that the electric field, magnetic field, and wave direction of electromagnetic waves were all at right angles

The proof was in the pudding. His model provided mathematical descriptions of electromagnetic waves that were not previously identified by experiment or common logic. Furthermore, the equations withstood the test of time. Modern scientists working on cell phone technology are still using them.

Maxwell understood light waves were transverse waves like electromagnetic waves. When he calculated the speed of electromagnetic waves through space using his mathematical equations, he came up with a number that was remarkably close to the number that had previously been determined for the speed of light. He concluded that light waves must be electromagnetic waves. This was a remarkable discovery, yet his colleagues were not that impressed. Even his close friend Cecil Monro wrote to Maxwell, *"The coincidence between the observed velocity of light and your calculated velocity of a transverse vibration in your medium seems a brilliant result. But I must say I think a few such results are needed before you can get people to think that every time an electric current is produced a little file of particles is squeezed along between two rows of wheels"* [16].

Maxwell took the criticism to heart and decided to develop a mathematical description of electricity and magnetism based on a hypothetical undulating medium in space discarding the rolling spheres and twisting vortices. He presented the mathematical formulation in a paper entitled *A Dynamical Theory of the Electromagnetic Field* at a Royal Society meeting in 1864. Maxwell explained, *"The theory I propose may be called a theory of electromagnetic field because it has to do with space in the neighborhood of the electric and magnetic bodies, and it may be called a dynamical theory because it assumes that in space there is matter in motion, by which the observed electromagnetic phenomena are produced...The energy of electrification resides in the dielectric medium whether that be solid, liquid, gaseous, dense or rare, or even what is called a vacuum, provided it is still capable of transmitting electrical action"* [17].

The hypothetical undulating medium Maxwell was referring to was *ether*, a substance that was everywhere present, in free space and in the interstices of observable matter.

He described the *ether* as *"an elastic, quivering, universal medium...the seat of storage and as the vehicle of transmission for the electrical and magnetic energy."* The concept of ether dates back to the ancient Greeks and was the foundation of Descartes' theory of planetary motion. Maxwell provided ether with elastic properties to store and transmit electromagnetic energy.

Maxwell initially used eight equations in his *Dynamical Theory of the Electromagnetic Field.* The equations described the behavior of electromagnetism in terms of the electric field, the magnetic field, and the currents. Maxwell used vector calculus with values of both magnitude and direction in three dimensions to express the mathematical formulas. Each vector equation was in triplicate, one equation for each of the three dimensions in space. Of the eight equations, six of them were vector triplets so that the total number of equations was 20. Over time, the equations were revised and reordered into four equations known as "Maxwell's Equations."

Maxwell had left his mechanical model of spinning spheres and vortices behind and replaced it with intricately linked quantities that varied in space and time represented by abstract mathematical symbols. Many in the audience at the Royal Society were frustrated with the lack of a mechanical explanation and most were overwhelmed by the mathematics of the new theory. Even William Thomson, a solid mathematician who could follow the equations, wandered whether Maxwell had "lapsed into mysticism" [18]. The Fellows in the Royal Society in the mid-1860s were just not prepared for this paradigm shift in the world of physics that would become the norm in the twentieth century.

In their book on Faraday and Maxwell, Science writers Nancy Forbes and Basil Mahon nicely summarized the significance of Maxwell's Dynamical Theory of the Electromagnetic Field. *"Here was a* ***physical*** *theory expressed solely by its equations, and it brought a profound change in physicists' concept of reality. The startlingly new idea was that reality exists* ***on two levels****. Beneath all the things that we can touch and feel (or make models of) lies a deeper reality that expresses itself in the language of mathematics. In this underlying layer are quantities, like electric and magnetic field intensities, that are quite different from anything we can assess with our senses. They must, in some sense, be real, because they give rise to all the mechanical forces that we* ***can*** *feel and represent in models, but the only way we can describe them is through abstract symbols in equations"* [6].

Maxwell continued to refine his dynamical theory, and in March of 1871, he returned to Cambridge as the first professor of experimental physics and the head of a new laboratory, the Cavendish Laboratory, that would open 3 years later. With all the responsibilities of starting a new laboratory and teaching graduate students, Maxwell never had a chance to verify his theory experimentally. In 1873, he published his famous book *Treatise on Electricity and Magnetism*, which became *the* textbook of physics for students and scholars and remarkably is still in print [19]. Maxwell was working on a major revision of his book at the time of his death in 1879. He died of stomach cancer at a young age like his mother.

Making Sense of Maxwell's Equations

There can be little doubt that Maxwell's dynamical theory of the electromagnetic field was a pivotal moment in the history of science. The atomic physicist Richard Feynman called it the most significant event of the nineteenth century. Yet, at the time, nearly all contemporary physicists had a difficult time following the mathematics and most were uneasy with the lack of a mechanical model that they could relate to. A small group of contemporary mathematicians and physicists, known as" Maxwellians," deciphered the mathematics and conducted experiments to verify the predictions of the theory [20].

Chief among the Maxwellians was a self-taught mathematician, Oliver Heaviside, who produced a simplified four-equation version of Maxwell's theory that remains in general use today. Heaviside was born into a poor family in London in 1850, the youngest of four sons. He suffered scarlet fever as a child leaving him with permanent hearing loss which no doubt contributed to his eccentricities and stubborn independence. Heaviside showed an early interest in science but received little encouragement at a local school with its old-fashioned teaching methods. Attending a university was financially out of the question but due to his passion for learning, he was allowed to spend 2 years studying at home while his older brothers worked to help support the family.

At age 18, Heaviside was fortunate to get a job as a telegraph operator for the Danish–Norwegian–English Telegraph Company through the intervention of Charles Wheatstone, an uncle by marriage. Wheatstone was an English scientist and inventor best known for developing the Wheatstone bridge, a device to measure electrical resistance. Wheatstone built the first commercial telegraph in Britain and was involved in developing the technology for suboceanic telegraphy. Heaviside quickly learned Morse code and became a star troubleshooter for problems in the North Sea cable after he was posted at the main Danish operating station at Frederica. At age 20, he was already known for his technical knowledge of electricity and was assigned as chief operator at the company's English headquarters in Newcastle.

In his spare time, Heaviside studied mathematics and the physics of electricity at the Newcastle public library. This was where he first saw Maxwell's famous *Treatise on Electricity and Magnetism.* He later wrote: "*When I saw on the table of the library the work that had just been published (1873) I browsed through it and was astonished! I read the Preface and the last chapter, and several bits here and there; I saw that it was great, greater and greatest, with prodigious possibilities in its power. I was determined to master the book and set to work*" [21].

Heaviside gradually became bored with the routine of working long hours for the telegraph company and at the age of 24, he decided to quit the job and return home to his parent's house in London and spend full time researching electromagnetism. His first task was to master Maxwell's theory which was not easy considering the complexity of the mathematics and the lack of a simple way to visualize the predictions. As he began to understand the equations, he saw ways to simplify the formulas without affecting the meaning.

To simplify the vector analysis, Heaviside represented each vector in three-dimensional space with a single letter [22]. This enabled him to represent three equations with a single equation that could be manipulated with simple algebra. Maxwell's equations contained calculations for field forces and potentials, but Heaviside concentrated just on the field forces which he considered were "real" whereas the potentials were "metaphysical" and could be neglected. With this strategy, he was able to reduce Maxwell's eight equations (twenty including the vector triplets) into four basic equations, two describing electrical field forces and two magnetic field forces.

According to Maxwell's theory, electromagnetic energy was transmitted in space so that at any moment in time each part of space contained a fixed amount of energy. Heaviside focused on how this energy flowed which he described with a simple formula:

$$W = E \times H$$

where E and H are the vector products of electric and magnetic field forces, respectively, and W is the resultant energy-flow vector. In other words, the energy-flow vector (W) was the product of the electric (E) and magnetic (H) field forces. The energy-flow vector was at right angles to the electric and magnetic field forces, and it was maximum when the electric and magnetic field forces were at right angles to each other.

Like Maxwell, Heaviside was fascinated by the mystery of how electricity flowed within a wire. According to many of his fellow electricians when he was working for the telegraph company, a telegraph wire was just a conduit through which a battery pumped a kind of fluid called electricity. But when Heaviside began to explore the consequences of Maxwell's theory with his new formulas, he came to the remarkable conclusion that no electrical energy passed along the wire itself, but the wire simply acted as a guide for the flow of energy. The energy was borne by the electric and magnetic force fields in the space surrounding the wire. The electric force radiated out from the wire like the spokes of a wheel and the magnetic force encircled the wire. The flow of energy was at right angles to these fields, running parallel to the wire. When the lines of force contacted the wire, they turned sharply inward and were converted into heat.

Heaviside was a recluse who rarely traveled away from home, so he had little direct communication with other scientists working in his area. Fortunately, he had developed a working friendship with C. H. W. Biggs, the editor of the weekly trade journal, *Electrician*. Biggs regularly published his work despite Heaviside's frequent satirical comments and his increasingly complicated mathematics which were beyond most readers of the journal. Heaviside particularly enjoyed taking jabs at pompous aristocrats and authorities in the church. For example, in one of his articles, he suggested that God created Ohm's law to make it easier for electricians to make their calculations.

Gradually, other mathematicians and physicists in Britain became aware of Heaviside's work and recognized his unique insight into Maxwell's theory. Two such scientists, George Francis FitzGerald and Oliver Lodge, began communicating with Heaviside, and together the three formed the core of the Maxwellians. According to Maxwell's theory, electromagnetic waves were constantly being beamed into space, and the Maxwellians decided that it should be possible to reliably generate and detect these waves.

Generating Electromagnetic Waves

In 1842, the American physicist Joseph Henry reported that the magnetization of a steel needle could be reversed by the spark discharge of a nearby Leyden jar. The effect of a single spark could be transmitted at least 30 feet from his cellar to the second floor. William Thomson, unaware of Henry's report, showed that a spark discharge from a Leyden jar oscillated until all the energy of the charge had been expended as sound, light, and radiation consistent with the law of conservation of energy. In 1875, the American inventor Thomas Edison reported that a spark from a telegraph key triggered a spark in a nearby metal plate with an attached wire when the plate was close to another metal object. Similarly, he observed a spark between the slightly separated tips of two carbon points when a nearby telegraph key sparked. Finally, in 1879, the Welsh scientist David Edward Hughes noted that electric sparks could activate a telephone receiver 500 yards away. These were hints that electromagnetic waves could be produced and detected although the researchers were not primed to understand the implications of their observations.

In 1883, George Francis FitzGerald, Professor of Natural and Experimental Philosophy at Trinity College in Dublin, predicted how to generate electromagnetic waves based on Maxwell's theory. Born into an affluent Irish Protestant family, FitzGerald was a gifted mathematician who after obtaining his degree from Trinity College received a prized fellowship. Although he tended to be somewhat disorganized in his research focus, he developed a reputation among British physicists for his interpretation of Maxwell's and Heaviside's equations. Heaviside called him *"the quickest and most original brain of anybody"* [23].

Based on Maxwell's theory, FitzGerald calculated that the energy of oscillating electric currents radiated into space with energy proportional to the fourth power of the frequency. The key to generating radiant energy that could be detected was to use a high-enough frequency. He suggested that the discharge of a capacitor like a Leyden jar should be high enough in frequency to be easily detected. He estimated a discharge frequency of several million cycles per second with a wavelength of a few meters. As a wave generator, he proposed a circular loop of wire with ends attached to the Leyden jar. The problem was detecting the waves. FitzGerald noted that the detector should somehow be tuned to the frequency of the waves, but he had to admit that he had no good idea how this would be achieved.

Oliver Lodge, a professor of physics and mathematics at Liverpool University College was in regular communication with FitzGerald. Both were highly impressed with Heaviside's insight into Maxwell's equations, and both shared a desire to find a way to detect the predicted electromagnetic waves in space. Lodge, a clay merchant's son from Staffordshire, worked his way to a doctorate degree from University College, London. He was a colorful character with a passion for science but also a tendency toward mysticism that, as we will see in the next chapter, led him to some bizarre ideas. Yet, he had a practical side that led to the development of the first reliable detector for radio waves.

Lodge made an interesting observation when he was experimenting with spark discharges from a Leyden jar. He noted that the spark could be made weaker or stronger by varying the length of the wires attached to the Leyden jar. After discussing the finding with a young colleague, he realized that he was seeing evidence of Maxwell's electromagnetic waves streaming along the wires. He was observing the effect of standing waves, reflected waves from the end of the wires combining with waves traveling with the wires. By changing the length of the wires, he was changing how the two waves combined peak to peak. He presented his findings at an 1888 meeting of the British Association for the Advancement of Science before leaving for a vacation of walking in the Alps. While on the train ride to the Alps, Lodge read the exciting yet devastating (to him) news that a German physicist, Heinrich Hertz, had already generated and detected electromagnetic waves in space and demonstrated that the waves could be reflected and refracted just like light waves.

Heinrich Hertz and His Electromagnetic Wave Detector

Heinrich Hertz was born in the port city of Hamburg in 1857 into an affluent family, the oldest of three brothers and a sister [24, 25]. His father was a lawyer and senator of the city. In the newly developing German empire, Hertz decided on a career in engineering and enrolled in the School of Polytechnics in Dresden. After graduating and completing a mandatory year in the army, he moved to Munich where he studied mathematics and natural science along with engineering. He finally decided to give up engineering and attend the University of Munich to focus on natural science. After studying mathematics and experimental physics in Munich, in 1878, Hertz moved to the Berlin Physical Institute, Germany's finest scientific research center. At the Institute, Hertz came under the tutelage of Hermann von Helmholtz, Germany's most prominent man of science.

As we have already seen, Helmholtz had a wide range of scientific interests with electromagnetism being at the center. In 1847, he was the first to note that the discharge of a Leyden jar was not a simple one-way flow but rather a very high-frequency oscillation. The oscillating energy produced radiation that could move out into space. He was aware of Henry's observation that the radiation could magnetize a steel needle, but this was before Maxwell's theory and there was no ready explanation for the radiant action.

When Maxwell's theory was published in the 1860s, Helmholtz was the first scientist on the continent to analyze Maxwell's equations and understand the potential significance regarding propagation of electromagnetic waves. At the time, there were rival electromagnetic theories by his German compatriots Wilhelm Weber and Ernst Neumann based on the traditional Newtonian assumption of direct forces acting at a distance as opposed to Faraday and Maxwell's forces moving at a finite speed. Helmholtz found that the equations of all three theories worked fine for predicting electromagnetic action in closed metallic circuits, but Weber's and Neumann's equations ran into problems when there was a dielectric (insulator) such as a capacitor in the circuit. Recall that Maxwell's theory predicted displacement currents in the dielectric that produced oscillating electromagnetic waves. As a research project, Helmholtz suggested that Hertz try to detect the predicted displacement currents in a dielectric. Although Hertz's initial efforts were unsuccessful (the currents were too weak to be recorded with the instruments he had available), this experience proved useful when he returned to the problem later.

After a few years, as an unpaid lecturer at Kiel University where he had time to study the theoretical aspects of Maxwell's theory, Hertz was offered the position of professor of experimental physics at the Hochschule in Karlsruhe. There in the summer of 1886, he met and, after a three-month courtship, married Elizabeth Doll, the daughter of a professor at the school. In November of that same year, working in his new well-equipped laboratory, he began his experiments on electromagnetic waves that would make him famous. Interestingly, just as a chance observation led to Oersted's discovery of electromagnetism, a chance observation led to Hertz's detecting electromagnetic waves in space. In each case, the researcher had been primed to understand the significance of the chance observation.

Hertz was preparing an exhibition of Faraday's principle of induction for a class using two coils that were magnetically linked. As expected, he was able to induce an electrical current in the second coil from the nearby first coil. To illustrate the current in the second coil, he produced sparks across a small gap between metal balls on the wire terminals of the second coil. By chance, at the same time, Hertz noticed sparks coming from the ends of a circular piece of wire lying on the table. Could this be evidence of electromagnetic waves traveling through space? Had he found a simple method to detect the waves, a loop of wire with the ends near each other so that sparks could jump across the gap? All he had to do was to make the detector wire loop the right size and shape to be tuned to the frequency of the waves (Fig. 7.4). The waves would generate enough electromotive force in the detector loop to make sparks jump across the gap.

Hertz began to experiment with the detector wire changing the length and the distance of the spark gap [26]. He carried a detector around the room and faint sparks appeared in different locations. Assuming he was detecting electromagnetic waves, he made a large thin sheet of zinc as a reflector and systematically mapped the sparks of the wire detector identifying areas where the sparks were strong and others where they were weak or absent. This was evidence for standing waves produced by combining traveling waves with their reflection from the zinc screen. He then showed that the waves could be reflected by any metal surface, yet they could pass through a thick wooden door.

Once he was certain that he was detecting electromagnetic waves, Hertz began a series of experiments comparing the detected waves to light waves. He showed that

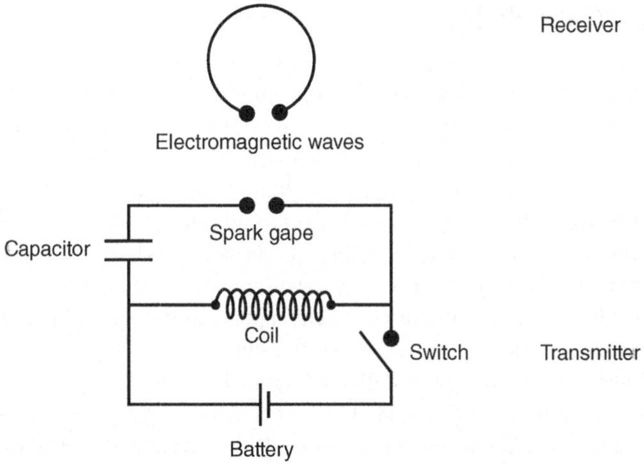

Fig. 7.4 Schematic drawing of Hertz's transmitter and receiver of electromagnetic waves

the waves moved at the speed of light, could be polarized like light, and could be refracted by a lens or prism just like light waves. In other words, they were light waves but at a lower frequency and longer wavelength than visible light.

When Hertz published his findings in 1888, there was relatively little enthusiasm in his German audience. Most German physicists accepted Weber's and Neumann's action-at-a-distance theories and even Helmholtz who studied Maxwell's equations was not entirely ready to accept the concept of electromagnetic waves moving through space at a defined speed. On the other hand, the English Maxwellians were ecstatic with Hertz's reports. He was their hero. As soon as Lodge returned from his vacation, he made a replica of Hertz's apparatus and reproduced his results. Fitzgerald lauded Hertz at a British Association for the Advancement of Science meeting calling his device "beautiful." Heaviside wrote to Hertz thanking him for dealing a "death blow" to the action-at-a-distance theories.

Soon, Hertz's international recognition soared, and he received offers from other institutions in Germany and abroad. The Berlin Physical Institute tried to lure him back, but he finally accepted the position as Professor of Physics at the University of Bonn where he would have less teaching responsibilities and more time for research. In 1890, Hertz published a paper entitled The Fundamental Equations of the Electromagnetics in which he elucidated and extended Maxwell's ideas on electromagnetics for bodies at rest and in motion. He lauded Maxwell's equations as being rich and comprehensive: *"It is certainly desirable that a system which is so perfect, as far as its contents are concerned, should also be perfected as far as possible in regard to its form"* [27]. Together, Hertz and Heaviside modified Maxwell's original equations into a form that would be widely accepted and pass the test of time.

Hertz and Cathode Rays

While at Bonn, Hertz began a series of experiments with cathode rays that are less well known but had the potential to be as important as his work on detecting electromagnetic waves. In 1875, while investigating the conduction of electricity in low-pressure gases, the English physicist William Crookes discovered what became known as Crookes tube (see Chap. 8). Called a cathode ray tube, the tube consisted of a sealed glass vessel with an electrical cathode at one end and the anode off to the side. When the pressure in the tube was lowered and electrical current was applied to the electrodes, a patch of fluorescent light appeared on the wall of the tube across from the cathode. Crookes later observed that the "cathode rays" that he assumed streamed out of the cathode were deflected with a magnet.

Shortly after moving to Bonn, Hertz decided to investigate whether the cathode rays were particles or electromagnetic waves. If they were particles, they should be deflected by a static electric field. To prevent ionization of gas particles in the tube, which could interfere with the electric field, Hertz created a vacuum in the tube. He did not see deflection of the cathode rays, but in retrospect, this was because he did not remove all the air from the tube. As we will see, in 1897, J.J. Thomson, working at Cambridge, was able to produce a better vacuum and showed that the rays were a stream of negatively charged particles which he called corpuscles, later named electrons. Ironically, with modern quantum theory, electrons are considered both particles and waves. Vacuum cathode ray tubes are the basis for a variety of image display inventions including oscilloscopes, radar screens, and televisions.

Unfortunately, Hertz's work on cathode rays was cut short after completing the initial studies. He became ill and was diagnosed with a form of bone cancer. After a series of painful operations, he died at the age of 37 in January 1894. The two most gifted mathematicians and physicists of the nineteenth century, Maxwell and Hertz died at a young age depriving the world of what likely would have been years of productive research.

Guglielmo Marconi and Wireless Telegraphy

After Oliver Lodge read about and replicated Hertz's experiments, it became apparent to him that Hertz's electromagnetic wave detector was functional but not very efficient. Lodge had previously investigated the principle of electrostatic cohesion and decided to develop an automatic decoherer for detecting electromagnetic waves [28].

The principle of electrostatic cohesion dates to the mid-nineteenth century when several investigators observed that finely powdered materials that were normally resistant to the passage of electrical current in their loose state would cohere and become a good conductor when subjected to electrostatic voltage. An English electrical engineer working for a telegraph company in Manchester, Samuel A. Varley made the first practical application of the principle of coherence developing a detector of lightning strikes. Charcoal powder in a small tube was a nonconductor until it was caused to cohere by a flash of lightning. The French physicist, Edouard Branly, studied a large variety of materials under different electrical influences and noted that metal filings such as iron, copper, brass, aluminum, and zinc were the best coherers. The coherence of the metal filings produced by an electrical discharge persisted but rapidly dissipated when the tube containing the filings was tapped.

For his automatic decoherer, Lodge used a tube filled with metal filings and an electric bell in the coherer circuit. Electromagnetic waves from a nearby spark caused the metal filings to cohere closing the circuit, ringing the bell, and tapping the tube to break the circuit. He demonstrated his coherer at a meeting of the British Association in 1894. The transmitter consisted of a battery-powered induction coil, a relay key, and a spark gap. When the key was pressed, a spark was triggered, and the coherer detected electromagnetic waves up to 150 yards away. The signal passed through walls and other barriers with ease.

Despite the obvious implications of Lodge's demonstration, he did not mention practical applications such as wireless communication. He would later try to justify his lack of foresight. "*I was too busy with teaching work to take up telegraphic or any other development, nor had I the foresight to perceive what has turned out to be its extraordinary importance to the Navy, the Merchant Service, and indeed land and war services too*" [29]. The man with such foresight was a young Italian inventor and entrepreneur, Guglielmo Marconi.

Marconi was born in Bologna, Italy, in 1874, the son of an Italian country gentleman and his Irish wife [28]. He showed an early interest in electricity and was enrolled at the Technical School at Leghorn where he became enamored with Hertz's electromagnetic waves. He followed the work of Augusto Righi, professor of physics at Bologna University who made several refinements in Hertz's transmitter. Righi found that the size of the emitted waves could be influenced by the size of the metal balls at the spark gap, the smaller the balls the shorter the wavelength. Hertz's waves were estimated in the meters range, whereas Righi could generate waves as short as a few centimeters. Righi also observed that he could make the waves more consistent by placing the spark gap in thick petroleum jelly.

While still a young student, Marconi decided that it should be possible to use electromagnetic waves for communication and he dreamed that he was the man to develop wireless telegraphy. He could not wait to get started, and at the age of 20, he began experimenting with wireless telegraphy at his father's country estate in Pontecchio near Bologna. As an emitter, he used Hertz's induction coil and Righi's brass balls separated by a small gap immersed in petroleum jelly. He controlled the sparking with a Morse key in the induction coil circuit. He moved his apparatus outside so he could increase the distance of transmission and soon discovered that he could improve transmission by connecting one of the terminals of the induction coil to a large cylinder mounted at the top of a pole and the other terminal to a metal plate on the ground.

For the receiver, he connected an identical cylinder on a pole to one end of a coherer and the other end to the ground. He began with coherers like the ones used by Branly and Lodge but found that they were not very sensitive, so he experimented with variations. Rather than a long tube with a large amount of coarse metal filings, he markedly reduced the size of the tube and the gap between the plugs in the tube to about 2 mm. The plugs were made of polished silver and connections to the plugs with platinum wires. After numerous trails, he found that the best metals for the filings was a mixture of 95% nickel and 5% silver sifted for fineness, filling about half the space between the two silver plugs. Air was removed from the glass tubes before they were sealed.

This trial-and-error approach was typical of Marconi's approach to experimentation. He was not driven by theoretical considerations but rather by results. Maxwell and Hertz had provided the theoretical basis for wireless telegraphy; someone had to work out the mundane details of a workable system. Marconi gradually extended the range of his wireless transmission from meters to kilometers by 1895. He exchanged the metal cylinders in the antenna with metal cubes and noted that doubling the height of the antenna quadrupled the distance of transmission. He found that he could concentrate the transmitted waves in the desired direction by placing the discharge balls of the transmitter and the receiving detector in line with parabolic reflectors, the predecessor of modern beam aerials.

In 1896, Marconi traveled to England and through family connections applied for and received a patent for his wireless telegraphy apparatus. In 1897, he formed The Wireless Telegraph and Signal Company Limited to hold his patents and erected his first wireless telegraph stations at Alum Bay on the Isle of Wight and at Bournemouth. The antenna height was 120 feet and the transmitter used a 10-inch spark induction coil. The first two paid wireless telegraphs were sent by William Thomson, by then Lord Kelvin. Marconi then switched his focus to the maritime uses of wireless telegraphy for ship-to-ship and ship-to-shore communication, which rapidly became commercially successful.

Oliver Lodge followed Marconi's achievements in wireless telegraphy with mixed emotions considering that he had failed to take advantage of his invention of the coherer. Lodge pointed out that if wireless telegraphy was to be widely used, there must be a way to tune the system to maintain privacy. He developed and patented a method to tune the transmitter and receiver circuits to the same frequency by balancing inductance and capacitance in the two circuits. His patent overlapped with later patents obtained by Marconi and to avoid legal battles, Marconi's company ultimately bought Lodge's patent and hired Lodge as a consultant.

Marconi rapidly became aware that he could only increase the distance of transmission to a limited degree by increasing the height of the antenna. The key to long-distance transmission was improving the electronics of the transmitter and receiver circuits.

In early 1900, Marconi took out a patent for "tuned or syntonic telegraphy," which removed the transmission and receiving circuits from the antenna circuits and joined them together through magnetic induction. This allowed him to tune the transmitter and receiver so that they oscillated in unison and the aerials could be used with multiple sets of transmitters and receivers. Even though multiple electromagnetic waves might be impinging on the aerials, the receiver would only respond to the wavelength to which it was tuned. Despite some overlap with Lodge's patent, Marconi's patent was ultimately upheld. As we will see in Chap. 9, there was also overlap with Nikola Tesla's patents of an electric oscillator in America and eventually Tesla's patents won out.

Marconi's next goal was to transmit across the English Channel and in early 1899, the French Government gave permission for Marconi's company to build a station at Wimereux near Boulogne. He built a matching station at the South Foreland lighthouse in Dover and in March 1899 transmitted the first wireless messages across the Channel. The public was enthralled although there were exceptions. Marconi's company received several letters from people who claimed they were made ill by the wireless waves passing through their body.

The most dramatic incident occurred at the Wimereux station. A gunman burst into the wireless room complaining that the wireless waves were causing him intense body pains and that he would shoot the engineer if he did not shut down the operation. The quick-thinking engineer, a Mr. Bradfield, expressed his concern for the stricken victim and indicated that he could take care of his problem by inoculating him against electricity. All the man had to do was to agree to receive a shock from the aerial. Of course, the man had to remove any metal coins from his pockets and lay down his metal revolver. The man agreed and happily accepted his electric shock convinced that his body pains were cured.

Transatlantic Wireless Transmission

Marconi's biggest challenge was to transmit wireless messages across the Atlantic Ocean. At the time there was a consensus among scientists in the field that it would be impossible to send such messages because of the curvature of the earth. Electromagnetic waves traveled in a straight line from the source so how would they bend around the earth to reach a receiving station in the Americas. But by the summer of 1899, Marconi had already shown that he could transmit signals from ship to shore over a distance of at least 150 miles so somehow the signals seemed to hug the surface of the earth.

The other big problem was to markedly extend the distance that electromagnetic waves could be transmitted. Although Marconi's company had shown that increasing the aerial height increased the distance of transmission, they had already reached the practical economic height of wooden masts to support the aerial of about 200 feet. The only option was to increase the power of transmission. It was estimated that the transmitter would need to be 100 times more powerful than the ones currently being used. An induction coil powered by batteries would no longer suffice. They would need an electrical plant to generate the power required. For help, the company hired Professor John Ambrose Fleming at University College London who was an expert in dealing with high-voltage alternating currents. Marconi and Fleming decided on an alternator driven by an engine with an output of about 25 W. High-tension transformers would charge a bank of capacitors that would discharge across a spark gap. Marconi left it up to Fleming to design the details of the plant and to devise the circuits to handle the high voltage.

For the English station, Marconi decided on Poldhu, on the west coast of Cornwall, relatively isolated from civilization and its rapidly expanding electric lighting. The conical aerial of 400 wires was supported by 20 wooden masts, each 200 feet high, arranged in a circle 200 feet in diameter. By early 1901, the power station at Poldhu was completed and Marconi and Fleming completed successful preliminary tests between Poldhu and the Isle of Wright. Marconi and his newly hired chief engineer, R.N. Vyvyan, sailed to America to select a sight for a corresponding station at South Wellfleet, Cape Cod, Massachusetts. After satisfying himself that the site was appropriate, Marconi returned to England leaving Vyvyan to organize and construct a replica of the Poldhu station.

The work at Cape Cod was completed by the summer of 1901, but before a transmission could be arranged, the aerial masts on the windward side were severely bent over by winds to a dangerous degree. Not long after the masts at Poldhu completely collapsed in a gale and a few weeks later the masts at Cape Cod also collapsed. Marconi was determined to forge ahead despite the setback, and he quickly rebuilt an aerial at Poldhu with 54 wires between two sturdy masts 150 feet in height. When he tested transmission with the new aerial, he received very strong signals at a station on the west coast of Ireland and felt sure that the signal was strong enough to travel across the Atlantic. To improve the chances of success, he decided to make the first test in a temporary station in Newfoundland, the closest part of the Western Hemisphere.

Marconi along with his equipment and two assistants arrived in Newfoundland on December 5, 1901, where he was warmly received by government officials. His first aerial attached to an inflated balloon immediately broke away and was lost, so he turned to a kite which he successfully flew at a height of about 400 feet. He had instructed the crew at Poldhu to send the Morse code for the letter S (three dots) daily from 3 to 6 PM. On December 12, 1901, Marconi and one of his assistants heard faint S signals. The reception occurred despite the kite rising and falling in the wind, altering the electrical capacity of the aerial.

Because of concerns of failure and its effect on public perception, Marconi had kept his goal of transatlantic wireless transmission a secret. When he first announced his success, there was worldwide enthusiasm. But skepticism from scientists soon followed. All they had was the word of Marconi and his assistant. Several suggested that he was likely recording atmospheric noise. In an interview, Thomas Edison noted *"Let us not forget that there are such things as strays"* [30]. Marconi's work in Newfoundland ended abruptly when the Anglo-American Telegraph Company notified him that they controlled all telegraphic communication from Newfoundland.

Marconi was immediately invited by the Canadian Government to build a wireless telegraph station in Canada. They offered 16,000 British pounds to build the station with the stipulation that the wireless rates would not exceed 10 cents per word. In the meantime, Marconi decided to see how far the Poldhu station could transmit to a ship at sea. He selected the S.S. Philadelphia of the American Line with an aerial atop a mast 170 feet high. Just married, Marconi and his bride sailed on the ship near the end of February 1902. Messages were recorded on tape at 1551 miles from Poldhu and the three dots of the letter S were recorded up to 2099 miles. Interestingly, they found that the electromagnetic waves traveled farther at night than in the day. Marconi realized that his demonstration in Newfoundland would have been more impressive if the wireless signals had been transmitted at night.

Later in 1902, Oliver Heaviside provided a possible explanation for the transmission of electromagnetic waves around the Earth. He suggested that there was a layer of the atmosphere that reflected the waves so that they could skim about the planet. Within a few years, the Canadian station was erected at Glace Bay on Cape Breton Island and wireless transatlantic messages were routinely transmitted. As we will see in Chap. 9, at the same time Marconi was working on long-distance wireless telegraphy, Nikola Tesla was using similar technology for his dream of transmitting high-energy electricity over long distances without the need for transmission wires.

Fessenden and Continuous Wave Transmission

When Reginald Fessenden was growing up in Fergus, Ontario, he was enthralled with the work of Alexander Bell whose family lived in nearby Brantford having emigrated from Scotland in 1870 [31, 32]. Alexander Bell had subsequently moved to Boston where he developed and patented the telephone sending his famous "*Mr. Watson—come here—I want to see you*" message to his assistant in March of 1876. In August of 1876, Bell returned to the family home in Brantford and made the first long distance telephone call, 113 km from Brantford to Paris, Ontario. When discussing this historical event with his mentor and uncle, Cortez Fessenden, young Reggie queried why Bell needed wires to transmit his voice messages. After all, thunder traveled through the air without the need for wires. His uncle who was a physics teacher explained that thunder traveled by sound waves different from the electrical waves that Bell transmitted in his wires. It was impossible to transmit words without wires. But Reggie was not convinced and would later prove his uncle wrong.

Fessenden had a patchwork of education in Canada in his young life where he was recognized as a prodigy in mathematics. He never formally graduated from a university but worked at Edison's New Jersey Laboratory and Westinghouse's Pittsburgh laboratory from 1897 to 1892 before he was offered the chair of electrical engineering at Purdue University. In 1893, the University of Pittsburgh recruited him to the same position with the financial support of George Westinghouse and other Pittsburgh millionaires. By the turn of the century, Fessenden had already accumulated nine patents and after experimenting with radio waves for several years made the first voice transmission over radio in December 1900.

At the time, wireless radio transmission as pioneered by Marconi and Fleming used a spark gap in an oscillatory circuit consisting of an inductance and capacitance in series to produce radio waves for transmitting Morse code signals. With each spark, a burst of damped oscillating radio waves went into space from the attached antenna. This was fine for transmitting dots and dashes but did not work for voice transmission. Fessenden recognized early on that what was needed for voice transmission were continuous waves not irregular bursts with rapid decay. But there was a consensus that the only way to generate powerful radio waves was with a spark gap generator.

A solution that immediately came to Fessenden was to increase the frequency of the spark rate above the voice bandwidth. In 1899, Fessenden began experimenting with a high-frequency interrupter to increase the spark rate to approach a continuous wave. Colleagues at the University of Pittsburgh constructed an interrupter that gave 10,000 breaks per second and by the end of 1900, Fessenden successfully transmitted a wireless voice signal over 1.5 km. His transmitter consisted of a carbon microphone in series with the antenna lead. The first wireless voice message was: "*Hello, one, two, three, four. Is it snowing where you are Mr. Thiessen? If it is, telegraph back and let me know*" [33].

Over the next 5 years, Fessenden worked to develop an improved transmitter that was more continuous wave like. He used an AC generator to provide the energy for the sparks produced by a spoked wheel that rotated in synchrony with the AC generator. But these stop-and-go spark systems were noisy and did not produce the quality of voice and music that Fessenden was seeking. By 1906, Fessenden was convinced that it was possible to apply a high-frequency alternating current directly to an antenna to produce the continuous waves he desired. His high-frequency alternator had a fixed armature in the form of a ring and a revolving fixed magnet with 300 teeth that rotated at a speed of 140 revolutions per second generating an alternating current of 50,000 Hz with a maximum power of 300 W.

Marconi and Fleming initially ridiculed Fessenden's idea that high-frequency AC current could be applied to an antenna for high-quality voice transmission. Everyone knew that wireless transmission required sparks. But the proof was in the pudding. In January 1906, Fessenden conducted the first two-way transatlantic wireless conversation from his station at Brand Rock, Massachusetts, with his chief assistant James Armor located at a station in Macrihanish, Scotland. Marconi tried to take advantage of his spark gap expertise to produce a transmitter that approached Fessenden's continuous wave technology, but the signal was noisy, and eventually, it was replaced by a high-frequency alternator.

On Christmas Eve, 1906, Fessenden made the world's first radio broadcast. It was broadcast from the station in Brant Rock, MA, and heard by radio operators on ships of the US Navy and the United Fruit company equipped with Fessenden's radio receivers. After reading a passage from the Bible, Fessenden played *O Holy Night* on the violin. Fessenden's high-frequency alternator had an output of about 50 kW and the General Electric Company soon after marketed a scaled-up version with an output of about 200 kW named the Alexanderson alternator after the man who supervised the work. Subsequently, it became known as the Fessenden–Alexanderson alternator.

Fessenden also needed to develop a receiver for his continuous wave transmitter. Fleming's coherer would be of little use for detecting continuous waves. After extensive experimenting, he developed what he called a barretter from the French word for exchanger, a devise to convert alternating to direct current, a rectifier. He used a platinum-coated wire that made point contact with an acid solution. He initially patented the devise in 1903 and a year later patented a sealed version for use on ship. Fessenden's rectifier was soon replaced by a galena crystal and eventually by a vacuum tube.

While Fessenden was working on his high-frequency alternator, an American inventor, Lee de Forest, was developing an arc transmitter that produced near-continuous waves that could be amplitude modulated for voice transmission. De Forest first broadcast music from his New York city laboratory in February 1907, and in July 1907, he broadcast ship-to-shore reports of the results of the Annual Inter-Lakes Yachting Association Regatta on Lake Erie. As discussed in Chap. 10, de Forest is best known for his development of a vacuum tube amplifier, the Audion, that made radio receivers readily available to the broad public and paved the way for neuroscientists to understand nerve transmission. This also marked the beginning of medical diathermy where electromagnetic waves in the radio to microwave range are used to heat body tissues to help with healing after injury. Diathermy is still commonly used by physical therapists and physicians.

In Brief

James Clerk Maxwell converted Faraday's lines of force theory into mathematical equations that provided the foundation of modern physics. The equations showed how electromagnetic waves pass through space as vibrations in an all-pervasive ether.

Maxwell began with a mechanical model of spinning spheres and vortices and ended with a model that linked quantities varying in space and time represented by abstract mathematical symbols. One of the few contemporary scientists that appreciated the significance of Maxwell's equations, Heinrich Hertz discovered a way to detect the ubiquitous electromagnetic waves. He showed that the waves passed through a thick door and were deflected and refracted by a metal surface. His work led to the development of wireless telegraphy by Marconi and voice radio by Fessenden.

References

1. Bordeau SP. Volts to hertz – the rise of electricity. Minneapolis: Burgess Publishing Company; 1982. p. 164–83.
2. Bordeau SP. Volts to hertz – the rise of electricity. Minneapolis: Burgess Publishing Company; 1982. p. 166.
3. Bell ET. Men of mathematics: the lives and achievements of the great mathematicians from zeno to poincaré. New York: Simon and Schuster; 2009. p. 218–69.

4. Bordeau SP. Volts to hertz – the rise of electricity. Minneapolis: Burgess Publishing Company; 1982. p. 169.
5. Foster GC. Wilhelm Eduard Weber. Nature. 1891;44:229–30.
6. Forbes N, Mahon B. Faraday, Maxwell and the electromagnetic field. Guilford: Prometheus Books; 2019. p. 127–240.
7. Bordeau SP. Volts to hertz – the rise of electricity. Minneapolis: Burgess Publishing Company; 1982. p. 184–207.
8. Forbes N, Mahon B. Faraday, Maxwell and the electromagnetic field. Guilford: Prometheus Books; 2019. p. 129.
9. Bordeau SP. Volts to hertz – the rise of electricity. Minneapolis: Burgess Publishing Company; 1982. p. 188.
10. Finger S, Wade NJ. The neuroscience of Helmholtz and the theories of Johannes Müller. J Hist Neurosci. 2002;11:240–2.
11. Forbes N, Mahon B. Faraday, Maxwell and the electromagnetic field. Guilford: Prometheus Books; 2019. p. 96–101.
12. Forbes N, Mahon B. Faraday, Maxwell and the electromagnetic field. Guilford: Prometheus Books; 2019. p. 155–7.
13. Simpson TK. Maxwell on electromagnetic field. New Brunswick: Rutgers University Press; 1997. p. 57.
14. Bordeau SP. Volts to hertz – the rise of electricity. Minneapolis: Burgess Publishing Company; 1982. p. 193.
15. Simpson TK. Maxwell on electromagnetic field. New Brunswick: Rutgers University Press; 1997.
16. Forbes N, Mahon B. Faraday, Maxwell and the electromagnetic field. Guilford: Prometheus Books; 2019. p. 196.
17. Simpson TK. Maxwell on electromagnetic field. New Brunswick: Rutgers University Press; 1997. p. 255.
18. Forbes N, Mahon B. Faraday, Maxwell and the electromagnetic field. Guilford: Prometheus Books; 2019. p. 212.
19. Maxwell JC. A treatise on electricity and magnetism. 3rd ed. Oxford: Clarendon; 1891. Reprinted by London, Oxford University Press, 1998.
20. Hunt BJ. The Maxwellians. Ithica: Cornell University Press; 1994.
21. Forbes N, Mahon B. Faraday, Maxwell and the electromagnetic field. Guilford: Prometheus Books; 2019. p. 244.
22. Heaviside O. Electrical papers. 2nd ed. Providence: Chelsea Publishing; 1970. Originally published 1893–1912.
23. Hunt BJ. The Maxwellians. Ithica: Cornell University Press; 1994. p. 187.
24. Appleyard R. Pioneers of electrical communication. London: Macmillan; 1930.
25. Bordeau SP. Volts to hertz – the rise of electricity. Minneapolis: Burgess Publishing Company; 1982. p. 220–43.
26. Hertz H. Electric waves. Originally translated by Jones DF in 1893. New York: Dover; 1962.
27. Bordeau SP. Volts to hertz – the rise of electricity. Minneapolis: Burgess Publishing Company; 1982. p. 239.
28. Vyvyan RN. Marconi and wireless. Yorkshire: EP Publishing Limited; 1974. First published by Routledge and Kegan Paul in 1933 under the title: Wireless Over Thirty Years.
29. Vyvyan RN. Marconi and wireless. Yorkshire: EP Publishing Limited; 1974. p. 9.
30. Vyvyan RN. Marconi and wireless. Yorkshire: EP Publishing Limited; 1974. p. 30.
31. Belrose JS. Fessenden and the early history of radio science. Radioscientist. 1994;5:1–16.
32. Fessenden HM. Reginald Fessenden: builders of tomorrow. New York: Coward-McCann; 1940.
33. Belrose JS. Fessenden and the early history of radio science. Radioscientist. 1994;5:9.

Chapter 8
Ether, Cathode Rays, and Electrons

> *In the very intense electric field in the neighborhood of the cathode the molecules of the gas are dissociated and split up, not into ordinary chemical atoms, but into these primordial atoms, which we shall for brevity call corpuscles.*
>
> *JJ Thomson, 1897*

Abstract William Crookes was one of the most enigmatic scientists of the nineteenth century. His pioneering work on spectroscopy led to the discovery of new elements and his cathode ray tube was a forerunner of several modern electronic devices. Yet, he actively participated in psychic research with charlatans and strongly believed in a psychic force that could be transmitted through space by the living and the dead. While working with Crookes cathode ray tube, Wilhelm Röntgen identified highly penetrating emissions that could pass through solid objects and produce an image on a photographic plate. There was immediate worldwide interest in the medical applications of Röntgen's X-rays. JJ Thomson used Crooke's tube to study the effects of cathode rays on gases and was able to measure the charge-to-mass ratio of the negatively charged corpuscles in cathode rays. The corpuscles (later called electrons) were several orders of magnitude smaller than a hydrogen atom. The discovery of the electron marked the beginning of the subatomic world of modern physics.

Keywords Vortex ring theory · Electrolysis · Radiant energy · Cathode rays · Corpuscles

To the ancient Greeks, ether (aether) was the fifth element, the material that filled the Universe beyond the terrestrial sphere. In the seventeenth century, the French philosopher Rene Descartes proposed that ether was composed of tiny atoms like the particles in a gas (see Chap. 1). The earth and planets were pushed around the sun as the ether moved in a large vortex around the sun. There was no such thing as a vacuum and any object that moved had to enter a gap left by another moving object. Descartes suggested that ether explained how light was able to radiate in all directions although his treatise on light was never completed.

It was only natural that Maxwell should turn to ether when he tried to explain the mysterious electrical and magnetic lines of force. Ether was thought to be near weightless yet rigid enough to transmit light waves. It was everywhere yet could not

R. W. Baloh, *Brain Electricity*, https://doi.org/10.1007/978-3-031-62994-5_8

be seen or measured with any known device. According to Maxwell, *"whenever energy is transmitted from one body to another in time, there must be a medium or substance in which energy exists after it leaves one body and before it reaches the other…we ought to endeavour to construct a mental representation of all the details of its action"* [1].

Relationship Between Ether and Matter

A key issue that nineteenth-century scientists who believed in ether needed to address was the relationship between ether and matter. Based on the historical concept of Lucretius' atom, the pervading view was that the world was made up of material atoms embedded in an all-pervading fluid-like ether. But there were other ways that ether and matter could be related. Ether could be made up of tiny particles of matter or matter could be part of the structure of ether.

A strong proponent of the concept of a material ether was William Thomson (Lord Kelvin), Maxwell's fellow Scotsman and mentor. His vortex ring theory of the atom became all the rage in British physics in the latter part of the nineteenth century [2]. The theory considered atoms to be vortex rings in a continuous fluid, ether. All the properties of matter could be explained by motion in this incompressible perfect fluid. The theory had the potential to be a theory of everything and could be described with the known mathematics laws of hydrodynamics. The fundamental entity of nature was an all-pervasive fluid and everything in the universe could be represented by the energy at any point in the fluid. Thompson set out to provide a mathematical description of the model that would include a description of electric and magnetic lines of force.

The beginnings of Thomson's vortex ring theory can be traced to the German physician and physicist, Hermann Helmholtz, and his work on hydrodynamics. We first met Helmholtz in Chap. 5 for his work on the speed of nerve conduction. Helmholtz described what he called vortex lines and vortex tubes that formed closed rings when set in motion in a frictionless incompressible fluid. With the mathematics of hydrodynamics, Helmholtz showed that vortex rings were strong and long-lasting, and he pointed out they might even relate to physical phenomena like electricity and magnetism. A crude example of Helmholtz's vortex rings are smoke rings that persist for seconds but then disperse since the air is not an incompressible perfect fluid.

Helmholtz's work was originally published in 1859, but Thompson did not come up with the vortex ring theory until January of 1867, when his friend Guthrie Tait demonstrated a simple apparatus for producing long-lasting smoke rings. Tait, who was also a close friend of James Maxwell, had read Helmholtz's paper on vortex rings and translated the material into English. On seeing Tait's smoke rings, Thomson had the sudden inspiration that the smoke rings could be an analogy for atoms in ether.

Thomson immediately wrote to Helmholtz outlining his theory: *"The absolute permanence of the rotation and the unchangeable relation you have proved between it and the portion of the fluid once acquiring such motion in a perfect fluid, shows that if there is a perfect fluid all through space, constituting the substance of all matter, a vortex-ring would be as permanent as the solid hard atoms assumed by Lucretius and his followers (and predecessors) to account for the permanent properties of bodies (as gold, lead, etc.) and the differences of their characters. Thus, if two vortex-rings were once created in a perfect fluid, passing through one another like links of a chain, they never could come into collision, or break one another, they would form an indestructible atom; every variety of combinations might exist"* [3]. Thompson publicly announced his new theory in February 1867 in a lecture at the Royal Society of Edinburgh.

Clearly, Thomson was not a fan of Lucretius' atomism. He felt that it was *a "monstrous assumption of infinitely strong and infinitely rigid pieces of matter"* [4]. His vortex theory of ether and matter saw atoms as hard elastic particles that were expansible yet highly stable. The vortex atoms were not just interlocking rings but could be complex interlocking structures called knots that looked like bizarre pretzels (Fig. 8.1). Thomson felt that the increasing complexity from a single ring to the highly complex knot structures could explain the increasing complexity of atoms in different elements.

In1887, Thomson finally published a detailed description of the vortex theory of ether and atoms. His goal was to provide a detailed description of wave transmission in the "*luminiferous aether*". The problem was the mathematicians were not providing a clear picture of wave transmission and Thomson was already becoming frustrated with his model. In the end, he would be the one to raise critical objections against the model.

Fig. 8.1 William Thomson's vortex ring theory. The complex knot structures explained the increasing complexity of atoms in different elements

Around the same time, another British mathematician and physicist trained at Cambridge, William Hicks, introduced what he called the vortex sponge theory of ether. Hicks saw ether as an incompressible fluid filled with closely packed tiny vortex rings. His fluid medium allowed for large vortex atoms as proposed by Thomson but in addition had the tiny vortex rings throughout (sponge-like) that accounted for the propagation of electromagnetic waves.

This new version of Thomson's vortex ring theory was received with great enthusiasm, particularly in the British Isles. Two physicists and associates of Thomson and Maxwell, George Francis FitzGerald and Oliver Lodge, introduced in Chap. 7, became strong proponents of the vortex sponge ether theory. Both believed in a universal medium, ether and each gave his own unique properties to the ether.

FitzGerald felt that there was no real difference between matter and ether. Matter resulted from any type of motion within the ether. He suggested *"a possible theory of ether and matter was that space was full of such (straight and hollow) infinite vortices in every direction, and that among them closed vortex rings represented matter threading its way through the ether…If it be true, ether, matter, gold, air, wood, brains, are but different motions"* [5].

Lodge, who regularly corresponded with FitzGerald, shared the notion that all of nature evolved from ether. In what he called the modern view of ether, Lodge envisioned *"One continuous substance filling all space: which can vibrate as light; which can be sheared into positive and negative electricity; which in whirls constitutes matter; and transmits by continuity, and not by impact, every action and reaction of which matter is capable"* [6]. In other words, the vortex theory of ether was a grand mechanical model of the universe. As we will see shortly, Oliver Lodge was a strong proponent of a role for ether in psychical phenomena such as telepathy and telekinesis and became a prominent member of the Society for Psychical Research. Recall that Lodge's automatic decoherer was a key discovery in the development of wireless telegraphy.

The Unseen Universe

In 1875, two Scottish physicists, Balfour Stewart and Peter Guthrie Tait, published a book entitled *The Unseen Universe: or Physical Speculations on a Future State* in which they attempted to prove that the universe as we know it, the "visible universe" is only a small part of a larger universe, the unseen universe, which was both eternal and infinite [7]. The book became a best seller and served as a springboard for scientists' fascination with mysticism and spiritualism in late nineteenth- century Victorian Society.

Stewart and Tait were both physicists trained at Cambridge, with close ties to William Thompson and James Maxwell, and both shared an interest in metaphysics and a dislike for the scientific trend toward materialism. They started with Thompson's vortex ring theory, which assumed ether to be a perfect primordial fluid in which vortex rings (matter) appeared through divine intervention. Stewart and

Tait felt that the unseen world of ether was infinite but that the visible world was finite, the result of vortex rings coming and going spontaneously within the ether. The visible world was transient *"just as the smoke-ring which we develop from air...is ephemeral, the only difference being in duration, these lasting only a few seconds, and others it may be for billions of years"* [8].

The obvious implication of their premise was that the visible world was finite and would come to an end. *"Immortality is therefore impossible in such a universe"* [9]. But how were the two universes interrelated? Stewart and Tait thought that the visible and invisible universes were interconnected by a transfer of energy from one to the other, secured by the law of conservation of energy. As energy in the visible universe traveled outward in space, it was gradually transferred into the invisible universe. The ether in space was not just a bridge between parts of the visible universe but rather *"a bridge between one order of things and another"* [9].

The *Unseen Universe* was at its core an attempt to unite science and religion. Keep in mind that Cambridge University was a highly religious institution in the nineteenth century and the vast majority of faculty had strong religious beliefs. To Stewart and Tait, the existence of a Creator was taken for granted. One of the goals of their book was to explain the immortality of the soul and religious miracles such as the resurrection of Christ. The Creator produced the present visible universe by transferring energy from the invisible to the visible universe. Miracles that were considered impossible from the prospective of the visible universe could be explained *"as the result of a peculiar action of the invisible upon the visible universe"* [10].

To Stewart and Tait, matter was not fundamental to the overall universe but only to the seen universe. They noted, *"It is only within the last thirty or forty years that there has gradually dawned upon the minds of scientific men the conviction that there is something besides matter or stuff in the physical universe"* [11]. That something was *energy* and it had *"as much claim to be regarded as an objective reality as matter itself"* [12]. Matter and energy were interchangeable as in Thomson's vortex atom theory. Stewart and Tait saw a close link between the science of energy and the existence of spirits and their combined action in the visible world.

William Crookes and the Bridge Between Science and Spiritualism

One of the most enigmatic scientists of the late nineteenth century was the English chemist, journalist, and entrepreneur, William Crookes. His work in spectroscopy revolutionized the field of chemistry and his development of the radiometer and cathode ray tube directly led to the discovery of the electron and much of modern electronics. He was a successful scientist and businessman, yet he regularly took advantage of colleagues and was notoriously self-promoting. He was elected a fellow in the Royal Society at a young age but was frequently disparaged for unscientific speculation. His lifelong foray into the fringes of science with the study of spiritualism caused many of his colleagues to question his scientific qualifications and even his sanity. Despite or because of all this, by the end of the century, he was a public celebrity and was awarded the Order of Merit by King Edward VII in 1910.

William Crookes was born in London in 1832, the son of a successful tailor and investor, Joseph Crookes [13]. Overall, Joseph had 21 children, 5 by a first wife, who died in childbirth, and 16 by his second wife, 8 of which survived into adulthood. William was the oldest child of the second wife. He grew up in an affluent household with multiple servants and his basic education was provided by a tutor. He showed an interest in science at an early age even though no one else in the family had any interest in mathematics or science. The Royal Polytechnical Institution with its scientific lectures, exhibitions, and demonstrations was just a short walk on Regents Street from the Crookes family home.

After a brief time at boarding schools, in 1848, at the age of 16, William enrolled at the newly opened Royal College of Chemistry (RCC) just a few miles from his home. The college was more of a technical school than an academic institute and most graduates entered industrial or commercial careers. Crookes later considered enrolling at Oxford University, but he lacked the academic qualifications to pass the entrance examination. In many ways, the RCC provided ideal training for his talents and temperament. He developed hands-on experience with chemical reactions that were crucial for his studies in photography and spectrometry, and he became a skilled glass blower that proved useful for developing the radiometer and cathode ray tube.

The Director of the RCC, the German chemist August Wilhelm Hofmann, had a special interest in coal tar and its derivatives. Hofmann visited the laboratory and talked with each student at least twice per day and he rapidly took an interest in Crookes. At the end of the first year, Crookes received the College's only prize, the Ashburton Scholarship, which allowed him to work directly with Hofmann. Crookes remained Hoffman's personal researcher up until he finished his training at the RCC in the summer of 1854. In that role, he had a chance to meet scientists that would play an important role in his later career.

Crookes attended several meetings at the Royal Society with Hoffman where he saw presentations by leading British scientists including his hero, Michael Faraday. The secretary of the Royal Institution, The Reverend John Barlow, decided to take a course in analytic chemistry and Hofmann assigned Crookes to be his tutor. Barlow subsequently invited Crookes to attend several Royal Institute sessions in his home where Crookes had the opportunity to meet some of the top British scientists that he would later call upon to help him with his investigations. At a session in 1850, he met the Cambridge mathematician George Gabriel Stokes, who at the time was working on ultraviolet light, a subject that Crookes was trying to investigate with photography. They developed a working relationship and corresponded regularly over the years. On numerous subsequent occasions, Crookes turned to Stokes to help him understand the mathematics in published papers or to provide a mathematical interpretation of his own work. Crookes had little interest or training in mathematics. Crookes' biographer William Brock referred to Stokes as one of Crookes' many *"invisible helpers"* who throughout Crookes' career provided important contributions to his work but did not receive proper recognition from Crookes [14].

It was at one of Barlow's home sessions in 1853 that Crooke met Michael Faraday and witnessed his famous table-turning experiment. At the time, so-called modern spiritualism was rapidly spreading from America to Britain and the European continent. Home seances and public lectures given by entranced mediums were becoming all the rage. One of the common demonstrations was so-called table turning. The medium and other participants sitting around a table placed their hands on the table and then called upon disembodied spirits to turn the table [15]. Miraculously the table would begin to turn without any apparent mechanical force. A leading skeptic of spiritualism, the physiologist, William Benjamin Carpenter, asked Faraday to help him prove that the force of table turning was not from spirits but the subjects participating in the experiment were involuntarily pushing the table in response to strong expectation that the table would turn.

Carpenter was a highly respected scientist who was able to convince Faraday to participate in the experiment even though Faraday preferred not to become involved in such public spectacles. Faraday's device consisted of a small tabletop made of two thin boards with glass rollers in between so that the upper board could slide a limited amount over the lower board. The boards were fastened together by rubber bands. A rod attached to the center of the upper board monitored the amount of turning. When in a staged seance the experimental subjects placed their hands on the table, the upper board began to turn. If Carpenter informed the participants that they were involuntarily turning the tabletop, the phenomena stopped. Carpenter and Crookes would later become embroiled in a heated dispute over the claims of several mediums that Crookes supported.

Spectroscopy and the Discovery of Thallium

Remarkably, it was Crookes' research in the physical sciences that provided the impetus for his venture into the spiritual world. Throughout his 6 years at the RCC, Crookes became enamored with the newly developing field of photography, both the chemistry and the physical optics. He became an avid photographer experimenting with different chemicals in the development process. In 1852, based on the interaction with physical scientists he had met at Barlow's home, Crookes decided to photograph the refraction patterns of different crystals placed in polarized light. In the process, he invented what he called, a spectrum camera. He published a description of the camera in 1856 and patented the process 5 years later. He used the sodium spectrum as a standard noting that sodium produced a bright double yellow line, but he failed to recognize the meaning of the lines in terms of absorption and emission of light. George Stokes would later develop the mathematical theory to explain the phenomena.

Around the same time in Germany, the chemist Robert Bunsen (of Bunsen burner fame) and the physicist Gustav Kirchhoff demonstrated how the color pattern given off by different elements in a flame were unique to each element [16]. The emission spectrum, a series of colored lines, provided a signature for each element. The process became known as spectroscopy, and it provided a technique to discover new elements not previously known. As soon as Crookes read of Bunsen's and Kirchhoff's finding, he made minor changes in his spectrum camera to make a spectroscope and began to search for new elements. He saw this as a quick way to establish his credentials as a bonified scientist. The problem was time to work on the project. Crookes was busy with other projects including compiling patents and editing articles for his new journal *Chemical News*. He sought help from a former colleague at the RCC, Charles Williams, who was working as an analytical chemist in Glasgow, Scotland. Crookes asked Williams to be his partner in the project noting that there were numerous new elements waiting to be discovered and that he personally had seen "*several suspicious looking spectra*" [17].

In early March 1861, Crookes examined deposits from the Tilkerode mining district in Germany that he originally received from Hoffmann while at the RCC and noticed a bright-green line on the spectroscope located between the lithium and sodium spectral lines. He immediately alerted Williams and indicated that he would send him some of the material so that they could work on the project together. To absolutely confirm that it was a new element, Crookes and Williams would have to isolate enough of the element from the raw material to establish its chemical reactions and determine its atomic weight. Based on confirmation of the new spectral line from Williams, Crookes published a report in *Chemical News* in late March 1861 describing his detection of the green line, its location between the lithium and sodium lines and speculation that it probably represented a new element [18]. There was no mention of William's contribution.

Crookes had to first be sure that his green line was not due to one of the 50 known elements at the time. When Crookes and Williams analyzed other mining deposits for the proposed new element, Crookes found that deposits from a Sulphur mine in Sicily had a bright green line and he wondered if the green line might be part of the Sulphur spectrum. After trying different methods, he developed a separation technique using zinc to precipitate a black powder from the deposits that showed the bright green line without lines of any other known element. He was convinced that he had discovered a new element which he named thallium.

Crookes' next challenge was to extract enough of the new element characterize its reactions and determine the atomic weight [19]. In his search for new deposits rich in thallium, he discovered a rich source of the new green line in copper pyrite ore from a mine in Belgium. After arranging for delivery of two tons of the ore, he sought the help of fellow chemist and photographer from his RCC days, John Spiller, and the two feverishly worked on extracting thallium from the pyrite ore in a specially designed furnace at Brook Green. While this was going on, Crookes became aware that a French chemist and physicist, Claude-Auguste Lamy was about to report to the French Academy of Sciences that he had independently discovered the green line of thallium in mineral deposits from a French factory and that he had extracted a large quantity of the new element for characterization. Lamy had a 14 g ingot of thallium which he noted resembled metallic lead and had chemical reactions resembling an alkali metal.

Crookes immediately contacted his friend William Allen Miller who was Vice-President of the British Royal Society informing him of the situation and that Lamy was likely ahead of him in characterizing the new element. They decided that it was critical to British interests that Crookes should present a paper before the Royal Society before Lamy's presentation at the French Academy. Crookes rapidly prepared a short paper to present at the Royal Society meeting on June 19, 1862, 4 days before Lamy's scheduled presentation. Crookes was the sole author on the paper. By this time, Crookes and Spiller had extracted enough thallium to better describe its physical and chemical properties. It was a metal like lead that was soft and easily drawn into a wire. Crookes made thallic acid and various salts and identified at least two thallic oxides. He sent samples to Michael Faraday who found that thallium was diamagnetic.

Not surprisingly, there was a battle over priority in the discovery of thallium. For his part, Crookes argued that he clearly had been the first to identify the characteristic green line of thallium on spectroscopy. Lamy noted that he had independently identified the green line and only later became aware of Crookes' March 1861 report in *Chemical News.* Furthermore, a chemist could only claim to have discovered a new element after having isolated a significant amount of the element. Lamy had a large quantity of thallium and was clearly the first to show that it was a metal like lead.

Adding to Crookes' frustration, Lamy was awarded a medal "*for the discovery of new and abundant sources*" of thallium at an International Exhibition held at Hyde Park in London the summer of 1862. Crookes wrote to the Commissioners of the Exhibition claiming that an injustice had been done and that he was the rightful discoverer of thallium. "*This my Lords and Gentleman, is what I protest against. I do not ask for a prize, but I do ask for justice. The honour of possessing an Exhibition medal would be insignificant in comparison to that which I have already gained by being the first Englishman since Sir Humphry Davy's time, who had discovered a new element; but I feel most keenly the injustice with which I have been treated. M. Lamy has received all the honours of the discovery, whilst I have been quietly snubbed*" [20].

The British commissioners were embarrassed. British honor was at stake. They were being accused of intentionally snubbing a British scientist while honoring a French scientist. Their response was to claim that Crookes had won a prize, but his name was inadvertently omitted from the prize list. In a revised edition of the prize list, Crookes was awarded a medal for the discovery of thallium and Lamy's medal was only for isolating a small ingate of the new element. Crookes then began a press campaign to solidify his claim to the discovery of thallium. He argued that he clearly had been the first to publish the discovery of the new element and that by the time of the International Exhibition, he had already isolated a significant quantity of thallium and demonstrated its physical and chemical qualities. Not surprisingly, the

French press supported Lamy's priority claim that he was the first to isolate and characterize the new element.

As a clear reward for his discovery of thallium, Crookes was elected a Fellow of the Royal Society. This was a remarkable achievement for a 30-year-old chemist and journalist without academic credentials. William Carpenter who was on the Royal Society council hinted that there was opposition to Crookes' election, but his supporters won out. A useful benefit of his Fellowship was it enabled him to obtain a government grant administered by the Society to determine the atomic weight of thallium. As we will see, it was during these studies that he made observations that ultimately led to the discovery of the radiometer and the cathode ray tube (Crookes tube). But before addressing these studies, we will take a side tour into Crookes' studies of spiritualism which also influenced these discoveries.

Crookes' Psychic Force

When Crookes attended the demonstration by Carpenter and Faraday at Barlow's home in 1853, he completely agreed that table turning was a hoax and could be explained by psychological and mechanical mechanisms. Subsequent comments in his journal *Chemical News* indicated that he was highly skeptical of spiritualists and their extravagant claims [21]. But this changed after the death of his youngest brother Philip in September 1867. Philip was the only member of the family other than William who showed an interest in science, planning a career in electrical engineering. He died from yellow fever while on a voyage to the Caribbean to lay telegraph cables. The entire Crookes family was deeply affected, and William had a period of deep depression. At the suggestion of friends, he and his wife attended a séance in search of some type of solace for their loss.

The birth of modern spiritualism can be traced to two American teenagers, Margaret and Kate Fox, who claimed to receive messages from the spirit world in the form of rapping noises in their home in Hydesville, New York, in 1848 [22]. Over the next several years, they along with a third sister began giving public demonstrations of the rapping as a form of communication between the living and the dead. Many thought that the mystery of electricity and electrical telegraphy provided an analogy for how such communications might occur. In 1855, two teenage American brothers, Ira and William Davenport developed the so-called dark séance in which they communicated with a spirit named John King who talked to them through a trumpet. The brothers were tied up and locked in a closet while musical instruments played and flew about the room, people were nudged, hats flew off, and John King spoke. The Davenports argued that darkness was necessary for the spirits to feel safe to communicate but it also provided a cover for deceit.

In the early 1860s, many American spiritualists came to Victorian England where interest in occultism was on the rise and people were willing to pay money for the experience. Even scientists such as Crookes began to see similarities between the mysteries of spiritualism and the mysteries of electricity and electromagnetism.

Crookes and his wife attended their first séance in 1868 given by a new English medium, 18-year-old Frank Herne. During the séance, Crookes' deceased brother Philip took control of Herne's body and spoke to Crookes so convincingly that Crookes was sure that it was Philip. Young mediums such as Herne, the Foxes, and the Davenports all had managers who took care of arrangements and provided support and information for the seances. There was a great deal of publicity in the press about Philip's death and the fact that William had considered suing the telegraph company for negligence. Herne was no doubt prepped in advance with information about William and Philip from his manager. Crookes was hooked on spiritualism, and he was convinced that it could be studied with scientific methods just like other unexplained phenomena. As we will see later, it was around this time that he was beginning to investigate the mysterious forces that led to his notion of a "psychic force" and the development of the radiometer.

Through Herne, Crookes met and attended seances of several other London mediums. Crookes began to search for a medium who he trusted enough to work with him on his scientific experiments. He chose Daniel Dunglas Home, a popular medium recommended by several of his colleagues. Beginning in early 1870, Home conducted a series of seances in the dining room of Crookes London house. He enthralled the Crookes family with floating objects, self-levitation, and handling hot coals. Unlike most other mediums he did not demand total darkness and allowed himself to be searched. Crookes became convinced that Home possessed an unusual force associated with his body that allowed him to perform gravity-defying fetes.

Home had an interesting background, to say the least [23]. He was born in Edinburgh and was adopted shortly after by an aunt who emigrated to America when he was age 9. After seeing an exhibition by the Fox sisters, at age 22, Home began "spirit rapping" on his own. Following the trend of the time, he decided to move to London and then to the salons of Europe and even Russia. During his travels, he was banned from Rome, declared a "sorcerer" by the pope, and married a wealthy Russian aristocrat who died shortly after of tuberculosis. He became involved in a lawsuit over her considerable fortune, but the family won out and he left Russia and returned to London in 1866.

In London, Home managed to become secretary of the Spiritual Athenaeum, founded by the wealthy Irish journalist and ardent Christian spiritualist, Samuel Carter Hall. In this esteemed position, Home met Jane Lyon, a wealthy widow who wanted to contact her dead husband. During a series of seances with Home over the next year, on instructions from her dead husband, Mrs. Lyon gave Home 60,000 pounds for his spiritualist work and made him the sole beneficiary in her will. However, rumors from disgruntled competitors that Home was a fraud reached Mrs. Lyon, and she sued for return of her money. In a highly publicized trial, Home was found to have taken advantage of the defenseless widow and had to return her money.

With this less-than-stellar history, how could Crookes have selected him for his experimental studies? Strong supporters of Home, even after the Lyon debacle, were Crookes' friends, the telegraph engineer Cromwell Varley and his clairvoyant wife, Ada. In the early 1850s, Varley suggested that table rapping was produced by some type of electric force. Another Crookes friend, the astronomer James Lyndsay, claimed that he had personally observed Home levitate out of one window and return through an adjacent window during a séance. These endorsements from respected colleagues convinced Crookes that Home was authentic.

In May 1871, Crookes began a series of experiments with Home in the dining room of his house. He invited a friend, the astronomer, and Vice-President of the Royal Society, William Huggins, to work with him and verify the results of the experiments. Home began by playing an accordion without touching the keys demonstrating his apparent ability to exert a force at a distance without mechanical aid. To ensure that there was no chance for deception, Crookes placed the accordion in a wire cage under the dining table and Home's hands were tied. In the second "scientific" experiment, Crookes attached one end of a thick mahogany plank to a spring balance hung from a tripod and the other end rested on the edge of the dining table. He had Home place his fingers on the end of the plank near where it touched the table and had him exert maximum force. Crookes and Huggins noted the spring balance registered from 1 to 4 kg. Crookes then stood with his entire weight on the plank at the exact location where Home had placed his fingers and the spring balance registered 2 kg.

Crookes was convinced that they had witnessed some new type of force generated by Home's body, a *psychic force*. He felt that it was his duty to publish the "experiments" and sent off a paper to the Royal Society in June 1871. He placed his old friend, the mathematician George Stokes, one of two Society secretaries, in a difficult position. Stokes was critical of the experimental conditions and questioned whether Home could have used a lever effect. There was no information about lighting conditions and the positions of the different participants. Based on Stokes' feedback, Crookes constructed a much more sophisticated apparatus with gears and pulleys and measured the force on a smoked-glass plate that moved horizontally by a watch spring mechanism [24]. He had Home hold his hand above the apparatus without touching it and placed a container of water between the hand and the apparatus. Home was still able to register a measurable force although less than in the initial experiment. But Stokes and other influential Society members "*had heard too much of the tricks of Spiritualists*" and even after revisions the paper was rejected by the Royal Society [25].

Crookes tried the British Association for the Advancement of Science, but the paper was also rejected so he finally published it in his own *Quarterly Journal of Science* in October 1871. In the paper, Crookes quoted his correspondence with the Royal Society noting that the established scientific establishment was being unreasonable. His friend Cromwell Varley who had just been elected to the Society, admonished his fellow members for exhibiting a double standard. "*If there is one thing that is clear about the progress of modern science, it is the tendency to reduce all scientific problems...to problems of molecular physics—that is, to attractions, repulsions, motions and co-ordination of the ultimate particles of matter. Yet these ultimate particles, molecules or atoms are creatures of the imagination and as pure assumptions as the spirits of the Spiritualists*" [26].

At the same time, Crookes' old nemesis William Carpenter heartily agreed with the skeptics in the Royal Society. He pointed out that Crookes' efforts were completely inadequate to deal with Home's trickery and that Crookes provided no objective evidence of a psychic force. He even went so far as to suggest that physical scientists like Crookes "*lacked the proper mental appliances*" to objectively investigate spiritualists. The physicist, Belfour Stewart, mentioned earlier for his speculation on an unseen universe, suggested that Home may have exerted an "electro-biological" influence (hypnosis?) over Crookes, causing him to think he had made an objective measurement of Home's psychic force [25].

Having taken a beating from the established scientific community regarding his psychic research, Crookes decided to turn to a more receptive audience, the spiritualistic community. From that point onward, he published his more traditional scientific work in established scientific journals and psychic research in spiritualistic journals. For the most part, he was able to separate the two, although he truly believed that his scientific and spiritualistic research would eventually come together.

Over the next 20 years, while Crookes was publishing new experiments regarding his radiometer and cathode ray tubes in scientific journals, he was publishing claims of witnessing ghosts and recording psychic forces in spiritualistic journals. Probably the most bizarre of all of Crookes' spiritualistic investigations occurred with the English medium Florence Cook. Crookes began working with Florence in the early 1870s and maintained contact with her up to the time of her death in 1904 [27].

In early 1872, Florence began conducting a series of seances in which the spirit Katie King would physically materialize. Katie was the daughter of John King, the spirit contacted by the Davenport brothers. Florence was tied up with tape in a cabinet from which Katie appeared in a white robe. After speaking with the guests, Katie reentered the cabinet, and after a short time, the cabinet was opened revealing Florence still tied with tape. Supporters of Florence approached Varley and Crookes to conduct a séance with Florence to document her authenticity. In the meantime, a bizarre event occurred in one of Florence's seances. A member of the audience, who became convinced that Katie Fox was Florence, grabbed the ghost, and after a struggle, Katie broke free and ran back into the cabinet. There was a delay before anyone entered the cabinet and when they did, Florence was still tied with tape. This highly publicized event was apparently arranged by a rival medium to discredit Florence.

Varley and Crookes agreed to a séance at the London home of Florence's manager in which they could control the events. At the onset, Crookes' wife Ellen and other ladies searched Florence and after she was tied up, the knots were sealed with wax and stamped with Crookes' signet ring. In place of a cabinet, Florence was tied up in a curtained adjoining room. A few minutes later Katie King appeared from behind the curtain and announced that her medium was not well. Crookes described hearing sobbing behind the curtain. After the spirit returned to behind the curtain, Varley and Crookes entered the room and examined Florence, noting that Crookes' seal was intact.

In a subsequent seance also at her manager's house, Crookes used an electrical test devised by Varley to prove that Florence and Katie could not be the same person. They hooked Florence into an electric circuit with telegraph wire attached to her wrists with moistened blotting paper. Florence's body resistance was measured with a mirror galvanometer (the latest technology) located in the séance room so that Florence could be monitored throughout the event. Katie appeared from the curtain without any obvious attached wires and except for a brief fall and rise in the galvanometer reading there was no break. Crookes was convinced of Florence's authenticity, and she became a regular guest at Crookes' home often remaining for a week at a time. In a subsequent seance, Crookes asked permission to grasp Katie's arm and he stated that she was a material being. In another session, he was allowed to enter the adjoining room to see Florence although he was never able to clearly see the two ladies together at the same time. In an article published in *The Spiritualist,* Crookes wrote that "*enough has taken place to thoroughly convince me of the perfect truth and honesty of Miss Cook, and to give me every reason to expect that the promises so freely made to me by Katie will be kept*" [28].

Much has been written about Crookes and Florence Cook both regarding their personal relationship and Crookes investigations of her spiritualistic powers. Some suggested that Crookes had an affair with the very attractive young lady and actively participated in the deceit but there is little concrete evidence for such an affair. Why didn't Crookes simply restrain Katie and check if Florence remained in the adjacent room rather than using the elaborate electrical circuit to monitor Florence. Crookes did note a similarity of the facial features of the two young ladies, the room was darkened, and Crooke had notoriously poor far vision. On the other hand, others noted that Florence and Katie did not look alike suggesting that Florence must have had an accomplice. Regardless, one must marvel at Crookes' gullibility.

The final phase of the Crookes and Katie King saga occurred in May 1874 when Crookes announced that he was permitted to photograph Katie in his laboratory. The week-long photographic session with Florence and Katie was so bizarre that one might wonder whether Crookes' entire family was involved in a grand conspiracy. Adding to the drama Katie announced that she would no longer materialize after the session. The photographic sessions took place in the laboratory in front of several invited guests and Crookes family members with an adjoining library serving as the medium's dark room. Only Crookes was allowed to move about and enter the medium room and Crookes reported that on a few occasions he saw both Katie and Florence together although usually only Florence. One condition imposed was that the medium's head was always covered by a shawl to prevent light from hitting her face and breaking the trance. In the only photograph of the two together, Katie was seated in front of Florence blocking any view of her face.

In one session Crookes took photographs of Katie and Florence standing with him and with his wife and on comparing the photos, he noted that Katie was taller and bigger than Florence. This suggested that there must have been an accomplice. But if so, she was not recognized by Crookes or members of his family. On one occasion, Crookes described Katie sitting with his children around her, telling amusing stories of her adventures while visiting India. He noted a wide range of facial expressions as the ghost told her story. Overall, Crookes took a total of 44 photographs, but unfortunately none of the originals and only a few copies have survived. The originals were destroyed after Crookes' death probably by a family member trying to protect his scientific integrity. The best reproduction of Katie's appearance is a watercolor portrait made by the Irish author J. Hawkins Simpson from one of Crookes' original photographs. In it, Katie is shown standing with her eyes closed in a dream-like state wearing a long white robe cut low to reveal her ample bosom. She held a bouquet of lilies.

To Crookes' dismay, his old friend Daniel Home wrote a book in 1877 that revealed the tricks spiritualists used to fool the public including many of the ones he had used with Crookes [29]. For example, he described how he used a small harmonica in his mouth easily concealed by his mustache and beard to produce the accordion music from afar. Like a ventriloquist, he directed attention to the accordion by looking that way. Indeed, he used many of the techniques and apparatus used by illusionists and magicians who were also popular at the time. Crookes admitted that there was fraud being practised by some spiritualists, but he remained convinced that his experiments proved the existence of a psychic force.

In a report made to the Society of Psychical Research in 1889, shortly after Home's death, Crookes described in detail more than 30 scientific sessions he conducted with Homes. He described lights flashing, objects flying across the room, and whispers in people's ears. In his experiments, he tried to block Home's psychic force by electricity, magnetism, and electromagnetic waves all without success. His report concluded: *"The evidence for the genuineness for the phenomena obtained by Home in my presence seems to me to be strengthened by the discussions of conjuring, and the exposures of fraud which have since taken place. The object of such discussions is to transform vague possibilities of illusions and deception into definite possibilities; so far as this has yet been done, it has I think, been made that certain of Home's phenomena fall quite outside the category of marvels producible by sleight of hand or prepared apparatus"* [30].

British Society of Psychical Research (SPR)

The Society of Psychical Research (SPR) that received Crookes' report was formed in 1882 with the stated goal of providing a scientific response to the proliferation of lay groups and associations interested in paranormal phenomena in Victorian Britain. The members were *"to examine without prejudice or prepossession and in a scientific spirit those faculties of man, real or supposed, which appear to be inexplicable in terms of any generalized hypothesis"* [31]. The SPR boasted as members many of the top academic mathematicians and physicists of the time. The members varied in their degree of skepticism and belief.

The first president of the SPR, Henry Sidgwick, was a Cambridge Trinity Graduate in Classics who eventually became the Knightsbridge Professor of Philosophy. He had a long-standing interest in psychic research through the Cambridge "Ghost Society" and his role as a philosopher and Anglican apostate. Sidgwick was generally considered fair and honest in his evaluation of psychic events and despite having attended numerous séances, he admitted that he never personally observed what he considered convincing evidence of paranormal phenomena. By contrast, William Crookes who later became president, was a strong proponent of spiritualism claiming to have witnessed many paranormal phenomena. Two other prominent members who were strong proponents were Frederic Myers and Oliver Lodge.

Myers became interested in the paranormal through Sidgwick who was his tutor at Trinity College in the 1860s. Myers went on to be a Lecturer at Trinity College and later a well-known poet, critic, and essayist. Myers used what he considered to be scientific evidence of the unseen universe to support the science of the human psyche and the existence of an immortal soul. An admirer of the eighteenth-century scientist, theologian, and mystic Emanuel Swedenborg, Myers believed that there was a clear relationship between the natural (physical), the spiritual, and the divine worlds. He likened Swedenborg's effort to bring scientific discovery into the realm of the supranatural to that of his contemporaries in the SPR.

Myers summarized his ideas on the role of the unseen world on the human psyche and soul in his book *Human Personality and Its Survival of Bodily Death*, published in 1903 [32]. The book, long and rambling at times, addressed normal and abnormal human psychology based on understanding of physics at the time. Myers introduced the terms *supraliminal* and *subliminal* to describe the mental life that goes on above and below the threshold of consciousness. On the subliminal end, consciousness merged into organic operations and on the supraliminal end, it merged into reverie or ecstasy. It was in these extremes, above and below the level of consciousness, where paranormal phenomena and the soul resided.

For an analogy of the human psyche, Myers turned to the spectrum of electromagnetic waves. Consciousness was represented by the visible spectrum of light rays coming from the sun. "*Beyond each end of that prismatic ribbon are ether-waves of which our retina takes no cognisance. Beyond the red end come waves whose potency we still recognise, but as heat and not as light. Beyond the violet end are waves* [for example x-rays] *still more mysterious; whose very existence man for ages never suspected, and whose intimate potencies are still but obscurely known. Even thus, I venture to affirm, beyond each end of our conscious spectrum extends a range of faculty and perception, exceeding the known range, but as yet indistinctly guessed*" [33]. Based on our current understanding, this statement is more prescient for its description of the spectrum of electromagnetic waves than for its insight into the human psyche.

Based on his analogy, Myers addressed the cause of hysteria and its treatment with hypnosis. Hysteria was essentially "*an undue permeability of the psychical diaphragm, or a confused interchange of elements which should lie above and elements which should lie below the threshold of waking consciousness*" [34]. As an example, he noted that hysterical anesthesia (numbness) was not due to a loss of sensation but rather a lack of attention to the affected area. The anesthesia did not follow nerve distributions but was the result of "self-suggestions" originating in the subliminal centers. With regard to hysterical weakness, he noted that a hysteric patient who had a feeble grip when asked to squeeze a dynamometer could produce great force when influenced by emotions originating in the subliminal centers. In other words, the barriers between the subliminal and supraliminal centers and waking consciousness broke down.

Myers saw sleep as a normal phase of the human personality part of the relationship between the supraliminal and subliminal self. Hypnosis was an intense form of sleep analogous to abnormal sleep states such as somnambulism. Suggestion was the best way to induce this unusual sleep state. *"Hypnotic suggestion intensifies the physical recuperation of sleep and aids the emergence of those supernormal phenomena which ordinary sleep and spontaneous somnambulism sometimes afford"* [35]. Myers observed that hypnosis although long opposed and ignored by orthodox science had become an accepted treatment for hysteria along with psychotherapy. As evidence, he quoted case reports by Janet and Freud to support this contention.

Finally, in the last half of his book, Myers addressed evidence for an immortal soul, messages from spirits after death, and transmission of thought from one living person to another. He felt that it was possible to study these phenomena with scientific observation just like any other hypothesis and that he personally had observed all three phenomena. They existed in the subliminal and supraliminal self, outside the realm of consciousness just like electromagnetic waves existed outside the spectrum of visible light.

Oliver Lodge and the Science of Immortality

In 1883, shortly after Oliver Lodge received his appointment to the University College, Liverpool, he was approached by two of the founding members of the Society of Psychical Research (SPR), Frederic Myers and Edmund Gurney, to help them with a psychic experiment [36]. Two young girls working at a Liverpool drapery were reported to have a remarkable ability to read minds. Myers and Gurney felt that Lodge's reputation as a bonified scientist made him an ideal participant. Lodge initially resisted because he was skeptical of the reports, and he was concerned for his reputation, but eventually, he gave in and agreed to participate.

Conditions for agreeing were that the young girls were brought to the college for the experiment and Lodge could enlist two friends, a headmaster and biologist, from the college, who were unknown to the girls. After completing a series of tests with the help of his friends, Lodge was convinced that the girls did have supranatural powers. Two examples of the tests were: (1) with one of the girls blindfolded and sitting behind a screen, Lodge's friends were asked to stare at two different objects, an open square and a large X, the girl drew a square with an X inside, and (2) his friends stared at the cutout shape of a teapot and the girl drew a picture of a silver duck that Lodge felt could be a crude reproduction of the teapot. One of his friends noted that he was thinking that the teapot looked like a duck.

At the time, Lodge started reading the work of Maxwell and Heaviside and was enamored by the notion of energy in space in the form of electromagnetic waves. In his report on the experiments with the mind-reading girls published in the Proceedings for the Society of Psychical Research in June 1884, Lodge wrote:

"That the brain is the organ of consciousness is patent, but that consciousness is located in the brain is what no psychologist ought to assert, for just as the energy of an electric charge, though apparently on the conductor, is not in the conductor, but in all space round it, just as the energy of an electric current, though apparently in the copper wire, is certainly not all in the copper wire, and possibly not any of it; so it may be that the sensory consciousness of a person, though apparently located in the brain, may be conceived of as also existing like a faint echo in space, or in other brains, though these are ordinarily too busy and pre-occupied to notice it" [37].

Lodge saw an analogy between "brain waves" and electromagnetic waves in the ether of space. Consciousness was not confined to the brain but existed in the surrounding space just like electromagnetic waves surrounded moving electric current. Some perceptive individuals had the ability to tune into the energy field of the consciousness of others. In other words, thoughts traveled through the ether just like electromagnetic waves. Having had this revelation, Lodge felt that it was important for scientists to study brain waves just like any other physical phenomena and he joined the Society for Psychical Research.

For the next few years, Lodge avoided any association with the Society's studies of spiritualism and the afterlife because of the potential *"odor of vulgar sensationalism"* [38]. In 1889, however, Frederic Myers became ill and asked Lodge to take his place at a séance to be given by an American woman, Mrs. Piper, sent from Boston by the famous American psychologist William James. Lodge's experience at the séance would change his perception of the afterlife.

When William James, the founding father of American psychology, began teaching at Harvard in 1874, Boston was a hotbed of spiritualism [39]. Séances were a popular form of evening entertainment. Like many of his colleagues, James was initially skeptical of psychics, and when attending his first séance, he quickly concluded that the medium was a fraud, moving a piano not by psychic power but with the aid of a "skillful knee". But this would change after he and his wife, Alice, attended a séance given by Leonora Piper shortly after the death of their son, Herman.

Piper was an ordinary middle-class housewife who apparently was not interested in fame although she did accept payments for her services. She allowed her sessions to be recorded by a stenographer and was open to psychical research. Piper would typically enter a trance and report messages from the dead often through a "Dr. Phinuit" but for James through a diseased former psychical research colleague, Richard Hodgson. James was impressed with the authenticity of Piper's séances which he attended on and off over a 13-year period. Piper became a friend of the James family, and she even spent a week with them at their summer home in New Hampshire. During a visit with his brother Henry in England in 1882, James was introduced to the group of scientists that founded the Society for Psychical Research and a few years later, he joined the Society and launched a branch in the United States.

Piper's English séance was held in Cambridge sponsored by the SPR with Lodge and multiple other members present to assure authenticity. While in a trance-like state, Mrs. Piper made connections with a variety of spirits that entered her range of reception. Then suddenly, Lodge heard the voice of his long-dead Aunt Anne speaking to him from someplace in the room. Lodge was sure he recognized her voice when she reminded him of her promise to come back if she could [38]. Mrs. Piper also reported communications from other dead relatives with personal information that Lodge felt only he could know. Lodge saw the snippet of his Aunt Anne's consciousness as evidence of an immortal soul. In his report for the Society Proceedings, Lodge wrote, *"[T]wo worlds, the living and the dead, had at last been introduced, and surely science could have no more important invitation, no course of investigation of greater human consequence"* [38].

As suggested earlier, Lodge was a strong proponent of William Thomson's vortex motion in ether theory [2]. He wandered whether just like electromagnetic waves human thoughts rippled through the ether. In 1891, Lodge built what he called an "*ether whirling machine*" in his laboratory in Liverpool consisting of two massive iron discs less than an inch apart, spun by a powerful dynamo to achieve speeds up to 4000 mph. The goal was to see if the rapidly revolving matter would drag some ether along with it. He used a beam of polarized light sent around the annular space to see if the moving ether would change the speed of the projected light. Despite numerous trials, he found no evidence of moving ether, but he was convinced that experiments like this were the only way to understand the physical nature of ether.

Remarkably, Lodge's best-known foray into the spirit world occurred shortly after he made his famous demonstration of wireless transmission before the British Association for the Advancement of Science in 1894. He along with Frederic Myers and a Polish philosopher, Julian Ochorowicz, sailed to a remote island off the coast of France at the invitation of the French physiologist and future Nobel Prize winner, Charles Richet [36]. The location was chosen since it was deserted except for Richet's small cottage at one end and a lighthouse at the opposite end. The secluded island was thought to be an ideal location to objectively assess the highly publicized supranatural powers of a Neapolitan laundry woman, Eusapia Palladino.

Lodge and his fellow scientists attended an "experimental" séance organized at the cottage late in the evening after arrival. Eusapia and the investigators sat around a table in the main room. Richet's private secretary sat just outside the window taking notes and making sure no one entered or left the room. To prevent Eusapia from manipulating the experiment, one man under the table clasped her legs and another behind her held her neck. Men on each side held her hands. Wires were run across her body to see if some form of induction might occur due to electromagnetic waves.

Despite these restrictions, Eusapia put on a show. According to Lodge's published report, large pieces of furniture flew about the room, curtains gathered into the form of a phantom face, and an accordion played as the walls of the cottage shuddered. While in a deep trance Eusapia excreted a white substance that congealed into what appeared to be phantom limbs protruding from her body. The investigators indicated that they could feel themselves slapped, pulled, and even pinched by the phantom limbs. Lodge speculated that somehow Eusapia used ether that was part of her own biological mass to project forms external to her body. Richet suggested that it might be "ectoplasm" the medium's own vital protoplasm projecting outside of her body.

It is hard to read the bizarre happenings at the séance without scratching one's head and wondering how could this elderly uneducated Neapolitan laundry woman have bamboozled such an elite group of nineteenth-century scientists.

Crookes' Radiometer

Around the time that William Crookes was performing his experiments with the medium Daniel Home, he was also conducting the tedious process of trying to determine the atomic weight of thallium [13]. In a process that took more than 8 years to complete, he discovered what he considered to be a new type of radiation that he called the "*fourth state of matter*".

A key participant in this endeavor was a new young laboratory assistant, Charles Gimingham, who moved into Crookes' home and ran the laboratory while Crookes was busy with his many business and journalistic activities. Gimingham served more like a modern-day graduate assistant than a laboratory technician. He soon became a talented glass blower, and he developed a new mercury vacuum pump apparatus that allowed them to obtain precise weight measurements of samples in a vacuum. In the process of weighing thallium samples in a vacuum, Crookes was surprised to find that warm samples were slightly lighter than cool samples. He assumed this was not due to air currents from the heat since it occurred in a near vacuum. In discussions with Gimingham, Crookes wondered whether heat produced some type of radiation that could alter gravity.

To answer this question, Crookes and Gimingham needed a device to measure the proposed radiation. They began with two tiny pith balls at the ends of a piece of straw balanced on a needle pivot at the center. The whole apparatus had to be small enough to fit inside a glass tube attached to the vacuum pump. When they brought a lamp below the apparatus, the pith ball nearest the lamp moved slightly down toward the lamp but as they gradually evacuated the air from the tube, the effect became less and then disappeared. But with further evacuation, they were surprised to see that the pith ball was vigorously repelled by the lamp. Even the warmth of a human finger was enough to repel the pith ball in the vacuum.

Crookes next decided to test the effects of sunlight. The repulsive effect of the sun's rays was so strong that it almost broke the apparatus. It seemed to them that the light rays were pushing the pith ball. Had they identified a new mechanical effect of light? Crookes then tried passing the light through a prism to test the effects of different colors. The maximum effect occurred with colors near the red end of the spectrum.

It is quite clear that in these early stages of development of his radiometer, Crookes was influenced by his work with Home and his desire to identify a new force that could modify gravity, a psychic force. He even suggested that his vacuum tube apparatus might be an instrument for testing the special powers of mediums. By the spring of 1872, Crookes was demonstrating his new instrument to fellow scientists and spiritualists. The biometrician Francis Galton, a cousin of Charles

Darwin, was enthralled by Crookes' demonstration. Although still confidential since Crookes had yet to publish, he gave Galton permission to tell Darwin about the instrument. Darwin was enthusiastic and suggested that Crookes should *"get some instrument-maker to sell it, then everyone could buy one and try for himself. This would settle the question at once, whether any power does come out of the human body of certain or many individuals"* [40].

Crookes and Gimingham experimented with different materials in the force-sensing apparatus. They tried replacing the pith with ivory, brass, glass, and copper. The ratio of the surface area to the weight was more important than the material. They improved the balance by replacing the straw with a thin glass rod. By the spring of 1874, Crookes decided that it was time to publish, and he sent off a paper to the Royal Society. As was his custom, he was the only author although he did acknowledge Gimingham for his technical support. Crookes had learned his lesson and made no mention of any spiritualistic implications. For the first time, detailed illustrations of the apparatus were displayed in the text rather than as lithographs at the end of the article, a journalistic milestone.

Crookes believed that the pith ball movement in his apparatus was driven by a new type of force, radiant molecular energy. In his Royal Society paper, he considered other possible explanations including residual gases in the vacuum tube, electricity from incidental radiation, and evaporation and condensation of vapors trapped in the materials but argued against each of them. When Crookes demonstrated his apparatus at a Royal Society meeting in April 1874, an electrical engineer from Manchester, Osborne Reynolds was skeptical of Crookes' interpretation and suggested that the pith ball movement could be explained by the kinetic theory of gases. He went back to his laboratory and produced his own apparatus like Crookes', and after a series of experiments, he was convinced that the pith ball movement resulted from evaporation and condensation on the surface. Since Crookes used a mercury vacuum pump, he suggested the pith balls had absorbed mercury vapor.

Reynolds read his paper at a Royal Society meeting in June 1874, and Crookes responded later in the month at a lecture before the Physical Society. Crookes remained convinced his radiation hypothesis was correct, arguing that the movement was a direct result of *"the impact of the waves* [radiation] *upon the surface of the moving mass, and not secondarily, through the interaction of air-currents, electricity or evaporation and condensation. Whether the aethereal waves actually strike the substance moved, or whether the mysterious boundary-surface separating solid from gaseous matter there are intermediary layers of condensed gas which, taking up the blow, pass it on to the layer beneath, are problems the solution of which must be left to further research"* [41]. In the subsequent debate, Reynolds questioned Crookes' understanding of the basic physical principles involved. A modified version of Reynolds explanation would eventually win out, but in the meantime, Crookes' incorrect radiation hypothesis would drive experiments that led to the development of his cathode ray tube.

If radiant matter was driving the pith ball movement, Crookes reasoned that the effect would be altered by the color of the pith ball. A white surface would reflect the radiation while a black surface would absorb it. He created an apparatus with the pith balls blackened on one side so that "an impetus given by a ray of light always acts in the same direction of movement" [42]. With this new instrument the pith balls rotated continuously when exposed to "radiation". The initial apparatus had two arms but they soon found that four arms were more stable and worked better (Fig. 8.2). Each arm had a flattened pith ball with the blackened sides all facing in the same direction. When the new "radiometer" was in uniform light, the arms rotated continuously and when in bright sunlight, they rotated at a high speed.

Being an avid photographer, Crookes suggested that his new instrument could be used as a photometer allowing photographers to determine if an object was adequately lit and if a darkroom was free of light. He showed that the speed of revolution was inversely proportional to the square of the distance between a light source and the instrument. Two candles at the same distance doubled the number of revolutions of one candle. Three candles tripled the effect.

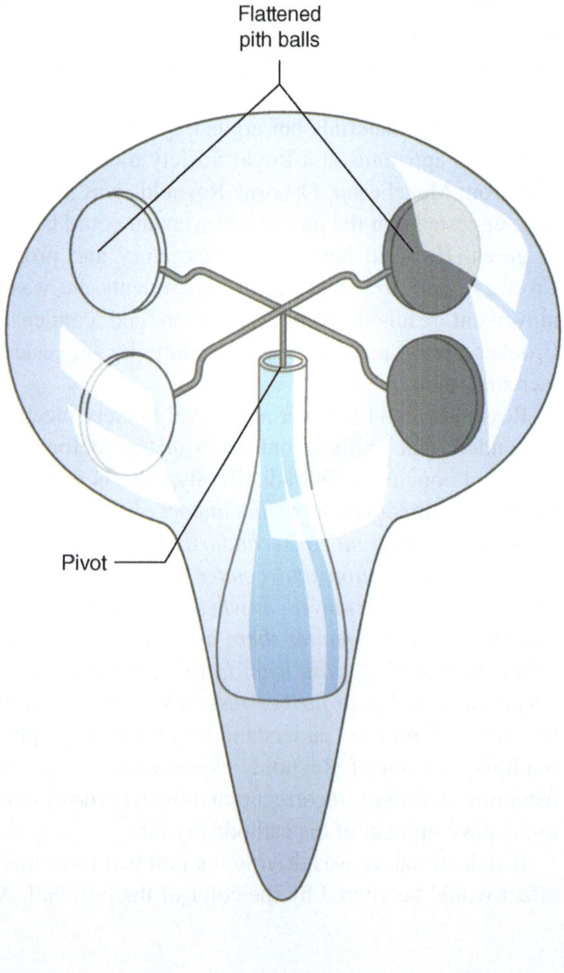

Fig. 8.2 William Crookes' radiometer. The flattened pith balls were painted white on one side and black on the other and fastened to the arms with the blackened sides facing in the same direction

In early 1876, Crookes gave a lecture at the Royal Institute on "weighing a beam of light". He estimated that when the sun's energy was equivalent to 10 candles 6 inches from the radiometer, the pressure of the sunlight was more than two tons per square mile. Crookes believed that he had forced matter into a new state, the fourth radiant state, the other three states being solid, liquid, and gas. He saw this fourth state of matter, radiant matter in a vacuum, as a possible link between the physical and spiritual worlds.

Crookes' radiometer was the scientific sensation of the late nineteenth century. Regardless of how it worked, everyone wanted one. No scientific exhibit was complete without a radiometer. Gimingham made a miniature version that Crookes gave to his friends. The English botanist and close friend of Charles Darwin, Joseph Hooker, suggested why the radiometer was so popular: *"It is the mystery attending the phenomenon that gives it its great importance. There is evidently some action going on with which we are not at present acquainted; and there is no saying what a thorough investigation into the cause of the phenomenon may lead to"* [43].

Our current understanding of the radiometer is based on the kinetic theory of gases. Heat energy from any source causes one side of the disc to be warmer than the other side. When gas particles strike the warmer side of the disc they rebound with greater velocity than those that strike the cooler side, causing a recoil. At atmospheric pressure, this effect is minimal since recoiling particles move only a fraction of a millimeter before colliding with particles moving in the opposite direction. In a partial vacuum, there are fewer particles to collide with and the recoil effect is markedly amplified. The key test of this theory is that if the vacuum is complete, which was not possible with the vacuum pumps used by Crookes and Gimingham, the radiometer should stop spinning. Crookes' theory predicted that the speed of rotation would continue to increase as the vacuum improved. As better vacuum pumps were developed, the test was conducted, and the radiometer decreased its speed of rotation and stopped in a near-complete vacuum. Crookes had to accept that residual gas explained the radiometer rotation but by then he had moved on to experiments on the electrical propulsion of molecules in a vacuum that led to the cathode ray tube and eventually to the discovery of the electron by J.J. Thomson.

Crookes' Tube

The vacuum discharge tube or Geissler tube was developed by the German glass blower, Heinrich Geissler in the early 1850s [44]. Geissler came from a long line of glass blowers in Bohemia and his skills were in demand at several German Universities. In 1852, he set up his own workshop and served as an instrument maker at the University of Bonn. It was there that he was asked to produce an apparatus for studying electrical discharges into evacuated glass tubes by the physicist Julius Plücker. In 1855, Geissler developed a hand-crank mercury pump to evacuate the tube and sealed platinum electrodes into the lead glass. Using Geissler's tube, Plücker observed that as the tube was evacuated, and electricity passed through the electrodes the tube glowed. In fact, Geissler showed that the quality of the vacuum could be assessed by the brightness of the glow.

Michael Faraday became interested in Geissler's tube in the late 1850s. He pointed out that there was a band of darkness between the glow surrounding the negative electrode, the cathode, and the striated glow that filled the remainder of the tube. The band became known as the Faraday dark space. Plücker showed that as the Geissler tube was further evacuated, the Faraday dark space grew, and the glass walls of the tube began to glow. Plücker then made the remarkable observation that the striated glow could be bent toward a magnet outside the tube. It was around this time that Crookes decided to seal electrodes into his radiometer tubes. Ironically, about the same time in Germany, Geissler saw a market for Crookes' radiometer and became the main supplier of radiometers throughout Europe, to the annoyance of Crookes who planned to exploit the market himself.

Clearly, Crookes did not invent the cathode ray tube, but he and Gimingham made important modifications and he was an excellent promotor and not above claiming priority for the work of others [45]. Crookes tubes had thicker glass and a better vacuum than Geissler tubes due to Gimingham's new mercury vacuum pump. Another important feature of Crookes tubes was a high-voltage power supply up to 100 kV.

Over time, Geissler's tube became known as Crookes' tube, and it was rare to see a photograph of Crookes without a Crookes' tube in his hands (Fig. 8.3). When Crookes and Gimingham first began to experiment with electric discharge vacuum tubes, they became fascinated by the glow like everyone else. Crookes identified a second smaller area of darkness between the cathode and the surrounding glow. He called this Crookes dark space in contrast to Faraday dark space located between the cathode glow and the striated glow in the rest of the tube. Crookes dark space also increased as the vacuum in the tube increased. Crookes became convinced that charged particles of gas were being driven off the cathode by the electric current and that the glow resulted from collisions of these particles with residual gas ions in the evacuated tube. As more gas ions were removed from the tube with the vacuum pump, the likelihood of collisions decreased, and Crookes' dark space increased in size.

By contrast, the German physicist Eugen Goldstein was convinced that a form of radiation emanated from the cathode which he called cathode rays. He and other German physicists argued that the cathode rays were a form of electromagnetic radiation moving in the ever-present ether. Crookes argued that charged particles (molecules) left the cathode at right angles forming a stream that produced a glow when it struck residual gas molecules or phosphorescence when striking the glass wall. To test this hypothesis, Crookes and Gimingham produced a vacuum tube with a concave-shaped cathode so that the beam of particles could be focused onto a solid object in its path. They noted that the object was heated by the impact of the particles and if the object was shaped like a Maltese cross, a shadow of the cross appeared on the wall behind. To further demonstrate the force of the stream of charged particles, Crookes and Gimingham made a tiny paddle wheel with its axle set on glass rails in the discharge tube. When the current was turned on, the paddle wheel rotated

Fig. 8.3 Caricature of William Crookes holding his famous vacuum tube published in 1902 edition of Vanity Fair. The caption read "ubi Crookes ibi lux"—Where there is Crookes, there is light

and moved along the rail even uphill when the tube was tilted. Although the mechanism of this mechanical action of the cathode discharge was later shown to be the same heat energy effect that drove Crookes' radiometer, Crookes was convinced that the paddle wheel was turned by corpuscular radiant matter.

Although the German physicists were impressed with Crookes' and Gimingham's technical skills with evacuated discharge tubes, they were not ready to accept his corpuscular radiant matter hypothesis. Based on mathematical calculations using the kinetic theory of gases, charged molecules were predicted to collide with residual gas molecules long before they could reach the glass wall. Furthermore, electrically charged molecules should be deflected by an electrical field outside the tube but Goldstein and Hertz had both tried to deflect cathode rays with an electrical field without success. Whereas, just like light rays, cathode rays were deflected by an outside magnet. When asked to review Crookes' paper on electrically charged particles, James Maxwell raised these same concerns and also questioned why if cathode rays were charged particles, they were not deflected by the positively charged anode in the tube. As we will see shortly, this apparent dilemma was solved by

J.J. Thomson when he suggested that the charged corpuscles (electrons) were several orders of magnitude smaller than a molecule. Crookes idea of a radiant matter, the fourth state of matter, in part driven by his spiritualistic beliefs turned out to be partially correct even though the underlying hypothesis was completely wrong.

Röntgen and His X-Rays

Another key piece in the puzzle that led to the discovery of the electron was the discovery of X-rays by Wilhelm Röntgen in 1895. No doubt Crookes and other scientists working with cathode ray tubes were producing X-rays on a regular basis. They just did not notice them.

Röntgen was born in Germany in 1845, the son of a cloth manufacturer [46]. The family moved to the Netherlands when he was a child and after attending a technical school in Utrecht, he enrolled at the Polytechnic Institute in Zurich, Switzerland (ETH Zurich). He received his PhD degree in mechanical engineering from the University of Zurich in 1869. He then returned to Germany, where he eventually became a professor of physics at Würzburg Physical Institute at the University of Würzburg. Röntgen became interested in cathode rays and decided to start by repeating some of the recently published studies conducted by Heinrich Hertz and his graduate student Philip Lenard.

Like his German colleagues, Hertz was convinced that cathode rays were ethereal waves just like other electromagnetic waves. Lenard, who carried on the work after Hertz's death in early 1894, questioned whether cathode rays could pass through the tube into the outside air. He found that cathode rays could not only pass through the glass but also penetrate a thin metal film consistent with them being waves and not moving atoms or molecules. Of course, at the time atoms were thought to be the smallest possible particle. Lenard heavily coated the end of a Crookes tube leaving only a small window in the center directly across from the cathode. He then covered the window with a thin piece of aluminum foil. He found that holding a florescent screen just outside the window showed fluorescence and a photographic plate darkened despite the absence of light.

Röntgen used a Crookes tube and simply covered the end with a sheet of black cardboard to block the cathode rays. To his surprise, he observed phosphorescence on a photographic screen well behind the cardboard. Some unknown type of radiation must have penetrated the cardboard. He went on to study the properties of these new rays produced by the cathode ray tube which he called X-rays or unknown rays. They penetrated books and papers on his desk and just about anything else he placed in their way. He was amazed to find that he could image the bones of his hand and a compass needle in an iron box. When he published the picture of his wife's hand including her ring in December 1895, there was worldwide excitement over the medical applications of this new imaging technique (Fig. 8.4). By early 1896, physicists were refining the production and focusing on the new rays for imaging different parts of the body.

Medical imaging turned out to be the first practical application of Crookes tubes. The key to the production of X-rays was the high voltage applied to the tube. At least 5000 V were required to generate the high-energy rays. The energy of the cathode rays also depended on the amount of residual gas in the tube. With Crookes' tubes, the amount of residual gas gradually decreased over time due to absorption into the glass wall. This process reduced the number of cathode rays produced but increased the energy of those that were produced. These high-vacuum Crookes tubes were said to be "hardened" because they generated more penetrating rays. Eventually, the vacuum became so complete that the rays stopped all together. To counter this problem, devices were developed to regularly inject small amounts of gas into frequently used X-ray tubes, a process called "softening".

Crookes and the many other investigators working with cathode ray tubes were obviously exposed to X-rays in their work. Although Crookes never published any studies on X-rays, he later commented that he personally had not suffered any ill effects from X-rays. Of course, long-term risks due to X-ray exposure were not known at the time. Röntgen received the first Nobel Prize in physics in 1901 and Lenard received the Prize in 1905.

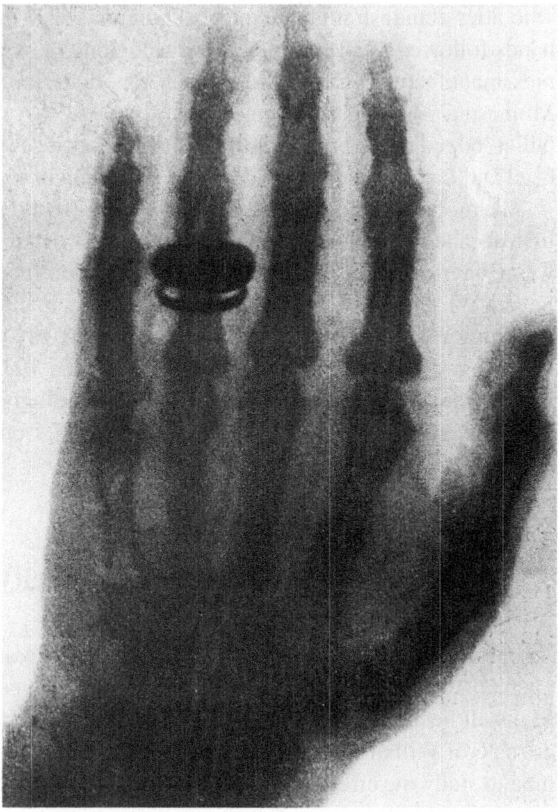

Fig. 8.4 Röntgen's X-ray of his wife Anna's hand taken on November 8, 1895. When she saw the photo she reportedly exclaimed "I have seen my death!"

Dangers of X-Rays

Just 2 weeks after Röntgen published the report of his discovery of X-rays, in December 1895, a dentist in Braunschweig Germany, Frederich Otto Walkhoff, took the first dental radiograph on himself. He wrapped a photographic glass plate in a rubber dam and held it in his mouth between his teeth. He then sat on the floor for 25 min with the X-ray beam focused on the plate. The only apparent ill effect of the exposure was a slight loss of hair and tanning of the face and hands. The pictures of his teeth were crude but proved proof of concept [47].

Later in 1896, Walkhoff teamed up with his friend and colleague Fritz Giesel to establish the first commercial X-ray laboratory in the world focusing on images of the jaw and head. The only ill effect of the X-rays that they noted was frequent loss of hair on the side exposed to the X-ray tube. Giesel died of metastatic carcinoma in 1927 thought to be due to radiation exposure while Walkhoff died at age 74 of presumed natural causes.

An English translation of Röntgen's original article appeared in the English journal *Nature* on January 23, 1896, followed shortly by the American journal *Science* on February 14, 1896. Reports of the immediate effects of X-ray exposure soon followed. In August of 1896, the *Electric Review* reported a doctor in New York who after demonstrating the medical potential of X-rays developed swelling of the hands followed a few days later by a sloughing of skin [48]. Employees of companies manufacturing X-ray tubes soon reported similar experiences. A patient in Minnesota who underwent one of the first skull X-rays to locate the position of a bullet lodged in his brain in July 1896, developed severe burns with swelling of the head and face and hair loss a day after a continuous exposure from 8 AM to 10 PM.

The more ominous long-term effects of X-rays did not become apparent until the first decade of the twentieth century. The first death attributed to X-ray exposure was Clarence Dally who died of metastatic carcinoma in 1904 at age 39 years. Several prominent physicians and X-ray tube manufacturers died of carcinoma including Wolfram Fuchs who opened the first X-ray laboratory in Chicago in 1896 and imaged the first brain tumor in 1899. In his 1936 book *American martyrs to science through Roentgen rays,* the Boston radiologist Perry Brown provided a litany of scientists who died of X-ray exposure [49]. Ironically, Brown himself died of cancer related to X-ray exposure in 1950.

JJ Thomson and Conduction of Electricity Through Gases

Joseph John (JJ) Thomson was another pivotal character in the history of electromagnetism. He idolized Michael Faraday, followed in the footsteps of James Maxwell becoming the third Cavendish Professor of Physics at Cambridge, was a devotee of William Thomson's (no relative) vertex ring theory, and used Crookes' tube to study the effects of electricity and X-rays on gases ultimately leading to the discovery of the electron. His measurement of the charge-to-mass ratio of the negatively charged corpuscles in cathode rays showed that they were several orders of magnitude smaller than a hydrogen atom. This marked the beginning of the subatomic world of modern physics. Ironically, JJ Thomson received the Nobel Prize for showing that the electron was a particle while his son GP Thomson later received the Nobel Prize for showing that it was an electromagnetic wave.

JJ Thomson was born into a Victorian middle-class family in 1856 in a suburb of Manchester, England [50]. His father was in the book publishing business, so JJ was exposed to the intellectual elite of Manchester growing up. But when JJ was 16 years of age, his father died and the family faced economic hardship. Fortunately, his father had sent JJ to Owens College in Manchester at age 13, where he made valuable contacts that assured his later enrollment at Cambridge University. Probably, most influential of his teachers at Owens College was the Scottish physicist Balfour Stewart who we met earlier in the Chapter for his work with Peter Tait on the "Unseen Universe". Stewart introduced JJ to Maxwell's famous book on electricity and magnetism and spent many hours working with JJ in the laboratory on a project measuring changes in weight associated with chemical reactions. Stewart worked at the boundary between physics and chemistry and influenced Thomson's lifelong notion that the two disciplines were closely intertwined.

JJ Thomson arrived at Cambridge at age 19 in 1876 having won a scholarship at the prestigious Trinity College. Like James Maxwell, Thomson went on to become second wrangler in the mathematical tripos. His undergraduate years were relatively uneventful spent studying and attending the expected religious and social events. His fellowship provided just enough money to get by and his main exercise was frequent 3-to-4-mile walks. Not surprisingly, Trinity college students, particularly those interested in mathematics, were in a unique position to learn Maxwell's theory on electromagnetism. William D Niven, a close friend of Maxwell, provided a series of lectures on Maxwell's *Treatise*. Thomson already had a good grasp of Maxwell's mathematical equations as an undergraduate student.

Being a top wrangler, Thomson had an inside track to a fellowship at Trinity College and he won a fellowship in the summer of 1880. He wrote two well-received papers on electromagnetic theory within months of graduating from Trinity. To Thomson, Maxwell's equations were gospel. He decided to apply the equations to the moving particles in a Crookes tube, Crookes' so-called fourth state of matter. He modeled the particles as charged spheres in a fluid medium under the influence of an electric field. According to Maxwell's equations, the moving charged particles must experience a resistance in the dielectric and the resistance could be mathematically modeled as an increase in mass. *"This additional mass is not in the sphere but in the space around it"* [51]. This notion would later reemerge to possibly explain the mass of his corpuscle (electron).

Thomson's interest in the origin of matter naturally propelled him to investigate William Thomson's vortex ring theory. His essay "On vortex rings" won the 1882 Adams Prize at Cambridge University. The essay contained more than 50 pages of mathematical equations that enabled the reader to work out *"a complete dynamical theory of gases"*. This was the beginning of a life-long interest in the physics of gases. He argued that explaining matter as a secondary structure of ether was a simple but powerful idea. The theory *"proposes to explain by means of the laws of hydrodynamics all the properties of bodies as consequences of the motion of this fluid. It is thus evidently of a very much more fundamental character than any theory stated"* [52].

Thomson worked on the mathematics of the vortex ring theory throughout the 1880s but gradually became frustrated with his progress and moved on to a variation of the theory based on what he called "Faraday tubes," in the early 1890s. The problem he faced was that electricity moved near the speed of light but charged atoms or molecules in discharge tubes were not energetic enough to move at such a high speed. What was needed was something analogous to a wire for conduction of electricity in gases. Thomson perceived Faraday tubes as channels in the ether like vortex tubes beginning and ending in matter to form a closed circuit. These tubes were direction fixed and the atoms or molecules that they began and terminated on had a unit of positive and negative electrification. Thomson noted, *"Before the electric field is intense enough to cause discharge, the induction in the field polarizes the gas. We may regard this polarization to being equivalent to the formation of chains of molecules analogous to the Grotthus chain in electrolysis"* [53].

In the early 1800s, the Baltic German natural philosopher, Theodor Grotthus, explained electrolysis as a moving chain of charged particles. He proposed that electric force set up a chain of charge exchanges in a solution such that positive charges moved step by step toward the cathode and negative charges step by step toward the anode like a two-way bucket brigade. This explained high-speed movement of the charge through the solution without the need for similar movement of the ions. The same mechanism was proposed for the movement of electric current through a wire. Thomson used his Faraday tubes to provide a general theory of electrodynamics. He argued, *"If we regard these tubes as having a real physical existence, we may, as I shall endeavor to show, explain the various electrical processes,—such as the passage of electricity through metals, liquids, or gases, the production of a current, magnetic force, the induction of currents, and so on,—as arising from the contraction or elongation of such tubes and their motion through the electric field"* [54].

Discovery of the Electron

As Thomson continued his work on the analogy of electrolysis with the movement of electricity through gases, he was puzzled by a key difference between ions in solution and charged molecules in gases. The sign of the electric charge of an ion in solution was a constant feature of the elements in solution whereas the sign of the charge of different elements in gases undergoing electrolysis could be different. For example, an atom of hydrogen could have a positive or negative charge. He surmised that the relationship between matter and electric charge was fundamental to understanding all chemical and electrical phenomena.

As the reader may surmise, the terms atom, molecule, and element were commonly used by scientists in the nineteenth century, sometimes interchangeably, but the meaning of the terms and how they were used was confusing. This was because there was no agreed-upon definition or a conceptual framework of what they represented. Based on its ancient Greek meaning, an atom was the smallest subdivision of matter, a body that could not be cut in two. A molecule was considered the smallest unit of a chemical compound. It could be made up of a group of atoms or a single atom. On the other hand, a chemical element was a chemical substance that could not be subdivided into other substances. Many thought that each chemical element was made up of different-sized atoms so that the number of different atoms corresponded to the number of chemical elements. Remarkably, at the beginning of the nineteenth century, the English chemist William Prout speculated that the masses of all elements were exact multiples of the mass of the hydrogen atom and therefore all elements were made up of aggregations of this basic atom which he called a protyle. It would take another century, however, before Prout's speculation was shown to be correct.

Röntgen's discovery of X-rays in 1895 provided a critical new tool for Thomson to use in his experiments on the conduction of electricity through gases. Working with his research student Ernest Rutherford, Thomson noted that the high-energy X-rays could ionize gases making them conductors of electricity when normally they were not. *"The passage of these rays through a substance seems thus to be accompanied by a splitting up of its molecules, which enables electricity to pass through it by a process resembling that by which current passes through an electrolyte"* [55]. Since the rate at which a charged plate emitted electricity after being radiated with X-rays was proportional to the intensity of the X-rays, he had a way to quantitatively assess the conductivity of a gas.

In 1897, Thomson published three papers that outlined an exciting new theory that cathode rays were composed of a stream of corpuscles with an extremely small mass. He estimated the mass-to-charge ratio of the corpuscle was 1.6×10^{-7} compared to the mass-to-charge ratio of a hydrogen atom of 10^{-4}. This estimate was based on measurements of the magnetic deflection and the heat given off by a substance impacted with cathode rays.

In the third and most talked about of the three papers published in the *Philosophical Magazine*, Thomson debunked a common argument made against the particulate nature of cathode rays, the fact that they could not be deflected by an electric field. He was able to deflect the rays with an intense electrical field and in the process obtained an alternate way of measuring the mass-to-charge ratio of the corpuscles (Fig. 8.5). The new measurement was remarkably close to his original measurement based on magnetic deflection and heat production strongly supporting his theory. He speculated that within a cathode ray tube *"in the very intense electric field in the neighborhood of the cathode the molecules of the gas are dissociated and split up, not into ordinary chemical atoms, but into these primordial atoms, which we shall for brevity call corpuscles"* [56].

Thomson went beyond cathode rays and suggested a role for the corpuscle as an elementary constituent of matter. *"Thus on this view we have in the cathode ray*

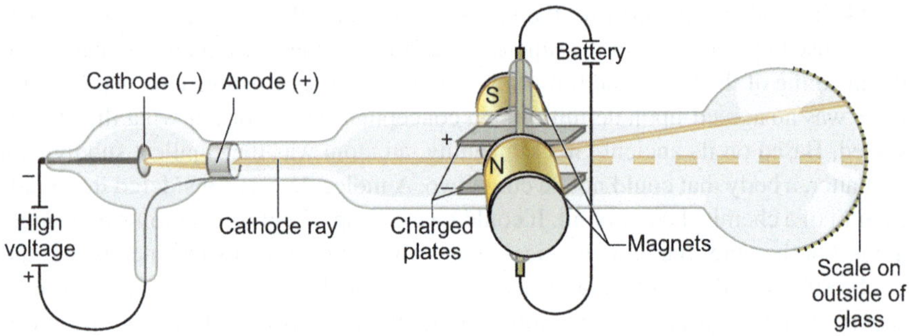

Fig. 8.5 Schematic drawing of JJ Thomson's cathode ray tube showing that cathode rays were deflected by both an electric and magnetic field. This allowed him to measure the mass-to-charge ratio of the corpuscles (electrons)

matter a new state, a state in which the subdivision of matter is carried very much further than in the ordinary gaseous state: a state in which all matter—that is, matter derived from different sources such as hydrogen, oxygen, &c.—is of one and the same kind; this matter being the substance from which all the chemical elements are built up" [57]. In other words, Thomson saw his corpuscle as being a basic state of matter, a building block of all atoms and molecules. He initially suggested that atoms were composed of corpuscles and holes, without defining what he meant by holes. Presumably, he was referring to holes in the ether (like a sponge?). He later refined his idea of the structure of an atom to corpuscles "arranged in electromagnetically stable configurations".

One aspect of electricity that baffled scientists since the time of Ben Franklin and his one fluid theory, was the prominent asymmetry between positive and negative electrification. Thomson's negative corpuscles provided a simple explanation. He saw "electric fluid" as a collection of corpuscles moving from one location where there is a gain in positive electrification to another location where there is a gain in negative electrification. *"These corpuscles may either remain free or get attached to molecules of matter with which they come in contact; thus positive electrification is always associated with ordinary matter, while negative electrification may or may not be, according as the corpuscles are or are not attached to molecules of ordinary matter"* [58].

How did Thomson's corpuscle become known as an electron? Wilhelm Weber had already considered the possibility of a small particle of unit electricity in 1846 when he noted that positive and negative charges moved with different velocities during electrolysis [59]. The term "electron" was first used to describe an elementary quantity of electricity by the Irish physicist, George Johnstone Stoney, in 1891. He went on to estimate the magnitude of what he called an "atom of electricity". Once Thomson published his work on the corpuscle, it was only natural that many scientists preferred the term electron to the more cumbersome and less descriptive term corpuscle. Thomson, however, continued to use the word corpuscle until 1911 when he gave in and began using the term electron.

Although Thomson had set out to understand the behavior of cathode rays, his discovery of the electron led to a general theory of electricity and the subatomic structure of matter. Thomson suggested that atoms were made up of an aggregation of corpuscles. *"In the normal atom, this assemblage of corpuscles forms a system which is electrically neutral. Though the individual corpuscles behave like negative ions, yet when they are assembled in a neutral atom the negative effect is balanced by something which causes the space through which the corpuscles are spread to act as if it had a charge of positive electricity equal in amount to the sum of the negative charges of the corpuscles"* [60]. As we will see in Chap. 12, the idea of combining negatively charged corpuscles with larger subatomic particles that had a positive charge to form an atom awaited studies on radioactivity carried out in the early twentieth century by Thomson's former graduate student Ernest Rutherford.

Crookes' Final Years

When Crookes read Thomson's work on corpuscles, he felt vindicated. Thomson's corpuscles were what he had called radiant matter or the fourth state of matter. Of course, Crookes originally based his theory on moving gas ions the size of molecules and had no concept of subatomic charged particles. In 1906, Crookes published a paper summarizing his studies on cathode ray tubes after Thomson discovered the electron [61]. The dark space (Crookes dark space) was a measure of the free path of electrons released from the cathode and was proportional to the atmospheric pressure in the tube. The better the vacuum the larger the dark space. Crookes used photographic film to detect electrons, positive ions, and X-rays released by cathode ray tubes. He showed that X-rays were best produced when electrons struck a metal surface and that the amount of X-rays produced varied with different metal surfaces.

Based on current understanding, Crookes' tube works by accelerating electrically charged ions and electrons by the electric field. At the onset, there are a small number of free electrons in the tube and these electrons collide with other gas molecules producing more electrons and positive ions in a chain reaction. As the positive ions are pulled toward and strike the cathode, they dislodge large numbers of electrons from the metal surface that are repelled by the cathode and attracted by the anode. This stream of electrons is the main component of cathode rays. In the near-vacuum condition, most of the electrons travel the length of the tube without striking gas molecules.

Because of the high voltage, the electrons are accelerated to speeds in the range of one-fifth of the speed of light. Although the electrons are attracted by the anode, they have so much momentum that they fly by the anode (typically located off to the side) and strike the end wall. Light (usually yellow-green) is produced when the electrons strike atoms in the glass and fluorescence occurs if the glass is coated with a fluorescent chemical such as zinc sulfide. X-rays are produced when the highest speed electrons hit the glass wall, or the anode made of a heavy metal such as platinum. Early X-ray tubes consisted of a platinum anode tilted at an angle to a concave cathode so that the X-rays generated at the anode were focused as a beam exiting the side of the tube.

Although Crookes' cathode ray tubes and early X-ray tubes had a much better vacuum than Geisinger's original cathode ray tubes, they required a small amount of air to work. Free electrons and ions were required to "prime the pump". These early tubes were called "cold cathode" tubes as opposed to "hot cathode" tubes invented in the early twentieth century. As we will see in the next chapter, these new tubes had a heated cathode that released electrons and worked best in a high vacuum.

Crookes remained an avid spiritualist up to the time of his death. In his later years, he played a prominent role in the Ghost Club serving as president from 1907 to 1912. The club boasted famous members such as writers Charles Dickens and Laurence Oliphant, poets Author Kennedy and William Butler Yates and financier Alexander Ionides junior. Unlike the more "scientifically oriented" SPR, the Ghost Club was an informal dining club where true believers could discuss their psychic experiences over dinner without fear of ridicule from non-believers. Typically, there were only a few dozen members present although this number was substantially increased by deceased members who regularly attended.

Alexander Ionides was a long-time friend of Crookes who had invested in several of Crookes' business adventures, including his unsuccessful light bulb business. Crookes biographer William Brock noted a conversation between Ionides and Crookes in which Crookes, in a rare moment of modesty, admitted to being "*a one-idea man, and all my discoveries have emanated from the idea that light must have a force of its own. This idea came to me as the result of some experience at a séance, and in trying to discover this force I hit upon the radiometer. In devising this, I discovered thallium, and so discoveries have branched out from one to another, all resulting from this one idea*" [62].

In Brief

William Crookes was one of the most enigmatic scientists of the nineteenth century. His pioneering work on spectroscopy led to the discovery of new elements and his cathode ray tube was a forerunner of several modern electronic devices. Yet, he actively participated in psychic research with charlatans and strongly believed in a psychic force that could be transmitted through space by the living and the dead. While working with Crookes cathode ray tube, Wilhelm Röntgen identified highly penetrating emissions that could pass through solid objects and produce an image on a photographic plate. There was immediate worldwide interest in the medical applications of Röntgen's X-rays. JJ Thomson used Crooke's tube to study the effects of cathode rays on gases and was able to measure the charge-to-mass ratio of the negatively charged corpuscles in cathode rays. The corpuscles (later called electrons) were several orders of magnitude smaller than a hydrogen atom. The discovery of the electron marked the beginning of the subatomic world of modern physics.

References

1. Bordeau SP. Volts to hertz – the rise of electricity. Minneapolis: Burgess Publishing Company; 1982. p. 206.
2. Kragh H. The vortes atom: a Victorian theory of everything. Centaurus. 2002;44:32–114.
3. Kragh H. The vortes atom: a Victorian theory of everything. Centaurus. 2002;44:38.
4. Navarro J. The history of the electron. J.J. and G.P. Thomson. Cambridge: Cambridge University Press; 2012. p. 37.
5. Kragh H. The vortes atom: a Victorian theory of everything. Centaurus. 2002;44:52.
6. Kragh H. The vortes atom: a Victorian theory of everything. Centaurus. 2002;44:51.
7. Stewart B, Tait PG. The unseen universe or physical speculations about a future state. London: Macmillan and Co; 1875.
8. Stewart B, Tait PG. The unseen universe or physical speculations about a future state. London: Macmillan and Co; 1875. p. 157.
9. Heimann P. The unseen universe. Physics and the philosophy of nature in Victorian Britan. Br J Hist Sci. 1972;6:77.
10. Heimann P. The unseen universe. Physics and the philosophy of nature in Victorian Britan. Br J Hist Sci. 1972;6:78.
11. Stewart B, Tait PG. The unseen universe or physical speculations about a future state. London: Macmillan and Co; 1875. p. 100.
12. Stewart B, Tait PG. The unseen universe or physical speculations about a future state. London: Macmillan and Co; 1875. p. 115.
13. Brock WH. William Crookes (1832–1919) and the commercialization of science. London: Routledge; 2008.
14. Brock WH. William Crookes (1832–1919) and the commercialization of science. London: Routledge; 2008. p. 11.
15. Forbes N, Mahon B. Faraday, Maxwell and the electromagnetic field. Guilford: Prometheus Books; 2019. p. 112.
16. Kirchhoff G, Bunsen R. Chemical analysis by spectrum observations. Philos Mag. 1860;20:89–109.
17. Brock WH. William Crookes (1832–1919) and the commercialization of science. London: Routledge; 2008. p. 63.
18. Crookes W. On the supposed new member of the calcium group of metals. Chem News. 1861;3:129–30.
19. Brock WH. William Crookes (1832–1919) and the commercialization of science. London: Routledge; 2008. p. 61–82.
20. Brock WH. William Crookes (1832–1919) and the commercialization of science. London: Routledge; 2008. p. 74.
21. Brock WH. William Crookes (1832–1919) and the commercialization of science. London: Routledge; 2008. p. 119–33.
22. Moore RL. Spiritualism and science: reflections of the first decade of the spirit rappings. Am Q. 1972;24:474–500.
23. Brock WH. William Crookes (1832–1919) and the commercialization of science. London: Routledge; 2008. p. 135–53.
24. Noakes R. Instruments to lay hold of spirits: technologizing the bodies of Victorian spiritualism. In: Morus IR, editor. Bodies/Machines. New York: Oxford University Press; 2002. p. 125–63.
25. Noakes R. Instruments to lay hold of spirits: technologizing the bodies of Victorian spiritualism. In: Morus IR, editor. Bodies/Machines. New York: Oxford University Press; 2002. p. 143.
26. Brock WH. William Crookes (1832–1919) and the commercialization of science. London: Routledge; 2008. p. 146.
27. Brock WH. William Crookes (1832–1919) and the commercialization of science. London: Routledge; 2008. p. 179–94.

28. Brock WH. William Crookes (1832–1919) and the commercialization of science. London: Routledge; 2008. p. 183.
29. Home DD. Lights and Shadoew of spiritualism. London: Virtue and Co; 1877.
30. Brock WH. William Crookes (1832–1919) and the commercialization of science. London: Routledge; 2008. p. 139.
31. Gauld A. The founders of psychical research. London: Rouledge and K Paul; 1968. p. 137.
32. Myers F. Human personality and its survival of bodily death. London: Longmans, Green and Co; 1903.
33. Myers F. Human personality and its survival of bodily death. London: Longmans, Green and Co; 1903. p. 117–8.
34. Myers F. Human personality and its survival of bodily death. London: Longmans, Green and Co; 1903. p. XXVII.
35. Myers F. Human personality and its survival of bodily death. London: Longmans, Green and Co; 1903. p. XXV.
36. Raia CG. From either theory to ether theology: Oliver Lodge and the physics of immortality. J Hist Behav Sci. 2007;43:19–43.
37. Lodge O. Experiments in thought transference. Proc Soc Psych Res. 1884;2:191.
38. Raia CG. From either theory to ether theology: Oliver Lodge and the physics of immortality. J Hist Behav Sci. 2007;43:33.
39. Simon L. Genuine reality: a life of William James. Chicago: University of Chicago Press; 1999.
40. Noakes R. Instruments to lay hold of spirits: technologizing the bodies of Victorian spiritualism. In: Morus IR, editor. Bodies/Machines. New York: Oxford University Press; 2002. p. 147.
41. Brock WH. William Crookes (1832–1919) and the commercialization of science. London: Routledge; 2008. p. 168.
42. Brock WH. William Crookes (1832–1919) and the commercialization of science. London: Routledge; 2008. p. 173.
43. Brock WH. William Crookes (1832–1919) and the commercialization of science. London: Routledge; 2008. p. 177–8.
44. Dahl PF. Flash of the cathode rays: a history of J.J. Thomson's electron. New York: CRC Press; 1997. p. 49–52.
45. Brock WH. William Crookes (1832–1919) and the commercialization of science. London: Routledge; 2008. p. 225–42.
46. Glasser O. Wilhelm Conrad Röntgen and the early history of the Roentgen rays. London: John Bale, Sons and Danielsson; 1933.
47. Forrai J. History of X-ray in dentistry. Rev Clín Pesq Odontol. 2007;3:205–11.
48. Frei GA. Deleterious effects of X-rays on the human body. Electr Rev. 1896;29:95.
49. Brown P. American martyrs in science through Roentgen rays. Springfield: Charles C. Thomas; 1936.
50. Navarro J. The history of the electron. J.J. and G.P. Thomson. Cambridge: Cambridge University Press; 2012. p. 6–28.
51. Navarro J. The history of the electron. J.J. and G.P. Thomson. Cambridge: Cambridge University Press; 2012. p. 35.
52. Navarro J. The history of the electron. J.J. and G.P. Thomson. Cambridge: Cambridge University Press; 2012. p. 38.
53. Navarro J. The history of the electron. J.J. and G.P. Thomson. Cambridge: Cambridge University Press; 2012. p. 62.
54. Thomson JJ. On the illustration of the properties of the electrical field by means of tubes of electrostatic induction. Philos Mag. 1891;31:150.
55. Thomson JJ. On the discharge of electricity produced by the Röntgen rays, and the effects produced by these rays on dielectrics through which they pass. Proc R Soc. 1896;59:276.
56. Thomson JJ. On cathode rays. Philos Mag. 1897;9:311.
57. Thomson JJ. On cathode rays. Philos Mag. 1897;9:312.

58. Thomson JJ. Conduction of electricity through gases. Cambridge: Cambridge University Press; 1903. p. 63.
59. Bordeau SP. Volts to hertz – the rise of electricity. Minneapolis: Burgess Publishing Company; 1982. p. 181.
60. Thomson JJ. On the masses of the ions in gases at low pressure. Philos Mag. 1899;48:565.
61. Crookes W. Experiments on the dark space in vacuum tubes. Proc R Soc. 1907;79A:98–117.
62. Brock WH. William Crookes (1832–1919) and the commercialization of science. London: Routledge; 2008. p. 441.

Chapter 9
The Battle of Electric Currents

> *One of the early observed and remarkable features of the high frequency currents, and one that was chiefly of interest to physicians, was their apparent harmlessness which made it possible to pass relatively great amounts of electrical energy through the body of a person without causing pain or serious discomfort.*
>
> *Nikola Tesla, 1898*

Abstract Thomas Edison's development of an affordable long-lasting incandescent light bulb led to demand for an efficient reliable system to deliver electricity to the public. Edison had a head start with his low-voltage direct current system but Westinghouse's high-voltage alternating current system eventually won out. The alternating current system had the major advantage of transmitting electricity long distances. A transformer lowered the voltage at the site of use. George Westinghouse teamed up with Nikola Tesla and his polyphase alternating current motor to produce an alternating current system that became the standard of the industry. In his later years, Tesla focused on wireless high-frequency electric power transmission. He demonstrated that humans could tolerate extremely high-voltage alternating current if the frequency was high enough. He speculated that the ultra-high-frequency current was good for one's health and even improved learning. This led to the development of the violet ray and widespread use of high-frequency electrotherapy in the United States and Europe.

Keywords Dynamo · Vacuum tube · Transformer · Polyphase current · Electrical oscillators

In the first half of the nineteenth century, other than the widespread medical use of electricity, the only other commercial use was for telegraphy which had rapidly expanded to major cities around the world. The successful completion of the Indo-European and trans-Atlantic telegraph cables marked the beginning of the modern era of high-speed communication. Electric telegraph systems were relatively simple, direct current usually from a battery passed along insulated telegraph cables to activate electromagnetic relays. The biggest challenge for electrical engineers was to maintain and repair the long cables that were notoriously subject to malfunction.

R. W. Baloh, *Brain Electricity*, https://doi.org/10.1007/978-3-031-62994-5_9

Electric lighting, a key development in the latter half of the nineteenth century, dramatically changed how electricity was generated, distributed, and perceived [1]. By the early twentieth century, electricity was no longer an exotic scientific idea but a practical necessity for most households in the developed world. With this widespread availability of electricity came increased exposure to electricity and risk of electrocution for electric workers and members of the public. At the same time, physicians had a reliable source of electricity that could be used for diagnosis and treatment of patients without the need for expensive and potentially dangerous equipment for generating electricity.

Early Electric Lighting

The first incandescent electric light was made by the English chemist Humphrey Davy in 1802. He used the powerful battery that he developed for his electrolyte experiments to pass current through a thin stripe of platinum in the open air. He chose platinum because of its high melting point. The light he generated was not very bright and lasted only minutes before the platinum strip burned up. Several years later, Davy showed that he could generate more powerful electrical lighting by passing current into two sticks of charcoal creating sparks across the tips a few inches apart. This early form of carbon-arc light was promising and with gradual improvements became commercially feasible.

The extremely bright carbon-arc light proved ideal for street lighting, lighthouses, train stations, and large department stores. But it required a higher voltage of electricity than that supplied by a battery. In 1858, Davy's protégé Michael Faraday oversaw the production of a steam-powered magneto-electric generator to activate a large carbon-arc light for the South Foreland Lighthouse near Dover. This was the first electric light that could be seen across the channel in France. Faraday spent a great deal of his time in his later years visiting lighthouses and taking sea voyages to test the visibility of the light from great distances. When commenting on a proposal for a new arc light design he was asked to review, Faraday wrote: "*There is no human arrangement that requires more regularity and certainty of operation than a lighthouse. It is trusted by the Mariner as if it were a law of nature, and as the Sun sets so it expects that, with the same certainty, the light will appear*" [2].

The Siemens Brothers and Electric Generators

Four German brothers led by the eldest, Werner von Siemens, took the lead in developing and manufacturing electric generators in the latter half of the nineteenth century [3]. Early on, Werner recognized that batteries were far too costly to provide the rapidly growing need for electric power and so he focused on mechanical generation of electricity. Later joined by three other brothers, the Siemens manufacturing and engineering operations developed an international reputation that has persisted to the present time.

Werner and his brothers were part of a family of ten surviving siblings who grew up in a small town near Hannover Germany. Their father was a successful farmer and estate manager. Werner showed an early interest in engineering, and at age 17, he entered an officer training program with the Prussian army. After studying at the Berlin artillery and engineering school for 3 years, Werner developed an interest in the application of electricity with telegraphy, a promising way for distributing information in the military. After he improved the speed and range of the army's existing systems, he received a commission in 1847 to lay a new underground telegraph system.

One of the biggest problems facing long-distance telegraph transmission was producing and maintaining insulation of the telegraph wires. Siemens decided to cover his transmission wires with a new type of coating, "gutta percha," a rubber-like substance made from the latex of trees in Malaysia. This natural plastic substance seamlessly coated the wires making them highly water resistant. The success of his new system led to other military projects and Siemens decided to leave the military and set up his own company. He asked a young instrument maker at the artillery school, Johann Georg Halske, to join him in constructing a plant to manufacture telegraph equipment.

The Siemens Halske Company rapidly was in demand for developing new telegraph systems for militaries throughout Europe. Since the Morse telegraph system developed in the United States became internationally accepted, Siemens began with that system making improvements in the relays by supplying a magnetic bias for long-distance transmission. The company specialized in laying undersea telegraph cables with the new gutta-percha insulation and they also developed a new technique for electroplating gold and silver.

Werner's younger brother Wilhelm was initially steered toward a career in business and banking by his father, but Werner convinced him that a career in engineering offered more possibilities, and he entered the University of Gottingen to study physics and mathematics. After serving as an apprentice at a steam engine factory, Wilhelm began to work for his brother initially focusing on his brother's new electroplating process. In 1843, Werner decided to send his brother to England to establish his patent for electroplating at a plant in Birmingham. Wilhelm proved highly successful and was soon acting as the British agent for the Siemens and Halske Berlin company. He decided to stay in England where he was joined by two younger brothers Karl and Friedrich. He eventually married an English woman and became a British citizen.

The Siemen brothers' cable factory was built near London in 1858 specializing in gutta-percha insulated cable specially designed for undersea use. They rapidly developed an international reputation and designed a special ship called the *Faraday* for laying undersea cables. The company went on to lay the Indo-European cable from London to Calcutta in 1870 and the trans-Atlantic cable to the United States in 1875.

With the rapid development of arc lighting and electroplating, there was a strong demand for cheap high-voltage electricity that could only be supplied by mechanical means, steam engines, and water wheels. The Siemens and Halske Company decided to branch out from telegraphy to develop mechanically driven electrical generators. Dating back to Faraday's original design, electric generators used disc-type armatures that moved through the magnetic field of permanent magnets (see Chap. 4). The basic idea was the armature wound with wire cut through the force lines of the magnetic field producing electricity. Siemens recognized that these armatures were bulky and inefficient, and he designed a simpler armature shaped like a weaver's shuttle so that the winding was exposed to a higher density of magnetic flux.

The next breakthrough was replacing the permanent magnets with more powerful electromagnets. With the initial versions, the coils of the electromagnets were excited by a separate power system. But in 1867, Siemens announced that their company had developed a new self-exciting system whereby the electromagnetic coils were energized by the armature of the generator (Fig. 9.1). The system worked because the iron cores of the electromagnets maintained some magnetism after the system was turned off so that when restarted there was enough residual magnetism

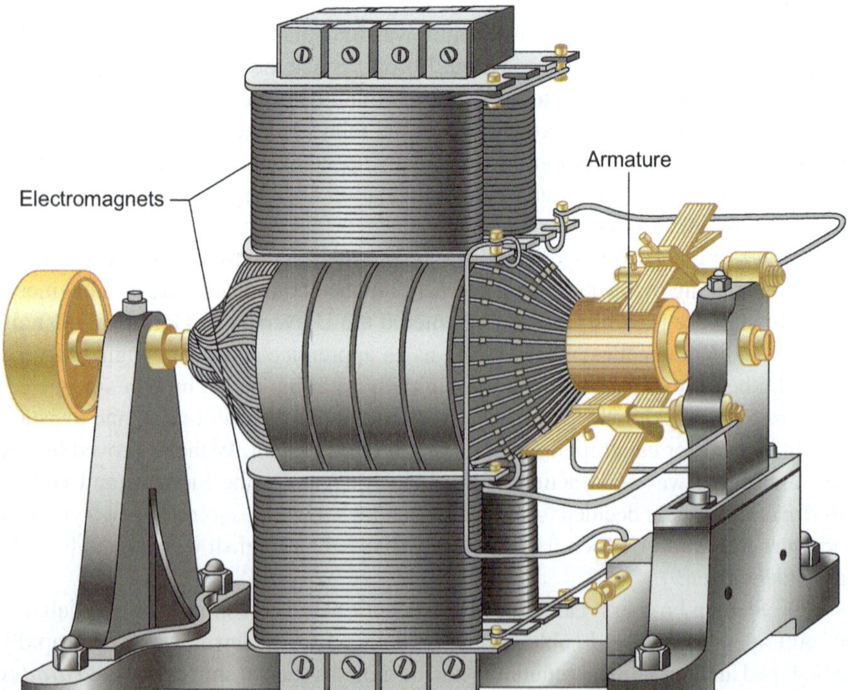

Fig. 9.1 Siemens self-exciting electric generator originally called the dynamo-electric machine later shortened to "dynamo"

to start producing electric current to "prime the pump" so to speak. Siemens called his self-exciting generator the "dynamo-electric machine," later shortened to "dynamo."

All the dynamos made by Siemens and competing companies at the time were direct current (DC) generators with a commutator to convert alternating current produced by the wired armature moving through the changing magnetic field into pulsed direct current. To deal with problems of overheating of the armature and sparking of the commutator, the Belgian engineer Zenobe T Gramme devised an armature wound with many separate coils connected to form a single continuous closed coil. He also developed an electric motor that was just the dynamo running in reverse. Siemens further improved the armature by winding the wire around a cylinder so that the entire length of the winding was equally cut by the magnetic flux, a design that persists to this day.

In the 1870s, the demand for electric power exponentially increased. Arc lighting, electroplating, and motors in factories were all big consumers of electricity. Electrical arc furnaces for metal processing were first developed in 1879. In the same year, Siemens and Halske designed the first electric railway and a year later the first electric elevator. These systems were supplied by Siemens and Halske direct current generators often backed up by a bank of batteries. By contrast, demand for alternating electric current was minimal although soon that would change.

Arc Lighting

Arc lighting became a big user of electric generator power when the Russian engineer, Paul Jablochkoff developed an efficient commercial version called the Jablochkoff candle in 1876 [4]. The "candle" consisted of thin carbon rods bound together by kaolin cement that also acted as an insulator. The rods were arranged so that when one burned down another was automatically lit. Jablochkoff's candles were longer lasting and produced a more pleasing light compared to earlier versions that lasted only a few hours and produced a glare. One problem was the candles did not work well with standard generators because direct current caused the carbon rods to burn unevenly. They worked much better with alternating current. The Siemens and Halske Company quickly developed a new line of alternating current generators to meet the demand. These new alternating current generators typically produced much higher voltage than the standard direct current generators.

By the late 1870s, the large cities of the world, Paris, London, and New York were inundated with electrical wires to service the burgeoning telegraph, telephone, and fire and burglar alarm systems that businesses relied upon. These wires carried low-voltage direct current but adding to the mix were high-voltage alternating current wires to supply the newly expanding arc light systems. Commercial sections of the cities contained an ugly maze of wires hanging from windows and roofs of buildings and carried over a forest of poles of different heights some higher than 100 feet. In some cases, the wires were simply hung in trees. Making matters even worse, many of the new electric companies soon went out of business, but their wires remained to fray and eventually drop from the poles.

The 1876 Centennial Exposition in Philadelphia featured a new arc light system powered by the first dynamo designed in the United States by the inventor Moses Farmer and his partner William Wallace, owner of a large brass and copper factory in Connecticut. A Cleveland chemist, Charles Bush built the first commercial arc light system for a department store in Boston in 1878. The Bush Electric Company constructed an arc lighting system (the Great White Way) on Broadway Avenue in New York in 1880 driven by more than 3000 V of alternating current. For this and several other nearby locations, Bush constructed three central power stations with his high-voltage wires strung among the jungle of functioning and nonfunctioning wires that currently existed. As we will see later in the chapter, this was an accident waiting to happen.

Thomas Alva Edison at Menlo Park

Thomas Alva Edison was an iconic American figure, revered by the public and respected by much of the international scientific community. But he was not without shortcomings. He had no formal training in physics or mathematics and cared little for hypothesis-driven research. He used a brute-force approach to solving scientific problems—repeat experiments until a solution was found. Regardless, he was a highly successful inventor and entrepreneur and ultimately became very wealthy.

Edison was born in Milan Ohio in 1847 and grew up in the small town of Port Huron, Michigan [5, 6]. His father worked at a variety of jobs including a grocery and truck farming business and his mother, a former schoolteacher, took in borders to help meet expenses. Edison had little formal education being taught mainly by his mother, a clever inventive woman. At age 12, he began noticing hearing loss. The cause of the hearing loss is disputed but probably resulted from repeat middle ear infections as a child. Later in life, he became completely deaf in one ear and could barely hear out of the other so that communication became difficult. At the same time, Edison noted that the hearing loss made it easier to concentrate on his work without distractions from the noisy environment.

At age 13, Edison became a newsboy for the Grand Trunk Railroad working for a dollar a day. He spent two dollars to join the Detroit Public Library where he embarked on a course in self-education, reading every book he could get his hands on. At age 16 in the middle of the Civil War, Edison taught himself Morse code and earned a position as a junior telegraph operator. He rapidly developed a reputation for hard work and enterprise and before long he was on a monthly retainer to Western Union for $400. His first patent was for an electric vote recorder in 1869, and in 1874, he patented a multiplex telegraph system with Franklin Pope that turned out to be very lucrative. He used the money from his patents and from wealthy investors to open the Edison Menlo Park Laboratory in 1876. Menlo Park, New Jersey, provided abundant space, peace, and quiet at an affordable cost. The focus of his laboratory was commercial utility. An invention had to be valuable from an industrial point of view? [7].

At the urging of a friend, Edison visited Wallace's brass and copper factory in Connecticut to see for himself the dynamo invented by Moses Farmer and first displayed at the Centennial Exhibition of 1876 [8]. The eight-horsepower electric dynamo that Farmer called the *telemachon* lit up a row of 8 arc lights in Wallace's factory. Edison was enthralled and immediately began to consider how he could improve on the system. By the time he left the factory, he bet Wallace that he could beat him in making practical electric lighting. Wallace who had a clear head start shook hands on the bet.

The Race to Develop a Durable Incandescent Light Bulb

When Edison returned to Menlo Park, he immediately began to work on the problem. "*It was all before me. I saw the thing had not gone so far but that I had a chance. I saw that what had been done had never been practically useful. The intense light had not been subdivided so that it could be brought into private houses*" [9]. Edison decided to focus on developing an incandescent light bulb that could light the homes in America more effectively and cheaper than gas lights.

But the task of creating a practical electric light bulb was more difficult than Edison had surmised. Scientists and inventors around the world had been working on the problem for several decades without much success. In England, William Crookes and his associate Gimingham had inserted a platinum coil into one of their radiometer bulbs and connected it to a battery (see Chap. 8). They were pleased to see a pleasant bright glow without the glare of an arc light. Even more important they demonstrated that the incandescence became greater as the vacuum in the bulb became greater. The problem was the platinum coil rapidly deteriorated. They published the results of their light bulb experiments in the journal *Engineering* in April of 1878. Crookes had already agreed on a contract with the Newcastle factory of Swan and Mawson to mass produce his radiometer and he would later work with Joseph Swan to produce incandescent lamps.

Edison experimented with a variety of filaments in his incandescent bulbs throughout 1877, trying carbon, boron, and silicon filaments before settling on a platinum filament by the end of the year. He tried several variations to prevent the platinum filament from melting. Edison was finally convinced that he had produced a practical incandescent light bulb (Fig. 9.2). He applied for a patent, had his attorney begin to identify potential investors, and invited a reporter from the New York Sun to come to Menlo Park and see and write about his triumph. He boasted to the reporter: "*I can light the entire lower part of New York city, using a 500 horse power engine*" [10]. The reporter was enthralled: "*There was the light, clear, cold and beautiful. The intense brightness was gone. There was nothing irritating to the eyes. The mechanism was so simple and perfect that it explained itself. The strip of platinum that acted as burner did not burn*" [11].

Fig. 9.2 Edison's original incandescent light bulb with a platinum filament from 1879. (From gettyimages.com)

What Edison failed to mention to the reporter was that his new light bulb only lasted an hour or two. When Edison read Crookes' report in *Engineering*, he realized that not only was platinum fragile but the better the vacuum, the longer the filament lasted. By early 1879, Edison was preoccupied with finding a better filament and improving the vacuum in his bulbs. At the same time, he recognized that Farmer's dynamo would not be adequate to provide the necessary electricity to power the thousands of light bulbs he envisioned he would need. He tackled these problems with his usual brute-force approach.

Throughout 1879, Edison and his team of workers tried many hundreds of different filaments to produce a longer-lasting light bulb. They started with a variety of metals and metal combinations and then moved on to baked carbon filaments. The thin filaments were carefully coiled and then carbonized in a furnace. Edison brought in a young German glass blower who had trained with the famous Hiernrich Geissler and improved the vacuum pumps for evacuating the bulbs (see Chap.3). The first breakthrough came with a carbonized cotton-thread filament that lasted 14.5 h. That lead to a mad dash to find an even longer-lasting carbonized filament. They tried silk, a variety of wood fibers, flax, parchment, and different hairs even those from the beards of the workers. Remarkably, a thin strip of carbonized cardboard won out as the best filament.

To generate the power needed to light the bulbs, Edison's team ordered the five best currently available dynamos and systematically made changes to improve their efficiency. They rewound the armatures and altered the shape of the electromagnets to concentrate Faraday's lines of magnetic force. The final product was affectionately called *long-legged Mary Anne* due to a pair of three-foot-long iron poles that supported the dynamo.

By the end of 1879, Edison and his team were ready to display their work to investors and the public. They would light up Menlo Park with a lighting display on New Year's Eve. This meant buying steam engines to power the dynamo and developing and manufacturing ancillary equipment including underground and building wiring, switches, fuses, and of course light fixtures. Western Union sent men to help with the wiring and Edison and his men oversaw the whole operation. Despite the daunting task, the team was successful, and Menlo Park was electrified by early December 1879.

Edison chose his favorite reporter, Marshall Fox, from the *New York Herald*, for an initial exclusive report of the remarkable Menlo Park achievement. The New York Herald was read by only the best and most influential New Yorkers. It cost three cents, an extravagant amount compared to the one- and two-penny newspapers read by the masses. Fox's story that appeared in the December 21, Sunday edition began: "*Edison's electric light, incredible as it may appear, is produced from a tiny strip of paper that a breath would blow away. Through this little strip of paper is passed an electric current, and the result is a bright, beautiful light, like the mellow sunset of an Italian autumn*" [12]. Edison boasted that his light was "*cheaper than that from the cheapest oil*" [12]. On New Year's Eve, 3000 people including future donors arrived at Menlo Park and marveled at the light display despite severe winter weather.

Lighting Up New York

Edison's next step was to take the system that he developed in Menlo Park and apply it to a New York business district effectively competing with the price for gas lighting. He would run the insulated copper wires from the power plant underground, and at the buildings, the wires could be run through existing gas pipes into gas lamps that would be replaced with light bulbs. A major limitation of Edison's direct current electrical system was the short distance that the current could be reliably transmitted, about half of a mile. He planned to have multiple power plants each serving an area of about one square mile. This was feasible in heavily populated downtown New York but would not work for suburbs and smaller towns.

In early 1880, Edison began a new phase of his project, lining up investors and mass-producing cost-effective, reliable equipment. He had a natural gift for selling his wares and soon had the big corporate money he needed for the project but there were still technical problems to overcome. His carbonized cardboard light bulbs, although longer lasting than earlier versions, still only lasted about 300 h. Furthermore, when he started the patent process, he found out that a rival lightbulb inventor, William Sawyer had already filed a patent for a bulb with a similar filament. Edison had to find a light bulb with a better filament. His team again embarked on the tedious process of trying a wide variety of filaments for a new bulb. They were looking for a natural substance with geometrically parallel fibers suitable for the filament. After literature searches and many tries and dead ends, in July of 1880, they settled on a carbonized bamboo fiber for the filament. It proved longer lasting and easier to work with.

A major problem that Edison and his team faced was laying and maintaining underground copper wires for distribution of the direct current. He had a bad experience with the Menlo Park system having buried the wires in narrow wooden conductor boxes that had to be repeatedly dug up and replaced because of water damage. They finally solved the problem by insulating the wires with layers of muslin covered with paraffin, tar, linseed oil, and asphalt. Then there was the rapidly increasing price of copper, which represented the single biggest cost of his proposed electrical system. By using a network of thinner multiple feeder wires instead of a single very thick and costly feeder wire, Edison was able to save a large sum of money. Every step of the process provided a new challenge. Edison lamented: "*I have to make the dynamos, the lamps, the conductors, and attend to a thousand details the world never hears about*" [13].

By February 1881, Edison's staff began to move to Manhattan to begin building the first direct current electrical lighting system [14]. They promised to have electric current flowing to the buildings by the fall. At the end of April, they received city permission to begin digging the subways for the underground wires. An Irish immigrant crew, working mainly at night, had to dig down two feet, which proved much more difficult than originally expected. Instead of the wooden boxes used at Menlo Park, the wires were run in iron pipes with connector boxes every 20 feet. It rained nearly every day during the dig and the company making the iron pipes abruptly stopped delivery. Adding insult to injury, Edison had to pay five city inspectors $25 dollars a week to monitor progress. Fortunately, they only showed up once a week to collect their pay.

Edison's team had been searching for a reasonably priced large building to locate their central generators and finally settled on a building on Pearl Street in a slum district at a cost of $65,000. It was located within a half mile of the all-important financial district of Wall Street and large enough to hold the massive dynamos and coal-fired steam engines. At every step in the process of building the generator plant and distributing the wires, there were unexpected delays. Making matter even worse the nation went into morning with the shooting of President Garfield and his survival for more than 2 months with a bullet in his spine. By December of 1881, even the press began to turn on Edison. The New York Times reported despite having laid a considerable amount of wire, Edison was no were near lighting up the Financial District. Edison admitted there was still a long way to go blaming delays in the delivery of supplies and problems with digging in the frozen ground.

In the meantime, Edison had successfully lit up the mansion on Madison Avenue of his chief investor, J.P. Morgan. This required a separate electric generator plant on the grounds since the home was too far north of the main plant on Pear Street to run a direct current line. The Morgan home project had its own list of delays and problems not the least of which was a major fire that destroyed Morgan's beautiful wood-paneled library and furniture. Despite all the problems Morgan was proud to be the first to have electric lighting and he had his house rewired on multiple occasions despite the repeat damage to the walls and floors. Morgan was reported to joke: "*I hope the Edison Company appreciate the value of my house as an experimental station*" [15].

Slowly but surely, Edison and his team overcame the obstacles, and in August 1882, they began to run tests on the system. A setback occurred when a steam-heating company, also digging in the area, accidentally broke one of Edison's iron pipes causing a short. Edison's team found out about the problem when horses passing by received a brief shock causing them to jump and run off. What would happen when they turned on the full power of the generators? Could pedestrians be in danger? These concerns just added to the suspense of the upcoming full-scale test. Finally, at 3 PM on September 4, 1882, his team flipped the main circuit braker at the Peal Street plant and Edison personally flipped the switch in J. P. Morgan's financial office to turn on the lights. The test was a complete success and the New York Times lauded Edison's incandescent light as a giant step ahead of the gas light. Edison had kept his promise to deliver an electric light system and there was a huge lucrative market for similar systems awaiting in the Unites States and throughout the world.

George Westinghouse in Pittsburgh

George Westinghouse and Thomas Edison had a lot in common. They were the same age, passionate for invention and discovery, and highly successful inventors and entrepreneurs. Both had hundreds of patents. Yet, their battle over the primacy of direct versus alternating current for distribution of electricity lasted over a decade and resulted in some of the most bizarre confrontations and deceptions in the history of science.

Westinghouse was born in a small town in New York state in 1846 the son of a machine shop owner [16, 17]. With the Civil War heating up, he left school at age 15 to join the New York National Guard. He eventually joined the Navy where he served as an assistant engineer on a gunboat through the end of the war. After returning home, he briefly enrolled at Union College in Schenectady but dropped out after only one semester. Westinghouse showed an early interest in machines in his father's shop and his first invention at the age of 19 was a rotary steam engine. He became interested in the rapidly expanding railroad industry and he invented a device to get derailed cars rapidly back on the tracks. His most important and lucrative early invention at age 22 was an air brake that allowed the engineer to rapidly stop all the cars at once for the first time. He cleverly refused to sell his patent to railroad companies, setting up exclusive production at his newly constructed Pittsburgh factory.

In defending his exclusive right to produce his air brakes, Westinghouse developed a reputation for threatening and carrying out lawsuits against any company that tried to infringe upon his patents. The large railroad companies, who were used to getting around patents by making slight modifications, soon found that Westinghouse was a business force to be reckoed with. Westinghouse first became involved in the electrical industry when in 1881 he began looking to purchase promising new patents for electrical devices to activate train signals. Oil lamps as signaling lights just were not very good. With his own and purchased patents, he set up the Union Switch and Signal Company in 1882.

Westinghouse was living in New York when Edison was installing his first electric lighting system in the financial district and like most everyone, he was impressed with the final product. If there was going to be a demand for centralized electric plants to provide direct current for every square mile of cities in America, why could not his company compete for building the plants. In 1884, Westinghouse decided to set up an electric lighting Company. He began by hiring a young inventor, William Stanley Jr. who had worked for the Swan Incandescent Light Company in England and H. M Byllesby, a mechanical engineer who was working for Edison. He paid the Swan Company $50,000 to use two of their patents, one a carbonized silk-filament light bulb and the other a self-regulating direct current dynamo. Stanley and Byllesby were charged with developing a commercial direct current lighting system for Westinghouse.

By chance in early 1885, Westinghouse came across an article in the English journal *Engineering* describing an alternating current system for lighting that could economically carry electricity over long distances. It used the novel approach of transmitting the electricity at high voltage and then stepping it down at the site using newly developed transformers. The stepping down in voltage occurred outside the buildings to be lighted minimizing the risk of exposure to high voltage. The new alternating current system had been developed by Lucien Gaulard and John Gibbs and at the time Gaulard was in Turin Italy building a test system to transmit electricity generated at a waterfall far from the city to the city for individual lighting. One of Westinghouse's young engineers happened to be in Italy visiting family and Westinghouse asked him to check out the Turin system.

At the time, many of the most prominent electrical researchers in the world had an aversion for alternating current. As we will see, Thomas Edison vehemently opposed the use of alternating current. He truly believed that it presented an unacceptable risk to the public. In England, Sir William Thomson (Lord Kalvin) was an ardent foe of alternating current throughout his long-distinguished career. Although the German Siemens Halske Company eventually produced an alternating current generator for arc lights, Werner von Siemens saw little practical long-term use for alternating current. One of the reasons was the lack of a motor that ran on alternating current. Fortunately, Westinghouse's engineer in Italy sought the advice of the Ganz Electrical Company in Budapest, Hungary, who were already working to improve the Italian alternating current system. Westinghouse was sold and he purchased an option on the American patent. He also bought one of the Gaulard and Gibbs transformers and a Siemens alternating current generator designed to run arc lights.

A reluctance to work with alternating current was not just on the outside but also within Westinghouse's own work force. His first task was to alleviate concerns and convince colleagues of the long-term benefits. When Westinghouse received the Gaulard and Gibbs transformer, it was disappointing, not even properly soldered, but after disassembling and reassembling it, Westinghouse saw how it could be markedly improved with proper craftmanship and a new design. The concept of the transformer as originally devised by Michael Faraday in 1831 was simple, when alternating current was passed through a primary coil of insulated wire the current was passed on to a separate secondary coil surrounding the primary coil. If the secondary coil had more turns of wire than the primary coil, it would jack up the voltage of the current in the secondary coil whereas if the secondary coil had less turns, it would decrease the voltage. The transformer only worked with alternating current since continuously changing magnetic flux was required.

Westinghouse Chooses Alternating Current

Westinghouse's new transformer was shaped like an H with the horizontal central bar machine wound with primary and secondary copper wire coils. He could mass produce the transformers to step up or step down the alternating current voltage as desired. Rather than making many different-size transformers, he could simply run several transformers in parallel to control the voltage change. Around this time, William Stanley Jr. was just about to test the first Westinghouse direct current lighting system that he and Byllesby had been working on. Stanley was one of the few Westinghouse employees who immediately recognized the advantages of an alternating current system, and after discussions with Westinghouse, they decided to scrap the direct current system and convert to alternating current.

Despite naysayers all around, Westinghouse incorporated his fifth company, the Westinghouse Electric Company, in January 1886 with stock worth one million dollars. His goal was to use alternating current generators to produce high-voltage current that could be transmitted long distances and stepped down to safe voltages at factories and homes. The Westinghouse direct current laboratory built and managed by William Stanley was in Stanley's hometown of Great Barrington, Massachusetts. While Stanley was in Pittsburgh consulting with Westinghouse, the Edison Lighting Company built their own conventional direct current system showcased with the lighting of one of Great Barrington's finest homes. The local newspaper was enthralled running the headline– *A Brilliant Spectacle* [18]. They marveled at how well the mansion and grounds were illuminated but also noted that a fire truck stood by ready to extinguish any fires that might occur.

A week later, Stanley powered up a 25-horse-power steam engine located in a barn next to the laboratory in Great Barrington to run the Siemens alternating current generator Westinghouse had purchased. The generator sent 500 V of alternating current along a copper wire running in a row of elm trees to the basement of the general store of Stanley's cousin, R. I. Taylor, where the current was passed through a new Westinghouse transformer and stepped down to 100 V. The lower voltage current was passed through interior wires to light the store's incandescent lights. Stanley reported to Westinghouse that all went well. Within a week, Stanley had signed up another general store, a doctor's office, and a drugstore and within a month several dozen new customers. At the same time, due to the distance restrictions, Edison's company enlisted only a dozen new customers. Remarkably, in this first head-to-head DC/AC competition between the Edison and Westinghouse companies neither the press nor the public was aware of the major difference between the two systems. Unlike Edison, Westinghouse tended to avoid interviews with the press, and in this case, he preferred to keep the details of his system a secret.

Stanley quickly designed and built a new alternating current generator to replace the Siemens' generator, and he carefully tested and improved all aspects of the new AC system. Westinghouse decided to install the new system in his Pittsburgh Union Switch and Signal Company plant and ran a test line three miles to the suburb of East Liberty. He tested transformers at each end of the line and systematically developed a working AC system. After being confident in the system's reliability, Westinghouse's first big customer was the Buffalo, New York emporium of Adam, Meldrum, and Anderson, a massive downtown department store that sold just about everything. In late 1886, Westinghouse's AC electricity lit both arc lights outside and incandescent lights inside and the public crowded the emporium to see the brilliant lights. Westinghouse soon had 27 new customers for his AC system from a wide range of locations.

If the public and news media were unaware of the differences between the Edison and Westinghouse electrical systems, Edison was aware and very concerned. Not only did he have a formidable new competitor, but the competitor was well respected and well-funded and had a revolutionary new system. Edison's immediate response was that Westinghouse's AC system was dangerous and it would not be long before someone was electrocuted. He could not understand how workers could possibly repair a damaged high-tension line in a busy city. With his DC system, there was no danger to the public or electric workers.

While Westinghouse was developing his new AC system in America, The Ganz Company in Budapest already had a system up and running that illuminated more than a thousand incandescent lights with 1300 V of alternating electric current. They were the first to use the term transformer to describe the Faraday induction coil that stepped the voltage up and down. The system was called the ZBD for its three Hungarian inventors (Zipernowsky, Blathy, and Deri). Ironically, the first customer for the new ZBD system was the Edison Company of Milan, Italy. The company was run by John Lieb, Edison's chief engineer at the Pearl Street plant during the first large DC lighting project. Lieb's group had successfully implemented an Edison DC system in Milan but they were unable to service the Theater dal Verm that was outside the reach of the DC system. They purchased the ZBD system and successfully lit up the theater.

The ZBD system was rapidly becoming a sensation in Europe and several of Edison's engineers suggested that he should consider buying the patents for the system. Edison sent his long-time associate Francis Jehl to Budapest to assess the system and Jehl was highly impressed [19]. The Ganz Company was enthusiastic in joining with Edison and sent one of the inventors, Otto Blathy to visit Edison and try to convince him to purchase the patents. Edison remained skeptical even after it became apparent that he could buy the patent rights for $20,000. Edison personally contacted Siemens and Halske in Germany for their opinion on the AC system, and as noted earlier, they were highly negative noting that the system was troublesome and dangerous. After ruminating about the problem for several months, Edison decided to stick with direct current. The technology of alternating current was simply too dangerous.

The Dangers of High-Voltage Alternating Current

In the spring of 1888, two accidental electrocutions in New York City reinforced Edison's warnings and brought the danger of high-voltage alternating current wires to public awareness [20]. As noted earlier, high-voltage AC wires were intermixed with the many low-voltage DC wires running on poles above the city. A Rumanian immigrant boy was skipping along East Broadway when he grabbed a wire that was dangling from one of the overhanging telegraph poles. There was a shower of sparks and the boy fell to the ground motionless. He was taken to the Chambers Street Hospital where he was pronounced dead. The US Illuminating Company whose high-voltage AC lines for arc lights ran along the street was charged with neglect and fined.

A few weeks later an even more dramatic episode occurred. An employee of the Brush Electric Company was high above the street clearing away old wires from the cornice of the second floor of a building when a worker inside the building heard a crackling sound and noticed smoke outside the window. On investigating, he saw a man sprawled on the wires with one of the wires partly cut through and the surrounding insulation burning. When rescuers tried to pull the body through the window, they received a shock. They had to use rubber sheets to remove the corpse from the wires. The newspapers mirrored Edison's concerns about high-voltage AC current. The wires were not only an eye sore but dangerous. At the very least, they needed to be removed and placed underground.

For his part, Edison used his considerable resources to try and block the use of alternating current systems. "*My personal desire would be to prohibit entirely the use of alternating currents. They are unnecessary as they are dangerous*" [21]. He embarked on a campaign to influence public opinion by exaggerating the danger of electrocution from high-voltage alternating current systems. He bought advertisement in leading newspapers emphasizing the risks of alternating current and lobbied powerful political friends against the use of high-voltage alternating current.

In Britain, the managing director and the resident physician of the Institute of Medical Electricity, Henry Newman Lawrence and Arthur Harries, called for better studies on the effects of electricity on the human body [22]. In their initial experiments, they compared the effects of alternating and continuous current on measured resistance in a group of ten human volunteers and noted a marked variability based on the moisture of the skin. The resistance of dry skin was about ten times greater than the resistance of saltwater-moistened skin. Overall, they noted that the resistance of the human body to continuous current was about seven to eight times greater than it was to alternating current regardless of skin moisture. These studies suggested that the human body was much more sensitive to the passage of alternating current as opposed to continuous current. Based on their data, they estimated that about 10 mA of continuous current could be passed through the body without discomfort whereas only about 2 mA of alternating current could be passed without discomfort.

But many practicing members of the Institute felt that Newman Lawrence and Harries had grossly overestimated the dangers of alternating current. George Forbes, the well-known Scottish electrical engineer who later played a major role in the Niagara Falls hydroelectric project, pointed out that their resistant measurements for alternating current were flawed. The sensation of electric shock did not depend on the current measured by a galvanometer but rather on the maximum current passing through the body. Other members noted that they had personally withstood shocks from much higher alternating currents than those suggested by Newman Lawrence and Harries. There was growing agreement that the human body was much more complex that an electrolyte solution and that its complicated structure and physiology prohibited any comparison with an inanimate conductor. Furthermore, different parts of the body responded differently to different types of electricity.

Electrocution of Criminals

In late 1887, Edison received a letter from a Buffalo dentist, Dr. Alfred Southwick asking him what he thought about electrocuting condemned criminals and if he could recommend the strength of current that would reliably produce death [23]. Southwick was one of three members of the New York State Death Commission. Edison immediately responded that he was against the death penalty and did not want to become involved. However, after thinking about it for a few weeks, Edison wrote back to Southwick that he had changed his mind and that electricity would be ideal for producing a rapid painless death. Furthermore, he recommended using an alternating current dynamo, like the ones manufactured in America by George Westinghouse. Even the slightest contact with Westinghouse's high-voltage AC current would lead to instantaneous death.

With Edison's endorsement, the New York State Legislature passed a law establishing electrocution as the means of capital punishment starting January 1, 1889.

Edison had cleverly established alternating current as a threat to human life and the Westinghouse dynamo a killer. What Edison needed was someone, preferably not involved with his company, to take up the torch and lead the crusade against alternating current. In stepped a little-known New York engineer and electrical consultant, Harold Brown. After the death of the Rumanian boy, Brown wrote a letter to the New York Evening Post condemning companies using alternating current for lighting. All they were interested in was *the almighty dollar*. They were just trying to save money on the heavy copper wires used by safe direct current lighting systems. Brown ranted about the dangers of high-tension alternating current and suggested safety rules that included outlawing alternating current above 300 V. He wanted to legislate alternating current out of business. Here was a man after Edison's heart.

Brown later appeared before the New York Board of Electricity Control and demanded that his proposed safety rules be published in the minutes of the meeting and sent to alternating current companies for comment. As usual, Westinghouse was averse to public controversy and tried to ignore the matter. But as Brown continued his attacks, Westinghouse felt that he had to respond. He described Brown's attack as "*more unmanly, discreditable and untruthful than any competition which has ever come to my knowledge*" [24]. He went on to note that if safety was the issue, there had never been a fire involving any of his alternating current systems he installed around the country whereas of the 125 Edison direct current systems installed, there had been several major fires with total destruction of three central stations and many user buildings including total destruction of a large theater in Philadelphia.

It was just a matter of time before Brown and Edison got together and when Brown asked Edison for a loan of some instruments to help prove his case, Edison invited him to use his private laboratory for his experiments. Brown accepted Edison's offer and was soon carrying out experiments late into the night at Edison's state-of-the-art laboratory. By mid-1888, Brown was ready to show the world the dangers of that "*damnable*" alternating current. He arranged a demonstration in a lecture room at the School of Mines, Columbia College on Madison Ave, and invited members of the Electrical Control Board, the press, and interested parties in the electrical community.

Brown started out by discussing the advantages of direct versus alternating current and noted that he had performed numerous experiments on live animals showing the creatures could tolerate direct current much better than alternating current. Brown then proceeded to bring a large black golden retriever dog onto the stage and after muzzling the dog he strapped it into a large wooden cage laced with copper wires. The poor dog who weighed about 75 pounds woofed through his muzzle. Not surprisingly, the crowd became restless, and murmurs were audible. Brown was assisted by Edison's chief electrician, Arthur Kennelly and a physician who specialized in electrotherapy, Dr. Frederick Peterson who held the struggling dog while Brown attached wires to its legs. Brown announced that the dog's resistance was 15,300 Ω and he applied 300 V of direct current. The dog was obviously uncomfortable and when Brown increased to 400 V it began to struggle and yelp. When he increased to 700 V, the struggling dog broke the straps and new straps had to be applied. Even though it was obvious that spectators were becoming dismayed with the cruelty, Brown went ahead and increased to 1000 V even as some left the room.

Brown finally turned off the direct current and sarcastically announced to the audience that the alternating current would be "less trouble" and even "make him feel better" [25]. He then connected the dog to 330 V of alternating current from a Siemens AC generator and the dog collapsed to the ground and was pronounced dead. Rather than convincing the audience of the safety of direct current, Brown had managed to make them angry. One reporter stood and demanded that Brown stop torturing dogs and a member of the American Society for the Prevention of Cruelty to Animals forbid Brown from executing any more dogs. Sensing the overall hostility Brown decided to end his demonstration but before he left the stage he remarked, "*the only places where an alternating current ought to be used were the dog pound, the slaughterhouse, and the state prison*" [25].

Not deterred from his mission by the unpleasantries of this initial demonstration, Brown decided to persevere. Less than a week later, he returned to the Columbia School of mines where he systematically electrocuted three large dogs with high-voltage alternating current. The audience was more select, mostly public health officials and newspaper reporters. Brown was assisted by Dr. Peterson but there were no Edison people present. After the display, Brown announced to the press that since high-voltage alternating current can kill a large healthy dog it can kill a man. The next day, he confidently wrote to Edison's right-hand man, Author Kennelly: "*Whatever action the Board of Electric Control may take, it is certain that yesterday's work will get a law passed by the legislature in the fall, limiting the voltage of alternating current to 300 volts*" [26].

Brown next focused his attention on convincing New York State officials that alternating current should be designated the official current for executions. The legislature asked the Medico-Legal Society of New York to set up a committee to recommend the best way to kill a condemned prisoner with electricity. The chairman of the committee was none other than Dr. Frederick Peterson, Brown's partner in the dog executions. Brown organized one final exhibition of the killing power of alternating current held at Edison's famous West Orange Laboratories. He invited all the key players including prominent physicians from the Medico-Legal Society, members of the New York State Death Commission, Buffalo dentist Dr. Southwick and most important of all, the great Thomas Edison. Brown wanted to show that alternating current could kill larger animals more akin to humans. He started with a 124-pound calf, moved on to a 145-pound calf and ended with a 1230-pound horse. All slumped to their death with a maximum of 700 V of alternating current.

The exhibition had the effect Brown hoped it would. The New York Times reported: "*The experiments proved the alternating current to be the most deadly force known to science, and that less than half the pressure used in this city for electric lighting by this system is sufficient to cause instant death*" [27]. At the next meeting of the Medico-Legal Society, the members unanimously accepted the recommendations of the electrocution committee for *death by alternating current* with the prisoner sitting or supine in a properly prepared chair or bed. The state of

New York was ready to execute its criminals with alternating current electricity. In the spring of 1889, the New York State Prisons hired Brown as their electrical expert.

The First Electrocution

They soon identified an ideal candidate for the first electrocution in William Kemmler, also known as John Hort. Kemmler, the son of German immigrants, was born in Philadelphia in 1860. After alcohol-related deaths in both parents at a young age, Bill, who was illiterate, moved to the slums of Buffalo, New York where he became a vegetable peddler. Known as Philadelphia Billy, his heavy drinking was legendary in the local saloons. While recovering from a drinking binge, he accused his girlfriend and common-law wife, Tillie, of stealing and planning to run away with a friend. He grabbed a hatchet and struck her, killing her instantly. He then calmly walked over to a neighbor's house and confessed to the murder.

The murder trial was over quickly. Kemmler was convicted of first-degree murder and 3 days later, sentenced to death. His attorneys appealed the execution via the electric chair as cruel and unusual punishment. Concerned about the adverse publicity that would accompany the execution, Westinghouse hired the prominent lawyer W. Bourke Cockran to help with the appeal.

One of the highlights of the appeal process was the deposition of Thomas Edison in Cochran's law office. This was a rare instance where Edison himself publicly voiced opposition to alternating current. By this time in July 1889, Edison was nearly deaf so that the yelled questions and his loud replies could be heard on the nearby street. Edison confidently stated that alternating current could reliably produce sudden painless death in human beings without exception. Cochran's detailed questioning about the resistance in ohms and current in amps required to execute a man revealed a surprising lack of basic knowledge about electricity by Edison, yet Edison's reputation won the day. Newspaper articles uniformly lauded Edison's expert testimony.

Despite Cochran's eloquent arguments against electrocution, the United States Supreme Court turned down the appeal. The execution took place on August 6, 1890. Kemmler was awakened at 5 AM; his head was shaved, and he was dressed in a suit with a neck tie and white shirt. Kemmler was ushered into the execution room by the warden, Charles Durston, where he was introduced to 17 male witnesses selected from the audience. After examining the chair, Kemmler calmy sat down and said: "*Gentlemen, I wish you all good luck. I believe I am going to a good place, and I am ready to go*" [28]. After a hole was cut in his suit to apply an electrode to his body, Kemmler was strapped to the chair, his face was covered, and a metal cap was placed on his bare head (Fig. 9.3).

After Durston was informed that everything was set, he ordered the electrical switch thrown. The alternating current generator was charged with 1000 V based on the successful test execution of a horse the day before. Kemmler sat quietly and did not scream or cry out. After current was passed for 17 s, it was turned off and

Fig. 9.3 Painting of the electrocution of William Kemmler on August 6, 1879. (From Alchetron, the Free Social Encyclopedia)

Kemmler was declared dead. But several of the witnesses noticed that Kemmler was still breathing. The attending doctors confirmed that he was still alive and a second attempt with 2000 V was quickly ordered. With the increased voltage, blood vessels under Kemmler's skin ruptured and bled and some witnesses claimed that parts of the prisoner's skin caught fire. The New York Times reported, "*an awful odor began to permeate the death chamber, and then, as though to cap the climax of this fearful sight, it was seen that the hair under and around the electrode on the head and the flesh under and around the electrode at the base of the spine was singeing. The stench was unbearable*" [28].

The whole procedure lasted only about 8 min, but to the witnesses, it seemed like an eternity. The headline of the New York Times article read: "*Far Worse Than*

Hanging" [28]. Westinghouse later commented that they could have done a better job using an axe. The English medical journal, *Lancet* described the execution as "*one of the saddest, histories that has ever been posted up in the historical minutes of a century now nine-tenths towards the close*" [29].

Alternating Current Wins

During the long appeal process of the Kemmler electrocution, another gruesome electrical accident occurred just a few blocks from city hall in New York [30]. A Western Union lineman, John Feeks, was cutting out dead wires high above the street packed with lunchtime workers. He was pulling apart a group of tangled wires when he received a sudden shock. He grabbed onto another wire to steady himself and onlookers noted that flames and smoke issued from his body as though he was on fire. He lost consciousness and was suspended in the tangled wires 40 feet above the ground. Shocked pedestrians called out for someone to help but his dead body remained suspended in the wires for 45 min before equipment arrive to extricate him. Mayor Hugh Grant had had enough and ordered all high-voltage alternate current lines shut down so that workers could repair and upgrade the lines and where possible place the lines underground.

Edison saw the Feeks' electrocution as another opportunity to publicize his war on alternating current. He wrote an article in the popular *North American Review* arguing that the Feeks' death demanded an end to high-voltage alternating current [30]. Burying the wires underground was not a solution since it would just transfer the risk to another group of people. He noted that he had personally observed a dog killed by as little as 178 V of alternating current and that he could not confidently name a safe voltage. The use of alternating current must be banned.

Interestingly, despite the initial outcry over the Feeks accident, once the arc lights were turned off, citizens began to complain about the blackness that cloaked the city. "*Darkness and gloom were everywhere*" according to the *New York Times* [31]. Due to city bureaucracy, little work was done on underground subways and residents lamented the loss of their brilliant lighting [32].

In the next month's issue of the *North American Review*, Westinghouse wrote a powerful reply to Edison's condemnation of alternating current. He noted that the struggle for the control of the electric lighting business was unparalleled in the history of commerce. "*Thousands of persons have large pecuniary interests at stake, and, as might be expected, many of them view this great subject solely from the stand-point of self interest*" [33]. To put the risk of electrocution in perspective, Westinghouse pointed out, in the last year five people in New York City were electrocuted whereas 64 were killed in streetcar accidents, 55 by omnibuses, and 23 by illuminating gas. Furthermore, the great majority of electrical engineers were choosing the more scientifically sound and safer alternating current plants over Edinson's revered direct current plants. Unless alternating current plants are somehow prohibited, Edison's plants ran the risk of becoming obsolete. He concluded with probably the most devastating argument of all, over the prior 3 years purchasers of electrical incandescent lighting systems chose alternating current systems more than five times as often as direct current systems. Even Edison's own people were selecting alternating current systems when given the chance.

In the end, first Edison and then Westinghouse were squeezed out of their positions as head of the Edison and Westinghouse Electric Companies. J. P. Morgan and the Wall Street money men eventually took over control. Both companies eventually merged into the General Electric Company, which did three-fourths of the country's electrical business.

Nikola Tesla and the Polyphase Alternating Current System

Nikola Tesla, a Serbian American inventor, had two monumental ideas in his long eventful life, a two-phase alternating current system and motor that operated without a commutator or brushes and an ultra-high-frequency electromagnetic power station that would allow long-distance wireless communication and the transmission of wireless electrical energy. In a collaboration with George Westinghouse, the first idea came to fruition and was the driving force behind the electrical revolution of the first half of the twentieth century while the second idea although generating important technical advances never reached a practical application. Although Tesla equalled and may even have surpassed the inventive abilities of Edison and Westinghouse, he lacked the common sense and business skills that propelled these men, and he spent his later years relying on the generosity of Wall Street moguls Morgan and Vanderbilt for basic support.

Tesla was born in 1856 to Serbian parents in a small town in Croatia, part of the then Austria-Hungarian Empire [34, 35]. As a child, it was expected that he would follow in the footsteps of his father, a clergyman in the Serbian Orthodox Church but he was much more like his mother who was known for her remarkable memory and inventive skills. In primary school, Tesla displayed a knack for mathematics and an incredible memory being able to recite long poems with ease. He became fascinated with languages and became fluent in multiple languages, including English. At age 21, he gained admittance to the Polytechnic Institute of Graz, Austria, one of the oldest and most prestigious in Europe. It was here that he was first exposed to the marvel of electricity, and he decided that this was the field he wanted to work on for the rest of his life.

It was during Tesla's second year at the Polytechnic Institute that the seed for one of his great ideas was planted, a motor without a commutator. He was in a physics class where a direct current motor was being demonstrated. The brushes on the commutator were sparking and Tesla naively commented to the professor it should be possible to make a motor without the sparking brushes. The professor ridiculed the suggestion as not unlike suggesting a perpetual motion machine. The embarrassed Tesla was determined to prove him wrong.

Tesla would later say: "*I undertook the task with all the fire and boundless confidence of youth. To my mind it was simply a test of will-power. I knew nothing of the technical difficulties. All my remaining term in Graz was passed in intense but fruitless effort, and I almost convinced myself that the problem was unsolvable. Indeed, I thought, was it possible to transform the steady pull of gravitation into a whirling force? The answer was an emphatic no. And was this not true of magnetic attraction? The two propositions appeared very much the same*" [36].

Tesla's Great Inspiration

After a few years in Graz, Tesla decided to complete his education at a university in Prague but shortly after beginning his father died and he felt he had to leave to earn a living and help support his family. He soon became the chief electrician at a new telephone company owned by a friend of his family in Budapest. His mind continued to ruminate about a brushless motor and in the summer of 1882 while walking through the Budapest City Park reciting poetry with a fellow electrician, Tesla had an inspiration. "*In an instant I saw it all, and I drew with a stick on the sand the diagrams, which Mr. Szigeti understood perfectly*" [36]. His idea was to create a spatially shifting electromagnetic field with alternating currents to generate the torque in an induction motor. The out-of-phase alternating currents produced a progressively rotating magnetic field. What Tesla did was the equivalent of adding cylinders to a one-cylinder gas engine. By using two circuits of alternating current of the same frequency but out of phase, as one began to wane the other would kick in. The beauty was there were no rubbing parts to wear out.

A few months after this epiphany, the telephone company in Budapest was sold and Tesla moved to Paris where he was introduced to Charles Batchelor, the head of Edison's European operations. He was offered a job as an engineer working on the incandescent lighting systems being installed under Edison patents. While working for Batchelor, Tesla had the opportunity to meet several American engineers and he was soon discussing his idea for an alternating current motor with them. They suggested that he should patent the motor and start a company to produce it, but Tesla found that this was much more difficult than he expected. Most wealthy Europeans were baffled by the mysterious phenomena of electricity. The notion of a machine operating without any apparent mechanical means seemed absurd. During a brief period in which he was sent to work in Strasbourg in the Alsace region, Tesla made a protype alternating current motor convincing himself that the principle was sound, but after returning to Paris, he still could not secure capital to move forward.

Tesla decided that there were greater opportunities in America, and he arranged for a transfer to The Edison Machine Works in New Jersey. He arrived in New York in June 1884 along with thousands of other Europeans searching for the "American Dream." When Tesla met the great Edison, he was in awe. "*When I saw this wonderful man, who had had no training at all, no advantages, and did it all himself, and see the great results by virtue of his industry and application. I felt mortified that I had squandered my life…*" [37]. But it did not take long for Tesla to realize that he was in the wrong place to achieve his dream of an alternating current motor. Edison was already entrenched in the battle for the supremacy of direct current and was totally opposed to the use of alternating current.

Tesla left the Edison works in 1885 and joined a new company organized by some friends to develop industrial arc lights. But he soon was at odds with these friends about the other possibilities for alternating current and he left the company. For the next few years, he worked at odd jobs, including ditch digging, to earn a living until he was able to attract financing based on two patents for arc lights. He established a small laboratory in New York and applied for patents on his alternating current motor. Tesla obtained the patents and shortly after he was invited to demonstrate his new motor to the New York Chapter of the American Institute of Electrical Engineers (Fig. 9.4). His presentation *A New System of Alternating-Current Motors and Transformers* established Tesla as a key player in the electrical systems business. When George Westinghouse heard about Tesla's presentation, he immediately saw the commercial possibilities. Tesla's alternating current motor was the missing link in his plans for a universal alternating current system. Just 2 months later, Westinghouse purchased the exclusive patent rights on the motor for a million dollars plus royalties on each motor sold. He also hired Tesla as an electrical consultant. Tesla was suddenly a rich man.

Westinghouse and Tesla Team Up

The first hurdle that Westinghouse and Tesla faced was that there was no currently available polyphase alternating current generator. Tesla improvised with a split-phase motor where he adapted single-phase alternating current to do the work of two phases, but this motor did not work well with the 133 1/3 cycles/s alternating current generally being used. Tesla calculated that the best frequency of alternating current for both his motor and incandescent lighting was 60 cycles/s (the frequency that would eventually become the world standard).

Westinghouse and Tesla took the opportunity to develop and display Tesla's new alternating current system at the great Columbian Exposition in Chicago in 1893 celebrating the 400th anniversary of the discovery of America. The Exposition was considered a great chance to showcase the remarkable industrial and cultural advances of Chicago and the United States. Westinghouse got into a bidding war with the new General Electric Company (that had consumed Edison's electric company) for providing electrical lighting for the Exposition and he took the risk of submitting a low bid to undercut General Electric for the opportunity to publicize the Tesla system. Tesla used 24 single-phase 500 horsepower engine-driven generators connected 90° out of phase to produce twelve 1000 horsepower two-phase alternate current generators. The 23,000 V 60-cycle two-phase output was transmitted and transformed to serve incandescent lights and induction motors. Tesla also designed a rotary converter for changing the two-phase alternating current into

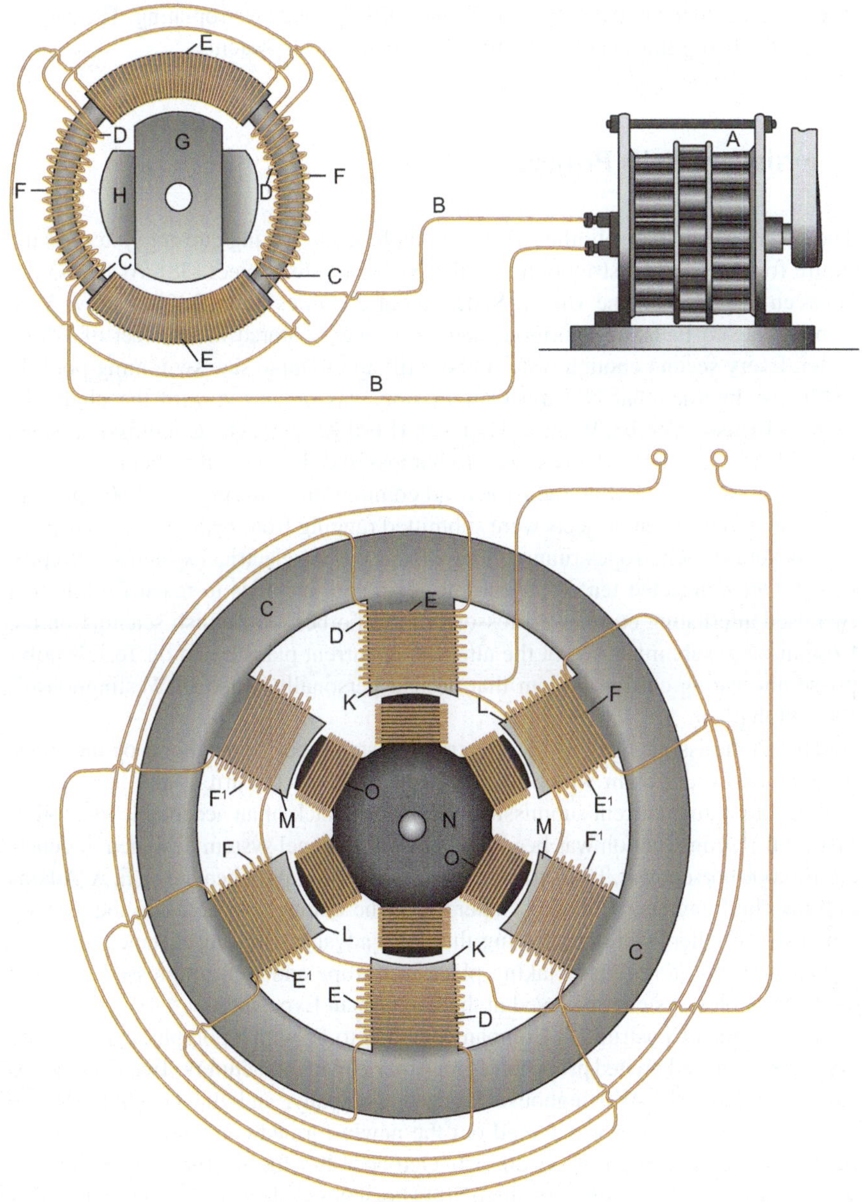

Fig. 9.4 Schematic drawing of Tesla's alternating current motor. (Redrawn from Tesla's 1896 patent application)

direct current that could be used for electric railways and electroplating. The system provided a first glimpse at the future of electrical power systems.

The Niagara Falls Project

The other project that solidified Tesla's polyphase alternating current system as the future for electricity distribution was the Niagara Falls project [38]. At the border between Canada and the United States about 22 miles from Buffalo, New York stands Niagara Falls, an enormous source of energy stored in 180 feet of falling water. Every second enough water falls to fill an Olympic size swimming pool. In 1890, the International Niagara Commission was organized with the chair, the famous British scientist, William Thomson (Lord Kelvin). The Commission began a worldwide search for the best way to harness and distribute the energy stored in the Falls to the surrounding industries and communities offering a $22,000 prize to the winner. Seventeen projects were submitted ranging from hydraulic and air pressure systems to wire ropes running long distances. Six electrical systems were proposed. One connected ten direct current generators in series to reach Buffalo and two used alternating current. Professor George Forbes, an English scientist on the Commission, submitted one of the alternating current plans based on Tesla's polyphase alternating current system that he had personally tested in Westinghouse's Pittsburgh plant.

The Commission favored the electrical systems but found none of the plans acceptable. Lord Kelvin was well known to favor direct current and he opposed Forbes alternating current submission. Despite the lack of an acceptable method of distribution, construction was begun on a massive tunnel system designed to generate 100,000 horsepower. The Cataract Construction Company headed by E. A. Adams and his chief engineer Coleman Sellers won the competition to build the system. Adams and Sellers were monitoring the rapid advances in long-distance alternate current transmission already taking place in Europe and were impressed with the Tesla polyphase system displayed at the Columbian Exposition.

The Cataract Construction Company decided to go with a polyphase alternating current system and invited proposals from American manufacturers. Two Companies submitted bids, the Westinghouse Electric Company, and the General Electric Company (once Edison was forced out the newly formed company rapidly began installing alternating current systems). In October 1893, the Westinghouse Company won out with its plan to produce three 5000 horsepower Tesla polyphase generators that would operate at 2250 V. It took almost 2 years to design, test, and install these high-powered generators with power delivered to the first customer in August 1895. The first to receive electrical power from Niagara was the Pittsburgh Reduction Company (now the Aluminum Company of America) located in a nearby industrial zone created for efficient use of Niagara power. Ironically, the Company required direct current for producing aluminum with its electrolytic process. Buffalo began receiving electric power from Niagara in November 1896 and soon seven more 5000 horsepower generators were installed.

With the successful Niagara model, alternating current rapidly became the accepted method for generating, transmitting, and using electric power in the United States. New York city installed massive steam engines to drive high-powered alternating current generators to provide electricity for all uses. Edison's direct current systems were run by converting alternating to direct current or completely changing over to alternating current. After costly litigation regarding Tesla patents, the Westinghouse and General Electric companies settled hundreds of patent infringement suites between them agreeing on cross-licensing that promoted alternating current power systems.

By 1890, Tesla was at the peak of his career with name recognition almost equal to that of the great Edison. He left the Westinghouse Company in 1889 and set up a laboratory in New York where he could work on his own ideas. He liked to be his own boss and never did well working for or with others. One cannot truly understand Tesla without knowing that he suffered from obsessive-compulsive disorder and recurrent bouts of depression. Based on his autobiography, I suspect he had bipolar disorder with periods of mania and delusions of grandeur. He was highly regimented, and he had a lifelong germ phobia. Everything he touched had to be thoroughly cleaned and he maintained his own separate toilet in the laboratory. He was very private and had few close relationships. He was generous and humble but did not like shaking hands or touching other people. Tesla greatly admired Westinghouse, and in the financial panic of 1893, Westinghouse convinced Tesla to give up his motor royalty agreement to save the Westinghouse company from bankruptcy. In effect, he gave up millions of dollars in royalties that later in his life he greatly needed.

The Tesla Coil

By the early 1890s, Tesla had achieved his goal of developing a low-frequency polyphase alternating current system and he was ready to move on to his other great idea, production of very high-frequency and high-voltage electricity as a form of radiant energy. As we have seen in Chap. 7, Hertz and Marconi recognized the potential of high-frequency electromagnetic waves for long-distance wireless communication. Tesla was aware of the potential of high-frequency electricity for communication, but he saw an even greater potential application in power transmission and lighting.

Tesla began his efforts to generate high-frequency alternating current using mechanical multi-polar alternators with large diameter armatures that rotated very fast. He produced frequencies between 20,000 and 30,000 cycles/s, at the low end of the radio frequency range. Tesla showed that the high-frequency alternating current could illuminate low-pressure gas tubes like the ones developed by Geissler and Crookes.

To generate even higher frequencies for his radiant energy lighting projects, Tesla realized that he had to move on from mechanical generators to electrical oscillators. Tesla had a lot of experience working with alternating current transformers and knew that they behaved differently at high frequencies and could be set into high-frequency oscillation by balancing the capacitance and inductance in the circuit. He developed a new type of transformer without an iron core driven by a resonant circuit to produce high-voltage, high-frequency alternating current that became known as a Tesla coil. The basic features of the coil were a high-voltage primary transformer, capacitor, and air core induction. He could alter the resonant frequency of the circuit by changing the capacitance and inductance. Tesla patented the coil in April 1891 and demonstrated its remarkable effects in a lecture *Experiments with Alternating Currents of Very High Frequency and Their Application to Methods of Artificial Illumination* before the American Institute of Electrical Engineers at Columbia College, New York in May 1891.

Showers of manmade lightning burst from his coil. Tesla sat amid the high-voltage *electric fire* without any concern for harm because of the ultra-high frequency, up to millions of cycles per second (Fig. 9.5). Evacuated or gas-filled tubes

Fig. 9.5 Photograph of Nikola Tesla reading while sitting next to his coil spewing out ultra-high frequency "electric fire." (From reddit.com)

or bulbs held in his hand were brightly lit by the electrical discharge. Tesla felt that the light resulted from molecular vibration in the tubes or bulbs set in motion by the rapidly alternating field of radiant energy generated by the coil. He even suggested that this could be a more efficient form of lighting than the hot filaments used in incandescent lights. He could illuminate a room without wires by electrifying plates in walls with the high-frequency electrostatic discharge of his coil.

As the last decade of the nineteenth century ran by, Tesla expanded his vision for high-frequency electrical oscillations from lighting to a general method of transmitting energy long distances. The high-frequency oscillations could pass through the Earth's stratosphere and core as efficiently as through wires. He obtained over 20 patents related to high-frequency electrical oscillations and their transmission through *natural media*. In one of his patents, he declared, "*It is to be noted that the phenomenon here involved in transmission of electrical energy is the one of true conduction and not to be confounded with the phenomena of electrical radiation which have heretofore been observed and which from the very nature and mode of propagation would render practically impossible the transmission of an appreciable amount of energy to such distances as are of practical importance*" [39].

One of his patents, an apparatus for wireless control of moving vessels or vehicles, was an immediate sensation because of its potential use in warfare. Tesla demonstrated his "*teleautomatic brain*" at a series of exhibitions at Madison Square Garden where he remotely controlled a metal boat several feet long floating in a large tank in the aquarium [40]. He announced that the boat was controlled by wireless electricity. In fact, he had invented the first radio remote control system. Radio waves generated by his *coil* were picked up by the boat's antenna and through a series of relays the operator could control the speed and direction of the movement.

Wireless High-Frequency Electric Power Transmission

Tesla had grandiose plans for multiple large electrical oscillators that could send communications and electric power around the world. His New York laboratory was not a good location for developing such devices, so he decided to move west to Colorado Springs based on the advice of a friend and financial supporter who had moved there for his health. In the summer of 1899, Tesla built a massive electrical oscillator, the largest, ever made, at the base of Pike's Peak. It consisted of a large wooden tower holding a 200-foot copper pole topped off with a 3-foot diameter copper ball. The coil had a 200-kilowatt input and could generate 100 million volts output. Around the large well-fenced property to keep out sightseers, Tesla located a series of receiving stations with resonant circuits.

On a warm summer evening, Tesla sat outside wearing thick rubber-soled shoes ready to test the system. He signaled for an assistant inside to turn on the power ready to watch the fireworks. Lightning bolts began to spray from the copper ball more than 100 feet in length followed by loud bursts of thunder. Inside the wooden structure, there was a blue aura with wild dancing lights. After a massive lightning bolt and crash of thunder, everything suddenly went black, and the pungent smell of ozone was powerful. Tesla initially thought his assistant turned off the power, but he then saw that all of Colorado Springs was dark. He had blown out the town's electric supply [41].

Tesla spent the rest of the year refining the electronics of his transmitting and receiving electrical circuits. He was particularly interested in studying the effects of lightning storms in the nearby mountains on his electronic equipment. Not only did he record electrical discharges as the storm passed by but after the storm there was a persistent waxing and waning in the intensity of discharge. His receiving circuits were recording the peaks and valleys of stationary waves. He became convinced that in addition to wireless communication, it should be possible "*to transmit power, in unlimited amounts, to any terrestrial distance and almost without any loss*" [42]. In practice, he was able to able to detect wireless signals at 1000 km and to power 200 incandescent lights at 30 km.

When Tesla returned to New York in 1900, he was the toast of the town basking in the accomplishments of his Colorado Springs venture. Pictures of Tesla sitting and reading next to his coil as it spit out giant lightning bolts were everywhere. He was an electrical wizard who would change the world. His grand plan was to build a series of the huge transmitter towers like the one in Colorado Springs for communication and transmission of electricity around the globe. Tesla stated that his towers would be "*capable of transmitting individual signals to the uttermost confines of the earth. Each of them will be preferably located near some important center of civilization and news it receives will be flashed to all points of the globe. A cheap and simple device that might be carried in one's pocket, will record the world's news or such special messages as may be intended*" [43].

Tesla's next hurdle was to obtain financial backing. The Colorado Springs venture had drained most of his personal wealth and building the proposed super towers would be costly. His proposal for wireless communication interested J. P Morgan, the banker of the American Telephone and Telegraph Company, and Morgan agreed to provide $150,000 for Tesla to build a tower for trans-Atlantic wireless telegraphy. As part of the deal Morgan received 51% ownership of Tesla's patents for wireless communication and wireless lighting. Tesla selected a 200-acre location that he called *Wardenclyffe* on Long Island just east of Brooklyn for an administrative building, a laboratory and the transmission tower with transformers and generators. At the top of a 57 m, high wooden tower sat a 60-ton metal half-dome. Below the tower, he tunneled deep into the earth. A receiving tower was planned for the coast of England in Cornwall.

Tesla had originally planned for transmission to England by the end of 1901 but cost overruns and delays in equipment delivery forced him to keep pushing back the date. In the meantime, in December 1901, the world received the astonishing news that Marconi had transmitted the first wireless trans-Atlantic telegram from Cornwall to Newfoundland. Furthermore, Marconi achieved the fete by transmitting relatively low-powered Hertzian waves and recording the waves with a simple coherer. When in early 1902, Tesla returned to Morgan for more money, Morgan refused. He saw no reason to continue backing the Wardenclyffe project and Tesla's grandiose plans for a gigantic transmitter using millions of horsepower. Marconi had achieved wireless telegraphy at a much lower cost. Morgan was also unwilling to return Tesla's patents so that he might obtain other financial backing. When Tesla made a last-ditch effort to change Morgan's mind in January 1904, he bitterly responded to Morgan's refusal, "*you fire me out like an office boy and roar so that you are heard six blocks away; not a cent. It is spread all over town, I am discredited, and the laughing stock of my enemies*" [44].

Tesla's fortunes only went downhill from there. Despite multiple efforts, he was unable to convince any of his influential friends to provide financial backing for the Wardenclyffe project. There were rumors that Morgan warned potential backers against supporting Tesla. By 2016, he was unable to pay his bills and he declared bankruptcy. He owed $20,000 to the Waldorf-Astoria where he had lived since 1898 and in the settlement the hotel received Wardenclyffe. They blew up the tower to improve the land sale. A few years later as Tesla's circumstance deteriorated even further, the Westinghouse Electric and Manufacturing Company (by then Westinghouse had died) began paying his monthly rent ($125) presumably due to concerns about bad publicity should Tesla become homeless. This was the company Tesla earlier saved from bankruptcy in 1893 when he agreed to give up the multimillion-dollar royalties on his alternating current motor.

Tesla and Electrotherapy

Being a self-described hypochondriac, Tesla had an early interest in the potential applications of high-frequency alternating current in the field of medicine. In his many demonstrations of the *Tesla Coil* in the United States and Europe in the early 1890s, he showed how the human body (his) could tolerate massively high voltages without any apparent evidence of damage. He later wrote, "*One of the early observed and remarkable features of the high frequency currents, and one that was chiefly of interest to physicians, was their apparent harmlessness which made it possible to pass relatively great amounts of electrical energy through the body of a person without causing pain or serious discomfort*" [45]. He regularly treated himself with high-frequency current and claimed that the treatments were highly effective. "*You see, electricity puts into the tired body just what it most needs – life force, nerve force. It's a great doctor, I can tell you, perhaps the greatest of all doctors*" [46].

When the human body was exposed to the ultra-high-frequency current from his coil, Tesla noted that "*small particles which may be adhering firmly to its surface, are torn off with violence.*" This could be anything from dirt to paint producing "*an instantaneous cleansing of the skin.*" Tesla speculated on the potential hygienic implications, "*an efficient and time-saving substitute for a water bath*" [45].

In 1894, Tesla began experimenting with passing ultra-high-frequency alternating current through Geissler and Crookes tubes. He noted a violet glow around the tubes and identified some type of invisible radiant energy that damaged photographic film in the laboratory. He had inadvertently discovered X-rays but did not recognize the significance. After learning of Röntgen's discovery of X-rays and X-ray imaging (see Chap. 8), in early 1896 Tesla developed a new type of vacuum tube with a single electrode attached to the output of the Tesla coil. He argued that his single terminal vacuum tube could produce X-rays of much greater power than those produced by standard devices.

Between 1896 and 1897, Tesla published a series of papers describing improved devices for generating X-rays and experiments on the biological effects of X-rays. He is credited with being one of the first to describe the harmful effects of X-rays on the human body. His single-electrode vacuum tube became the prototype of the vacuum electrode and violet ray electrotherapy that as will see became widely used by practising physicians [47].

Tesla noted that high-frequency currents can produce the potentially noxious gas, ozone. Electricity in the air converts O_2 to O_3 (ozone), a pale blue gas with a distinct pungent smell. Ozone is much less stable than O_2 breaking down to O and O_2. Ozone can irritate the skin and respiratory tract causing a burning sensation and difficulty breathing. It can also kill bacteria and be used as a disinfectant. Tesla patented a device for producing ozone in 1896 and later started the Tesla Ozone Company to market ozone products including therapeutic gels made by passing ozone through different oils and ozone room sanitizers for hospitals. In 1899, newspaper reports that Tesla had cured a patient with pulmonary tuberculosis with high-frequency currents were later shown to be bogus.

High-Frequency Electricity to Improve Learning

Tesla was convinced that exposure to ultra-high-frequency electrical oscillations improved brain function and overall human health. He felt refreshed and able to think more clearly after exposure. This led to the idea that such exposure might improve learning. In 1912, he came up with a plan "*for making dull students bright by saturating them unconsciously with electricity*" [48]. Details of his plan were published in the December 1912 edition of *Popular Electricity Magazine.* Tesla proposed to begin with a trial in mentally defective children by installing insulated wires in the walls of the classroom. The wires were to be concealed and the students completely unaware of the investigation. The equipment for generating millions of volts of high-frequency current in an adjoining room was turned on by a switch when class began.

William H. Maxwell, the superintendent of the schools of New York approved the plan after a heated debate at the Board of Education. The main concern was whether there might be unknown long-term consequences of the exposure. A Professor of physiology at the University of Illinois warned that it could take years to determine if the electrical experiment was safe. Tesla and other electrical experts reassured the board that based on their extensive experience, ultra-high-frequency electrical current was perfectly safe. Furthermore, they pointed out that a similar experiment had already been conducted in Stockholm, Sweden, and had proved successful without any observable side effects. No concerns were raised about experimenting on children without them or their parents being aware. It would be many years before the concept of informed consent was addressed.

Tesla had little understanding of human anatomy or physiology, and of course, overall understanding of brain physiology was in its early stages. He used simplistic arguments to support his claim for a beneficial effect of high-frequency current, increased motion of molecules like a *molecular massage* and increased oxidation to burn up waste products. To support his notion that electricity could improve learning in children he quoted the studies performed in Europe showing that electricity improved plant growth (see Chap. 3). In his grand scheme, Tesla saw a future where every home would be bathed in high-frequency electrical current from equipment built into the walls. People within would live healthy lives, saved from the effect of common illnesses and everyday aches and pains.

High-Frequency Electrotherapy in Europe

In Europe, several investigators assessed the effect of ultra-high-frequency electrical currents on living tissue. Oliver Lodge, professor of physics at the University of Liverpool teamed up with Francis Gotch, professor of Physiology to study the effects on a nerve-muscle preparation in a frog [49]. Potentials of several thousand volts that produced visible sparks did not stimulate the nerve. By contrast, ultra-low potential currents readily produced muscle contraction.

Even before Tesla's experiments on human subjects, the French physiologist Arsenne d'Arsonval began working on the effects of high-frequency electrical currents on the human body [50]. He used a resonant circuit like the one designed by Hertz to show that the human body was relatively insensitive to these high-frequency oscillations. When Tesla toured Europe in 1892, d'Arsonval was impressed with his high-voltage alternating current circuit design so he made his own version with modifications. D'Arsonval was not a practicing physician but he worked with physicians to test his new equipment. He used both animals and human subjects.

D'Arsonval developed two common methods for coupling the subject to the high-frequency oscillations. In one, he had the subject stand in the center of a circular cage wrapped with wires that he called an "auto-conduction cage," essentially a human-sized Faraday induction coil (Fig. 9.6). The high-frequency alternating current in the wires was transferred to the subject by magnetic induction. With the other device, the subject sat on a recliner holding an iron bar connected to one of the electrodes with the other electrode placed on a metal surface under an insulating cushion. He called this "auto-condensation" with the subject serving as one of the conducting plates of a condenser (capacitor), the cushion serving as the dielectric and the metal surface of the recliner the other conducting plate.

With these devices, d'Arsonval reported increased body temperature, metabolism, oxidation, secretions, and glandular activity. He also claimed that blood pressure was lowered, and the nervous system was soothed. He noted that it was up to practising physicians to determine how these effects might be used in their therapies.

Fig. 9.6 A subject stands in the center of Arsenne d'Arsonval's auto-conduction cage. (From Strong [52])

One of d'Arsonval's main collaborators was Paul Oudin, a French physicist-physician best known for developing a range of hand-held devices for application of high-frequency currents to patients. Oudin applied the currents locally through the air using the terms *spark*, *brush*, and *feather* to describe the nature of the electrical discharge [51]. A single pointed electrode produced sparks while an instrument with multiple points produced a spray or brush discharge which he called an *effluve*. He used a technique that was first used by Tesla, coating the ends of the metallic electrodes with a dielectric substance, now called dielectric barrier discharge. He also used the vacuum electrode developed by Tesla.

These treatments were an early version of what has become known as plasma medicine. It uses gases containing ions and electricity (physical plasma) for medical applications. The simplest plasma device used room temperature air at normal atmospheric pressure. It contains multiple ion species and free electrons with an overall neutral charge. Because of its electrical conductivity, the plasma couples to electromagnetic fields producing electrostatic and electromagnetic waves. Electrostatic waves are purely longitudinal while electromagnetic waves are longitudinal and transverse. Plasma within a vacuum tube responds differently to an applied electromagnetic field. Since the number of gas molecules is much less than in room air, free electrons can be accelerated to much higher velocities producing more powerful electrostatic and electromagnetic waves.

Oudin was most impressed with the ability of his effluve to relieve pain. In a series of 32 manual workers with low back pain, he reported that the pain was eliminated in all with from 1 to 3 daily sessions lasting 2–10 min. Most striking was his description of pain relief during tooth extractions. He used moistened cotton between the powered electrode and the tooth and a return electrode covered with rubber outside the cheek. After 5 min of high-frequency current application, the tooth was extracted, and the patient interviewed regarding pain. Twenty of twenty-four patients reported minimal or no pain, and in the four patients who reported pain, he determined that there were technical difficulties in applying the current. Oudin went on to use his effluve for treating localized lesions of the skin and mucous membranes including infections although he did not think that the high-frequency currents were bactericidal but rather, they improved local blood circulation. As for pain relief, he was not sure of the mechanism but since it worked why not use it.

The Violet Ray

An American physician, Fredrick Strong, is credited with popularizing the use of the vacuum electrode in electrotherapy or as it became known, the violet ray or violet wand. In his 1908 book entitled *High Frequency Currents*, he claimed to have invented the vacuum electrode in 1897 but as we have seen Tesla developed a vacuum electrode several years earlier [52]. When the high-frequency current supplied vacuum electrodes were held near the skin, there was a visible discharge between the glass and skin surface that varied as the vacuum in the tube was increased.

Based on the discussions in his book, Strong was clearly not an electrical engineer and had limited understanding of electric fields and electromagnetic waves. He did however assess the effect of lowering the pressure in the vacuum electrode on the appearance of the discharge. At intermediate pressures in the vacuum electrode, violet rays streamed from the glass filling the space between the glass and the skin, thus the terms violet ray or violet wand. Strong was generating violet and ultraviolet electromagnetic waves in addition to the plasma effect on the air between the electrode and skin. As Tesla had shown, with very high voltages and markedly reduced pressure vacuum electrodes produced X-rays that could damage the skin. Strong claimed that his vacuum electrode could cure a wide variety of diseases by improving circulation and metabolism and he suggested that it could kill bacteria by producing ozone.

The violet ray apparatus became extremely popular with physicians in the United States marketed to treat a variety of conditions from anxiety and neurasthenia to rheumatism and inflammatory disorders. In the 1920s, the Fitzgerald Manufacturing Company in Connecticut marketed the STAR-RIGHT Violet Ray consisting of two glass electrodes, an insulated handle and a cord connected to a transformer with Tesla's high-frequency resonant circuit. The transformer cord could be plugged into the standard 60-cycle alternating current outlet. The entire apparatus fit into a 10 by 6-inch box that a private practitioner could carry with him on house calls making electrotherapy available to just about anybody.

What exactly were physicians accomplishing with their violet rays beyond the obvious placebo effect? Based on our current understanding, the only definite effect was heating the tissues beneath the glass electrode with electrostatic and electromagnetic waves, a type of diathermy. This in turn led to reflexive vascular dilatation and increased metabolism no different from that associated with other heat sources. If the electromagnetic waves produced were nonionizing (shorter frequencies than ultraviolet and X-rays), they were relatively safe. The effects of the electric field on the plasma outside the electrodes could produce ozone and other potentially toxic substances like nitric acid. By producing exfoliation of the skin, these substances might have contributed to the wide use of violet rays for treating skin disorders such as acne, eczema, and lichen planus.

The German physician Karl Franz Nagelschmidt was the first to regularly use radio frequency (RF) electromagnetic waves for deep heating tissues in patients. He coined the term diathermy and wrote the initial textbook on diathermy in 1913. Physicians used the therapy for a wide range of medical conditions from arthritis to infectious diseases such as polio and pneumonia. In the early stages, the procedure was limited to "longwave diathermy" in the range of 0.1–2 MHz generated by spark gap discharge circuits such as Tesla's coil. But with the development of Fessenden's high-frequency alternator and de Forests amplifying vacuum tube (see Chap. 7), it became possible to produce so-called short waves in the range of 10–300 MHz, on the low end of the microwave range. Initially, there were dramatic claims of curing tuberculosis and cancer with the short waves, but these claims proved false. In the 1940s, microwaves were introduced for diathermy, but microwaves had relatively poor depth penetration and were subsequently largely replaced by infrared radiation. Currently, diathermy with radio waves, microwaves or infrared waves is primarily used by physical therapists to apply heat to deep tissue injuries, particularly muscle injuries.

Danger to the Brain of Radio Waves and Microwaves

With the earliest radio wave transmissions, people expressed concerns about the possibility of health risks. When the newly developed short waves were found to kill flies, the press called them "death rays" and raised concerns about possible damage to the brain. But over time it became apparent that the only consistent effect of radio waves was to heat tissue with the amount of heat proportional to the energy in the electromagnetic waves and inversely proportional to the distance from the source (inverse square law). In the usual circumstance where people were distant from the transmitting antenna, there was no risk at all. In the rare occurrence where people were accidently too close to the source, they initially felt extreme heat and perspired diffusely.

Not surprisingly, one of the early descriptions of the effect of high-energy radio wave exposure on the brain was provided by Nikola Tesla. Recall that Tesla spent hours sitting next to his famous coil that was spewing out extremely high-energy radio waves (see Fig. 9.5). Within minutes of exposure to the high-energy radio waves, Tesla described feeling very warm and began to perspire. He wrote: "*I have repeatedly, in demonstrations to friends, exposed myself longer to the actions of the oscillations, and each time, after the lapse of an hour or so, an immense fatigue, of which it is difficult to give an idea, would take hold of me. It was greater than I experienced on some occasions after the most straining and prolonged bodily exertion. I could scarcely make a step and could keep the eyes open only with the greatest difficulty. I slept soundly afterward, and the after-effect was certainly beneficial, but the medicine was manifestly too strong to be used frequently*" [45].

Interestingly, Tesla compared his symptoms to those experienced after extreme bodily exertion which also causes brain heating and potentially heat stroke. If blood circulation and other compensatory mechanisms fail and heat buildup persists, heat stroke can occur with permanent brain damage. The most temperature-sensitive brain cellular elements are the mitochondria and plasma membranes. With temperature above 40 °C proteins can be denatured and form aggregates that disrupt normal cellular functions and ultimately lead to neuronal death. Fortunately, Tesla had the common sense to limit his exposure when he felt the early symptoms of heat stroke.

With the development of microwaves and microwave ovens in the mid-twentieth century, there were renewed concerns about the dangers of these high-frequency radio waves on the brain. There is significant overlap between radio and microwaves. Typically, microwaves are defined as frequencies in the low end of the radio frequency range between roughly 300 MHz and 30 GHz (see Table 12.1). A few victims developed skin burns and nerve damage in the hands after accidental exposure to high-energy microwaves from leaking microwave ovens [53]. The skin and skull are a major barrier to these relatively low penetrating electromagnetic waves. The skin would be burned before the brain is severely affected.

The eyes and testicles are at the greatest risk to the heat effect of electromagnetic waves since they have the least shielding and relatively low blood flow. Development of cataracts and transient impotence have resulted from accidental exposure to high-energy radio waves and microwaves. We will return to this discussion at the end of Chap. 12.

In Brief

Thomas Edison's development of an affordable long-lasting incandescent light bulb led to demand for an efficient reliable system to deliver electricity to the public. Edison had a head start with his low-voltage direct current system but Westinghouse's high-voltage alternating current system eventually won out. The alternating current system had the major advantage of transmitting electricity long distances. A transformer lowered the voltage at the site of use. George Westinghouse teamed up with Nikola Tesla and his polyphase alternating current motor to produce an alternating current system that became the standard of the industry. In his later years, Tesla focused on wireless high-frequency electric power transmission. He demonstrated that humans could tolerate extremely high-voltage alternating current if the frequency was high enough. He speculated that the ultra-high-frequency current was good for one's health and even improved learning. This led to the development of the violet ray and widespread use of high-frequency electrotherapy in the United States and Europe.

References

1. Jonnes J. Empires of light. Edison, Tesla, Westinghouse, and the race to electrify the world. New York: Random House; 2003.
2. Forbes N, Mahon B. Faraday, Maxwell and the electromagnetic field. Guilford: Prometheus Books; 2019. p. 120.
3. Bordeau SP. Volts to hertz – the rise of electricity. Minneapolis: Burgess Publishing Company; 1982. p. 208–19.
4. Bordeau SP. Volts to hertz – the rise of electricity. Minneapolis: Burgess Publishing Company; 1982. p. 147–8.
5. Josephson M. Edison: a biography. New York: McGraw-Hill; 1959.
6. Israel P. Edison: a life of invention. New York: Wiley; 2000.
7. Lanier CD. Two giants of the electric age. Rev Rev. 1893;8:48.
8. Jonnes J. Empires of light. Edison, Tesla, Westinghouse, and the race to electrify the world. New York: Random House; 2003. p. 48.
9. Brock WH. William Crookes (1832–1919) and the commercialization of science. London: Routledge; 2008. p. 245–6.
10. Jonnes J. Empires of light. Edison, Tesla, Westinghouse, and the race to electrify the world. New York: Random House; 2003. p. 56.
11. Jonnes J. Empires of light. Edison, Tesla, Westinghouse, and the race to electrify the world. New York: Random House; 2003. p. 57.
12. Jonnes J. Empires of light. Edison, Tesla, Westinghouse, and the race to electrify the world. New York: Random House; 2003. p. 65.
13. Jonnes J. Empires of light. Edison, Tesla, Westinghouse, and the race to electrify the world. New York: Random House; 2003. p. 71.
14. Josephson M. Edison: a biography. New York: McGraw-Hill; 1959. p. 252–64.
15. Satterlee H. J. Pierpont Morgan: an intimate portrait. New York: Macmillan; 1940. p. 216.
16. Prout HG. A life of George Westinghouse. New York: Scribner; 1926.
17. Jonnes J. Empires of light. Edison, Tesla, Westinghouse, and the race to electrify the world. New York: Random House; 2003. p. 117–39.

18. Jonnes J. Empires of light. Edison, Tesla, Westinghouse, and the race to electrify the world. New York: Random House; 2003. p. 134.
19. Jehl F. Menlo Park reminiscences, vol. II. Dearborn Park: Edison Institute; 1938. p. 832–3.
20. Jonnes J. Empires of light. Edison, Tesla, Westinghouse, and the race to electrify the world. New York: Random House; 2003. p. 142–3.
21. Bordeau SP. Volts to hertz – the rise of electricity. Minneapolis: Burgess Publishing Company; 1982. p. 255.
22. Morus IR. The measure of man: technologizing the Victorian body. Hist Sci. 1999;27:269–71.
23. Essig M. Edison and the electric chair. Stroud: Sutton Publishing; 2003.
24. Jonnes J. Empires of light. Edison, Tesla, Westinghouse, and the race to electrify the world. New York: Random House; 2003. p. 168.
25. Died for sciences sake. The New York Times, July 31, 1888. p. 8.
26. Jonnes J. Empires of light. Edison, Tesla, Westinghouse, and the race to electrify the world. New York: Random House; 2003. p. 174.
27. Surer than the rope. The New York Times, December 6, 1888. p. 5.
28. Far worse than hanging. The New York Times, August 7, 1890. p. 1.
29. Capital punishment by electricity. Lancet. 1891;138:943–4.
30. Met death on wires. The New York Times, October 12, 1889. p. 1.
31. Edison TA. The danger of electric lighting. N Am Rev. 1889;149:625–33.
32. Like a city in morning. The New York Times, October 16, 1889. p. 1.
33. Westinghouse G. A reply to Mr. Edison. N Am Rev. 1889;149:653–64.
34. Tesla N. My inventions: the autobiography of Nikola Tesla. Williston: Hart Bros; 1982.
35. Bordeau SP. Volts to hertz – the rise of electricity. Minneapolis: Burgess Publishing Company; 1982. p. 244–75.
36. Bordeau SP. Volts to hertz – the rise of electricity. Minneapolis: Burgess Publishing Company; 1982. p. 250.
37. Tesla N. My inventions: the autobiography of Nikola Tesla. Williston: Hart Bros; 1982. p. 71.
38. Jonnes J. Empires of light. Edison, Tesla, Westinghouse, and the race to electrify the world. New York: Random House; 2003. p. 277–99.
39. Bordeau SP. Volts to hertz – the rise of electricity. Minneapolis: Burgess Publishing Company; 1982. p. 265–6.
40. O'Neill JL. Prodical genius: the life of Nikola Tesla. New York: David McKay; 1944. p. 167.
41. Jonnes J. Empires of light. Edison, Tesla, Westinghouse, and the race to electrify the world. New York: Random House; 2003. p. 359.
42. Bordeau SP. Volts to hertz – the rise of electricity. Minneapolis: Burgess Publishing Company; 1982. p. 246.
43. Bordeau SP. Volts to hertz – the rise of electricity. Minneapolis: Burgess Publishing Company; 1982. p. 247.
44. Jonnes J. Empires of light. Edison, Tesla, Westinghouse, and the race to electrify the world. New York: Random House; 2003. p. 362.
45. Tesla N. High frequency oscillators for electro-therapeutic and other purposes. IEEE. 1999;87:1282–92.
46. Rhees DJ. Electricity – "The greatest of all doctors": an introduction to "High frequency oscillators for electro-therapeutic and other purposes". IEEE. 1999;87:1280.
47. Graves DB. Lessons from Tesla for plasma medicine. IEEE. 2018;2:594–607.
48. Gilliams EL. Tesla's plan of electricity treating school children. Popular Electricity Magazine, December 1912. p. 813–4.
49. Lodge O. Past years: an autobiography. London: Hodder and Stoughton; 1931. p. 188–90.
50. Reif-Achermn S. Jacques Arsene d'Arsonval: his life and contributions to electrical instrumentation in physics and medicine. Part III: high-frequency experiences and the beginnings of diathermy. IEEE. 2017;105:394–404.
51. Graves DB. Lessons from Tesla for plasma medicine. IEEE. 2018;2:599–602.
52. Strong FF. High-frequency currents. New York: Rebman; 1908.
53. Guy AW. History of biological effects and medical applications of microwave energy. IEEE Trans Microw Theory Tech. 1984;32:1182–200.

Chapter 10
Electrophysiology in the Twentieth Century

> *A small action potential was recorded from the upper end of the axon, and this gradually increased as the electrode was lowered, until it reached a constant amplitude of 80–95 mV, at a distance of about 10 mm from the cannula. In this region the axon appeared to be in a completely normal condition, for it survived and transmitted impulses for several hours.*
>
> *Allan Hodgkin and Andrew Huxley, 1939*

Abstract Based on Walther Nernst's equation describing the potential difference between two salt solutions of different concentrations separated by a semipermeable membrane, Julius Bernstein developed his theory of cell membrane polarization at the turn of the twentieth century. The negative electrical potential inside excitable nerve and muscle cells was due to selective permeability of the cell membrane to potassium ions. Bernstein speculated that during an action potential, there was a brief reversible breakdown in the selective permeability allowing potassium ions to enter the cell. Keith Lucas and Edgar Adrian demonstrated the "all-or-nothing principle" of nerve transmission and suggested that the action potential moved down the nerve by exciting the nerve region just ahead enough to propagate the signal. Development of thermionic vacuum tubes for rectifying and amplifying electronic signals along with improved electrometers allowed researchers to accurately record action potentials for the first time. By taking advantage of the giant squid axon that could be penetrated with a glass electrode and a new voltage-clamp technique, Allan Hodgkin and Andre Huxley determined that the action potential resulted from independent changes in permeability for sodium and potassium ions with the conductance changing as a function of time and membrane potential. They went on to develop a mathematical model based on opening and closing of ion channels in the axon membrane that provided a remarkably accurate prediction of the threshold, shape, and propagation of the action potential. Finally, the patch-clamp technique allowed investigators to measure the current in a single ion channel.

Keywords Neurotransmitters · Membrane potential · Ion channels · Thermionic emission

R. W. Baloh, *Brain Electricity*, https://doi.org/10.1007/978-3-031-62994-5_10

At the turn of the twentieth century, several factors came together to produce a leap forward in the basic understanding of the electrophysiology of the nervous system. On the physiology side, the discovery of the semipermeable membrane and the theory of electrolyte dissociation provided an explanation for why charged free ions accumulate on each side of a cell membrane and an electromotive force is associated with the ion separation. On the electronic side, the invention of the hot cathode vacuum tube, the amplifying vacuum tube, and ultimately, the multistage amplifier allowed precise recordings of tiny electrical potentials in the millivolt range.

At the same time, physiologists and pharmacologists began probing the events that occurred at the neuronal synapse when an electrical impulse arrived. Recall that after Sherrington coined the term synapse controversy arose as to whether electricity jumped across the synaptic cleft (sparkers) or chemicals were released and crossed the synapse (soupers). We will first address the early evidence for chemical transmission at the synapse and then focus on the electrical events that occur as an action potential moves down the axon to a synapse.

The Autonomic Nervous System and Chemical Neurotransmission

For centuries, it was known that the autonomic nervous system modulates the internal organs of the body controlling heart rate, respiration rate, and digestion. The system was divided into two parts, the sympathetic and the parasympathetic whose actions tended to oppose one another. For example, electrical stimulation of sympathetic nerves increased heart rate and blood pressure while electrical stimulation of parasympathetic nerves decreased the heart rate and blood pressure.

The first hint that synaptic transmission in the autonomic nervous system was chemical came from an English physician and inventor, George Oliver [1]. To test a devise, he invented to measure constriction of arteries under the skin, he gave his young son an extract of the adrenal gland from a sheep and cow provided by a local butcher and observed a marked constriction of the arteries and an increase in blood pressure. Armed with this information, he sought out a London physiologist, Edward Schäfer and convinced him to conduct experiments on dogs to test the effect of the adrenal extract on blood pressure. The two went on to show that injecting adrenal extract into dogs produced the same effect as stimulating the sympathetic nerves with electricity [2].

John Newport Langley, a physiologist at Cambridge who specialized in the autonomic nervous system, was intrigued with Oliver's finding and suggested a student in his laboratory, Thomas Elliot, investigate the matter further. In 1904, Elliot showed that a chemical extracted from the adrenal gland, adrenaline, stimulated the smooth muscles in the walls of arteries comparable to electrical stimulation of the sympathetic nerves. He speculated that adrenaline or an adrenaline-like substance might be released at the synaptic junction between the sympathetic nerves and the smooth muscles in the walls of arteries causing them to constrict [3].

Around the same time, a Cambridge pharmacologist, Walter Dixon was studying the parasympathetic nerve supply to the frog heart that ran in the vagus nerve [4]. After showing that stimulation of the vagus nerve slowed the heart rate as expected, he removed the heart and produced an extract from it. When he applied the extract to another frog heart, the rate of beating slowed considerably. Dixon then set out to identify the chemical in the extract that caused the heart slowing and found that it resembled muscarine, a chemical found in a mushroom that was known to inhibit the heart rate. Furthermore, he noted that the vagal effect on the frog heart rate was blocked by the drug atropine which was known to block the effect of muscarine. Dixon concluded that a muscarinic-like substance was stored in the parasympathetic nerve synapses and released by electrical stimulation of the vagus nerve.

Although Elliot and Dixon's findings were intriguing, the neurophysiology community was not ready to accept chemical transmission at the nerve synapse. The evidence was indirect, and the transmitter substances were not conclusively identified. A young English physician and pharmacologist, Henry Dale provided the next major breakthroughs in the discovery of chemical neurotransmission [5].

Dale, who also trained at Cambridge with Langley, obtained his medical degree at St. Bartholomew's Hospital in London. In 1904, Dale accepted an offer to work at the Wellcome Physiological Research Laboratory a research component of Burroughs Wellcome and Company founded by two American pharmacologists, Silas Burroughs and Henry Wellcome, in London in 1880.

At the suggestion of Wellcome, Dale began to study the physiology of ergot, a fungus that grew on rye and was known to have both toxic and pharmacological properties. It was responsible for causing outbreaks of St. Anthony's fire in the Middle Ages (causing severe burning pain and convulsions) and had been used extensively in obstetrics to induce uterine contractions. Ergot was known to contain several chemical substances and Dale noted that one extract had adrenaline-like features but with much stronger stimulating effects than adrenaline. He worked with a chemist who synthesized chemical compounds at the Wellcome laboratory, and they identified noradrenaline, a chemical like adrenaline that would later be shown to be the neurotransmitter released by sympathetic nerves.

In 1913, Dale found another extract from ergot that could decrease the heart rate in a cat after intravenous injection. Aware of Dixon's work, Dale suspected that a muscarine-like substance was in the extract. When he and a colleague, Arthur Ewins, isolated the substance from the extract, it was clearly different from muscarine even though it had similar actions. They were aware of a prior report by an American pharmacologist, Reid Hunt, of a compound called acetylcholine that had a powerful effect on lowering blood pressure like muscarine. On comparing synthesized acetylcholine with the muscarinic-like substance in the ergot extract, they found it to be physiologically and chemically identical. In 1914, Ewins reported the chemical features of acetylcholine and Dale reported the physiological features [6, 7].

In his paper, Dale identified two separate actions of acetylcholine. It had a muscarinic action mimicking activation of the parasympathetic nervous system such as lowering blood pressure and heart rate and it had a nicotinic action mimicking the effect of low doses of nicotine on activating the nerve ganglia in the autonomic nervous system and the nerve-skeletal muscle junction. Solidifying the separate actions, the muscarinic action was blocked by atropine while the nicotinic action was blocked by curare, a poison used by South American Indians on their arrowheads known to block transmission at the neuromuscular junction.

Over a few short years, Dale and his colleagues had identified the neurotransmitter released at the sympathetic nerve synapses, *noradrenaline*, and the neurotransmitter released at the parasympathetic nerve synapses, the nerve synapses in the autonomic ganglia and the nerve synapses on skeletal muscle, *acetylcholine*. Dale was cautious about declaring that he had proved chemical synaptic transmission and that these were the transmitters since the substances were isolated from ergot extracts and not from animal tissue. He did speculate that if they are the natural transmitters, it might be difficult to prove because they were likely rapidly broken down by other naturally occurring chemicals at the synapses.

At the same time, a German pharmacologist, Otto Loewi, was performing similar experiments and arriving at similar conclusions [8]. Leowi was convinced of "neuro-humoral transmission" at the neural synapse, but like Dale, he had difficulty proving that neurotransmitters existed in living tissue. He used the term *Vagusstoff* (vagus substance) for the inhibiting substance released by the vagus nerve rather than acetylcholine because of prior criticism for speculating by his German colleagues. Like Dale, Leowi wondered if there was a naturally occurring substance to break down the Vagusstoff after it was released at the synapse. Working with frogs, Leowi and coworkers identified such an enzyme that they called cholinesterase, since *Vagusstoff* was an ester of choline, acetylcholine.

For years, scientists had been aware that an alkaloid extract from the African calabar bean could slow and even stop the heart of animals including humans. The active ingredient in the extract was found to be an anticholinesterase, physostigmine. Leowi showed that administering physostigmine to frogs could prolong the action of *Vagusstoff* (presumably acetylcholine) by inhibiting the breakdown at the synapse. This explained the slowing and even stoppage of the heart after administering physostigmine. By blocking the cholinesterase that breaks down acetylcholine, the action of the acetylcholine persisted unchecked.

After a delay in his research because of the war, Henry Dale moved to the National Institute for Medical Research in Hampstead and restarted work on neurotransmitters. He was intrigued by Loewi's speculation on acetylcholine but remained bothered by the fact that neither acetylcholine nor cholinesterase had been identified in animal tissue. He teamed up with a chemist, Harold Dudley, to isolate acetylcholine from the spleens of horses and oxen killed at a local slaughterhouse [9]. From 71 pounds of spleen minced in alcohol, they were able to extract 1/3 of a gram of acetylcholine. Dale and Loewi were now convinced that *Vagusstoff* was acetylcholine.

Regarding the neurotransmitter at the sympathetic nerve synapse, Dale and Loewi were unsure but agreed it was an adrenaline-like substance. In the early 1930s, Walter Cannon at Harvard Medical School found slight differences in the physiological effects of intravenous adrenaline and stimulating the sympathetic nerves in cats. But it was not until 1946 that the Swedish physiologist and pharmacologist Ulf von Euler first documented noradrenaline in the sympathetic nerve synapses of mammals.

Neuromuscular Transmission

Although Dale had identified a muscarinic-like action at the nerve-skeletal muscle synapse, Loewi felt that direct electrical transmission was more likely since the nerve endings seemed to have intimate contact with the muscle fibers. Furthermore, it seemed that synaptic transmission from nerve to muscle was much faster than between autonomic nerves and the internal organs. Efforts to extract acetylcholine from muscle had so far been unsuccessful. In 1932, while attending a meeting of the German pharmacology Society, Dale heard a presentation by a young German pharmacologist Wilhelm Feldberg who described a leech muscle model for detection of acetylcholine after electrical stimulation. As Loewi had previously described, Feldberg found that administering physostigmine increased the amount of acetylcholine making it easier to detect.

Just as Feldberg was developing his leech muscle acetylcholine assay, Hitler came to power and Feldberg of Jewish descent was dismissed from his position at the Institute of Physiology in Berlin. When Dale heard about the dismissal, he arranged for Feldberg to join him in Hampstead to work on the leech assay for acetylcholine at the neuromuscular junction. They studied the nerves to the leg muscles of frogs, dogs, and cats and obtained clear evidence that acetylcholine was released at the nerve-skeletal muscle synapse [10]. Once Loewi saw these results, he jumped on the bandwagon and agreed that neuromuscular and probably all synaptic transmission was chemical. Charles Sherrington, the father of the synapse, had already come to that conclusion.

Recall that although sympathetic nerves release an adrenaline-like substance at their contact with the smooth muscle of internal organs, Dale had shown that there was a muscarinic effect in the sympathetic ganglia. Feldberg went on to show that the nerves entering the sympathetic ganglia from the spinal cord released acetylcholine to activate postganglionic sympathetic neurons [11]. Dale attempted to simplify this potentially confusing situation by introducing the terms adrenergic for nerves that release adrenergic-like substances and cholinergic for those that release acetylcholine, terms that remain in general use today. Preganglionic sympathetic nerves are cholinergic and postganglionic sympathetic nerves are adrenergic.

It did not take long for clinicians to recognize the potential importance of these pharmacological discoveries [12]. In 1932, a German physician, Lazar Remen briefly reported that a patient with the muscle weakening disorder, myasthenia gravis, showed transient improvement after receiving neostigmine, a synthetic drug with effects like physostigmine. Unaware of Remen's report, an English physician working at St. Alfege's Hospital in Greenwich, Mary Walker, published two papers in 1934 and 1935 reporting successful treatment of patients with myasthenia gravis using anticholinesterase drugs. Walker got the idea when reading about the muscle paralysis produced by the South American arrow poison, curare. She reasoned that the anticholinergic drug physostigmine that reverses the weakness associated with curare poisoning might reverse the muscle weakness in patients with myasthenia gravis.

In what was called the "miracle at St. Alfege's," Walker reported a dramatic though transient improvement in muscle strength in a 56-year-old woman with myasthenia gravis after a single large dose of physostigmine. In the second report, she used the synthetic drug neostigmine with less side effects on another patient and reported similar success. The effect peaked about an hour after injecting the drug and lasted about 6 h. Walker suggested that anticholinesterase drugs might also be useful in treating patients with botulism and cobra bites, other curare-like poisonings associated with blockage at the neuromuscular synapse.

We now know that myasthenia gravis results from autoantibodies directed at the acetylcholine receptors located on the muscle side of the synapse blocking neuromuscular neurotransmission. Anticholinergic drugs increase the amount of acetylcholine at the synapse and help to partially overcome the block. Although these drugs are still used in the treatment of myasthenia gravis, modern treatment focuses on suppressing the immune system's production of blocking antibodies.

The Electron Microscope and Brain Neurotransmission

Despite the convincing evidence for chemical synaptic transmission in the autonomic nervous system, and at the neuromuscular junction in the early twentieth century, there was still a heated debate about chemical transmission in the central nervous system that persisted well into the mid-twentieth century [12]. The "sparkers" led by the Australian neurophysiologist John Eccles were not ready to concede the brain to the "soupers." As with many other advances in neurophysiology, improved understanding of chemical synaptic transmission occurred with the development of a new technology, the electron microscope.

Based on the work of Crookes and Thomson (see Chap. 8), it was known that a beam of electrons could be focused with a magnet analogous to light waves with a lens. As it became apparent in the early twentieth century that electrons had properties of electromagnetic waves (see Chap. 11), several researchers had the idea that a beam of electrons could be used for imaging just like a beam of light waves in a standard microscope [13]. The advantage of electrons was that their wavelength could be several orders of magnitude shorter than that of visible light. While light microscopes have a maximum resolution of about 200 nm, an electron microscope theoretically could have a resolution smaller than 0.1 nm.

The first electron microscopes were developed in the early 1930s by multiple researchers in Germany, the United States, and Canada. The first commercial electron microscope was produced by the Siemens Company in 1938. Researchers at the Rockefeller Institute in New York and at Washington University in Seattle simultaneously described electron microscopic pictures of neuronal synapses in the brain in 1954. The nerve terminal exhibited a localized swelling filled with tiny vesicles (Fig. 10.1). There was a clear gap of about 20 nm between the presynaptic and postsynaptic membranes visualizing the synaptic cleft for the first time. A year later Bernard Katz and colleagues at University College in London first speculated that the vesicles stored the neurotransmitters and released them by a process called exocytosis. The vesicles fused with the cell membrane releasing their contents into the synaptic cleft.

The 1950s was marked by the discovery of a wide variety of neurotransmitters in the central nervous system [14]. John Eccles and colleagues at the Australian National University in Canberra were the first to show that acetylcholine was released at the synapses between collateral branches of motor neuron axons and interneurons in the spinal cord. They used micropipettes to deliver the acetylcholine to the synaptic junctions of interneurons in the spinal cord in anesthetized cats. Discovery of the catecholamine neurotransmitters noradrenaline, serotonin, and dopamine in the brain soon followed. The final nail in the coffin for the sparkers was the finding that synaptic transmission could be excitatory or inhibitory. It would be another two decades before the main excitatory and inhibitory neurotransmitters in the brain were confirmed, the amino acids glutamate, and gamma-amino-butyric acid (GABA), respectively.

The next great challenge to researchers was identification and characterization of the neurotransmitter receptors located on the postsynaptic neuronal membranes. As we will see later in the chapter, this required an understanding of transmembrane

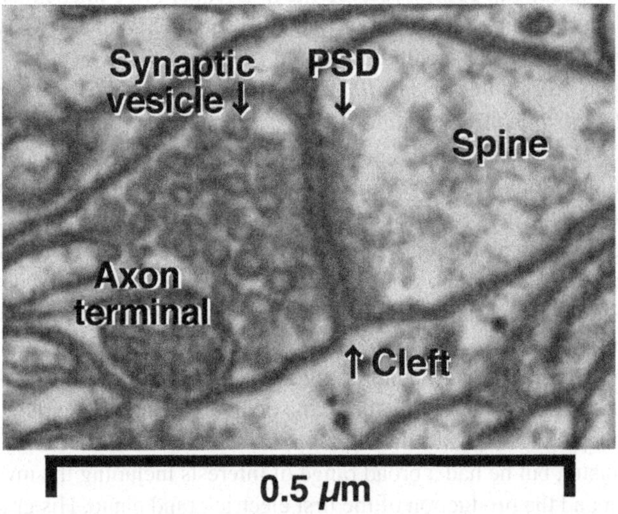

Fig. 10.1 The junction between a nerve axon terminal and a nerve cell spine showing synaptic vesicles, and the cleft between the axon terminal and spine and the post-synaptic density (PSD). (From https://blogs.zeiss.com/microscopy/en/brain-circuits/)

protein channels that developed later in the century. The number and diversity of these receptors were much greater than anyone could have anticipated.

Cell Membranes

The beginnings of physical chemistry and cell biology can be traced to the German chemist, Moritz Traube and his development of a mechanical semipermeable membrane in the mid-1860s [15]. By combining aqueous copper sulfate and potassium ferrocyanide solutions, Traube was able to produce a semipermeable membrane that grew by intussusception. He immediately saw the potential analogy with living cell membranes. In 1877, the German botanist Wilhelm Pfëffer combined Traube's delicate precipitate membrane with a stiff porous material to construct an artificial cell wall. Pfëffer showed that the artificial cell wall behaved like a plant cell wall.

In 1887, The Dutch chemist Jacobus Van't Hoff published *Studies in Chemical Dynamics* describing his research on the dynamics of chemical reactions [16]. He noted that the laws of thermodynamics could be applied to chemical reactions and the same mathematical laws could be used to describe the behavior of dilute solutions and gases. He developed an osmometer with a semipermeable membrane and found a remarkable similarity in the osmotic and elastic pressure of solutions and gases. "*The mechanism which, according to our current concepts, causes the elastic pressure of gases and the osmotic pressure of solutions is basically one and the same: in the first case gas molecules hit against the walls of the container, in the second the molecules of the dissolved body strike the semipermeable membrane*" [17]. For this work, Van't Hoff received the first Nobel Prize in chemistry in 1901.

The Baltic German chemist and polymath, Friedrich Wilhelm Ostwald recognized that semipermeable membranes could act as an "ionic sieve" allowing only certain ions to cross, "*in this case there is a separation of the electricity, and the resulting electrostatic forces inhibit further passage whenever the osmotic pressure of the negative ions equalizes*" [18]. He was one of the first scientists to recognize that separation of ions across a semipermeable membrane might explain the electrical potential in nerve and muscle cells. Ostwald set up the first department of physical chemistry in Leipzig and one of his early recruits was a young German chemist, Walther Hermann Nernst who would become famous for his equation describing the electrical potential generated by unequal concentrations of ions across a semipermeable membrane.

Nernst was born into an affluent family in Briesen, West Prussia in 1864 [19]. He was educated at several of the leading Universities in Europe (Zurich, Berlin, and Graz) before joining Ostwald in Liepzig. His focus was on thermodynamics and electrochemistry, but he had a broad range of interests including the invention of an electric light and the production of the first electric grand piano. His graduate thesis at Graz studied the thermoelectric phenomenon that became known as the Nernst effect. If a material that allows electrical conduction (best with a semiconductor) was exposed to a magnetic field and a temperature gradient perpendicular to each other, an electric field was induced normal to both.

When Nernst joined Ostwald, the impressive group of scientists that Ostwald had recruited were known as the "ionists" because of their work on ions and chemical reactions. Recall that Faraday had shown that salts in solution form ions when exposed to electric current thus the name cation for ions that move toward the cathode and anion for ions that move toward the anode. A Swedish chemist, Svante Arrhenius working in Ostwald's laboratory, proposed that salts separate into ions in solution even without an electric field. He further proposed the law of independent migration, each ion carries the same amount of electricity at its own speed independent of other ions moving in the same or opposite direction.

With these assumptions, Nernst developed his theory of electrolytic dissociation and an equation to calculate the potential difference between two salt solutions of different concentrations separated by a semipermeable membrane. The salt passes from the concentrated solution to the diluted solution until a state of equilibrium is reached. An electrical potential is established as electrically charged free ions accumulate on each side of the membrane. To calculate the potential difference (*E*), Nernst relied on Van't Hoff's theory that salt solutions behave like gases assuming a complete dissociation of the molecules into ions.

$$E = KT / F \ln C_{\text{out}} / C_{\text{in}} \text{V}$$

where K (0.86×10^{-4}) is Nernst's electrolyte constant of gases, T the absolute temperature, F the Faraday electric charge/mole, and ln $C_{\text{out}}/C_{\text{in}}$, the natural log of the ratio of molar concentrations on the outside and inside of the membrane. As we will see later, based on the concentration of potassium and sodium ions outside and inside the cell, the estimated transmembrane potential of nerve cells is about −70 mV.

Near the end of the nineteenth century, Nernst began working on a problem that intrigued many of the top scientists of the time, electrical lighting. Unlike Edison and many others who were using carbon filaments in evacuated glass tubes, Nernst developed a ceramic element that produced a clear white light when heated in normal air. His lamp was twice as efficient as a carbon filament bulb, but it had to be preheated before it could conduct electricity. Nernst patented the "Nernst lamp" and immediately sold the patent to the German company AEG for a lump sum up front payment of a million marks making him a wealthy man. The lamp was an initial success in lighting the German pavilion at the 1900 Paris exhibition, but it soon lost out to the less expensive and equally efficient tungsten filament light bulb. If he had sold his patent for royalties as most others had done, he would have received only small fraction of the million marks.

When Nernst was later touring America, he described visiting an elderly Thomas Edison who lectured him on the importance of business sense and the irrelevance of academic achievement. Nernst shouted into Edison's ear trumpet (he was near deaf at the time), "*How much did you get for your light bulb patent?*" After Edison admitted that he got virtually nothing, Nernst shouted, "*I got a million marks for mine! The trouble with you Edison is that you are not a good businessman!*" [20].

Membrane Theory of Bioelectrical Potential

Julius Bernstein, the protégé of du Bois-Reymond and Helmholtz introduced in Chap. 5, had been following the work of the "ionists" in Ostwald's laboratory and agreed with Ostwald that the separation of ions across the cell membrane explained the electrical potential in nerve and muscle cells [16, 21]. By 1900, he had formulated his theory on cell membrane polarization, but it was not until 1902 that he published a detailed version of the theory. The basic premise was that the electrical potential between the inside and outside of excitable cells such as nerve and muscle cells was due to selective permeability of the cell membrane to potassium ions and the higher concentration of these ions inside the cell compared to outside the cell. According to Nernst equation, the positive potassium ions will move from the compartment with higher concentration (inside the cell) to the compartment with lower concentration (outside the cell) until an equilibrium of charge and concentration is achieved. Therefore, an excess of positive charge builds up outside the cell membrane and an excess of negative charge inside the membrane. This separation of charge cannot be compensated by the movement of other ions because of the exclusive permeability of the membrane for potassium ions.

Bernstein's theory explained the negativity of the current measured from an injured nerve and the negative variation or action potential as it moved along a nerve or muscle (Fig. 10.2). When a nerve was injured, the recording electrode was exposed to the negative charge. Bernstein assumed that during the action potential there was a brief reversible "rupture" of the membrane separating the inside and outside of the cell. Interestingly, although as shown in Fig. 5.2, Bernstein had identified an overshoot of the baseline in his recording of an action potential, he chose to

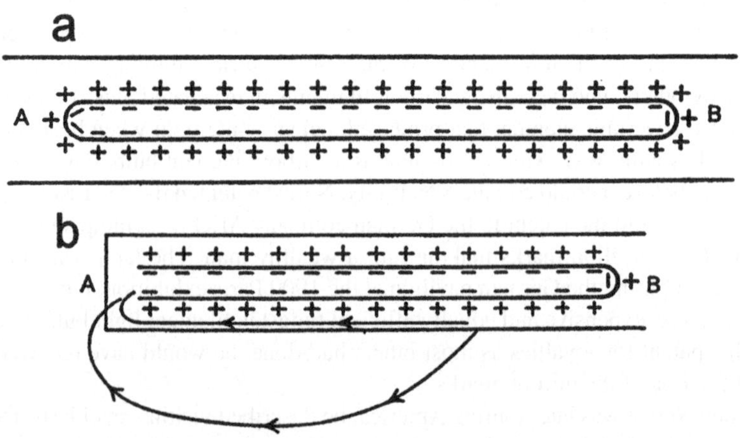

Fig. 10.2 Bernstein's theory of nerve and muscle cell excitation. In (**a**), there is an excess of negative charges inside the cell membrane at rest due to selective permeability of potassium, and in (**b**), damage to the cell membrane allows current to flow from the outside to inside of the cell. A similar phenomenon could occur during excitation due to transient changes in membrane permeability. (From Bernstein J. *Elektrobiologie*. Braunschweig, Vieweg and Sohn, 1912)

ignore the overshoot (assuming it was an artifact) in formulating his theory that predicted the action potential only reverted to baseline zero.

Around the same time, the British physiologist and biologist Charles Ernest Overton was working on membrane physiology first in Zurich and then at the University of Würzburg, Germany [22]. After studying plant and animal cell membranes testing permeability to hundreds of solutes, he concluded that a lipid-impregnated membrane accounted for the osmotic properties of living cells. He came to this conclusion based on the observation that lipid-soluble solutes such as anesthetics, narcotics, and many alkaloids penetrated cell membranes the fastest. Some such as ethanol crossed the membrane at a rate comparable to that of water.

The ability of a solute to pass through the membrane barrier did not primarily depend on the size of the molecule (as others had suspected) but rather its solubility in the lipid membrane. Surprisingly, solutes such as glucose and amino acids and inorganic ions such as sodium and potassium ions did not readily pass through the membrane suggesting some type of selective permeability to these important cell constituents. Overton showed great prescience in suggesting an active absorptive process for these substances across the membrane associated with the expenditure of metabolic energy. In particular, he proposed that the unequal distribution of cations such as Na^+ and K^+ between cells and their environment depended on an active process that could move ions in the opposite direction of that predicted by the laws of diffusion.

Just as Bernstein was preparing his potassium theory to account for the nerve and muscle action potential, Overton moved to Würzburg to work under Max von Frey in the Department of Physiology in 1901. Overton decided to focus on the excitability of muscle fibers in vertebrates. He began by testing muscle fibers bathed in isosmotic solutions of brown sugar devoid of salts. To his surprise, the muscle was completely unexcitable to electrical stimulation. Adding just a tiny amount of table salt (0.1% sodium chloride) was enough to fully restore the muscle excitability. He then tried other isosmotic solutions (glucose, lactose, alanine, etc.) with the same result. Without the table salt, the muscle was unexcitable. After trying other salts, he concluded that only sodium ions were crucial for "*the process of conduction of excitation and muscle contraction*" [23].

Overton was certainly aware of Bernstein's potassium theory, and he speculated on a theory that combined a role for sodium and potassium ions in the generation of nerve and muscle action potentials. "*With muscle contraction or the conduction of excitation a certain exchange [Austausch] [must] occur between the cations inside the muscle fibers (very probably an exchange of potassium ions) and the sodium ions in the solution bathing the muscle fibers*" [24]. The outward movement of the potassium ions accounting for the return to baseline and the influx of sodium ions for the overshoot of the baseline observed in the recorded action potential. It would be several decades before Hodgkin and Huxley would definitively prove that Overton was correct.

We now know that nerve and muscle cells possess potassium and sodium leakage channels that allow these cations to diffuse down their concentration gradient. Since there are many more potassium leakage channels than sodium leakage channels, potassium diffuses out of the cell faster than sodium leaks in causing the interior to be negatively charged relative to the exterior.

Improved Electronics

In the late nineteenth century, electrophysiologists still did not have the ability to record the tiny electrical change associated with action potentials moving along nerves or muscle fibers. Bernstein's rheotome allowed him to reconstruct the profile of the action potential using an averaging technique but it was not possible to measure the action potential in real time. The first breakthrough came with the discovery of the capillary electrometer in the 1870s [25]. After reports that a drop of mercury in an acid solution changed shape when a small amount of electricity passed through it, the French physiologist Etienne-Jules Marey made an electrometer by placing mercury and acid in a thin tube. When shining a bright light on the tube, he could record small movement at the surface of the mercury on film behind the tube. The instrument was first used to record the heartbeats of a frog and tortoise.

In 1888, the British physiologist Francis Gotch used the capillary electrometer to detect the action potentials in peripheral nerves and the spinal cord [26]. The action potentials appeared as featureless blips on the moving film that seemed to vary in shape and spacing. The distortions occurred because the mercury continued to move from inertia after the stimulus was complete. Gotch later noted there was always a small interval of time between action potentials suggesting a refractory period in the nerve after firing. This was the first indication that signals in nerves were conveyed by a series of discrete impulses and not by a single lingering impulse.

At the University of Leiden in Holland, the physiologist Willem Einthoven tried to use a capillary electrometer to record the electrical events associated with heart contraction, electrocardiography [27]. But like others he was frustrated with the poor resolution of the devise, so he began looking for other potential recording instruments. Einthoven heard a report that a thin wire, stretched between the poles of a strong electromagnet, vibrated when a small oscillating electrical current passed through it. This gave him the idea for his so-called string galvanometer. The instrument, which took several years to develop, weighed several tons and needed a massive amount of cold water to cool the electromagnet. Einthoven would eventually receive the Nobel prize for the first electrocardiographic recordings, but the recordings were still crude, and even with improvements, efforts to record the action potentials in nerves with a string electrometer were unsatisfactory.

What was needed was a paradigm shift in technology and as is often the case in the history of science, the new technology came from another field of research, wireless telegraphy, and radio transmission. In 1904, while working with the Marconi wireless telegraph company, the British electrical engineer and physicist, John Ambrose Fleming invented the thermionic two-electrode vacuum tube for wireless reception. Two years later the American physicist and inventor Lee de Forest added a third electrode, a control grid, to the Fleming vacuum tube creating the first electronic amplifier, a critical invention for long-distance telephone and radio communications. These vacuum tubes were key to the rapid advances in electrophysiology in the first half of the twentieth century eventually superseded by solid-state electronic technology of transistors in the latter half of the century.

Thermionic emission, also called the "Edison effect," has a long, convoluted history dating back to the mid-nineteenth century. The first clear description of the phenomenon was by the French physicist, Edmond Becquerel [28]. Becquerel is best known for creating the first photovoltaic cell when in 1839 he noted that an electric current was generated if he shined a light on a platinum electrode coated with silver chloride or silver bromide. This so-called photovoltaic effect or "Bacquerel effect" became the basis for modern solar cells. As we will see in Chap. 12, Bacquerel's son Henri was a pioneer in the field of radioactivity.

As described in Chap. 9, in 1880 Thomas Edison was puzzled by the blackening of his vacuum light bulbs particularly near the positive terminal of the filament. He made a light bulb with a metal foil surrounding the filament and found that this prevented the blackening. To try and understand the process, Edison used the metal foil as an electrode and connected it to a galvanometer. There was current flow through the vacuum from the heated filament to the foil if the foil was positive relative to the filament but not if the foil was negative relative to the filament. Edison noted that the one-way current emitted by the hot filament increased with increasing voltage and that there was enough current passing through the vacuum tube to operate a telegraph sounder. The hot filament was emitting electrons that were attracted to the positively charged foil but repelled by a negatively charged foil.

Edison exhibited his devise at the International Electrical Exposition in Philadelphia and filed for a patent as a voltage-regulating device in November 1884. This was the first patent for an electronic devise in the United States. The Welsh electrical engineer and inventor, William Henry Preece was visiting the United States at the time and Edison gave him several of his two-electrode vacuum tubes to take back to England. Preese published a paper in 1885 entitled, *On a peculiar behaviour of glow lamps when raised to high incandescence* in the *Proceedings of the Royal Society* in which he called thermionic emission, the *Eddison effect* [29].

Preece was a major player in the field of wireless telegraphy in England and became Engineer-in-Chief of the British Post Office in 1892. He was an ardent supporter of Marconi and helped him gain financial support from the Post Office for his transatlantic wireless project. Preece maintained a longstanding feud with the eccentric electrical engineer and mathematician, Oliver Heaviside. Recall from Chap. 7, Heaviside suggested that there was a layer in the upper atmosphere that reflected electromagnetic waves so that they could travel long distances about the planet. Preece disagreed and argued that the Earth's magnetic field was critical for long-distance wave transmission. With typical bluntness, Heaviside dismissed Preece as an incompetent mathematician.

In 1899, when Marconi began working on his transatlantic wireless project, he needed technical support and hired John Ambrose Fleming to design a new high-powered radio transmitter and receiver [30]. Fleming obtained his doctoral degree from the University of London and later became the first professor of electrical engineering at the University. As part of his remuneration for the transatlantic project, Fleming was promised 500 shares of Marconi stock if the endeavor was successful. But he had to sign an agreement that if successful only Marconi would receive credit for the success. Fleming had been frustrated with the inefficiency of Oliver Lodge's coherer for detecting radio waves and found that Edison's two-electrode vacuum tube could be a sensitive detector of radio waves. Fleming improved the design of a thermionic vacuum tube that he called a *diode* or *oscillation valve* and obtained a patent in November 1904 (Fig. 10.3).

The final piece to the thermionic emission puzzle came from a young American inventor, Lee de Forest and his triode vacuum tube, the first practical electronic amplifier [31]. Although he was expected to follow in the footsteps of his father, a Protestant Minister, de Forest dreamed of being a famous inventor from a young age. He obtained his undergraduate degree from Yale University's Sheffield Scientific School in New Haven in 1896, and after several years of postgraduate work, he earned his doctorate degree at Yale's Sloane Physics Laboratory in 1899. His doctoral thesis was *Reflection of Hertzian Waves from the Ends of Parallel Wires.*

On obtaining his degree, de Forest was convinced that wireless communication was the future, and he was determined to part of that future. After working for a variety of companies and starting his own company, de Forest developed a wireless system that unlike Marconi's discontinuous pulses produced by spark transmitters, used continuous wave signals that could be amplitude modulated (AM) for voice transmission. With this first system, he could transmit voice signals across his

Fig. 10.3 Photograph of Fleming's thermionic vacuum tube, the oscillating valve

laboratory. Like Fleming, he was frustrated with the available detectors for wireless signals, and he wondered if ionized gases heated in a partially evacuated glass tube would be a better detector. He began experimenting with partial vacuum tubes that he called *Audions*. He started with a diode tube-like Fleming's oscillation valve, and in late 1906, he added a third electrode in between the other two that he called a *grid*. Current in the grid amplified the signal passing through the tube. When he tested the new grid-controlled Audion or *triode*, he was convinced that it was superior to the diode for detecting radio signals.

De Forest filed for a US patent for the grid Audion and it was granted in February 1907. Fleming immediately accused de Forest of infringing on his patent. Marconi who owned Fleming's oscillating valve patent took de Forest to court arguing that de Forest simply added a grid electrode to Fleming's vacuum tube. De Forest countered that his invention not only rectified signals like Fleming's tube but also amplified the signals. The court eventually ruled in favor of Marconi. In the meantime, the two sides agreed to share a license so that both could manufacture triodes in the United States. Interestingly, the United States Supreme Court later invalidated Fleming's oscillating valve patent concluding that the technology in the patent was based on generally known information at the time (the Edison effect).

De Forest's initial three-valve Audion was not a very good amplifier or signal detector. To strengthen the audio frequency output from the grid, De forest fed the signal from one tube into a second for additional amplification. His final *cascade amplifier* consisted of three Audions joined together. But de Forest's premise that the tubes should only be partially evacuated was flawed.

At the same time, researchers at the American Telephone and Telegraph Company (AT&T) were looking for ways to amplify long-distance telephone signals and they invited de Forest to come and demonstrate his Audions. They were disappointed to find that the tubes did not work well with their low-voltage telephone lines. Dr. Harold Arnold and his research team at AT&T soon discovered that if they altered the tube design and fully evacuated them, the flow of electrons was better than in a partially evacuated tube. These new triode tubes operated well on existing telephone lines and became the prototype of modern electron-discharge amplifying vacuum tubes. In 1913, AT&T working through a third party to disguise the link bought de Forest's Audion patent rights for $50,000. De Forest badly needed money at the time.

The All-or-Nothing Principle

An important question facing physiologists at the turn of the century was how the electrical signals in nerves resulted in graded muscle responses. Did each muscle fiber gradually increase its strength of contraction or were more and more fibers recruited to increase the strength of contraction? Was the amplitude of the action potential gradable or was it an all-or-nothing phenomenon.

The first to address this question was Keith Lucas, a Cambridge-trained physiologist [32]. The son of the director of the British Telegraph Construction and

Maintenance Company, Lucas had a longstanding interest in electrical instruments. To address the graded muscle response question, he focused on a single frog muscle with about 200 fibers and dissected a small strip of the muscle containing about 20 fibers. He reasoned that if he applied electrical current to the strip and slowly increased the intensity, the recording of the muscle contraction would either show a smooth curve as the strength of all fibers gradually increased or there would be steps in the recording as new fibers were recruited. There was a clear staircase pattern to the recording indicating that the graded muscle contraction was due to some fibers contracting fully while others did not respond at all. The number of steps in the staircase never exceeded the number of fibers in the muscle strip.

Over the next few years, Lucas went on to develop a nerve muscle preparation so that he was able to show the same staircase muscle contraction pattern by stimulating the nerve to the muscle. In a paper published in the Journal of Physiology in 1909, he concluded that individual muscle fibers always contract maximally regardless of the strength of nerve stimulation [33]. He ended the paper by speculating that nerve fibers probably also had an all-or-nothing response although this was yet to be definitively proven.

In 1911, a bright young Cambridge postgraduate student, Edgar Douglas Adrian joined Lucas' laboratory to work on the nerve conduction question [34]. Adrian had just received a degree in physiology from Trinity College. Even though Lucas had some of the best electronic equipment available at the time including a newly designed capillary electrometer, Adrian began with an indirect method to address the nerve question just as Lucas had originally used for the muscle contraction question.

Adrian came up with the idea of an alcohol nerve block, exposing a frog nerve to just enough alcohol to weaken the impulse but not enough to completely block all the action potentials. If the all-or-none principle was valid, the action potentials that made it through the block would be at full strength. To test this hypothesis, he used a second alcohol block further down the nerve and found that it had to be a strong as the first block to stop the action potentials that got through the first block. Adrian further noticed that the strength of the electrical stimulus had little to do with the strength of an action potential. Even if triggered by a weak stimulus, the action potential shot down the nerve at full strength. Adrian and Lucas described the action potential as a "*propagated disturbance*" that moved down the nerve by exciting the nerve region just ahead enough to generate a continuous propagation.

At this time, research in electrophysiology took a back seat to the First World War, the war to end all wars. Lucas used his engineering background to develop new instruments for the Royal Air Force at a factory in Farnborough. He died tragically in a mid-air collision while testing a new instrument. Adrian completed his medical training in London and went on to treat soldiers with nerve injuries and to study shell shock and hysterical paralysis. A side benefit of the war was the new technologies driven by the need to develop wireless radio receivers and radio detection finders (radar) by the military. Improved thermoelectric vacuum tubes (de Forest's Audions) were used to amplify radio signals with little distortion. An American physiologist, Alexander Forbes, was actively involved in developing these technologies during the war, and immediately after the war, he obtained some of these vacuum tubes for his research [35]. In 1919, he used a string galvanometer and the new vacuum tubes in a circuit and was able to amplify an action potential 50 times in size.

Forbes, the son of the president of the Bell Telephone Company, attended Harvard where he developed an interest in physiology. He received a master's degree in 1905 and an MD degree in 1910. Forbes stayed on at Harvard to work with the physiologist, Walter Cannon of "fight or flight" fame. After reading the work of Charles Sherrington, Forbes received permission for a sabbatical to visit and work with Sherrington in Liverpool. Before returning home, Forbes visited the physiology laboratories in Cambridge where he met Lucas and Adrian. He was so impressed that he decided to delay his return to the United States for a few weeks and perform a series of experiments with them. The tickets he canceled were on the new luxury liner, the *Titanic*.

After the war, Adrian returned to Cambridge and Trinity College where he was offered the basement laboratory of his former mentor, Keith Lucas. As he began planning new experiments, he contacted Forbes regarding his vacuum tube amplifiers. Forbes agreed to a visit and brought some of his radio tubes in 1921. Forbes developed an amplifying circuit, and they performed a few experiments, but the results were not impressive. After Forbes returned to the United States, Adrian read about a new multistage amplifier developed by the physiologist Herbert Gasser at Washington University in St. Louis. Gasser was remarkably generous when contacted by Adrian offering to send him detailed instructions for building the multistage amplifier.

Gasser followed his mentor Joseph Erlanger from Wisconsin to St. Louis where despite Erlanger's background in cardiac physiology, Gasser convinced him of the potential in neurophysiology. Aware of the need for improved amplification of nerve currents, Gasser teamed up with a talented technician H. S. Newcomer and electrical engineers at the Western Electric Company to develop a multistage amplifier like de Forest's but with better newer vacuum tubes. They published their initial findings in 1921 using a sensitive string galvanometer and the new amplifier to view action potentials with remarkable clarity.

What Gasser needed was a way to display the tiny signals without the distortions due to the inertia of moving mechanical parts of standard recording instruments. For this, he turned to the cathode ray oscilloscope developed at the turn of the century by Karl Braun, a German electrical engineer and inventor. Braun's cathode ray tube shot a tiny stream of electrons through a vacuum tube where the current in an electromagnet deflected them onto a fluorescent screen for display (Fig. 10.4). Braun went on to start Telefunken, the pioneering television company. Combining Braun's cathode ray oscilloscope with the new amplifiers, Gasser and Erlanger were able to amplify and display nerve signals with high precision (amplified several thousand times) [36].

When Gasser and Erlanger began recording the impulses shooting down peripheral nerves in a variety of animals, they found a series of oscillations rather than a single up-and-down blip. The further apart the stimulating and recording electrodes,

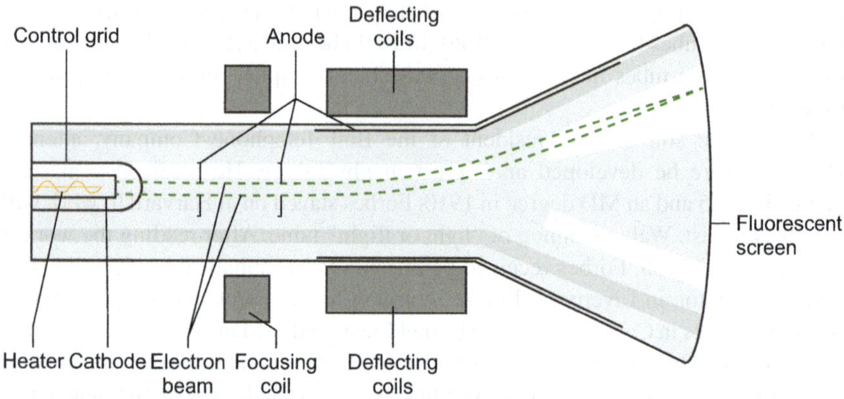

Fig. 10.4 Schematic drawing of Karl Braun's oscilloscope

the greater the oscillation. This suggested that peripheral nerves contained a variety of nerve fibers each with different conduction velocities. They assumed that action potentials moved faster in thicker fibers than in thinner fibers. They subsequently proved this assumption by selectively blocking different size fibers with different techniques such as anesthetics and freezing. Sensory fibers were thinner than motor fibers and fibers carrying different sensations (touch, pain, and temperature) could be selectively blocked.

Recordings from Single Nerves

While Gasser and Erlanger were making their pioneering recordings of peripheral nerves, in Cambridge Adrian followed Gasser's instructions and built his own new amplification system. Using Keith Lucas' old capillary tube electrometer with photography, he began recording nerve impulses in a frog nerve muscle preparation. He hung the frog leg on a hook and recorded from the exposed nerve. To his dismay, the recordings seemed to be noisy with frequent irregular blips. Was there a problem with his amplifying circuit? But the "noise" disappeared when he removed the frog leg from the hook and laid it flat on the laboratory counter only to return when it was replaced on the hook. It suddenly occurred to him that the "noise" might be the firing of proprioceptive sensory nerves previously described by Sherrington (see Chap. 9). The stretch receptors in the muscle were being activated when the muscle was hung from the hook.

When Adrian reported his findings in 1926, he hinted that it should be possible to record the signal from a single sensory receptor [37]. For this effort, he teamed up with Yngve Zotterman, a young Swedish Rockefeller Foundation Fellow who was previously a medical student at Cambridge. After trying to dissect out a single nerve fiber without success, Zotterman decided to try slicing off pieces of muscle until only a single stretch receptor remained. After multiple trials, they finally were successful. Zotterman later recalled: "*Adrian ran in and out, controlling the recording apparatus in the dark-room and developing the photographic plates. We were excited, both of us quite aware that what we now saw had never been observed before and that we were discovering a great secret of life, how the sensory nerves transmit their information to the brain*" [38].

The action potentials they recorded from the single fiber were uniform and traveled at the same speed consistent with the all-or-none law. The firing rate of the single nerve fiber varied with the tension on the muscle, a quarter gram of weight produced 21 action potentials per second whereas a gram produced 33 action potentials per second. The sensory information from the muscle to the brain was coded in the frequency of action potentials. Finally, they observed that the increase in action potentials after a change in the tension on the muscle was not sustained suggesting that sensory nerves responded to a change in tension, not to the steady state condition. This phenomenon that became known as *adaptation* could explain why sensory perceptions wane even though a sensory stimulus persists.

A few years later Adrian was able to record from a single motor neuron, this time with a visiting American physiologist, Detlev Bronk [39]. They selected the phrenic nerve to the diaphragm in rabbits and after anesthetizing the animals, they carefully dissected the tiny nerve so that only a few axons remained. In this case, they did not need to stimulate the nerve but rather relied on the regular impulses generated from the brain for breathing. With each breath, they recorded a burst of action potentials from a single axon all with uniform features. For the first time, they hooked up the output from the amplifiers to a loudspeaker so that they could listen to the nerve firing as well as recording it. This technique rapidly became standard in electrophysiology laboratories around the world.

Adrian shared the Nobel prize for "discoveries regarding the function of neurons" with Sherrington in 1932. In his Nobel address, he acknowledged the support of Lucas and the importance of the technological advances of Forbes and Gasser that allowed him to make his discoveries. It was another 12 years before Gasser and Erlanger received their Nobel Prize.

Sensory Representation in Cerebral Cortex

Having identified individual sensory nerve fibers and the signals they carried; Adrian became interested in where these signals went in the cerebral cortex. Using his sensitive amplification system, he mapped the representation of the skin senses in the brain of a variety of animals, including cats, dogs, and monkeys [40]. Adrian used natural stimuli such as rubbing or stroking the skin with a feather or glass rod or simply pressing the skin with his finger. He first noted that the sensation from one side of the body was mapped on the opposite parietal lobe but also noted a secondary representation just below the primary area.

When he first identified the secondary area in cats, Adrian thought it was just for the claws, but subsequently, he and others found that there was a secondary representation of the entire body in all animals. In the larger primary area, the animal's body was represented upside-down and facing forward and in the small secondary area the animal was lying flat on its back. Adrian assumed that the two reception areas must have different functions. Possibly the primary area was involved with finer tactile discriminations than the secondary area or the primary was concerned with active sensorimotor exploration while the secondary area delt with passive sensory stimuli.

One observation that impressed Adrian from the start was that although there were common features to the sensory maps in different animals there were also major individual differences. The size of the cortical representation for a given body part depended on its importance to the animal during its natural behavior. A pig's snout and the nose of a Shetland pony took up much of the somatosensory cortex roughly equal to all the other body parts combined. Whereas in monkeys, the hands and lips covered most of the parietal somatosensory cortex. Not surprisingly, the representation of the human parietal cortex was later shown to be close to that of the monkey.

Adrian spent his later research years focusing on the sense of smell (olfaction) across a range of animals. He applied a large variety of odors and recorded the olfactory activity in the cerebral cortex. He found that each animal responded best to odors that had the most significance in their natural habitat. For example, dogs and cat's olfactory systems best responded to foul odors such as urine and feces whereas rabbits responded best to grass and fruity odors. He then tried to tackle the difficult question of how olfactory receptor cells are programmed to respond to odors. When he recorded from nerves leaving the olfactory bulb there was some specificity for different odors but as the odorant increased above the threshold more and more nerves responded, and the picture became complicated. Adrian concluded that the sense of smell was a diverse system that depended on specific receptors, the area of the olfactory epithelium stimulated and the pattern of discharge resulting in remarkably large number of odors that could be distinguished [41].

The Giant Squid Axon

Not long after Adrian received his Nobel Prize for his work on how nerves transmit motor and sensory signals, a young undergraduate at Trinity College, Allan Hodgkin decided to revisit Adrian's nerve block studies to see if the arrival of an electrical impulse near the block resulted in a local decrease in membrane resistance as predicted by Bernstein's membrane theory. He induced a block by cooling a short piece of the sciatic nerve of a frog, stimulated the proximal nerve, and recorded from the nerve above and below the block. With this technique, he was able to show that the nerve segment beyond the block had increased excitability [42].

Hodgkin's experiments suggested that the nerve impulse resulted from a passive spread of current along the nerve fiber, but he was unable to definitively show an accompanying decrease in nerve membrane resistance. For this, he would need a new model system and better electronics. In 1936, Herbert Gasser, who had moved from St. Louis to New York to direct the Rockefeller Institute, invited Hodgkin to work in his laboratory for a year. During the year, he spent the summer at the Woods Hole Marine Biological Laboratory where he worked with the American biophysicist Kenneth Stewart Cole learning to dissect the squid giant axon. This would provide him with the model system he would need for his future groundbreaking research.

The squid giant axon is a single very large (0.5–1.5 mm diameter) axon that controls the squid's water jet propulsion apparatus for quick movements. It was identified in 1909 but first studied in detail by the English zoologist and neurophysiologist John Zachary Young. Trained in zoology at Oxford, Young went on to be one of the most influential biologists of the twentieth century. He conducted a series of studies regarding the giant axon's function in the 1930s first at the Marine Biological Association in Plymouth and then the Marine Biological Laboratory in Woods Hole, Massachusetts [43]. Kenneth Cole learned about the giant axon from Young and in late 1938 collaborated with another colleague H.G. Curtis to be the first to show a large increase in membrane conductance (or decrease in resistance) as an action potential moved along the giant axon [44].

In the spring of 1938 before returning to Cambridge, Hodgkin visited Joseph Erlanger in St. Louis who showed him the latest electronic equipment and encouraged him to continue to pursue his studies on the nature of the action potential. On returning to Cambridge, Hodgkin began by measuring the effect of altering the extracellular medium on nerve conduction velocity in crab and squid nerves. He found the speed of conduction could be decreased by immersing the nerve in oil and increased with metallic conductors. Hodgkin published his results in the *Journal of Physiology* (London) in 1939 [45].

At the same time, another young physiologist, Andrew Huxley finished his undergraduate work at Cambridge and accepted an offer from Hodgkin to work with him at the Plymouth Marine Biology Association on conduction in the squid giant axon [46]. After joining Hodgkin in Plymouth, the two young physiologists waited for several weeks before local fishermen started to bring in catches of squid. They first attempted to measure the viscosity of the intracellular compartment by monitoring the movement of small mercury droplets placed on top of a cut axon. But the droplets just sat there refusing to move. Then, they jointly came up with the idea to insert a thin capillary tube electrode inside the nerve axon (each later suggested that the other came up with the idea).

The electrode was a glass capillary tube about 100 μm in diameter and 20 mm long. The tube was filled with sea water in contact with an inserted 20 μm diameter silver wire whose tip was coated with silver chloride. The capillary tube was pushed through the cut end of the nerve and tied into the nerve with a thread of silk. By illuminating from behind the hanging cut nerve, the position of the electrode inside the nerve could be monitored with mirrors and a microscope.

The result of their experiment was a remarkably clear recording of the action potential between electrodes inside and outside the axon (Fig. 10.5). They noted: "*A small action potential was recorded from the upper end of the axon, and this gradually increased as the electrode was lowered, until it reached a constant amplitude of 80–95 mV, at a distance of about 10 mm from the cannula. In this region the axon appeared to be in a completely normal condition, for it survived and transmitted impulses for several hours*" [47]. They emphasized two important conclusions (1) the action potential arises at the surface of the nerve and (2) the magnitude of the recorded potential in the range of 90 mV was larger than previously recorded potentials from external electrodes due to the short-circuiting effect of the extracellular fluid.

Several months later, the Americans Cole and Curtis who had independently developed an intracellular recording electrode reported similar findings at the Marine Biological Laboratory in Woods Hole [44]. In the meantime, Hitler had invaded Poland and England declared war bringing the exciting work of Hodgkin and Huxley to a complete stop. Hodgkin spent most of the war working on radar for use in aircraft and Huxley worked on improving targeting for long-range weapons. After the war, Adrian arranged for Hodgkin and Huxley to restart their collaborative work, and in addition, he suggested that the German-born British physician and biophysicist Bernard Katz join them. The group spent the summers of 1947, 1948, and 1949 working with the squid giant axon where they developed the voltage clamp technique that led to the famous Hodgkin-Huxley equation.

Voltage Clamp and the Theory of Membrane Excitability

There is some debate as to who first came up with the idea to voltage clamp across an axonal membrane, but the first to use and publish the technique were the Americans Kenneth Cole and George Marmont. By using electronic feedback, they were able to "clamp" the voltage across the giant squid axon so that the electrical currents across the membrane could be investigated ("*taming the axon*" as Cole described it). They began using the technique at Woods Hole in 1947 and each published a paper describing the technique in 1949. According to Cambridge physiologist Christof Schwiening, Hodgkin and Huxley independently discussed the concept of voltage clamping before the end of the war [46]. Both had been working on electronic feedback circuits during the war, so they had the expertise for making such a system. Regardless, Hodgkin and Huxley improved on the design of Cole and Marmont and in their landmark papers published in 1952 they determined the kinetics of the underlying ion channels that led to the Hodgkin-Huxley equation.

Hodgkin and Huxley conducted the key voltage clamp studies on the squid giant axon in July and August of 1949. Their voltage clamp circuit using a dual-electrode system, one inside and one outside the axon, allowed them to avoid the problem of electrode polarization and to measure the ionic currents moving across the axonal

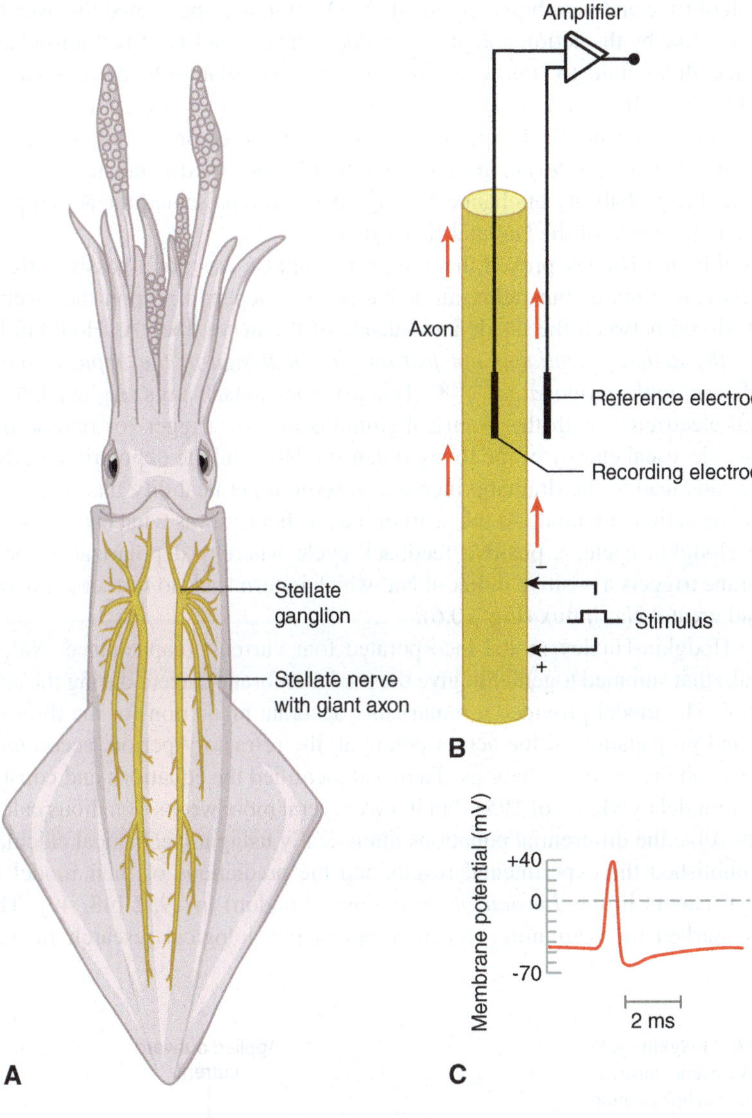

Fig. 10.5 Hodgkin and Huxley record an action potential in the squid giant axon. (**A**) Position of giant axon in the squid, (**B**) recording from inside to outside the axon, (**C**) time course of the action potential

membrane without any change in membrane potential. Bernard Katz helped with studies on the kinetics of the underlying ion channels.

The team showed that applying a step depolarization to the squid axon resulted in an inward current followed by an outward current. Using voltage clamping and ionic substitution, they determined that the two currents resulted from independent changes in permeability for Na^+ and K^+ ions with the conductance changing as a function of time and membrane potential. As Overton had speculated the overshoot of the baseline by the action potential was due to a large influx of Na^+ ions causing the intracellular fluid to transiently become positive relative to the extracellular fluid (Fig. 10.5C). Hodgkin and Huxley went on to develop a quantitative model that explained the smooth changes in electric current based on opening and closing of pores or channels (*gates*) in the axon membrane. They used a statistical approach to predict the probability of channels being open and represented the action potential as mainly a sum of the Na^+ and K^+ currents.

Hodgkin and Huxley proved that the nerve impulse was not a passive effect of the electrical stimulus but rather an active process depending upon the electrical energy stored between the inside and outside of the nerve fiber. As Hodgkin later noted: "*the action potential is not just an electrical sign of the impulse, but the causal agent in the propagation*" [48]. This gets back to Galvani's original theory of "animal electricity" with the electrical stimulus being a trigger for release of an intrinsic electrical energy of the living organism. How did the depolarizing electrical stimulus lead to the dramatic increase in sodium permeability that marked the onset of an action potential? At the heart of the mechanism was what became known as the Hodgkin cycle, a positive feedback cycle whereby depolarization on the membrane triggers a positive influx of Na^+ which in turn leads to further depolarization and greater Na^+ influx (Fig. 10.6).

The Hodgkin-Huxley model incorporated four currents (capacitance, Na^+, K^+, and leak) that summed together to give the total membrane current during the action potential. The model provided a remarkably accurate prediction for the threshold, shape and propagation of the action potential, the refractory period, accommodation, and subthreshold oscillations. They had identified the equations and constants for their model by March of 1951, but it took several more weeks of tedious calculations to solve the differential equations numerically using a mechanical calculator. They published the experimental results and the predictions of their model in a series of papers in the *Journal of Physiology* (London) in 1952 [48, 49]. These papers marked the beginning of a new era in physiological research in which

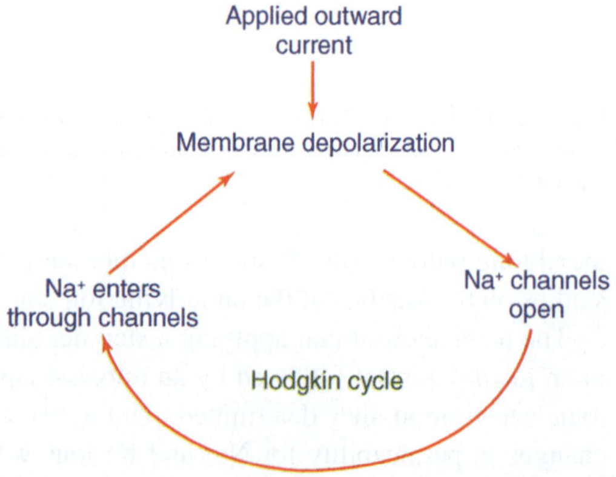

Fig. 10.6 Hodgkin cycle. Applied current causes membrane depolarization that initiates an influx of sodium ions that lead to further depolarization

qualitative descriptions are replaced with quantitative models that can reliably predict the function of complicated biological systems.

Development of Patch-Clamp Technique

The study of single-cell electrophysiology depended on the development of glass microelectrodes with tiny tips that allowed them to penetrate cells without injury [50]. In 1919, the American physiologists Frederick Pratt and John Eisenberger developed a pointed capillary pore electrode a few microns in diameter allowing them to stimulate a single skeletal muscle fiber. A few decades later, Gilbert Ling and Ralf Gerard first obtained an intracellular recording using a similar microelectrode to penetrate and measure the membrane potential in the frog sartorius muscle. Before long microelectrodes were developed to penetrate and record from a variety of cell types including neurons in a living brain.

Based on the work of Hodgkin and Huxley and many others, it was generally accepted that the electric current flowing across the cell membrane resulted from the opening of channels or pores within the lipid membrane. But how the channels worked was a mystery. The intracellular recordings of the current flow were too noisy to identify the individual channel components. Even though noise was only about one ten-billionth of an ampere, it was still 100 times greater than the current flowing through a single channel (about 1 pA, 10^{-12} A). The patch-clamp technique provided a way to measure these tiny currents.

The first to use the patch-clamp technique was the American neurophysiologist, Alfred Strickholm. After receiving his PhD in neurophysiology from the University of Chicago in 1960, Strickholm served as a postdoctoral fellow in membrane biophysics at the University of Uppsala, Sweden, where he began extracellular recordings of muscle cells in the frog with micropipettes having tip diameters of several micrometers. In the early 1960s, he developed a technique for isolating a patch of the external muscle cell membrane by pressing the micropipette against the membrane producing a seal [51]. He was able to measure the impedance and monitor the currents flowing across the small patch of muscle membrane isolated by the tip of the micropipette.

By the end of the 1960s, the technique was improved by Erwin Neher, a young German doctoral student in membrane biophysics at the Max Planck Institute for Psychiatry in Munich. In his doctoral thesis, Neher used patch clamp to record neuronal membrane currents in the subesophageal ganglia in snails. He measured the currents with high input impedance operational amplifiers and introduced a gentle suction to the pipette interior (2–10 mm of mercury) that caused the tip to clamp onto the cell membrane [52]. The microelectrode tip was about 100 μm in diameter, so he still could not resolve individual channel currents in these early studies.

While working on his PhD thesis in Munich, Neher met and became friends with another young biophysics' researcher, Bert Sakmann and the two later teamed up to perfect the patch-clamp technique for single ion channel recordings that ultimately led to a joint Nobel Prize in 1991. After leaving Munich in 1970, Sakmann studied biophysics at University College, London, before entering medical school at the University of Göttingen where he received his MD degree in 1974. Neher moved to the Max Planck Institute for Biophysical Chemistry in Göttingen in 1972, where he was joined by Sakmann in 1974. Their collaboration continued even with Neher's move to the University of Washington, Seattle, and later to Yale University.

Neher and Sakmann first tackled the neuromuscular junction of a leopard frog [53]. The goal was to monitor single-channel currents with the opening of a nicotinic acetylcholine receptor in the postsynaptic muscle membrane. They constructed an extracellular glass pipette with a tip diameter of a few micrometers that when pressed against the surface of an isolated skeletal muscle fiber covered a patch of less than 10 μm^2. They ran into difficulties in achieving a tight seal to prevent electrical leaks but by cleaning the cell surface, optimizing the size and shape of the pipettes, and producing just the right amount of suction they were able to able to see individual channel currents in response to applying acetylcholine.

> To measure these elementary currents individually, one can press the tip of a patch pipette onto a muscle fiber at the end plate. This region of muscle surface contains the acetylcholine receptor channel. When low concentration of acetylcholine is in the pipette solution, the current recorded with the pipette switches between two levels. At one level, effectively no current flows because all the ion channels in the membrane patch are shut…The opening of the receptor channel is triggered by its binding to acetylcholine; the unbinding of acetylcholine causes the channel to close again. The randomness of the lengths of time for which the channel stays open and closed reflects the probabilistic nature of the interactions between acetylcholine molecules and their receptors. [54]

Over the next several years, Neher and Sakmann improved the patch-clamp technique to develop a high-resistance seal between the micropipette tip and the cell membrane that they called a giga-ohm seal (more than a billion ohms). The latest low-impedance transistor amplifiers minimized distortion in the recorded signals. They found that by gently pulling with the attached pipette, they could excise microscopic patches of the membrane for study in isolation (Fig. 10.7). By exposing the cytoplasmic mouth of the channel, they could study the effect of altering the constituents of the living cell's cytoplasm. These techniques would revolutionize the study of ion channel physiology.

The patch-clamp technique was used to investigate the electrophysiology of transmitter-gated and voltage-gated ion channels. Neher and Sakmann noted that the elementary event in recording ion channels was not a single current pulse, but a series of brief pulses separated by gaps. At the neuromuscular junction, the acetylcholine receptor channel opened and closed several times before the acetylcholine dissociated from the channel mouth. In collaboration with David Colquhoun at University College in London, they used probability theory to study the rate of the transitions from opening to closing and concluded that each receptor channel must have two binding sites for acetylcholine. When both sites were occupied, the channel had nearly a 100% chance of being open. Thus, with high concentrations of acetylcholine in the synaptic cleft, these channels were essentially "*rapid, chemically gated current switches*" [54].

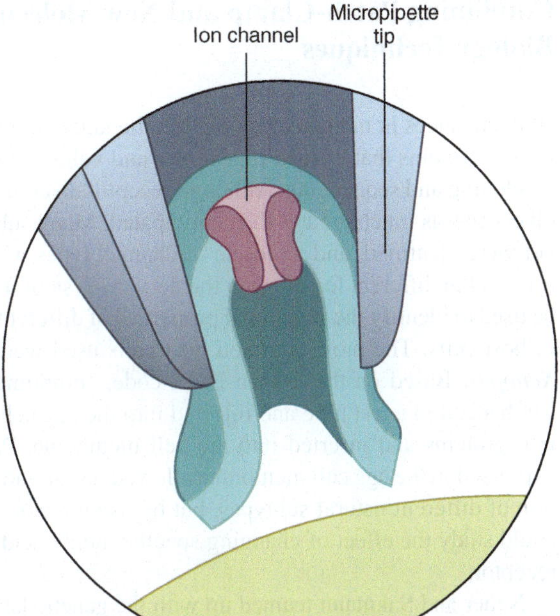

Fig. 10.7 Patch clamp technique. A small piece of membrane containing a single ion channel was clamped by the micropipette tip and pulled away so that the channel properties could be studied in isolation

Patch-clamp studies of the main transmitter-gated ion channels in the central nervous system (glutamate and GABA) showed the same random opening and closing like the acetylcholine receptor channels in the muscle postsynaptic membrane [55]. However, there was much more complexity at the neuronal synapses. Channels could be partially open or closed and there was a wide variety of receptor subtypes so that the same transmitter could have markedly different effects in different brain regions.

Although there are similarities between transmitter-gated and voltage-gated channels, there are important differences. Since the transmission of signals to and from the brain and the body must be rapid for sensory-motor responses to be effective, the voltage-gated channels in axons must open and close more rapidly than transmitter-gated channels. Patch-clamp studies of voltage-gated channels show that they rapidly respond to a difference in electrical potential between the inside and outside of the cell. As noted earlier in the chapter, voltage-gated sodium channels underlie the rapid rise of the action potential as it moves down the axon. As with other ion channels, voltage-gated sodium channels that are randomly opening and closing have a high probability of opening shortly after the onset of a voltage change. The action potential is a summation of near-simultaneous opening of tens of thousands of sodium channels. A segment of the transmembrane protein of the channel is specialized to sense the change in voltage.

Combining Patch-Clamp and New Molecular Biology Techniques

With advances in molecular biology in the latter twentieth century, the DNA codes for the proteins that formed transmitter and voltage-gated receptors were identified by cloning and sequencing. Although receptor diversity was suspected the degree of diversity was much greater than anticipated. Many subtypes of each receptor channel were identified, and a mosaic of channel types was expressed in different neurons and in different locations in the nervous system. Patch-clamp techniques could be used to identify the functional properties of different receptor subtypes expressed in host cells. The most common host cells used were eggs from the clawed frog *Xenopus*. Based on the known DNA code, complimentary RNA sequences were synthetized in a test tube and injected into the egg cells where they were translated into proteins and inserted into the cell membrane. Patch clamps of the channels expressed in the egg cell membrane allowed investigators to not only study the function of different natural subtypes, but by using recombinant DNA techniques, they could study the effect of changing specific amino acids at different locations in the receptor.

Neher and Sakmann teamed up with the genetic laboratory of Shosaku Numa at Kyoto University in Japan and used the egg cell model to study the acetylcholine receptor [55]. They wanted to identify the amino acid sequence in the receptor that formed the inner wall and controlled the ion permeability. The channel of the receptor was lined by four membrane-spanning segments, like the staves of a barrel. By directed mutations that caused substitutions in amino acids in these regions, they could see the effect of the changes on ion transport through the channel. They identified a cluster of negatively charged amino acids at the intracellular and extracellular mouths of the channel. These negatively charged rings of amino acids determined the ions that could cross the channel and the rate of current flow through the channel.

These same techniques were later used to study other ion channels including voltage sensing sodium, potassium, and calcium channels [56]. The voltage-gated sodium channel critical for generating nerve action potentials has four transmembrane domains like the acetylcholine channel each with six transmembrane segments. The fourth segment of each domain, the voltage sensing part of the channel, has positive amino acids at every third position. The gate of the channel opens with depolarization allowing sodium ions to pour into the cell for a few milliseconds and then abruptly shuts. The voltage-gated calcium channel that senses the action potential at the presynaptic membrane and triggers the release of neurotransmitter into the synaptic cleft has a similar transmembrane structure and similar voltage-sensing segments. On depolarization, calcium ions enter the presynaptic membrane and set in motion a cascade of proteins that cause the transmitter vesicles to fuse with the presynaptic membrane.

Mutations in the genes that code for neurotransmitter and voltage-gated ion channels cause some of the most common genetic neurological diseases. Clinical researchers have been able to use these same techniques to study the physiological consequences of different mutations on channel function. The mutated genes are cloned and sequenced and then expressed in frog egg cells where they can be studied with patch-clamp methods. Not surprisingly, the mutations that cause the most severe symptoms involve highly conserved regions of the genes that code for critical channel functions such as voltage sensing and pore selectivity.

Voltage Clamping with Patch Clamping

Traditional voltage clamp analysis as performed by Hodgkin and Huxley required implanting an axial wire or at least two microelectrodes into a nerve cell so it could only be used with large nerve cells such as the giant squid axon. The tiny nerve cells in a mammalian brain measuring less than 30 μm were not amenable to such a procedure. Neher and Sakmann found that on occasion they penetrated the cell membrane with their pipette without breaking the giga-ohm seal between the pipette and the cell membrane providing access to the inside of the cell [55]. This allowed them to apply a voltage clamp to much smaller cells than the traditional method and record synaptic currents with a resolution of 10–50 times higher than with traditional microelectrodes. Whole-cell recordings with this procedure called voltage clamp analysis were ideal for studying neuronal signaling at the cellular level.

Although it is possible to penetrate and record from nerve cells in the intact mammalian brain, because of the complexities of whole-cell voltage-clamp analysis much of it has been performed on intact thin slices of brain tissue [56]. The connections between neighboring nerve cells in the slices remain functional. Neher and Sakmann found that when a neuron in a brain slice releases glutamate from the presynaptic membrane and it binds with the postsynaptic receptor in a neighboring neuron, it induces excitatory postsynaptic currents (EPSCs) in the neuron increasing the probability of firing an action potential and releasing neurotransmitter of its own. By contrast, if the neuron releases GABA from the postsynaptic membrane, it induces inhibitory postsynaptic currents (IPSCs) in neighboring neurons decreasing the probability of firing an action potential and releasing neurotransmitter. Release of a single presynaptic vesical containing a quantum of neurotransmitter opened about 20–30 postsynaptic receptor ion channels. This was an order of magnitude smaller than that seen at the neuromuscular synapse indicating that there was a finer tuning at brain neuronal synapses than at neuromuscular synapses.

Whole-cell recordings had the additional advantage of allowing researchers to manipulate the intracellular environment through the opening of the pipette. With this technique, it became apparent that ion channels that responded to neurotransmitters or voltage changes were modulated by a variety of intracellular substances. These so-called second messengers which included a wide variety of proteins and ions provided feedback for controlling the function of the receptor ion channels. For example, many potassium ion channels expressed in neurons are controlled by both the membrane voltage and the concentration of calcium ions in the intracellular fluid. If calcium concentrations are high, the potassium channels open with only a slight change in voltage whereas when calcium concentrations are low the channels require high voltage changes across the membrane to open. Furthermore, some calcium channels are regulated by the concentration of intracellular calcium. These mechanisms provide an inhibitory feedback system for ion channel function.

Nerve Signals and Learning

To survive, all living organisms must be able to adapt to their surroundings. Even single-cell organisms can respond to noxious stimuli with evasive behavior. Animals with a nervous system alter their behavior by modifying the connectivity within the nervous system. In the late nineteenth century, the Russian physician and physiologist Ivan Petrovich Pavlov conducted a series of experiments that revolutionized our understanding of brain plasticity [57]. Based on his simple observation that dogs in the laboratory would drool when they saw a lab coat, Pavlov surmised that the dogs associated the lab coat with food because they always received food from someone wearing a lab coat.

To prove his theory, Pavlov developed a method to quantify saliva production and changed the "conditioning stimulus" from a lab coat to a loud noise, ringing a bell. If he rang a bell each time that food was presented after multiple trials the animals would salivate when the bell rang even without food. This learning to associate the bell with salivation dissipated over time after the training stopped but was reinstated with a brief period of retraining. Pavlov also studied the animals' responses to a noxious stimulus such as a painful shock to the foot. Normally animals rapidly withdrew the extremity, the pupils dilated and the heart rate and breathing rate increased. If a benign stimulus such as a touch or a bell ring was presented immediately after the painful stimulus, the animals had the same response to the benign stimulus as to the noxious stimulus as though the nervous system had somehow rearranged itself.

Pavlov surmised that there must be changes in the brain associated with these different types of learning but knowledge regarding neuronal mechanisms was rudimentary at the time. Jerzy Kornorski, a Polish neuropsychologist who spent 2 years in Pavlov's laboratory, was the first to use the term plasticity to describe the changes that occurred in the brain with learning. He proposed that neuronal systems were plastic and could change permanently based on the number and combination of incoming signals.

The work of the Canadian psychologist Donald Hebb on the synaptic basis of learning not only changed the course of psychology but it has influenced neuroscience well into the twenty-first century. Hebb was born in Nova Scotia in 1904, the son of two physicians [58]. His mother was enamored with the theories of Montessori and schooled him at home until age 8. Hebb was a brilliant student in mathematics and science, but he originally wanted to be a novelist and his writing skills served him well throughout his career in psychology. As a graduate student in psychology at McGill University, Hebb focused on the nature-nurture controversy regarding intelligence, and he speculated that intelligence was influenced by early learning.

After McGill, Hebb trained with the American psychologist, Karl Lashley, at the University of Chicago and then followed Lashley to Harvard where they conducted early experiments on learning and memory. Hebb then returned to Canada to work with the neurosurgeon, Wilder Penfield, at the Montreal Neurological Institute for 2 years followed by a faculty position at Queen's University in Kingston, Ontario. Here, he developed IQ tests for rats and humans and provided experimental support for his earlier notion that one's early experience influences intelligence, a radical idea at the time since most thought that intelligence was innate.

When Lashley became director of the Yerkes Laboratory of Primate Biology in Florida in 1942, he invited Hebb to join him and study the effects of localized brain lesions on emotions and behavior. While in Florida Hebb had the time to read and think about a question that had dogged him since his graduate school days—How does the brain store thoughts and concepts? The simple model of a brain where sensory nerves synapse on motor neurons to produce reflex behavior left little room for complex learning and behavior.

At that time Hebb read a paper by the Spanish neurophysiologist, Raphael Lorente de Nó describing how sensory signals originating from the semicircular canals of the inner ear were passed on to chains of interneurons in the brainstem that modulated and prolonged the signals on their way to higher centers. Lorente, who trained with the famous neuroanatomist Ramon-y-Cajal, drew beautiful pictures of the feedback loops of neurons that received the incoming sensory information. Hebb immediately saw these loops of reverberating neurons that he later call "cell assemblies," as a potential substrate for storing information. He systematically questioned how "cell assemblies" could generate ideas, memories, images, thoughts, and behaviors. His famous book, *The Organization of Behavior* published in 1949, outlined his theory of the "cell assembly" and even speculated on how the theory could explain various aspects of human behavior [59]. A basic feature of the Hebb's cell assembly was the "Hebb synapse"—when two interconnected nerve cells within the assembly repeatedly fired together their synaptic connectivity increased over time. Hebb speculated that growth or metabolic change occurred at the level of the synapse providing the basic building blocks for learning (plasticity) within the brain. Thoughts and concepts were stored within chains of interneurons whose connectivity was constantly being modified and updated by new sensory experience.

Further breakthroughs in understanding brain plasticity required improvements in technology and molecular biology that only became available in the latter half of the twentieth century. Viennese-born Eric Kandel performed a series of key experiments on neural plasticity in the giant marine snail, *Aplysia*, which ultimately led to his receiving the Nobel Prize [60]. In 1939 at age 9, Kandel fled Austria with his family and came to the United States. He wanted to become a psychoanalyst and became interested in learning and memory early in his career. While working as a postdoctoral fellow in the National Institute of Health (NIH) in Bethesda, Maryland, he conducted a series of experiments recording from neurons in memory centers in the temporal lobes of cats. Although he and his colleagues were able to obtain basic information about neuronal firing patterns, Kandel rapidly became convinced that the system was much too complex to clarify the basic mechanisms of learning. There are billions of neurons in the cat brain, and relatively little was known about connectivity between the individual neurons. He concluded that he would need to study a much simpler nervous system to be able to understand how learning occurred. His basic premise, which fortunately proved to be true, was that mechanisms identified in primitive nervous systems would generalize to more complex nervous systems including the human brain.

The brain of the sea snail, *Aplysia*, consists of about 20,000 neurons located in nine separate groupings or ganglia. With time, Kandel was able to map out the neurons and their connections in one of the ganglia, the abdominal ganglia consisting of about 2000 neurons. He focused on one reflex mediated by the abdominal ganglia, the gill-withdraw reflex. Stimulating a point on the skin of the siphon caused the gill to withdraw to protect it from damage. The reflex was mediated by six sensory neurons that relayed the sensation to two interneurons and to six motor neurons that innervated muscles that caused the gill to retract. One cell in the ganglia (R2) was so large (about 1 mm in diameter) that it was easily identified by the naked eye without the need for magnification. He was able to penetrate the neurons with a glass pipet and maintain recordings for many hours. By comparison the largest neurons in a cat's brain could only be seen with a high-power microscope and intracellular recordings could only be maintained for a few minutes. By electrically stimulating axons synapsing on R2 from other neurons in the ganglia while recording from R2, he was able to show that all of Pavlov's paradigms could be replicated at the synaptic level. The synapses were physically changing with experience.

Kandel studied the cellular basis for learning at the synapse between the sensory and motor neurons in the gill-withdraw reflex. The neurotransmitter released at the sensory-motor synapse was glutamate, the main excitatory neurotransmitter in the mammalian brain. Shocking the tail of the animal enhanced glutamate release and synaptic transmission at the synapse for several minutes after the shock—the classical paradigm for sensitization. The tail shock activated interneurons whose axon terminals released the neurotransmitter serotonin (known to be involved in behavioral responses) at synapses on the axon terminals of the sensory neurons. Serotonin receptors in the presynaptic membrane then activated a cascade of intracellular "second messengers" that increased the amount of glutamate released at the sensory-motor synapse.

If the animal received a long series of shocks to the tail, it produced a much longer sensitization associated with sprouting of new synaptic terminals between the sensory and motor cells. In this case, a second messenger entered the nucleus of the sensory cell activating specific genes to produce proteins for forming new synapses. This process called long-term potentiation (LTP) is the most studied model of neuronal synaptic plasticity. Early-phase LTP does not require new protein synthesis and lasts for up to 3 h and late-phase LTP requires new protein synthesis and lasts for more than 3 h and can last for as long as the animal lives. LTP in the brain has subsequently been shown to be a key factor in learning and memory in the synapses of Hebb's cell assemblies.

In Brief

Based on Walther Nernst's equation describing the potential difference between two salt solutions of different concentrations separated by a semipermeable membrane, Julius Bernstein developed his theory of cell membrane polarization at the turn of the twentieth century. The negative electrical potential inside excitable nerve and muscle cells was due to selective permeability of the cell membrane to potassium ions. Bernstein speculated that during an action potential, there was a brief reversible breakdown in the selective permeability allowing potassium ions to enter the cell. Keith Lucas and Edgar Adrian demonstrated the "all-or-nothing principle" of nerve transmission and suggested that the action potential moved down the nerve by exciting the nerve region just ahead enough to propagate the signal. Development of thermionic vacuum tubes for rectifying and amplifying electronic signals along with improved electrometers allowed researchers to accurately record action potentials for the first time. By taking advantage of the giant squid axon that could be penetrated with a glass electrode and a new voltage-clamp technique, Allan Hodgkin and Andre Huxley determined that the action potential resulted from independent changes in permeability for sodium and potassium ions with the conductance changing as a function of time and membrane potential. They went on to develop a mathematical model based on opening and closing of ion channels in the axon membrane that provided a remarkably accurate prediction of the threshold, shape, and propagation of the action potential. Finally, the patch-clamp technique allowed investigators to measure the current in a single ion channel.

References

1. Dale HH. Natural chemical stimulation. Edinb Med J. 1938;45:461–80.
2. Oliver G, Schäfer EO. On the physiological action of extract of the suprarenal capsule. J Physiol. 1895;18:230–76.
3. Elliott TR. The action of adrenalin. J Physiol. 1905;32:401–67.
4. Dixon WE. The mode of action of drugs. Med Mag. 1907;16:454–7.
5. Dale HH. Autobiographical sketch. Perspect Biol Med. 1958;I:125–37.
6. Ewins A. Acety-choline, a new active principle of ergot. Biochem J. 1914;8:44–9.
7. Dale HH. The action of certain esters and ethers of choline, and their relations to muscarine. J Pharmacol Exp Ther. 1914;6:147–90.
8. Dale HH. Otto Loewi, 1873-1961. Biogr Mem Fellows R Soc. 1962;8:67–89.
9. Dale HH, Dudley HW. The presence of histamine and acetylcholine in the spleen of the ox and the horse. J Physiol. 1929;68:97–123.
10. Dale HH, Feldberg W, Vogt M. Release of acetylcholine at voluntary motor nerve endings. J Physiol. 1934;86:353–80.
11. Feldberg W, Gaddum JH. The chemical neurotransmitter at synapses in the sympathetic ganglia. J Physiol. 1934;82:121–8.
12. Finger S. Minds behind the brain. A history of the pioneers and their discoveries. New York: Oxford University Press; 2000. p. 276–8.
13. Calbick CJ. Historical background of electron optics. J Appl Phys. 1944;15:685–90.

14. Eccles JC. Excitatory and inhibitory synapses. Ann NY Acad Sci. 1959;81:431–60.
15. Sourkes TL. Moritz Traube, 1826-1894: his contributions to biochemistry. J Hist Med Allied Sci. 1955;10:379–91.
16. De Palama A, Pareti G. Bernstein's long path to membrane theory: radical change and conservation in nineteenth-century German electrophysiology. J Hist Neurosci. 2011;20:306–37.
17. De Palama A, Pareti G. Bernstein's long path to membrane theory: radical change and conservation in nineteenth-century German electrophysiology. J Hist Neurosci. 2011;20:321.
18. De Palama A, Pareti G. Bernstein's long path to membrane theory: radical change and conservation in nineteenth-century German electrophysiology. J Hist Neurosci. 2011;20:322.
19. Barkan DK. Walter Nernst and the transition to modern physical science. Cambridge: Cambridge University Press; 1999.
20. Stone AG. Einstein and the quantum. Princeton: Princeton University Press; 2013. p. 144.
21. Seyfarth E-A. Julius Bernstein (1839-1917): pioneer neurobiologist and biophysicist. Biol Cybern. 2006;94:2–8.
22. Kleinzellar A. Charles Ernest Overton's concept of a cell membrane. In: Deaner DW, Kleinzellar A, Fambrough DM, editors. Membrane permeability. 100 years since Ernest Overton. New York: Academic Press; 1999. p. 1–22.
23. De Palama A, Pareti G. Bernstein's long path to membrane theory: radical change and conservation in nineteenth-century German electrophysiology. J Hist Neurosci. 2011;20:327.
24. De Palama A, Pareti G. Bernstein's long path to membrane theory: radical change and conservation in nineteenth-century German electrophysiology. J Hist Neurosci. 2011;20:327–8.
25. Lenoir T. Models and instruments in the development of electrophysiology. Hist Stud Phys Biol Sci. 1986;17:1–54.
26. Gotch F, Horsley V. Observations upon the mammalian spinal cord following electrical stimulation of the cortex cerebri. Proc R Soc Lond. 1888;45:18–26.
27. Finger S. Minds behind the brain. A history of the pioneers and their discoveries. New York: Oxford University Press; 2000. p. 240–1.
28. Becquerel E. Researches on the electrical conductivity of gases at high temperatures. Philos Mag. 1853;6:456–7.
29. Preece WH. On a peculiar behaviour of glow lamps when raised to high incandescence. Proc R Soc Lond. 1885;38:219–30.
30. Eccles WH. John Ambrose Flemming, 1849-1945. Obit Not Fell R Soc. 1945;5:231–42.
31. Chipman RA. De Forest and the triode detector. Scientific American, March 1965. p. 93–101.
32. Finger S. Minds behind the brain. A history of the pioneers and their discoveries. New York: Oxford University Press; 2000. p. 241–2.
33. Lucas K. The "all-or-none" contraction of the amphibian skeletal muscle fibre. J Physiol. 1909;38:113–33.
34. Finger S. Minds behind the brain. A history of the pioneers and their discoveries. New York: Oxford University Press; 2000. p. 239–58.
35. Eccles JC. Alexander Forbes and his achievements in electrophysiology. Perspect Biol Med. 1970;13:388–404.
36. Gasser HS, Erlanger J. A study of the action current of nerve with the cathode ray oscillograph. Am J Physiol. 1922;62:496–524.
37. Adrian ED. The impulse produced by sensory nerve endings. J Physiol. 1926;61:49–72.
38. Zottermannn Y. Touch, tickle and pain. Oxford: Pergamon Press; 1969. p. 220.
39. Adrian ED, Bronk DW. The discharge of impulses in motor fibres. Part I J Physiol. 1928;66:81–101. Part II J Physiol. 1929;67:119–51.
40. Adrian ED. Afferent areas in the brain of ungulates. Brain. 1943;66:89–103.
41. Adrian ED. The action of the olfactory organ. J Laryngol Otol. 1956;70:1–14.
42. Hodgkin AI. Evidence of electrical transmission in nerve. Part I and II. J Physiol. 1937;90:183–232.
43. Young JZ. Structure of nerve fibres and synapses in some invertebrates. Cold Spring Harb Symp Quant Biol. 1936;4:1–6.

44. Cole KS, Curtis HJ. Electric impedance of the squid giant axon during activity. J Gen Physiol. 1939;22:649–70.
45. Hodgkin AL. The relationship between conduction velocity and the electrical resistance outside a nerve fibre. J Physiol. 1939;94:560–70.
46. Schwiening CJ. A brief historical perspective: Hodgkin and Huxley. J Physiol. 2012;590: 2571–5.
47. Hodgkin AL, Huxley AF. Action potentials recorded from inside a nerve fibre. Nature. 1939;144:710–1.
48. Piccolono M. Animal electricity and the birth of electrophysiology: the legacy of Luigi Galvani. Brain Res Bull. 1998;46:381–407.
49. Hodgkin AL, Huxley AF. A quantitative description of membrane current and its application in conduction and excitation in nerve. J Physiol. 1952;117:500–44.
50. Verkhratsky A, Parpura V. History of electrophysiology and the patch clamp. In: Martina M, Taverna S, editors. Patch-clamp methods and protocols, Methods in Molecular Biology, vol. 1183. New York: Humana Press; 2014. p. 1–19.
51. Strickholm A. Impedance of a small electrically isolated area of the muscle cell surface. J Gen Physiol. 1961;44:1073–88.
52. Neher E, Lux HD. Voltage clamp on Helix pomatia neuronal membrane; current measurement over a limited area of the soma surface. Pflugers Arch. 1969;311:272–7.
53. Neher E, Sakmann B. Single-channel currents recorded from membrane of denervated frog muscle fibres. Nature. 1976;260:799–802.
54. Neher E, Sakmann B. The patch clamp technique. Sci Am. 1992;266:44–51.
55. Sakmann B, Neher E. Patch clamp techniques for studying ionic channels in excitable membranes. Annu Rev Physiol. 1984;46:455–72.
56. Edwards FA, Konnerth A, Sakmann B, Takahashi T. A thin slice preparation for patch clamp recordings from neurons of the mammalian central nervous system. Pfugers Arch. 1989;414:600–12.
57. Pavlov IP. Conditioned reflexes. An investigation of the physiological activity of the cerebral cortex (translated by Anrep GV). London: Oxford University Press; 1927.
58. Milner OM, Milner B. Donald Olding Hebb: 22 July 1904-20 August 1985. Biogr Mem Fellows R Soc. 1996;42:193–204.
59. Hebb DO. The organization of behavior. New York: Wiley; 1949.
60. Kandel ER. In search of memory. New York: WW Norton & Co; 2006.

Chapter 11
Brain Waves and Brain Stimulation

> *There are in the temporal lobe innumerable neurone patterns which constitute records of memory. The electrode [electrical stimulation] causes the patient to have a psychical experience, like the memory of some past event and he can describe it as he lies on the operating table.*
>
> *Wilder Penfield, 1954*

Abstract Although spontaneous and evoked electrical activity was identified in the brains of animals and human subjects in the late nineteenth century, it was not until the electronic advances of the early twentieth century that the clinical application of electroencephalography (EEG) was appreciated. Hans Berger first identified rhythmic alpha activity in human subjects and Adrian later concluded it was a default rhythm that occurred when subjects were inattentive. EEG proved most useful for diagnosing seizure disorders and neurosurgeons soon began using it to localize the seizure discharge prior to surgical resection. William Penfield revolutionized the surgical treatment of epilepsy at the Montreal Neurological Institute using electrical stimulation of the cortex in awake patients to decide which areas were safe to resect. Although initially used to treat schizophrenia, electroconvulsive therapy proved most effective for treating severe depression. Major concerns about safety remain despite continued use. Because of its rapid action compared to antidepressant medications, it is most used for treating patients at high risk for suicide. Deep brain electrical stimulation has a checkered past regarding both science and ethics, but it has established benefit for treating movement disorders such as dystonia and Parkinson disease. The benefits for treating psychiatric disorders are promising but to be established.

Keywords Brain waves · Oscillograph · Psychomotor seizures · Polygraph · Microelectrodes

After peaking in popularity at the turn of the twentieth century, the use of electrotherapy in neurology began to gradually decline until by the mid-1930s it was largely abandoned as an accepted treatment for any neurological disease. This decline correlated with improved understanding of how neurons transmit signals in the central and peripheral nervous system and realizing the futility of bombarding

R. W. Baloh, *Brain Electricity*, https://doi.org/10.1007/978-3-031-62994-5_11

the nervous system with electricity in a generalized fashion. At the same time, neurologists saw potential uses for electricity in the diagnosis of a range of diseases involving the peripheral and central nervous system.

Electrical stimulation of peripheral nerves could localize disease to nerves or muscles and recording the spontaneous electrical activity of the brain could identify seizure foci and areas of impaired function. The introduction of electroconvulsive therapy provided a new treatment for a range of psychiatric disorders previously considered untreatable, but side effects were worrisome, and the mechanism of action was ill-defined.

In the latter half of the twentieth century with the technological advances in electronics, surgeons began to implant electrodes in localized regions deep within the brain to treat a variety of neurological and psychiatric diseases. As the electrodes, electronics, and batteries became smaller surgeons recorded from and stimulated multiple neurons at the same time making it possible to alter communication between different brain structures. Feedback loops between the frontal cortex and the deep brain structures were found critical for much of normal human behavior and altering these feedback loops was used to treat anxiety and depression.

Early Evoked Potentials and Electroencephalography

In the late nineteenth century as researchers identified electrical activity in peripheral sensory and motor nerves carrying signals to and from the brain, others began searching for electrical activity originating in the brain [1]. One of the first to address this issue was a young physiologist at the Royal Infirmary School of Medicine in Liverpool, Richard Caton. Having received a small grant from the British Medical Association, Caton used William Thomson's mirror galvanometer and Du Bois-Reymond's non-polarizing electrodes to observe both spontaneous and evoked electrical activity in the brain of a variety of animals, mostly rabbits. In 1875, he made a report entitled *The Electric Currents of the Brain* before the Association in Edinburgh and published a note in the proceedings: "*In every brain hitherto examined, the galvanometer has indicated the existence of electrical currents…Feeble currents of varying directions pass through the multiplier when the electrodes are placed on two points of the external surface, or one electrode on the grey matter, and one on the surface of the skull…Impressions through the senses were found to influence the currents of certain areas,* e.g., *the currents of that part of the rabbit's brain which Dr. Ferrier has shown to be related to movements of the eyelids, were found to be markedly influenced by stimulation of the opposite retina by light*" [2].

Unfortunately, Caton had no way to record the electrical activity, so his results relied on written descriptions. Over the next several years, he attempted to induce evoked potentials in many animals including a few monkeys but reported only success with sensory stimuli to a limb and bright lights but no response to sounds and odors. The clear visual-evoked responses helped convince skeptics who suggested he was simply recording muscle activity.

In 1890, a Polish graduate student at the University of Cracow, Adolph Beck decided to use Caton's work as the focus of his doctoral thesis under the supervision of Professor Napoleon Cybulski. Like Caton, Beck did not have a recording devise but made detailed maps of the rabbit and dog cerebral cortex where evoked potentials were recorded and drawings of the evoked potentials. In addition to good responses to light, Beck identified an area on the temporal cortex that responded to loud noise. Like Caton, he described spontaneous oscillations in electrical activity onto which the evoked potentials were superimposed.

The most localized early evoked potentials were reported by a young Russian neurophysiologist Vladimir Efrimovich Larionov in his doctoral thesis under the supervision of the famous Russian neurophysiologist Vladimir Mikhailovich Bechterev. Larionov focused on auditory evoked potentials using tuning forks with different frequencies for his auditory stimuli. He borrowed a galvanometer from the physics department and carefully sketched the evoked potentials. He made maps of the cat temporal cortex showing the highly localized responses to different frequency sounds in an article published in *Pflügers Archiv* in 1899 (Fig. 11.1).

The first photographic records of the electroencephalogram and evoked potentials were provided by the Ukrainian neurophysiologist Vladimir Vladimirovich Pravdich-Neminisky in 1913. Pravdich-Neminisky was born and educated in Kiev, but after his first year of medical school, he transferred to the University of Kazan where Larionov had conducted his evoked potential work. But just as his career in neurophysiology was beginning to develop, he got into trouble with the Russian government for taking part in student demonstrations and he had to return to Kiev. It was there that he made his famous recordings.

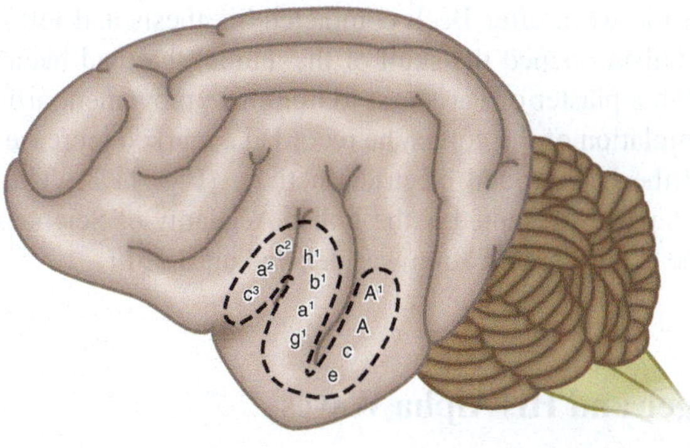

Fig. 11.1 Vladimir Larionov's sketch of the cat's brain showing the points on the temporal lobe that responded to different auditory tones. (From Brazier [1] with permission)

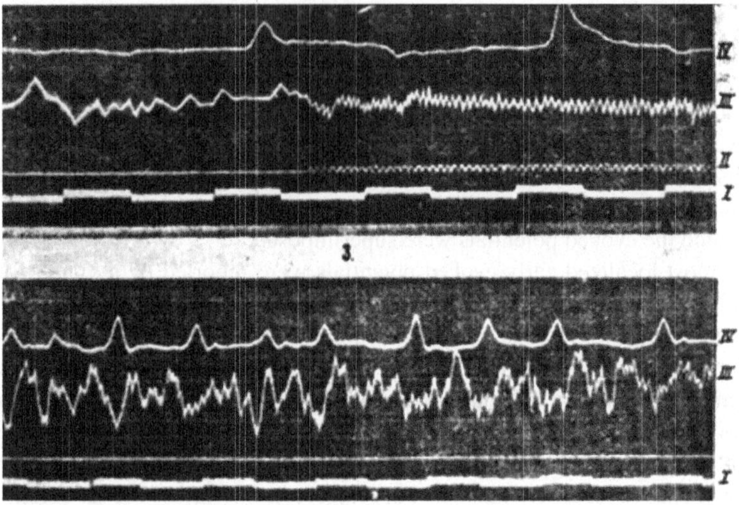

Fig. 11.2 The first published electroencephalographic (EEG) recording of an induced seizure in a dog. The traces run from right to left, I—time (seconds), II—electrical stimulus, III—cortex, IV—heart rate. The lower strip follows the upper one. (From Brazier [3] with permission)

Pravdich-Neminisky used an Einthoven string galvanometer with moving photographic paper to record the brain electrical activity. He prepared the anesthetized animals (mostly dogs) by injecting curare to block all muscle activity that might interfere with the brain electrical activity. The recordings showed a high-frequency background oscillation with distinct slow wave responses to electrical stimulation of the sciatic nerve in the region of the motor cortex.

The first electroencephalographic recording during an induced seizure was made by Napoleon Cybulski the mentor of Anton Beck at the University of Cracow, Poland [3]. Impressed with Beck's findings, Cybulski decided to purchase a camera and continue the work after Beck completed his thesis and left Cracow. With his assistant, Cybulski opened the skull of the animal that had been anesthetized and immobilized in a plaster of Paris cast to minimize movement artifacts. After direct electrical stimulation of the cortex, he recorded a marked increase in amplitude and frequency in the background electrical activity (Fig. 11.2). The recordings were published in a paper presented at the Polish Academy of Sciences in 1914 but due to the war, the general publication was delayed until 1917.

Hans Berger and His Alpha Waves

These early studies of evoked potentials and electroencephalography focused on animal experiments with minimal emphasis on possible clinical applications. In 1924, a little known German psychiatrist and neurologist working in Jena, Hans

Berger, attempted the first electroencephalographic recording in a human patient with a skull defect. According to neuroscience historian Mary Brazier, "*Although initially unsuccessful (probably because his wire electrodes had too high an impedance for this string galvanometer) this experiment proved to be the seed of the enormous growth of knowledge that today constitutes clinical electroencephalography*" [4].

Berger was born in Neuses which is now Coburg in Bavaria, Germany. He was initially interested in mathematics and wanted to become an astronomer but after a semester at the University of Jena, he enlisted in the cavalry of the military service. During a training exercise, he fell from his horse and was nearly crushed by a horse-drawn cannon but fortunately escaped without injury. Many miles away his sister had a premonition that he was in danger and convinced their father to telegraph Hans. This event would have a profound effect on Berger for the rest of his life. Berger later wrote: "*It was a case of spontaneous telepathy in which at a time of mortal danger, and as I contemplated certain death, I transmitted my thoughts, while my sister, who was particularly close to me, acted as the receiver*" [5]. Like many scientists of the era, Berger became a true believer in the psychic world and telepathy which turned out to be a major motivation for his studies of *brain waves*.

According to his diaries, Berger began working on evoked potentials and electroencephalography in animals in 1910 but he worked in isolation and kept his findings to himself. He began recording in human subjects with skull defects in 1924 using an Edelmann string galvanometer that gave him inconsistent results [6]. He purchased a new Siemens double-coil galvanometer in 1926 that could be used with low-impedance electrodes, and he began to have more consistent reliable results. He recorded the output, up to 3 s with a photographic technique. He began with needle electrodes made of platinum or zinc-plated steel but later turned to large plate surface electrodes fastened to the front and back of the head. He recorded spontaneous activity with the subject's eyes open and closed and noted the effect of mental attention and the response to loud noises and painful stimuli.

It was not until 1929 that Berger began publishing his findings. He observed brain waves of about ten cycles per second when subjects were relaxed with eyes closed which he called *alpha waves* (Fig. 11.3). When the subjects opened their eyes, the rhythmic alpha waves were replaced by high-frequency low-amplitude activity, *beta waves*. Since he could record alpha waves from any part of the skull, he felt that they must be the result of rhythmic discharges from brain cells over the entire cortex. He saw the rhythmic alpha waves as the fundamental activity of the cerebral cortex that was blocked by visual input with eyes open.

Berger's initial efforts at electroencephalographic recording in an epileptic patient were marred by prominent movement artifacts, but he did note a general loss of amplitude during the stupor that followed the seizure. He also reported that patients with long-standing uncontrolled seizures exhibited overall slowing in their baseline recordings. He was the first to recognize that intermittent sharp waves on interictal recordings indicated a propensity for developing seizures. Berger noted that electroencephalographic recordings in young children were different from those in adults and he was the first to record three per second large amplitude waves

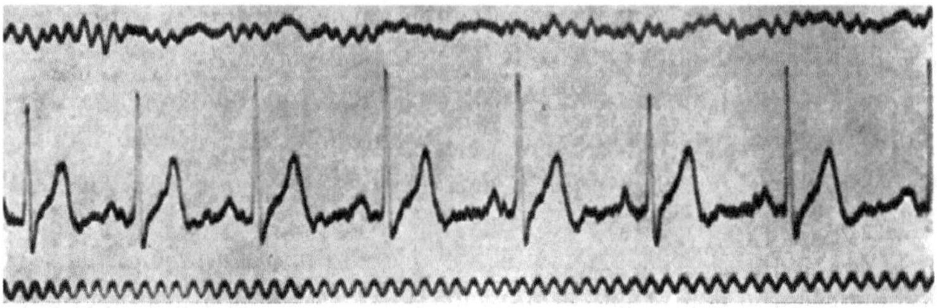

Fig. 11.3 Early Berger recording of alpha waves in a normal human subject. Upper trace—EEG, middle trace—EKG, lower trace—10 cycle/s signal. (From İnce et al. [62] with permission)

associated with absence spells in children, petit mal seizures. His recording did not have the resolution to see spikes associated with the waves as will be described later by Gibbs and associates (see below).

Over the next several years, Berger published more than 20 reports on electroencephalography, but the reception was not very enthusiastic particularly by his fellow Germans in the medical and scientific establishment. One of his problems was he had an almost complete lack of knowledge regarding the technical aspects of electricity and electronics. It was only near the end of his career that he sought help from an electrical engineer. Although Berger initially came into conflict with the Nazi regime in the mid-1930s, his relationship with the Nazis was complicated and he later participated in some questionable activities. He had a history of depression, and, in 1941, he became despondent and committed suicide by hanging.

One scientist who was impressed with Berger's work was Edgar Adrian at Cambridge [7]. After reading Berger's publications, Adrian decided to try some recordings on his own. With Adrian's electronic experience and the technical expertise of a colleague, Brian Mathews, the team began recording brain waves of anesthetized rabbits and cats. Mathews designed a new oscilloscope for recording the brain waves. They showed that the rhythmic activity was made up of the synchronized discharge of individual nerve cells. Adrian used the equipment to record sensory evoked potentials and localize the sensory cortex in the animals as described earlier in Chap. 10.

The team then moved on to human subjects where their oscilloscope recordings were a major improvement over Berger's photographic recordings. They showed that Berger's alpha rhythm that was abolished with opening the eyes could be reinstated if the subject stared at an empty visual space. On the other hand, they noted that the alpha rhythm that occurred with eye closure disappeared if the subject performed a mental task such as imagining a visual seen. Adrian interpreted these findings to suggest that alpha rhythm was a default rhythm that occurred when subjects were inattentive and not due to lack of sensory input. "*We regard the effect as due to a spontaneous rhythmic activity of a group of cortical cells in some part of the occipital lobe. These tend to beat synchronously when they are undisturbed, but visual activity or widespread non-visual activity in the brain breaks up the rhythm by exposing cells to a mosaic of excitations which makes synchronous action impossible*" [8].

Clinical Electroencephalography (EEG)

An English psychologist who was part of the Adrian team, W. Grey Walter was instrumental in introducing electroencephalography into clinical practice in England. In 1936, Walter reported that focal slow waves (delta waves) could be useful for localizing brain tumors [9]. He went on to describe EEG findings in seizures and a variety of mental disorders. Walter also participated in the early use of electrocorticography (ECoG) during brain surgery.

In America, the neurophysiologist Alexander Forbes, mentioned in Chap. 10 for his collaboration with Adrian before and after the war, played a major role in developing electrophysiology at the Harvard Medical School in Boston. In 1918, obviously unaware of Berger's early studies, Forbes and the neuropsychiatrist Donald MacPherson used a string galvanometer Forbes brought back from Cambridge to make photographic recordings of rhythmic waves from the exposed cerebral cortex of cats. These were never published since Forbes thought they were due to tremors from nearby construction [10]. Forbes later studied the effects of anesthetic agents on spontaneous and evoked electrical activity, but his main work of clinical relevance was conducted with a young Harvard physician, Hallowell Davis.

After graduating from Harvard Medical School in 1922, Davis spent a year working with Adrian at Cambridge before returning to work with Forbes on nerve physiology. In 1929, Forbes and Davis decided to focus on the electrophysiology of the auditory system. They developed a state-of-the-art laboratory with the latest vacuum tube amplifiers and a cathode ray oscilloscope. They also developed specialized sound-generating equipment and a sound-insulated animal room. Davis took the lead and along with a variety of graduate students recorded auditory evoked potentials in the cochlea, brainstem, and cortex of the cat. As an electrode for localized unit recordings, they used an insulated silver wire with just the tip exposed. In collaboration with the neurosurgeon Harvey Cushing, Davis and colleagues recorded auditory, somatosensory, and visual-evoked potentials in patients undergoing brain surgery.

Davis and Forbes first became aware of Berger's recordings of human electroencephalography in 1933. Their initial reaction was that the recordings must be an artifact. Based on their prior evoked potential work, they thought that it was extremely unlikely that cortical electrical activity could be synchronized to produce such a slow, regular rhythm particularly without exposing the brain. But their equipment was certainly capable of recording the 10 per second alpha waves if they were real so he suggested that two young colleagues give it a try. Davis later noted that after they tried to record Berger's waves for a few weeks they became frustrated and reported back: "*You were right, chief. We have stuck needles in each other's scalps (vertex and occipital). The baseline is unsteady, but we can't see anything like the*

10/sec rhythm on the scope. But come and see if we are doing it right before we say anything about it" [11].

Davis went to the laboratory with his young colleagues and, indeed, when one of them was hooked up with needle electrodes in the scalp, no clear alpha waves were seen on the oscilloscope. But just to be sure Davis asked them to hook him up and to everyone's surprise Davis had very clear 10/s alpha waves when he closed his eyes. When they tested other members of the laboratory staff, about half had strong alpha waves (*Bergers'*) and half had weak alpha waves (*non-Bergers'*). Davis speculated that those with strong alpha waves might have been more relaxed than those with weak waves [12].

A major problem that these pioneer electroencephalographers faced was the shortcomings of the photographic oscillograph recordings. There was no "real time" feedback and no memory capability other than with tedious photographs. By the fall of 1933, Davis was convinced that an ink-written trace was critical if the field was to advance, and he turned to the Physiology Department's engineer E. Lovett Garceau to make such a device. Garceau took a single-channel Western Union Morse code graphic recorder and over the next 6 months converted it into an ink-writing single-channel oscillograph. He replaced the old magnets and springs with stronger magnets and springs to increase the frequency response and on July 12, 1934, the first recording of a human electroencephalogram on an ink recorder was made from the head of Hallowell Davis.

Two epileptologists working at the Harvard Medical School, Frederic Gibbs and William Lennox, observed a demonstration of the ink-writing oscillograph in Davis' laboratory and immediately saw potential clinical applications [13]. Gibbs obtained his medical degree from Johns Hopkins University in Baltimore and moved to the Harvard affiliated Boston City Hospital Neurological Unit where he met Lennox who shared an interest in epilepsy. In 1930, Gibbs married Lennox's research assistant, Erna Leonhardt and the three, Gibbs, Gibbs, and Lennox worked together on epilepsy research for decades. When Frederic Gibbs approached Davis regarding the clinical applications of his new ink-writing electroencephalogram, Davis was supportive and had one of his graduate students show Gibbs how to make recordings. In October 1934, Gibbs received a formal appointment as a fellow in the Physiology Department at Harvard.

The research teams first breakthrough came in December 1934. Lennox had selected some children with brief absence spells (petit mal seizures) to record because they would be still during the seizures so no problem with movement artifacts. To everyone's surprise, they all exhibited a highly consistent 3/s spike and wave pattern throughout the absence spells (Fig. 11.4). This finding proved to be pathognomonic of this common seizure disorder in young people. The team then went on to study the effects of sleep, nitrogen breathing, and hyperventilation on the human EEG and recorded several grand mal epileptic seizures.

Some of the most frightening and bizarre seizures are psychomotor seizures or now commonly called partial complex seizures. Patients experience strange auras manifested by odors or tastes, sudden emotions such as fear or joy, feelings that what's happening has happened before (déjà vu) and even vividly recalled prior

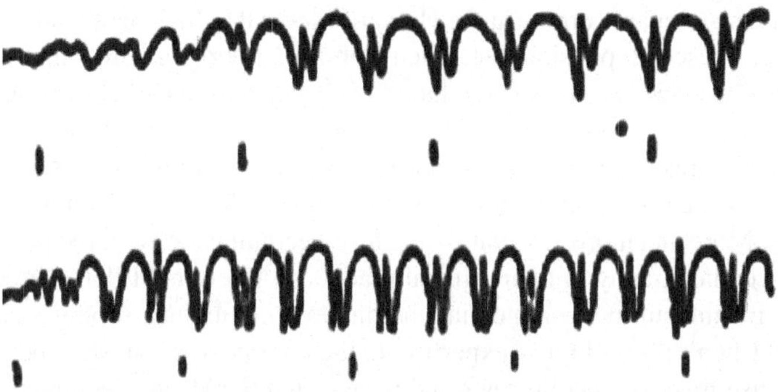

Fig. 11.4 Early EEG recordings 3/s spike and wave discharge during petit mal seizures in two different patients. Vertical lines mark seconds. (From Gibbs et al. [63] with permission)

recollections. The auras are followed by clouding of consciousness and automatisms such as lip smacking, chewing, swallowing, and repetitive finger motions. Hughlings Jackson first described these dreamy states calling them *uncinate fits* and localizing them to the temporal lobe because this was where the sense of smell and taste had been localized in animals.

In 1936 Gibbs, Lennox, and Gibbs reported the first EEG recordings in patients with partial complex seizures and they coined the term "psychomotor seizures" based on the combination of psychic and motor symptoms [14]. They noted a high rate of psychiatric disorders in their patients with psychomotor seizures and described what they considered the characteristic EEG discharge, incorrectly localized to the medial frontal cortex because of the lack of active temporal electrodes and routine use of a reference ear electrodes. It was almost a decade later and after Gibbs had moved to the University of Illinois at Chicago that he and colleagues recognized that the epileptic discharge originated from the temporal lobe. They also made the important observation that the discharge with psychomotor seizures was activated by sleep EEG studies, markedly improving the diagnosis in these patients.

The Grass Instrument Company and Foundation

In their early years, Gibbs and Lennox had been recording on the single-channel ink-writing recorder of Garceau and it became obvious that more channels were needed if they hoped to localize epileptic discharges. In the May of 1935, Gibbs contracted with a young electrical engineer, Albert Grass to build a 3-channel pen-writing electroencephalographic recorder for clinical use. Grass had just graduated from MIT and stayed on to work on simulated earthquakes. Albert and his brother Everett built the new electroencephalographic machine in their father's basement over a 3-month period. As a result, Albert was hired as a part time Research Instrument Engineer for the Department of Physiology at Harvard.

When Grass began working at the Harvard Medical School in the fall of 1935, he met a young research physiologist, Ellen Robinson, working with Hallowell Davis and the two married less than a year later [15]. Ellen was in a postgraduate program in the Department of Biology making recordings of the auditory cortex in rabbits. As word of Grass' three channel pen-writing electroencephalographic machine spread in the neuroscience community, the demand for such machines increased markedly. More and more clinicians were also becoming aware of the potentials for electroencephalography in their clinical practices. Albert and Ellen started a small business manufacturing commercial instruments combining Alberts engineering skills and Ellen's physiological expertise. In these early years, most of their employees and customers were colleagues and friends and they loaned equipment to neurophysiologists throughout the world. They formed a close relationship with the Marine Biological Laboratory at Woods Hole that persisted when they formalized the Grass Instrument Company and later the Grass Foundation.

The Grass Instrument Company dominated the market for research and clinical electroencephalographic equipment throughout the twentieth century and the Grass Foundation provided key support with grants and fellowships in electrophysiology.

I have fond memories of the Grass 4 channel ink-pen recorder that I used when I began working on the neurophysiology of the vestibular system in the early 1970s. The machine had remarkable versatility and reliability.

Electrocorticography and Epilepsy Surgery

The German neurologist and neurosurgeon Otfrid Forester is often considered the father of epileptic surgery not only for his own pioneering work but because he trained several of the later giants in the field (including the North Americans Wilder Penfield and Percival Bailey) [16]. Forester was born in Breslau, Germany (now Wroclaw, Poland) in 1873. He received his medical degree in Breslau in 1897, and after studying neurology in Paris with Joseph Dejerine for 2 years, he returned to Germany to work with Karl Wernicke on functional localization in the brain. They published their famous anatomical atlas of the brain in 1903. In addition to his academic endeavors, Forester spent 1922–1924 in Russia as Lenin's chief physician after his stroke, so his name is prominent in Russian history with his signature on all the documents related to Lenin's death and postmortem examination.

Forester became interested in epileptic surgery during and after the First World War. In 1915, he described innovative techniques for treating gunshot wounds to nerves, the spinal cord, and brain. Several soldiers with post-traumatic epilepsy were referred to Forester in Breslau for possible surgical treatment, but at the time he had no way to localize the seizure other than the chance occurrence of a focal onset. In 1924, Forester reported that hyperventilation could induce a seizure in a patient with a localized lesion. He also showed that electrical stimulation around a post-traumatic scar in the brain or gentle traction of the scar could induce a seizure. He used local anesthesia during surgery so that the patient remained conscious and could report experiences with electrical stimulation of the exposed human cortex. This allowed him to protect intact nearby sensory and motor cortex as he extracted the scar tissue.

In 1928, a young American neurosurgeon, Wilder Penfield spent 6 months working with Forrester and learning his surgical techniques. They published a paper in the journal *Brain* in 1930 summarizing their experience operating on patients with post-traumatic epilepsy and speculated that the scar tissue and encased blood vessels could cause a vasomotor reflex that somehow initiated the seizure [17]. Penfield would go on to develop and become the first Director of the Montreal Neurological Institute and Hospital at McGill University where he revolutionized the surgical treatment of epilepsy.

Penfield was born in Spokane Washington in 1891 and spent much of his young life in Wisconsin [18, 19]. He attended Princeton University where he excelled in American football, and after graduating, he received a Rhodes Scholarship to study neuropathology with Charles Sherrington at Oxford. After surgical training in Boston and New York, he obtained a position at the Neurological Institute of New York where he performed his initial epilepsy surgeries. After returning from training with Forrester, Penfield accepted a position at McGill University in Montreal, Quebec, the first neurosurgeon in the city. Six years later in 1934, with support from the Rockefeller Foundation, Penfield founded the Montreal Neurological Institute and became a Canadian citizen.

The Montreal Neurological Institute

In 1938, Penfield was joined in Montreal by the American neurophysiologist Herbert Jasper who along with Gibbs, Gibbs, and Lennox was a pioneer in American electroencephalography. Penfield was initially skeptical of the benefits of electroencephalography, but Frederic Gibbs convinced him of its importance and suggested that Jasper would be the ideal person for him to recruit and join him in Montreal. Jasper obtained his Ph.D. in neurophysiology from the University of Iowa in 1931 and after studying abroad for a few years he established a neurophysiology laboratory at Brown University in Rhode Island [20]. Working with the psychologist, Leonard Carmichael, Jasper confirmed many of the findings of Berger and Adrian and the two published the first major article on electroencephalography in America in 1935. They used a Westinghouse mirror galvanometer, direct current amplifiers with a frequency response about 2 cycles/s and photographic recordings. Jasper later developed a technique to localize brain electrical activity, by a process of *triangulation* using bipolar electrodes.

With Jasper, Penfield used the so-called *Montreal Procedure* to treat patients with medically intractable seizures by selectively destroying the epileptic focus in the brain where the seizures originated. After exposing the suspect area, they would stimulate the brain with electrical probes with the patients' conscious under local anesthesia and record their verbal responses. They always began stimulating with an alternating current of 60 cycles/s at a half volt. With this technique, they localized the seizure focus and preserved nearby regions that were functioning normally. Penfield also developed the *Penfield Dissector,* a scalpel with a long thin handle and a flattened blunt blade at the end for scooping out the damaged tissue.

With the many electrical stimulations in patients over the years, Penfield and Jasper were able to map the human cerebral cortex in detail. Not only did they map the motor and sensory areas but also higher-level functions such as auditory and visual recollections. Stimulation of the temporal lobes produced the most varied and bizarre responses. Patients reported hearing specific songs or seeing familiar visual scenes and experiencing a range of emotions from fear and loneliness to a sense of strangeness or familiarity. Recall of a memory experience was often tied to a specific emotion.

They published their landmark book *Epilepsy and the Functional Anatomy of the Human Brain* in 1951 and a revised more complete 2nd edition in 1954 [21]. The book is a remarkable compilation of case histories, electroencephalographic recordings, detailed descriptions of electrical stimulation results, and anatomical localizations. Penfield wrote the clinical chapters describing case histories and surgical procedures and Jasper wrote the chapters on electroencephalography and electrocorticography. Scalp and cortical electroencephalograms were recorded on an 8-channel ink-writing oscillograph. For routine scalp recordings, they used 16 sliver cup-like electrodes filled with electrolyte jelly fastened with collodion and distributed roughly equally over the cerebral cortex. For cortical recordings, they developed an array of eight flexible electrodes mounted on a horizontal bar that was clamped to the skull at the edge of the craniotomy. The electrodes consisted of silver wires insulated except at the tips.

In a joint chapter entitled *Centrencephalic or Highest Level Seizures,* Penfield and Jasper speculated that some common generalized seizure types such as petit mal and grand mal seizures that begin with loss of consciousness may result from discharge within a subcortical neuronal system that projects diffusely throughout the cortex. The neurophysiologists Horace Magoun and Giuseppe Moruzzi at Northwestern University in Illinois had just described an ascending reticular activating system (RAS) consisting of groups of interconnected neurons from the low brainstem to the midline thalamus receiving collateral sensory input and projecting throughout the cortex [22]. When they electrically stimulated the RAS, it caused the animals to be alert and abolished rhythmic alpha activity on electroencephalography replacing it with low-amplitude fast activity just as mental activity abolished alpha activity in human subjects. Destroying the RAS in animals caused them to enter a persistent coma-like state. Penfield and Jasper suggested that an electrical discharge in the RAS could explain the immediate loss of consciousness and the generalized nature of the so-called centrencephalic seizures.

The most common surgery performed at the Montreal Institute was temporal lobectomy (removal of a part or all of a temporal lobe) for treating intractable psychomotor seizures [23]. They operated on 68 patients with psychomotor epilepsy between January 1939 and April 1949. There was one operative death, seven were explored without a surgical procedure, and five were found to have an occult tumor that was removed. Of the remaining 55, 53 had localized atrophy (scarring, sclerosis) and 3 hemangiomas (benign tissue growths) that were excised. As to the cause of the localized scarring of the mesial temporal lobe found in most patients, Penfield speculated that they may have had a compromise in the blood flow to this region due to a difficult birth or childhood injury. Fifty-one of the 55 were contacted regarding outcome 1 to 10 years after the procedure. Twenty-two percent reported little or no improvement, 25% reported at least 50% improvement, and 53% were cured. These were remarkable results considering the innovative nature of the operations.

Based on his electrical probing of the temporal lobe, Penfield developed a particular interest in the mechanism of memory storage. He concluded that "*there are in the temporal lobe innumerable neurone patterns which constitute records of memory. The electrode* [electrical stimulation] *causes the patient to have a psychical experience, like the memory of some past event and he can describe it as he lies on the operating table. The hallucinations thus produced may be auditory or visual or both, but it is neither a single sound or frozen picture…It may be the reproduction of a song he has heard. If so, he 'hears' it from beginning to end, not all at once. The experience may seem to the patient a dream, rather than an accurate memory. But it is a dream in which familiar places are seen and well-known people speak and act. Such hallucinations, or memories, or dreams continue to unfold slowly while the electrode is held in place. They are terminated suddenly when the electrode is withdrawn*" [24].

Penfield emphasized that the patient not only had a playback of prior experiences and events but also a reproduction of the thinking (interpretation) related to the event and the emotions it evoked. This suggested a coordination or integration of a wide range of nerve impulses from both hemispheres. He speculated that such neuronal coordination might take place within the centrencephalic system, "*the formation of memory patterns depends upon neuronal currents that pass outward from the centrencephalic system to the temporal cortex of the two hemispheres*" [25]. Penfield noted that fortunately only a few of his patients operated on for temporal lobe seizures developed postoperative memory loss. At the time, it was well known that damage to both temporal lobes in monkeys and humans resulted in severe memory loss while one-sided damage had only subtle effects on memory. Penfield speculated that in his patients with postoperative memory loss recurrent seizures on one side might somehow damage the other side over time.

With the good results reported by Penfield and Jasper, neurosurgical centers using electrocorticography for treatment of psychomotor seizures began popping up around the world. In 1944, Frederic and Erna Gibbs left Boston and moved to the University of Illinois at Chicago and teamed up with neurosurgeon Percival Baily to set up an epilepsy surgery program [26]. Unlike Penfield who based his surgery on identifying structural pathology, Baily was willing to base surgery on finding localized spike activity on electroencephalography. He typically removed 5–6 cm from the anterior tip of the temporal lobe. Most pathological specimens contained sclerosis and success rates were comparable to those reported by Penfield and Jasper.

Early Fever and Convulsive Therapy

Electroconvulsive therapy developed completely independent of electrotherapy. It became popular as interest in electrotherapy waned. The rationale for the use of electroconvulsive therapy was based on the clinical observation that some patients with mental illness seemed to improve if they suffered an epileptic seizure. The idea of treating one disease with another disease can be traced to ancient times when Hippocrates and Galen noted that certain illnesses were cured if the patient developed a high fever from another illness. Typhoid, erysipelas, and cholera were all used for the so-called fever therapy.

In the late nineteenth century, a young Austrian psychiatrist, Julius Wagner-Jauregg observed that a woman in his clinic diagnosed with severe psychosis was cured when she developed a high fever associated with erysipelas, an acute bacterial skin infection [27]. After writing a review article on the effect of fever on psychosis, he abandoned the subject for several decades only to return in 1917 when he treated a series of patients diagnosed with general paresis of the insane by injecting them with blood from a patient with malaria. General paresis of the insane is a form of psychosis associated with end-stage neurosyphilis characterized by delusions, paralysis, and dementia. At the time, it represented up to 10% of psychiatric admissions to mental institutions.

Wagner-Jauregg's fever treatment proved to be the first effective biological treatment for psychiatric symptoms. Of the first nine patients treated, six showed significant improvement and three were reported cured. By the end of 1921, he reported results in over 200 patients with 50 recovering enough to return to work. After 7–12 days of recurrent fevers, he administered quinine for the malaria symptoms and Neosalvarsan, the main drug for treating syphilis at the time. For this work, Wagner-Jauregg was the first psychiatrist to receive a Nobel Prize in Medicine in 1927. His findings were replicated by psychiatrists in Europe and America, but his reputation suffered when late in his life he joined the Nazi Party and endorsed the National Socialist agenda.

The success of Wagner-Jauregg's fever treatment paved the way for other biological treatments of psychosis. Another Austrian psychiatrist who later moved to America, Manfred Joshua Sakel, introduced insulin shock therapy for psychosis in the late 1920s. He injected large doses of insulin into to the bloodstream to produce hypoglycemic coma. Some of the patients developed convulsions but Sakel felt that the treatment worked by decreasing glucose in the parts of the brain that produced psychosis [28].

After graduating from the University of Budapest School of Medicine in 1921, Ladislas Meduna became interested in neuropathology working under Professor Karoly Schaffer first at the Hungarian Interacademic Institute for Brain Research and then at the Psychiatric Institute in Budapest [29]. Meduna focused on the study of glia cells, the supporting cells of neurons in the brain. He observed that the brains of patients with epilepsy appeared to have excessive glia cells whereas those from patients with schizophrenia had fewer glia cells than normal. He was aware of sensational reports of patients with schizophrenia being cured after a seizure and that the incidence of epilepsy in hospitalized patients with schizophrenia was lower than expected so he decided to try to treat patients with schizophrenia by inducing seizures. His hypothesis for the treatment was that seizures would increase glia cells in the brain.

After trying a variety of chemicals in animals to induce seizures, he chose camphor dissolved in oil for his first studies in patients. Meduna began his patient experiments with camphor in January 1934 noting that only about a third of patients experienced a seizure. But there were a few dramatic responses in those who experienced a seizure particularly those with catatonia. Meduna described the response of a 33-year-old man with severe catatonia: "*...on the morning of February 10, 1934 (two days after the fifth injection) for the first time in four years, he got out of bed, began to talk, requested breakfast, dressed himself without any help, was interested in everything around him, and asked how long he had been in the hospital. When we told him he had spent four years, he did not believe it*" [30].

Meduna soon switched from camphor to Metrazol (Cardiazol) since it more rapidly and reliably induced seizures. After treating more than 100 patients with schizophrenia with medically induced seizures, Meduna reported that about half had a good result [31]. The treatment seemed to work better in patients with new onset schizophrenia than in patients with long-standing disease. Metrazol did have bothersome side effects. Many patients complained of severe anxiety and a sense of smothering prior to losing consciousness with a seizure.

The Beginnings of Electroconvulsive Therapy

Ugo Cerletti, the chair of Neuropsychiatry at the University of Rome, had witnessed Meduna's pharmacologically induced seizures in Vienna and began treating some of his patients with schizophrenia using Metrazol. He was concerned about the frightening symptoms reported by the patients before seizure onset, and he began to look for another method of inducing seizures. He may have gotten the idea to use electricity after seeing pigs electrocuted at a local slaughterhouse. The electricity was used to induce an epileptic coma prior to slaughtering them by slashing their throats.

Cerletti enlisted the help of a young research technician, Lucino Bini, to develop an apparatus to safely induce seizures in his patients with electricity [32]. Bini began by experimenting on laboratory dogs using the traditional Viale technique for inducing seizures with the electrodes in the mouth and rectum. With this method, some dogs died as the current traveled through the heart inducing a fatal cardiac arrhythmia. By using electrodes on each side of the head, Bini found that seizures could be reliably induced without the likelihood of death. A unique feature of the device was a circuit to measure the head resistance by applying a weak calibration current to the bitemporal electrodes to determine the voltage used with the stimulation. Bini applied for a patent of his electrical apparatus in Italy in October 1938, and it was granted in March 1939 with Bini as the sole inventor.

Cerletti used Bini's apparatus to perform the first electroconvulsive treatment on a patient in April 1938. The patient was a 40-year-old man diagnosed with schizophrenia who spoke nothing but "*incomprehensible gibberish.*" He described the tension associated with this initial experiment in an article published in the *American Journal of Psychiatry* in 1950: "*Bini and I fixed the two electrodes* [bitemporal] *well wetted in salt solution, by an elastic band to the patient's temples. As a precaution, for our first test, we used reduced tension (70 volts) with a duration of 0.2 seconds. Upon closing the circuit, there was a sudden jump of the patient on the bed with a very short tensing of all his muscles; then he immediately collapsed on to the bed without loss of consciousness*" [33].

In addition to Bini, several members of Cerletti's staff were present at the treatment session including a neurologist Lothar Kalinowsky, who monitored neurological function during and after the seizure. Kalinowsky was a German emigrant of Jewish origin who moved to Rome after he lost his position at the psychiatric hospital of the Charite in Berlin when the National Socialists came to power in 1933. According to Cerletti, there was a heated discussion among those present about proceeding and despite being in the minority he decided to go ahead and increase the stimulus to 110 V for 0.5 s.

> *The immediate, very brief cramping of all the muscles was once again seen; after a slight pause, the most typical epileptic fit began to take place. True it is that all their hearts in the mouths and were truly oppressed during the tonic phase with apnea, ashy paleness, and cadaverous facial cyanosis—an apnea which, if it be awe inspiring in a spontaneous epileptic fit, now seemed painfully never ending—until at the first deep, stertorous inhalation, and first chronic shutters, the blood more freely in the bystanders veins as well: and lastly, to the immense relief of all concerned, was witnessed a characteristic, gradual awakening by step. The patient sat up of his own accord, looked about him calmly with a vague smile, as though asking what was expected of him. I asked him 'what has been happening to you?' He answered with no more gibberish: 'I don't know, perhaps I have been asleep'* [33].

Over the next 2 months, the man received 13 more electroconvulsive treatments and he was discharged in complete remission. He was reported to be symptom free for 2 years before he was lost to follow-up. Cerletti gave the first public display of electroconvulsive therapy in May 1938 before the Medical Academy in Rome and news of the exciting new treatment rapidly spread throughout Europe.

Spread of Electroconvulsive Therapy

The journey of Lothar Kalinowsky and his promotion of Bini's electroconvulsive apparatus while fleeing Hitler and the Nazis paralleled the spread of the new electroconvulsive treatment amid the chaos enveloping Europe at the time [32]. Kalinowsky had an unpaid position at Carletti's clinic and to generate some income for his family he developed an arrangement with Bini to help sale his electroconvulsive apparatus that was manufactured by the Arcioni Company in Milan. Bini was concerned that other European companies would copy his machine without paying royalties, so he and the company were very secretive about releasing information.

Kalinowsky fled Rome to exile in France in 1939 when the alliance was formed between Fascist Italy and Nazi Germany annulling the diplomas of non-Aryan foreigners in Italy. Anxious to get his electroconvulsive apparatus distributed throughout Europe, Bini gave his plans to Kalinowsky as he left for Paris with the goal of convincing the well-known French medical engineering company, Duflot, to produce his patented apparatus. But the company procrastinated and even after Bini and Kalinowsky agreed to lower the patent fee from 20% to 15% of the sales price, Duflot delayed signing a contract. Kalinowsky suspected that they planned to build their own apparatus.

Kalinowsky left Paris for the Netherlands where he helped introduce electroconvulsive therapy to Dutch psychiatrists. In July 1939, a psychiatric institute in the small city of Heiloo received Bini's electroshock apparatus purchased from Arcioni and Kalinowsky used the apparatus to demonstrate Cerletti's method for electroconvulsive therapy to the local psychiatrists. He then moved on to England and approached the Solus Electric Company convincing them to build an electroshock apparatus according to Bini's patented plans. Psychiatrists at Warlingham Park Hospital were the first to use the Solus machine and acknowledged Kalinowsky's help when publishing the results of their initial treatment trials.

In England, Kalinowsky encountered competition with another electroshock apparatus commissioned by Frederick Gola and Grey Walter manufactured by the Edison Swan Electric Company. It sold for about half of the price of Bini's Arcioni apparatus. Recall that Grey Walter was a highly respected epileptologist and electroencephalographer who originally worked with Edgar Adrian at Cambridge. The Gola and Walter apparatus was based on the principle of current constancy rather than voltage constancy used by the Bini apparatus. They also used much smaller electrodes, about a quarter of the area of the Bini electrodes, that were applied to the front of the head on each side by an elastic band rather than the clip construction used by Bini.

In an article published in the *Proceedings of the Royal Society of Medicine in 1940*, Walter and colleagues described his initial experience with electroconvulsive therapy describing the benefits of electrical- versus chemical-induced seizures [34]. Walter began by noting that the electroencephalographic signal is attenuated about a hundredfold by the overlying tissue, the scalp, skull, and the lining of the brain, so that the current required to stimulate the brain from scalp electrodes was about a hundredfold greater the current required to stimulate an exposed brain (in the range of 500–1000 mA). A potential of between 50 and 150 V and a duration of one-tenth to one-fifth of a second were typically effective for inducing a seizure. The very brief duration accounted for the relative safety compared to the long duration used for electrocution.

> *The instantaneous effect of applying a current is twofold: firstly, a short sharp jerk of every part of the body indicating direct stimulation of the pre-central gyrus* [frontal lobe], *and a secondary immediate loss of consciousness. These two facts together indicate the profound and widespread effect of the stimulus. The direct motor jerk is probably comparable to the*

> *so-called first myoclonic stage of the Cardiazol* [Metrazol] *convulsion, and it is probably this jerk which is responsible for those few accidents to bones and joints which have been reported. The instantaneous loss of consciousness is a valuable feature of the treatment since it is of course associated with complete amnesia. This is both retrograde, covering a few seconds before the stimulus, and anterograde eliminating from the memory events occurring during the early recovery stages* [34].

Walter went on to describe how a stimulus just below threshold could produce a brief loss of consciousness lasting seconds while a stimulus just at the threshold could produce a seizure delayed in onset up to 30 s. Patients were typically given a series of trial shocks before a convulsion was induced and because of the amnesia any distress from these trial shocks was wiped from their memory. Finally, Walter noted that the electrically induced seizures were a shortened version of a spontaneous seizure with a total duration of about 45 s.

Considering the political alliance between Italy and Germany, Bini expected to have little trouble getting his apparatus patented in Germany. In 1939, he approached the SRW electro-medical research laboratory of the Siemen's Company to manufacture and distribute the apparatus. But after receiving blueprints of Bini's apparatus from their Dutch affiliate, Siemens decided they could make a cheaper better version. They found a way to circumvent Bini's Italian patent using the National Socialist patent reform of 1936 claiming that a German, Erich Schilf working at the Psychological Institute at Berlin University, developed electroshock technology before Bini. As early as 1922, Schilf produced epileptic seizures in dogs by passing brief duration alternating current through electrodes attached to the dog's eyes. The animals immediately lost consciousness, seized, and then recovered without any apparent ill-effects. SRW argued that there was nothing new in Bini's apparatus compared to Schilf's.

The German "Konvulsator" made by SRW excluded the resistant measurement circuit of the Italian "Telettroshock" made by Arcioni but included a switch to directly adjust the current in milliamps like the English apparatus of Golla and Walter. It also included the safety protection of a cut-off relay and a surge arrester. Also like the English apparatus, the electrodes of the Konvulsator were held against the head by a band rather than the large clip electrodes of the Telettroshock. Bini and Cerletti argued vehemently that their apparatus was safer and more effective than the German version but ultimately, they did not receive a German patent for their devise.

Kalinowsky's finally reached the United States in his flight from Hitler and his promotion of Bini's electroconvulsive apparatus in 1940. The first Arcioni apparatus had already been introduced in New York by another exiled associate of Cerletti the year before. Kalinowsky joined the research team at the New York Psychiatric Institute at Columbia University where he convinced them to exchange a model of Golla and Walter for a new model built according to Bini's plans. Despite Kalinowsky's efforts just like in England and Germany, Bini's plans were not accepted by the US patent office.

As in Europe, Meduna's chemical convulsive shock therapy was already in general use in America before electroconvulsive therapy was introduced. A psychiatrist,

Abram Elting Bennett, working at the Bishop Clarkson Memorial Hospital and the University of Nebraska in Omaha began using Metrazol-induced convulsions to treat psychiatric patients in the mid-1930s and he summarized his results in a paper published in the *American Journal of Medical Science* in 1939 [35]. Bennett noted that although chemical electroshock therapy was initially used to treat schizophrenia, the results were not encouraging since on follow-up most had relapsed. On the other hand, the results were much more encouraging for treating chronic affective disorders, both the depressive and manic types. Patients with midlife and presenile depression showed the best response. Schizophrenia patients who did respond typically were those with strong affective components.

Bennett addressed a major problem that became more apparent as convulsive shock therapy was more widely used—traumatic complications particularly fractures of the spine and lower extremities [36]. Even though Meduna reported only about a 1% incidence of such complications others were finding much higher rates in the range of 30% when patients were systematically assessed. In 1940, Bennett described a series of patients treated with Metrazol-induced seizures who he paralyzed with curare just before receiving Metrazol. Curare blocks the neuromuscular junction like in patients with myasthenia gravis. Bennett worked out the safe dose of an alcohol extract from crude curare by determining the lethal dose of each batch on mice and then injecting between one-fourth to one-half the estimated lethal dose into the patients. This markedly reduced the possibility of injury from violent jerks and allowed him to safely use a slightly higher dose of Metrazol which improved the therapeutic outcome. Although suppression of respiration was an infrequent complication, patients had to be monitored carefully during and after the procedure and the physician had to be ready to administer artificial respiration if necessary.

Through the 1940s, electricity gradually replaced chemicals for inducing seizures in psychiatric patients primarily because of the amnesia associated with electroconvulsive therapy. By the early 1950s, curare was replaced by a synthetic neuromuscular blocker, succinylcholine, of much shorter action than curare [37]. The drug is rapidly broken down into natural metabolites succinic acid and choline. It was typically administered in combination with a barbiturate about 20 s before the electric shock and normal respiration resumed within 1–2 min. The barbiturate reduced the anxiety associated with the procedure.

Risks of Electroconvulsive Therapy

As electroconvulsive treatment became widely used, it became apparent that it had limitations and potential complications. In 1954, the English neuropsychiatrist Hugh Jarvie reported the results of a 3-year follow-up of 97 patients treated for depression with electroconvulsive therapy at the Cheadle Royal Hospital [38]. He estimated that of the approximately 60,000 people admitted to mental hospitals in England and Wales annually about a third received electroshock therapy at some time during their hospital stay. Of the 97 patients treated for depression at Cheadle,

71 (73%) were discharged recovered and 26 (27%) discharged improved. Of the 73% discharged recovered, 35% had a relapse during the relatively short period of follow-up. Regarding complications, a few complained of back pain and a woman who received the most shock treatments (60) developed epilepsy. None complained of persisting amnesia or intellectual impairment although this was not specifically asked about on the questionnaire.

Cognitive side effects were recognized by others, however, and a one-sided stimulation technique developed in the early 1950s was shown to produce generalized seizures with decreased risk of cognitive complications. There was a clear relationship between the number of treatments and the risk of permanent cognitive impairment. Ultrashort pulse stimulation techniques further decreased the risk of cognitive impairment. Another major development in the 1950s was the introduction of effective neuroleptic and antidepressant medications. Psychotropic drugs were easier to administer and overall, better accepted than electroconvulsive therapy by patients and their family. After a period of gradual decline in use, electroconvulsive therapy had a resurgence in the 1980s particularly in patients resistant to medical treatment.

Electroconvulsive therapy administered with muscle relaxation and brief anesthesia is still widely used for treating depression, acute mania, catatonia, and drug-resistant schizophrenia. Because it acts much quicker than antidepressant medications, it has been recommended as the first-choice treatment for patients with very high risk of suicide. In the two most populous countries India and China, it is still regularly used to treat schizophrenia in part because it is the cheapest treatment option [39].

Despite the many years of use there still is no generally accepted mechanism of action for anticonvulsive therapy. Meduna's original idea that seizures increase depleted brain glial cells proved incorrect. The analogy of rebooting a computer is commonly used but this is an oversimplification. As we have seen throughout this book, the brain does not work like a computer. A leading theory is that the sudden widespread release of neurotransmitters, neurotrophic factors, and hormones causes a sudden massive stimulation of brain function. Another popular theory is that electroconvulsive seizures stimulate proliferation of neural stem cells generating new neurons in select regions of the brain. Likely there are multiple mechanisms of action including the ubiquitous placebo effect.

Despite the recent resurgence of interest in electroconvulsive therapy, it remains one of the most controversial treatments in all of medicine. Even if one has not personally witnessed an epileptic seizure, the frightening image of loss of consciousness followed by violent jerking of the extremities has been widely shown in the media. The portrayal of electroconvulsive therapy in movies such as Milos Forman's *One Flew over the Cuckoo's Nest* as a putative means of controlling psychiatric patients has influenced perception of the therapy in a large part of Western society. As a neurologist, I am aware of the risk of repeated epileptic seizures on brain health. With every seizure, there is a loss of some brain cells.

Early Deep Brain Electrical Stimulation

The origins of deep brain stimulation are rooted in the work of Forrester and Penfield and early epilepsy surgery. These pioneer neurosurgeons localized and ablated areas of the brain that were malfunctioning and causing seizures. In the 1940s, a stereotactic apparatus was developed that allowed neurosurgeons to localize and ablate regions deep within the brain for the first time, marking the beginning of deep brain electrical stimulation [40]. The stereotactic apparatus, fixed to the skull, delineated the brain as a three-dimensional Cartesian coordinate system so that any point within the brain could be defined by a set of three numbers.

A key element for successful stereotactic surgery was a method to image the brain so that the coordinate system could be overlaid onto recognizable brain features [40]. In these early times of stereotactic surgery, the only imaging technique available was X-ray pneumoencephalography. Since routine X-ray images do not distinguish between water and brain tissue, it was necessary to introduce a contrast material (air) into the spinal fluid system so that the ventricles at the center of the brain stood out. With the patient on the side, a needle was introduced into the dural sac in the back. Spinal fluid was removed and air introduced. By positioning the patient appropriately, the air moved upward into ventricles so that they were outlined by X-rays of the brain. This was a painful procedure that patients dreaded. Not only did they suffer severe headaches after the procedure, in some patients it was very painful getting the needle into the correct spinal position. Even with a successful procedure, the images of the brain were crude at best.

Early stereotactic deep brain stimulation followed Penfield's model using electrical stimulation to identify potential targets deep within the brain thought to be involved in the condition and then ablating the targets. As surgeons hunted for new and better target areas, a new body of information developed regarding brain function and connectivity ultimately leading to detailed stereotactic atlases. Electrodes could be left in place protruding from the skull for several weeks as the surgeon evaluated and decided on the area to be ablated.

In the late 1950s, surgeons began using stereotactic surgery to ablate deep brain areas for treating tremor particularly tremor associated with Parkinson disease [41, 42]. At the time, there was no effective medical option for treating this common progressive neurodegenerative disorder manifested by shaking and general slowness of movements. The initial target areas were parts of the thalamus and later the subthalamic nucleus. Early studies found that low-frequency stimulation could exacerbate the tremor while high-frequency stimulation (100–200 Hz) ameliorated the tremor. By the mid-1960s, deep brain stimulation and ablation became widely used for treating Parkinson disease, but this dramatically changed with the discovery of levodopa therapy in 1968. For the first time, a medication was effective in controlling Parkinson symptoms, and within a short time, stereotactic surgeries were almost abandoned only to reappear later in the century as it became apparent that the response to levodopa waned with time.

Another more controversial use of early stereotactic deep brain stimulation and ablation was in the field of psychosurgery. As with Parkinson disease, there were no good medical options for treating psychiatric illnesses in the early 1950s. The two most common treatments were electroconvulsive therapy (see Spread of Electroconvulsive Therapy) and frontal lobotomy. In 1936, a Portuguese neurosurgeon, Antonio Moniz performed the first frontal lobotomy on a patient with schizophrenia, and by 1949, when he received the Nobel Prize for his operation, more than 20,000 Americans had undergone frontal lobotomies. The procedure was performed by passing a sharp probe like an ice pick into the frontal lobes and reaming them out. The surgeon either drilled a small hole in the skull on each side or passed the probe through the eye socket and drove it through a thin layer of bone into the frontal lobe with a mallet. Although touted as a miracle cure, the procedure did not cure patients, it made them docile and easier to care for.

After the Second World War, more than half of Veterans Hospital beds in the United States were occupied by psychiatric patients. There was political pressure to release these veterans from the overcrowded hospitals and many were treated with frontal lobotomies. Critics pointed out that the procedure had frequent bleeding complications and vital areas of the brain were being destroyed without clear evidence of benefit. One of the leading critics, Fredrich Mettler, a neurologist at Columbia University in New York, teamed up with the Veterans' Administration to evaluate a more selective frontal lobectomy procedure called a topectomy where only a small part of the frontal lobes was removed. The Columbia-Greystone Project identified 28 patients at Greystone State Hospital with either schizophrenia or major depression and half underwent a topectomy and the other half were controls.

Robert Heath and the Septal Region

Mettler selected a young up and coming psychiatrist at Columbia, Robert Heath, to evaluate the mental condition of the patients before and after surgery [43]. Heath had completed his neurology training at Columbia and then switched to psychiatry. After working on the project for several months, Heath concluded that topectomy was no better than lobotomy with the same side effects including an overall flattening of emotions. Heath became convinced that the frontal cortex was not the problem in schizophrenia but rather one needed to look deeper within the brain. After reading the literature and discussing the problem with colleagues, he became enamored with a small region at the base of the frontal lobes just above the midbrain called the septal region. Heath believed this region was the focus of emotions such as pain and pleasure and possibly the key to treating schizophrenia. With the rapid advances in stereotactic neurosurgery, Heath was convinced that it was possible to electrically stimulate the region without damaging it.

At the same time, the dean of the medical school at Tulane University in New Orleans, Louisiana, was looking to recruit a new head of Psychiatry with the goal of making Tulane the Harvard of the South. With his background in neurology and psychiatry and his reputation of being an innovative researcher, Heath seemed perfect for the job. With an offer to recruit his own research team and the availability of a large psychiatric population at the affiliated Charity Hospital, Heath accepted the position in 1949. He brought along several colleagues from New York including the neurosurgeon from Greystone Project, Francisco Garcia. After some preliminary experiments in cats and a few monkeys, Heath and his team began performing deep brain electrical stimulation in psychiatric patients within a year.

By 1952, they had stimulation data from 22 patients, the majority with a diagnosis of schizophrenia. Heath decided it was time to present his material to the psychiatric community, so he set up a research symposium at Tulane to present his preliminary data. He invited some of the most prominent researchers in the field of psychiatry including his former mentor from Columbia, Fred Mettler. The presentations provided a detailed assessment of the patients before and after surgery.

After drilling a hole in the skull, the surgeon, Garcia, passed a thin silver wire electrode down deep through the frontal lobe near the base of the brain into what they called the septal region. He guided the electrode using a series of anatomical markers outlined with pneumoencephalography. For chronic recordings, the wire was held in place by a bone plug covering the hole with the skin pulled tight around it and bandaged to hopefully prevent infection. Typically, patients underwent multiple stimulations over several days with a range of stimulus parameters from 2 to 4 mA at 4–8 V. The patients remained fully alert during the stimulations so they could respond to questions and report any discomfort.

Overall, the results of the stimulation were mixed with only a few having a dramatic response. Even in those cases, there were potential complicating factors. For example, a teenage girl with catatonia and anorexia who had been hospitalized for a year was able to return home 4 months after treatment and resume her college education. After being stimulated on the operating table, the girl had several seizures and then slipped into a coma for 2 days before gradually improving over the next 4 months. A 26-year-old woman with catatonia who had responded to electroconvulsive therapy in the past began to manifest delusions and hallucinations. After electrical stimulation of the septal region, she began to speak and when returning to the ward she talked with the nurses for the first time in months. A few others showed good improvement, but four patients had no response, and one had a worsening of symptoms after stimulation. More bothersome, there was no explanation for why some patients responded and others did not.

Three of the patients presented at the symposium were considered controls since they had no mental illness. The patients agreed to the procedure since they suffered from medically intractable pain, one from severe arthritis and the other two from terminal cancer. Heath felt that stimulating the "pleasure" centers in the septal region might block the sensation of pain. The team reported that deep brain stimulation suppressed the pain in all three patients. One of the cancer victims who lived for 7 months after surgery reported that a 15-min stimulation session relieved her pain for a week. Even more intriguing the "controls" described being more alert and having a general sense of well-being after stimulation.

Heath's team presented their material to the esteemed guests expecting a reception that matched their enthusiasm. But what they observed was overall skepticism. Wasn't the operation and the interest and attention shown the patients before and after the surgery a strong placebo? Heath's old mentor Fred Mettler even hinted that the zeal exhibited by Heath and his team may have influenced interpretation of the data to conform to their hypothesis.

At the time little was known about the connectivity and function of the septal region of the brain. The region is at the midline deep in the brain below the septum pellucidum, the thin membrane that separates the lateral ventricles. The region gained fame a few years later in 1954 when Canadian researchers James Olds and Peter Milner implanted stimulating electrodes in the brains of rats looking for possible reward centers [44]. To their amazement, rats with electrodes implanted in the septal area repeatedly pressed on a foot plate in their cage that activated the electrodes. They pressed the plate 80–90% of the time at the exclusion of other activities including eating and sex. Olds and Milner speculated that the septal area was at the center of a reward system within the brain, a "pleasure" center.

Subsequent research showed that the septal region contains several nuclei that were potentially involved in reward processing. The ventral tegmental area rich in dopamine neurons proved to be closely interconnected with other reward centers in the brain. Interestingly, around the same time researchers discovered that Parkinson disease was associated with degeneration of dopamine neurons which ultimately led to an effective treatment, levodopa, a drug that increased dopamine production in the brain. It soon became apparent that dopamine played a role in behaviors such as gambling, and addiction and the pleasurable sensations associated with these behaviors. Patients with Parkinson disease have a blunted affect and are at high risk of developing depression. On the other hand, when treated with levodopa to increase dopamine production, some developed risk-taking behaviors such as compulsive gambling.

The 1950s was the "wild west" for deep brain stimulation. There was no such thing as "informed consent" and few researchers raised ethical concerns. Robert Heath not only implanted electrodes in the septal region but in several patients, he implanted electrodes throughout the brain to assess the response to septal stimulation. It was hard to justify these additional electrodes on a therapeutic basis. In one of his most controversial experiments, Heath attempted homosexual conversion by stimulating the septal region in a gay man while the man was having intercourse with a prostitute.

After hearing about Olds and Milner's rats, Heath decided to try self-stimulation with one of his patients. He had his technicians use the latest transistors to make a small box with buttons so that the patient could control the electrical stimulation. The device had three buttons, each one connected to a different electrode so the patient could stimulate up to three brain regions at a time with a half second pulse of current.

In her book *The Pleasure Shock*, science writer, Lone Frank described the bizarre results of Heath's experiment on a patient named Roy who had seventeen electrodes implanted in different regions of the brain [45]. In a 6-h "marathon" session, Roy self-stimulated different electrodes and electrode combinations while describing the sensations he experienced. The only location that triggered an unpleasant sensation was the right hippocampus that made him feel "*sick all over.*" When he stimulated the septal region, he described a "*wonderful feeling*" *with* "*some sexual overtones.*" But the region he stimulated the most, about five hundred times an hour, was the left central thalamus that produced an irritated and peeved sensation. He reported that when stimulating this region, "*it was as if he were always on the verge of grasping a particular recollection, a memory that he felt was important but kept getting away.*" He finally resolved the frustration by stimulating two other regions, the septal region and the high midbrain at the same time as the thalamus. After the experiment, Roy joked: "*I might want to buy this little box from you, Dr. Heath, and take it home with me.*"

Development of an Implantable Neurostimulator

Early attempts at deep brain stimulation had the obvious limitation of wires sticking out of the scalp attached to a bulky power source for generating electrical stimuli. Treatment outside of a hospital setting was not feasible. Furthermore, infections were not uncommon and a few deaths from brain abscesses occurred. As with other new medical therapies, technological advances were necessary before deep brain stimulation could become more widely used.

In this case, the technological advances occurred in another medical subspecialty, cardiology. In the early 1960s, the medical device maker, Medtronic, manufactured the first commercially available implantable cardiac pacemaker. Since batteries were still not good enough to be implanted in the body, these early pacemakers used a radio frequency system to pass energy in the form of radio waves through the skin to an implanted receiver. These small implantable devices were a huge success, and by the end of the 1960s, Medtronic was one of the fastest growing companies in the United States. It was only natural that they would expand into deep brain stimulation.

The initial trials of implantable neurostimulators were for treating chronic intractable pain. A Wisconsin neurosurgeon, Norman Shealy worked with Medtronic to adapt their implantable cardiac pacemaker into an implantable neurostimulator of the spinal cord [46]. When Shealy presented promising results in treating chronic pain with the neurostimulator, neurosurgeons around the country soon began implanting the devices. Other pacemaker companies started to manufacture implantable neurostimulators, and in 1975, Medtronic established a neurological division to market devices for treating chronic pain and trademarked the term "DBS" for deep brain stimulation.

How did stimulating the spinal cord inhibit pain? The Canadian psychologist, Ronald Melzack and British physiologist, Patrick Wall introduced the notion of a pain "gate" in the spinal cord in their famous 1965 *Science* article "Pain Mechanisms: A New Theory." [47] Their gate control theory proposed that small unmyelinated or thinly myelinated pain nerves and large heavily myelinated touch, pressure, and vibration nerves carried information from the periphery to the spinal cord where they activated two types of neurons, transmission neurons whose axons carry signals up to the brain and inhibitory interneurons whose axons feedback onto the transmission neurons to impede their activity.

Both the small and large sensory nerves excite transmission neurons, but the small nerves inhibit the inhibitory cells (allowing the transmission cells to fire) while the large nerves excite the inhibitory neurons (decreasing the firing of transmission cells). In other words, the inhibitory neurons act as a "gate" to control pain transmission in the spinal cord. The greater the large nerve (touch, pressure, vibration) activity relative to small nerve (pain) activity at the inhibitory cell, the less pain is felt. The theory explained why touching or applying pressure or vibration to a painful extremity eases the pain.

Melzack and Wall also speculated that signals carried by the pain transmission neurons to the brain could activate neurons that sent signals back down the spinal cord to modulate the activity of the inhibitory cells (the "gate"). Thus, they provided an explanation for how the brain might also influence pain transmission in the spinal cord and how stimulating certain regions of the brain could inhibit pain. More recently, this brain pain control system that includes multiple cortical and the brainstem centers became known as the descending pain modulatory system (DPMS). Many of the neurons in the DPMS produce endogenous opioids and the system is activated by opioid medications.

Based on the gate theory, surgeons implanted plate electrodes over the dorsal columns at the back of the spinal cord where large heavily myelinated sensory nerves predominate. The electrodes were connected to an electrical pulse stimulator that the patient could control. In the early trials, bone was removed to gain access to the spinal cord and the electrical stimulators were large and cumbersome but many patients reported good results. Over time, it was found that tiny electrode arrays implanted via a percutaneous needle worked just as well as the large plate electrodes and these electrode arrays could be placed outside the dura or near spinal nerves entering the dura often with equally good results.

The procedure was performed in two steps. First, temporary electrodes were implanted through a large needle and the electrodes connected to an electrical stimulator. The patient felt tingling in the region of the pain if the electrodes were in the proper location. The stimulator sent electrical pulses of different duration and amplitude to the electrodes to achieve the best pain control. During the trial period, that could last a few days to a few weeks, different pulse durations and amplitudes were tried, and pain relief and functional activities were monitored with the goal of obtaining at least 50% relief of pain. If the trial was successful, the electric pulse stimulator was implanted under the skin with local anesthesia where it could be programmed for the best results. It was not until the 1980s that lithium batteries were developed so that stimulators could be implanted for several years, and the devises could be programmed with a wireless console programmer.

Electrical stimulation of the spinal cord for pain control was not without risks. About a third of people had some type of complication although most were minor.

The most common complication was migration of the implanted electrode so that it no longer stimulated the correct area. This typically occurred within the first few days of implantation if it occurred. Not surprisingly, percutaneous electrodes were more likely to migrate than electrodes surgically implanted. Serious complications such as damage to a spinal nerve or the spinal cord by the percutaneous needle or bleeding into the space outside the dura occurred in only one to two percent of patients. Infection requiring a course of antibiotics occurred in up to 5% of people.

Implantable Neurostimulators for Deep Brain Stimulation (DBS)

With the improved implantable Medronic neurostimulator, there was a renewed interest in DBS at neurosurgical centers in the United States and Europe. The neurostimulator was tried in a variety of conditions from movement disorders such as Parkinson disease, dystonia, and cerebral palsy to psychiatric illnesses such as schizophrenia and severe depression. The reversible nature of neurostimulation seemed preferable to earlier ablation procedures.

In Germany, Fritz Mundinger used the Medronic stimulator to stimulate the thalamus in patients with dystonia reporting good results in a previously untreatable condition [48]. In the United States, New York neurosurgeon Irving Cooper reported high success rates in treating patients with dystonia with the Medronic stimulating electrode implanted on the surface of the cerebellum [49]. In Louisiana, Robert Heath reported excellent results in several of the patients with depression and schizophrenia using the Medronic neurostimulator to stimulate the cerebellum [50]. But what constituted a good or excellent result? This was largely based on the opinion of the treating physicians.

The 1970s were also marked by a backlash in public and physician opinion regarding DBS. As the disturbing aspects of lobotomy and early DBS for psychiatric illnesses became widely known, the public expressed major ethical concerns about continuing such treatments. With the initial good results of levodopa, antidepressants, and antipsychotic medications, physicians questioned whether the risk and costs of DBS were warranted. More than 700 deaths associated with medical devises had been reported between 1972 and 1975. In 1976, the US Congress passed the Medical Devise Amendment giving the Food and Drug Administration (FDA) authority over all medical devises. The "wild west" of medical devises was to be regulated.

Although it might not be possible to perform double-blind placebo-controlled trials that were required for introduction of new medications, new medical devises would be evaluated based on "*well controlled investigations, including clinical investigations where appropriate, by experts qualified by training and experience*" [51]. The manufacturers would have to convince an FDA-appointed committee of experts who had no relationship to the people conducting the treatment trials.

A major problem facing potential clinical investigators and the companies that manufactured the neurostimulators was the lack of verifiable objective measures of efficacy. For example, despite the widespread use of neurostimulators for treating chronic pain, it was well known that quantifying pain was highly subjective. Pain is influenced by both observer and patient related influences. Studies of pain medications consistently found a high placebo rate that could be influenced by many factors including the size, color, and even taste of the pill. Elaborate surgical procedures had an even higher placebo rate that was influenced by the enthusiasm of the surgeon and the expectations of the patient. The descending pain modulatory system (DPMS) mentioned earlier provided a neurological substrate to explain at least part of the placebo effect.

Initially, the companies that manufactured neurostimulators were offered the time to conduct the necessary clinical trials but after assessing the problems they would face they declined the offer. Clinicians in the field recognized that movement disorders such as Parkinson disease and dystonia offered the best possibility of developing an objective rating scale and by the 1980s rating scales were developed and clinical trials carried out. But it was not until near the end of the twentieth century that the FDA approved a Medtronic neurostimulator for suppression of medically intractable tremor in patients with essential tremor and Parkinson disease.

Although there is continued interest in using deep brain stimulation in psychiatric patients, the FDA approval process has provided a high bar that has yet to be reached. Early studies using open-loop deep brain stimulation for treating obsessive-compulsive disorder (OCD), depression, and post-traumatic stress disorder (PTSD) produced mixed results. Open loop means that the stimulus parameters for the electrode are pre-programmed at the time of insertion and then adjusted at follow-up visits (usually monthly) based on the patient's subjective report of benefits and side effects. Numerous potential problems explained the inconclusive results including choosing the wrong location for the electrodes, slight misplacement of the electrodes, difficulties programming the electrodes based on the patient's subjective reports and the lack of any objective measure that the target circuits were being stimulated. At the present time, the only DBS psychiatric treatment available outside of research protocols in the United States is open-loop DBS for treating patients with OCD, which has a Humanitarian Devise Exemption from the FDA [52].

There has been a great deal of interest in the use of DBS for treating PTSD probably because of the large number of people with this debilitating illness and the lack of effective treatments. An almond shaped nucleus deep within the temporal lobe, the amygdala is a key part of the brain's emotional network, the limbic system. In animal studies, the amygdala appears to be important for fear memories and persistent fear responses. The prefrontal cortex normally controls the amygdala modulating fear responses and a simple working hypothesis to explain PTSD is that loss of prefrontal inhibition of the amygdala causes fear memories to dominate emotional responses [53].

Studies in a rodent model of PTSD showed that stimulation of an electrode implanted in the amygdala improves fear conditioning and fear-related behaviors and electrodes were implanted into the amygdala of a few veterans suffering from refractory PTSD [54]. Although preliminary results with open-loop stimulation in the amygdala are encouraging there are problems with programming and measuring effectiveness of the devises. Without a biological marker for symptom severity, the stimulation parameters are set based on periodic reports from the patient which are highly subjective and constantly changing. By nature, PTSD is an episodic disorder so that continuous stimulation during periods when the patient is not having symptoms may be inappropriate.

The future of deep brain stimulation is probably some form of closed-loop stimulation. Closed loop means neuronal activity in one area of the brain controls the firing pattern of stimulating electrodes in another area. For example, in the case of PTSD, a logical closed-loop system might be to monitor activity in the prefrontal cortex to control stimulation in the amygdala. With recent advances in microelectrode technology, it is possible to record and stimulate from hundreds of microelectrodes in both locations so that one could fine tune the connections between the prefrontal cortex and the amygdala electrodes. Even with these multi-electrode arrays, however, the communication between brain centers will still be crude compared to the normal functioning brain and unintended consequences of the chronic stimulation are of concern. Even more important the ethics of placing electrodes in key emotional centers of the brain require more investigation and political discourse.

Magnetic Stimulation of the Brain

According to Faraday's law, the magnetic field in an excitation coil is proportional to the rate of change of the current in the coil. Conversely, only a time-varying magnetic field can induce a current in a nearby conductor (an eddy current). The first person to use this principal to stimulate the brain with a magnetic field was the French physicist and physician d'Arsonval introduced in Chap. 9 for his work on high frequency, high voltage alternating current. In addition to his body-encircling coil, d'Arsonval constructed a head-encircling coil and noted "*that an alternating magnetic field with an intensity of 110 volts, 30 amperes with a frequency of 42 cycles per second, gives rise to, when one places the head into the coil, phosphenes and vertigo, and in some persons, syncope*" [55].

As alternating current gradually replaced direct current, some workers in power plants perceived faint flickering lights (phosphenes) when standing near transformers with high intensity alternating currents. Unaware of d'Arsonval's report, the British physicist and electrical engineer, Silvanus Phillips Thompson, who designed commercial power plants in his early career, decided to investigate the effect of electromagnetic fields on the brain. Thompson, a polyglot Quaker Yorkshireman, was one of the last true renaissance men of Europe [56]. He began by studying music and art and then moved to science where he mastered chemistry, mathematics, and physics. He is probably best known for his book *Calculus Made Easy* and his extensive writings on the history of science.

In 1910, Thompson constructed a 32-turn copper wire coil just large enough to insert the head and he applied up to 180 A of power line alternating current (50 cycles/s) to the coil [57]. He was the first to stick his head into the coil and he reported experiencing flickering lights brightest in the periphery of vision. Other subjects reported a similar experience that occurred with eyes open or closed and in the light or dark. A few reported a strange taste sensation.

Other similar reports soon followed in America noting that the flickering lights were brightest with frequencies in the range of 20–30 Hz where the flickering seemed to follow the current frequency. At higher frequencies, the flickering lights were dimmer even when the current was increased. As expected, when direct current was used the subject only experienced a visual sensation when the current was turned on or off. When the current was turned on the subject experienced a luminous horizontal bar that move downward and when the current was turned off the bar moved upward. Initially, it was assumed that the visual perceptions resulted from eddy currents in the retinal from the changing magnetic field, but, with time, it became apparent that visual perceptions could occur with stimulation of the retina, optic nerve, and visual cortex.

It was not until 1959 that eddy currents produced in a volume conductor by a moving magnetic field were conclusively shown to stimulate a nerve. Biophysicist Alexander Kolin and colleagues at UCLA insulated an electromagnet with plastic and placed a frog nerve-muscle preparation in a dish filed with salt solution [58]. When they placed the dish on the pole face of the electromagnet and alternating current was applied to the coil, the muscle vigorously contracted. A few years later, researchers at UC San Diego used pulsed magnetic stimulation (from the discharge of a capacitor bank) to produce skeletal muscle contractions in live frogs, rabbits, and human subjects. Brief (about 300 μs) pulses with a peak field strength of 2–3 Tesla induced action potentials in the muscles supplied by the stimulated nerve.

The next major advance in magnetic stimulation of the brain occurred at Sheffield University in England. Anthony Barker investigated the possibility of using pulsed magnetic stimulation for clinical purposes as part of his PhD thesis published in 1976 [59]. Barker and colleagues built a practical magnetic stimulator that delivered a peak magnetic field of about 2 Tesla and showed how the devise could be used to activate peripheral nerves and measure conduction velocity with electromyography just as with electrical stimulation. They then moved on to the brain and showed that pulsed electromagnetic stimulation over the motor cortex in volunteer subjects induced action potentials in muscles in the hand on the opposite side recorded with electromyography [60]. The stimulation coil was approximately 10 cm in diameter with a peak current of 4000 A.

At the time, it was common to induce motor-evoked potentials with electricity by placing an electrode on the scalp over the motor cortex and recording in the opposite extremity. The procedure was uncomfortable and sometimes painful. With pulsed electromagnetic stimulation, there was no pain or discomfort. Clinicians immediately saw the potential for stimulating localized areas of the brain without any discomfort to the patient. Several companies in Europe and America began manufacturing transcranial magnetic stimulation (TMS) devises based on the

Sheffield model. Typically, the stimulation coil was placed on the scalp over the motor cortex and the intensity of the magnetic field was gradually increased until it produced muscle contractions of the hand on the opposite side. The coil was then moved to the part of the brain to be treated and the stimulus intensity was maintained or slightly decreased.

For the remainder of the twentieth century, TMS was mainly evaluated in research laboratories as investigators tried to understand what was happening in the brain. There was clear evidence that TMS activated the brain and depending on the location of the coil subjects experienced muscle twitching, visual disturbances and even altered cognition and speech arrest. Depending on the intensity of the stimulus and the configuration of the coil, the depth of stimulation ranged from 2 to 4 cm below the surface.

In the early twenty-first century, TMS has been tried and reported successful in treating a variety of psychiatric illnesses [61]. There was initial enthusiasm that it might be as effective as electroconvulsive therapy for treating depression without all the adverse consequences. The prefrontal cortex appeared to be a good target for TMS treatment in depression. In 2008, based on controlled trials, the devise was approved for treating depression by the FDA. The problem is that although it was significantly better than sham stimulation for treating depression it was not nearly as effective as electroconvulsive therapy. Furthermore, it was cumbersome with 4–6 weeks of hourly daily treatments compared to a few minutes' electroconvulsive treatment. Insurance companies have been resistant to reimburse for the treatments and it is still unclear where it will fit in the future of neurologic and psychiatric treatments.

In Brief

Although spontaneous and evoked electrical activity was identified in the brains of animals and human subjects in the late nineteenth century, it was not until the electronic advances of the early twentieth century that the clinical application of electroencephalography (EEG) was appreciated. Hans Berger first identified rhythmic alpha activity in human subjects and Adrian later concluded it was a default rhythm that occurred when subjects were inattentive. EEG proved most useful for diagnosing seizure disorders and neurosurgeons soon began using it to localize the seizure discharge prior to surgical resection. William Penfield revolutionized the surgical treatment of epilepsy at the Montreal Neurological Institute using electrical stimulation of the cortex in awake patients to decide which areas were safe to resect. Although initially used to treat schizophrenia, electroconvulsive therapy proved most effective for treating severe depression. Major concerns about safety remain despite continued use. Because of its rapid action compared to antidepressant medications, it is most used for treating patients at high risk for suicide. Deep brain electrical stimulation has a checkered past regarding both science and ethics, but it has established benefit for treating movement disorders such as dystonia and

Parkinson disease. The benefits for treating psychiatric disorders are promising but to be established.

References

1. Brazier MAB. Pioneers in the discovery of evoked potentials. Electroencephalogr Clin Neurophysiol. 1984;59:2–8.
2. Caton R. The electric currents of the brain. Br Med J. 1875;2:278.
3. Brazier MAB. The EEG in epilepsy. A historical note. Epilepsia. 1959/1960;1:328–36.
4. Brazier MAB. The EEG in epilepsy. A historical note. Epilepsia. 1959/1960;1:332.
5. Berger H. Psyche. Jena: Gustav Fischer; 1940. p. 6.
6. Stone JL, Hughes JR. Early history of electroencephalography and establishment of the American Clinical Neurophysiology Society. J Clin Neurophysiol. 2013;30:28–44.
7. Finger S, Edgar D, Adrian. In minds behind the brain. A history of the pioneers and their discoveries. New York: Oxford University Press; 2000. p. 239–58.
8. Adrian ED, Mathews BHC. The Berger rhythm: potential changes from the occipital lobes in man. Brain. 1934;57:356.
9. Walter WG. The location of cerebral tumors by electro-encephalography. Lancet. 1936;2:305–8.
10. Davis H. Alexander Forbes 1882–1965. J Neurophysiol. 1965;28:986–8.
11. Stone JL, Hughes JR. Early history of electroencephalography and establishment of the American Clinical Neurophysiology Society. J Clin Neurophysiol. 2013;30:32.
12. Davis H. Crossroads on the pathways to discovery. In: Worden FG, Swazey JP, Adelman G, editors. The neurosciences: paths of discovery. Cambridge: MIT Press; 1975. p. 311–21.
13. Niedermeyer E. Historical aspects. In: Niedermeyer E, Lopes da Silva FF, editors. Electroencephalography. Basic principles, clinical applications, and related fields. 5th ed. Philadelphia: Lippincott Williams and Wilkins; 2005. p. 1–15.
14. Gibbs FA, Lennox WG, Gibbs EL. The electro-encephalogram in diagnosis and localization of epileptic seizures. Arch Neurol Psychiatr. 1936;36:1225–35.
15. Zottoli SJ. The origins of the Grass Foundation. Biol Bull. 2001;201:218–26.
16. Piotrowska N, Winkler PA. Otrid Foerster, the great neurologist and neurosurgeon from Breslau (Wroclaw): his influence on early neurosurgeons and legacy to present-day neurosurgery. J Neurosurg. 2007;107:451–6.
17. Foerster O, Penfield W. The structural basis of traumatic epilepsy and results of radical surgery. Brain. 1930;53:00–119.
18. Feindel W. William Penfield (1891-1976) The man and his work. Neurosurgery. 1977;1:93–100.
19. Eccles J, Feindel W. Wilder Graves Penfield 26 January 1891-5 April 1976. Biogr Mem Fellows R Soc. 1978;24:472–513.
20. Andermann F. Herbert Henri Jasper 1906-1999. Epilepsia. 2000;41:113–120.
21. Penfield W, Jasper H. Epilepsy and the functional anatomy of the human brain. Boston: Little, Brown, and Co.; 1954.
22. Moruzzi G, Magoun HW. Brain stem reticular formation and activation of the EEG. Electroencephalogr Clin Neurophysiol. 1949;4:455–73.
23. Penfield W, Jasper H. Epilepsy and the functional anatomy of the human brain. Boston: Little, Brown, and Co.; 1954. p. 739–817.
24. Penfield W, Jasper H. Epilepsy and the functional anatomy of the human brain. Boston: Little, Brown, and Co.; 1954. p. 142–3.
25. Penfield W, Jasper H. Epilepsy and the functional anatomy of the human brain. Boston: Little, Brown, and Co.; 1954. p. 144.

26. Vannemreddy P, Stone JL, Vannemreddy S, Slavin KV. Psychomotor seizures, Penfoeld, Gibbs, Baily and the development of anterior temporal lobectomy: a historical vignette. Ann Indian Acad Neurol. 2010;13:103–7.
27. Tsay CJ. Julius Wagneer-Jauregg and the legacy of malarial therapy for the treatment of general paresis of the insane. Yale J Biol Med. 2013;86:245–54.
28. Sakel M. The methodologic use of hypoglycemia in the treatment of psychosis. Am J Psych. 1937;94:111–29.
29. Fink M. Convulsive therapy: theory and practice. New York: Raven Press; 1981.
30. Fink M. Meduna and the origin of convulsive therapy. Am J Psych. 1984;141:1034–41.
31. Meduna LJ. New methods of medical treatment of schizophrenia. Arch Neurol Psych. 1936;35:361–3.
32. Rzesnitzek L, Lang S. A material history of electroshock therapy. NTM. 2016;24:251–77.
33. Cerletti U. Old and new information about electroshock. Am J Psych. 1950;107:87.
34. Gola F, Walter WG, Fleming GWTH. Electrically induced convulsions. Proc Roy Soc Med. 1940;33:261–7.
35. Bennett AE. Metrazol convulsive shock therapying affective psychosis: a follow-up report of results obtained in sixty-one depressive and nine manic cases. Am J M Sc. 1939;198:695.
36. Bennett AE. Preventing traumatic complications in convulsive therapy by curare. JAMA. 1940;114:322–4.
37. Holmberg G, Thesleff S. Succinyl-choline-iodide as a muscular relaxant in electroshock therapy. Am J Psych. 1952;108:842–6.
38. Jarrie HF. Prognosis of depression treated by electric convulsive therapy. Br Med J. 1954;1:132–4.
39. Gazdag G, Ungvari GS. Electroconvulsive therapy: 80 years old and still going strong. World J Psych. 2019;9:1–6.
40. Spiegel EA, Wycis HT, Marks M, Lee AJ. Stereotaxic apparatus for operations on the human brain. Science. 1947;106:349–50.
41. Gildenberg PL. Neuromodulation: a historical perspective. In: Krames ES, Peckham PH, Rezai AR, editors. Neuromodulation. Amsterdam: Elsevier and Academic Press; 2009. p. 9–20.
42. Gardner J. A history of deep brain stimulation: technological innovation and the role of clinical assessment tools. Soc Stud Sci. 2013;43:707–28.
43. The FL, Shock P. The rise of deep brain stimulation and its forgotten inventor. New York: Penguin Random House; 2018.
44. Olds J, Milner P. Positive reinforcement produced by electrical stimulation of septal area and regions of the rat brain. J Comp Physiol Psychol. 1954;47:419–27.
45. The FL, Shock P. The rise of deep brain stimulation and its forgotten inventor. New York: Penguin Random House; 2018. p. 88–9.
46. Shealy CN, Mortimer JT, Reswick JB. Electrical inhibition of pain by stimulation of the dorsal columns: preliminary clinical report. Anesth Analg. 1967;46:489–91.
47. Melzack R, Wall PD. Pain mechanisms: a new theory: a gate control system modulates sensory input from the skin before it evokes pain perception and response. Science. 1965;150:971–9.
48. Mundinger F. New stereotactic treatment of spasmodic torticollis with a brai stimulation system. Med Klin. 1977;72:1982–6.
49. Cooper IS, Upton ARM. Use of chronic cerebellar stimulation for disorders of disinhibition. Lancet. 1978;311:595–600.
50. Heath RG. Modulation of emotion with a brain pacemaker. Treatment of intractable psychiatric illness. J Nerv Ment Dis. 1977;165:300–17.
51. Gardner J. A history of deep brain stimulation: technological innovation and the role of clinical assessment tools. Soc Stud Sci. 2013;43:717.
52. Greenberg BD, Gabriels LA, Malone DA, et al. Deep brain stimulation of the ventral internal capsule/ventral striatum for obsessive-compulsive disorder: worldwide experience. Mol Psychiatry. 2010;15:64–79.

53. Bina RW, Langevin J-P. Closed loop brain stimulation for PTSD, addiction, and disorders of affective facial interpretation: review and discussion of potential biomarkers. Front Neurosci. 2018;12:300.
54. Gouveia FV, Gidyk DC, Giacobbe P, et al. Neurostimulation strategies in post-traumatic stress disorder: from preclinical models to clinical applications. Brain Sci. 2019;9:45.
55. Geddes LA. History of magnetic stimulation of the nervous system. J Clin Neurophysiol. 1991;8:3.
56. Gooday G. Silvanus Phillips Thompson (1851-1916): an introduction to the spotlight section. Centaurus. 2021;63:453–8.
57. Thompson SP. A physiological effect of an alternating magnetic field. Proc R Soc. 1910;B82:396–9.
58. Kolin A, Brill NQ, Broberg PJ. Stimulation of irritable tissues by means of an alternating magnetic field. Proc Soc Exp Biol Med. 1959;102:251–3.
59. Barker AT. Determination of the distribution of conduction velocities in human nerve trunks. University of Sheffield, UK: PhD Thesis, 1976:53–58.
60. Barker AT, Jalinous R, Freeston IL. Noninvasive magnetic stimulation of human motor cortex. Lancet. 1985;1:1106–7.
61. Chail A, Saini RK, Bhat PS, et al. Transcranial magnetic stimulation: a review of its evolution and current applications. Ind Psychiatry J. 2018;27:172–80.
62. İnce R, Adanır SS, Sevmez F. The inventor of electroencephalography (EEG): Hans Berger (1873–1941). Childs Nerv Syst. 2021;37:2723–4.
63. Gibbs FA, Davis H, Lennox W. The electroencephalogram in epilepsy and in condition of impaired consciousness. Arch Neurol Psychiat. 1935;34:1133–48.

Chapter 12
Quantum Theory and Modern Neuroscience

When a light ray is spreading from a point, the energy is not distributed continuously over ever-increasing spaces, but consists of a finite number of energy quanta that are localized in points in space, move without dividing, and can be absorbed or generated only as a whole.

Albert Einstein, 1905

Abstract JJ Thomson's electron and Henri Becquerel's alpha particle were key to Ernest Rutherford developing his model of atomic structure. Disparities between the charge and mass of the atomic nucleus soon led to the discovery of the neutron. Niels Bohr's application of the new quantum theory of Max Planck and Albert Einstein to atomic structure solved the problem of electrons remaining in orbit and not crashing into the nucleus. Electrons had dual properties of particles and waves. Quantum theory provided new rules for understanding electricity and magnetism that led to a range of modern electronic devices from transistors and lasers to cell phones and modern computers. The quantum prediction that spinning protons and neutrons generate a magnetic field like the spinning electrons in orbit around the nucleus led to the discovery of nuclear magnetic resonance. This in turn led to the discovery of magnetic resonance imaging revolutionizing the medical specialty of neurology.

Keywords Crystals · Atomic nucleus · Phosphorescence · Photon · Semiconductors

When we left JJ Thomson in Chap. 8, he had identified a tiny negatively charged particle that he called a corpuscle (later called an electron) to explain the behavior of cathode rays. A stream of corpuscles was accelerated by the electrical potential in the vacuum tube and the corpuscles could be deflected by an external electric or magnetic field. When the corpuscles struck a glass or metal surface, electromagnetic waves (e.g., radio, light, and X-rays) were produced, spreading outward through the all-pervading ether. Over the next several years at the beginning of the twentieth century, Thomson developed a general theory of electrical conduction and subatomic structure based on the electron.

R. W. Baloh, *Brain Electricity*, https://doi.org/10.1007/978-3-031-62994-5_12

Electrons and Electrical Conduction

Thomson became convinced that electrons were a fundamental building block of all matter. He was able to measure the absolute value of the charge of a corpuscle and found that it was roughly the same as the positive electrification of a hydrogen ion even though it was three orders of magnitude smaller than a hydrogen ion. If negatively charged corpuscles were the building blocks of atoms, how did they assemble into a system that was normally electrically neutral? This led to his "plum pudding" model of atomic structure with raisins being the negatively charged corpuscles surrounded by a mortar of positive charge, the pudding.

After his cathode ray experiments, Thomson was convinced that corpuscles were key to understanding electrical conduction in gasses and solids. In 1899, he showed that corpuscles could be pulled from atoms in a process he called "ionization." "*From what we have seen, this negative ion must be a quantity of fundamental importance in any theory of electrical action; indeed, it seems not improbable that it is the fundamental quantity in terms of which all electrical processes can be expressed*" [1].

Based on Thomson's observations, in 1900, the German physicist Paul Drude formulated a new theory for electrical conduction in metals [2]. Drude was the chief editor of *Annalen der Physik,* the most prestigious journal of physics at the time. He postulated that although the atoms in metals were fixed in a solid array, an electron from each atom was free to move about. These "free" electrons could form a gas of charged particles that moved through the crystalline array when driven by an electrical potential or temperature gradient. Although Drude's theory which was later modified by Sommerfeld to be consistent with quantum theory remains popular and is often taught in the classroom to explain conduction of electricity in metals; based on modern statistical mechanics several aspects are now known to be false (discussed later in the chapter).

Thomson initially liked Drude's model of electrical conduction and suggested that it should be possible to calculate the number of electrons per unit volume and the speed of conduction based on the kinetic theory of gases. But when he began making the calculations, he found inconsistencies and suggested modifications to better fit the data. Rather than a gas of freely moving electrons, he proposed a model where electrons were passed from atom to atom reminiscent of the old bucket brigade, Gothius chain theory of electrolysis (see Chap. 8). This could explain the rapid movement of electricity without the need for a massive number of free electrons moving along the conductor. Of course, Thomson was still struggling with the question of the basic structure of the atom and its relationship to his corpuscle (electron).

Becquerel and Radioactivity

The discovery of radioactivity in 1896 by the French physicist Henri Becquerel would eventually lead to a completely new model of atomic structure proposed by the English physicist Ernest Rutherford—light negatively charged electrons orbiting a nucleus composed of much heavier positively charged particles called protons.

The phenomena of phosphorescence (objects that glow in the dark) had been observed for centuries but there was little understanding of the underlying mechanism. When some substances were exposed to bright sunlight they would glow in the dark as though the light had been stored and then released. Becquerel studied phosphorescence as part of the doctoral thesis and after hearing of Rontgen's discovery of X-rays in January 1896, he decided to test whether some of the phosphorescent materials in his laboratory might emit X-rays after being illuminated with bright sunlight [3, 4].

Becquerel selected uranium salts whose phosphorescence spectra he had previously tested. He enclosed unexposed photographic film inside of a black envelope, placed crystals of uranium salt on the envelope, and set the envelope on a window ledge in the bright sunlight. When he later opened the envelope and developed the film, there was an image of the uranium crystals. He then repeated the experiment placing other barriers between the crystals and the film and the image remained indicating that the light-exposed uranium salt was producing a penetrating radiation.

As with many of the great discoveries in the history of science, Becquerel made a serendipitous observation that changed everything. He prepared some envelopes with film inside, and uranium crystals glued to the outside and stored them in a dark cabinet since there was no sunlight for several days. When he developed the film in these envelopes along with others that were later exposed to sunlight, he was surprised to find that the images of the uranium crystals stored in the dark were brighter than the ones later exposed to sunlight. It appeared that the uranium salt crystals were spontaneously emitting penetrating irradiation of some type. He reported his finding to the French Academy of Sciences in March of 1896.

Becquerel went on to show that his newly discovered radiation contained electrically charged particles that could ionize gases and that pure uranium metal produced more penetrating radiation than uranium salts. Even more remarkable, he used a magnetic field to deflect the charged particles and noted two types: positive and negative, later named alpha and beta radiation, respectively. A few years later in 1900, Becquerel showed that the more penetrating negatively charged beta particles from radium had the same small mass as JJ Thomson's corpuscles (electrons). In the same year, the French physicist, Paul Villard identified an even more penetrating radiation emanating from radium, gamma rays.

At the time of Becquerel's initial reports, a young Polish graduate student Marie Sklodowska was fascinated by Becquerel's findings and decided to study the phenomena for her doctoral thesis. Marie met her future husband, Pierre Curie when she was completing her studies at the Sorbonne in Paris. They married in 1895 and when Pierre moved to the Municipal School of Industrial Physics and Chemistry, he arranged for a shed to be equipped as a laboratory for Marie to conduct her research. A former medical school dissecting room, the shed was badly ventilated, and the ceiling leaked. Marie would use the laboratory to make several groundbreaking discoveries in radioactivity.

Pierre and Jacques Curie and the Piezoelectric Effect

Years before Pierre met Marie, he and his brother Jacques Curie while working as research assistants at the Sorbonne made the exciting discovery that exerting pressure on certain crystals produced an electric charge across the crystal, one end became positively charged and the other negatively charged [5]. Jacques was working with the mineralogist Charles Friedel who was studying pyroelectricity, the ability to generate electric current by heating certain crystals.

A crystal is made up of atoms, ions, or molecules arranged in a highly ordered structure called a crystal lattice that extends in all directions. The spectrum of crystals ranges from table salt to diamonds. It was long known that heating a tourmaline crystal caused it to attract ash and glowing coals. Initially, this was thought to be a type of magnetism but later it was shown to result from an electrical polarization of the crystal. Friedel heated a variety of crystals and found several that developed a pyroelectric effect that seemed to depend on how the crystal was heated.

After discussing Friedel's findings, Jacques and Pierre thought that a temperature gradient must somehow alter the structure of the crystal. They wondered whether mechanical distortion with pressure would produce a similar electrical effect on the crystal. To test their idea, they cut crystals along their axis in thin slices and placed the slices between two pieces of tin foil connected to an electrometer. To exert pressure on the crystal slice, they used a clamp with insulation between the heads of the clamp and the tin foil. They found that a variety of crystals including tourmaline, boracite, topaz, and quartz exhibited a piezoelectric effect. Increasing the pressure on the crystal had the same effect as lowering the temperature while decreasing the pressure had the same effect as increasing the temperature. They then went on to show that each type of crystal had a unique piezoelectric constant that defined the relationship between the applied pressure and the electrical potential developed. For example, the piezoelectric constant for quartz was 5.4×10^{-8} coulombs per dyne.

After the Curie brothers published their findings, the Luxembourg-born French physicist, Gabriel Lippmann, known for developing the capillary electrometer, suggested that based on his thermodynamic calculations a reverse piezoelectric effect must also exist. Introducing an electrical current to the ends of a crystal should lead to a physical shift in the crystal lattice. The Curie brothers took up the daunting challenge of measuring less than a thousandth of a millimeter expansion or contraction in a centimeter thin slice of a crystal. After numerous tries, by using a lever effect and a microscope they were able to magnify the minute structural change in the crystal that occurred after applying electrical current.

We now know that a slight physical shift in the positive and negative atoms in the crystal explains the piezoelectric effect. For example, with quartz (silicon oxide), the silicone atoms have a positive charge and the oxygen atoms have a negative charge. When force is applied to the quartz crystal, the oxygen atoms move slightly to the left relative to the silicon atoms causing the left side of the crystal to be negative and the right positive.

More than 30 years later during the First World War, Paul Langevin a former student of Pierre Curie used the piezoelectric effect to make a sonar system to track German submarines. He glued small steel plates to both sides of a thin quartz crystal and applied very high-frequency alternating current to the plates. This caused the crystal to contract and expand at the frequency of the current producing ultrasonic sound waves. A reverse piezoelectric element was used to receive ultrasonic signals and convert them into alternating current signals. Quartz crystals were later used in a variety of electronic devices including early radios. In 1883, Jacques left Paris to teach mineralogy in Montpellier while Pierre who stayed in Paris met and married Marie and the two began their pioneering work in radioactivity.

Marie and Pierre Curie and Ionizing Radiation

In her new laboratory (shed), Marie Curie focused on the penetrating radiation described by Becquerel [3, 6]. She used a sensitive electrometer to measure electric charge that her husband and his brother had developed in their piezoelectric studies. Her first key observation was that the number of charged particles released by uranium compounds depended only on the amount of uranium present in the compound. This suggested that the radiation was originating from the uranium atoms themselves and not from some chemical reaction of the compounds. She then studied the uranium minerals, pitchblende and torbernite and the electrometer indicated that pitchblende had four times the activity as uranium alone and torbernite twice the amount. This suggested that these minerals likely contained other elements more radioactive than uranium. She began a search for these substances, and, by early 1898, she isolated the element thorium that also spontaneously released charged particles.

Keenly aware of the importance of rapid publication, Marie Curie sent off a paper summarizing her work on thorium to the French Academy of Sciences that was read for her by one of her professors, Gabriel Lippmann on April 12, 1898. Unfortunately, a month earlier a German physicist, Gerhard Carl Schmidt, had already reported that thorium was radioactive. By mid-1898, Pierre Curie became fascinated with Marie's discoveries and decided to set aside his work on crystals and join her on radioactivity. It was around this time that the husband-and-wife team began using the term "radioactivity" for the first time.

By the end of 1898, Marie and Pierre Curie published papers announcing the discovery of two other radioactive elements in pitchblende, polonium, and radium. Polonium was named for Marie's home country Poland and radium from the Latin word for "ray." They then began the tedious process of extracting appreciable quantities of these new elements from pitchblende. It took several years to separate a tenth of a gram of radium from a ton of pitchblende and they were never able to isolate polonium since it had an extremely short half-life of only140 days. Marie Curie was awarded a doctoral degree from the University of Paris in June 1903 and in December of that year she, her husband, and Becquerel shared the Nobel Prize in physics for their research on radioactivity becoming the first woman to win the prize.

At the time, it was rare for a woman to be involved in scientific research and most of the scientific community assumed that Pierre Curie was the driving force behind the couple's research. Initially the Nobel committee planned to award the prize to Pierre Curie and Becquerel but when Pierre was alerted, he complained to the committee, and they eventually added Marie's name to the award. She continued to face sexual bias throughout her career and after the tragic accidental death of Pierre in 1906 despite being the first woman to become a professor at the University of Paris she failed to be elected to membership in the French Academy of Sciences.

Marie Curie was not only the first woman to receive a Nobel Prize but the first person to receive two Nobel Prizes. In 1911, she received the prize in chemistry for her discovery of the elements radium and polonium. At the time Marie was involved in a romantic relationship with the physicist Paul Langevin, her husband's former student, who although married was separated from his wife. The tabloids accused her of home wrecking and preying on the younger Langevin. Marie was 5 years older than Langevin. A mob formed outside her home, and she had to flee with her two daughters to the home of a friend. The chair of Nobel Committee, Svante Arrhenius, tried to prevent Marie from attending the Nobel ceremony because of questionable morals but she ignored his comments and made her Nobel presentation.

The damaging effects of ionizing radiation were largely unknown in the early twentieth century when the Curie's and later Marie Curie alone was performing research on radioactive isotopes. The Curies did publish a paper noting that tumor cells were destroyed faster than normal healthy cells when exposed to radium but the use of radiation for treating cancer would come much later. It was not unusual for Marie Curie to carry test tubes with radioactive elements in her coat pocket and store them without precaution openly throughout the laboratory. Marie died at age 66 from complications of aplastic anemia probably related to her long-time exposure to ionizing radiation.

Rutherford and Atomic Structure

When Becquerel reported that uranium spontaneously emitted radiation in the form of electrified bodies, a young New Zealand research student, Ernest Rutherford, working in the Cavendish Laboratory at Cambridge with JJ Thomson, decided to investigate the nature of the electrified bodies [7]. In his initial experiment, he concluded that there were two basic types of discharge one much more penetrating than the other based their ability to penetrate layers of aluminum foil. In a paper published in 1899, he noted: "*These experiments show that the uranium radiation is complex and that there are present at least two distinct types of radiation—one that is very readily absorbed, which will be termed for convenience the alpha radiation, and the other of a more penetrative character, which will be termed the beta radiation*" [8]. Rutherford would go on to focus on the alpha radiation and show that it was a positively charged particle about twice the size of a hydrogen atom. As noted earlier, Becquerel showed that the highly penetrating beta particle had the same negative charge and small mass of an electron.

Ernest Rutherford was born in Nelson, New Zealand, in 1871, the son of a Scottish emigrant father and an English emigrant mother. His father was a wheelwright and his mother a schoolteacher. He received a scholarship to the University of New Zealand in Wellington where he excelled in mathematics and physics and after graduating in 1894, he obtained a scholarship to go to Trinity College where he met JJ Thomson. In 1898, the Macdonald Chair of physics at McGill University in Montreal, Canada, became available, and he took the position.

In Montreal, Rutherford teamed up with the chemist Frederick Soddy to develop their "transformational theory" of radioactivity. After studying a variety of radioactive elements, they concluded that the elements were in a constant state of transformation such that several new substances were being produced each with distinct chemical and physical characteristics different from the parent element. The transformation was atomic and not molecular, and each lost its radioactive properties according to specific laws. They wrote: "*The results obtained so far point to the conclusion that the beginning of the succession of chemical changes taking place in radioactive bodies is due to the emission of the alpha rays,* i.e., *the projection of a heavy charged mass from the atom. The portion left behind is unstable, undergoing further chemical changes which are again accompanied by the emission of alpha rays, and in some cases also beta rays*" [9].

As to the source of energy responsible for the release of the alpha rays with high velocities, Rutherford could only speculate that the atoms of these radioactive substances must consist of rotating or oscillating positively charged bodies that escape by some unknown internal or external force. In 1903, the Curie laboratory reported that radium generated heat and was always above the temperature of the surrounding air presumably due to the heating effect of the released alpha particles. Rutherford studied the phenomenon and found that radium produced heat proportional to the number of alpha particles that it released. The amount of energy released, about 80,000 horsepower in a day by a kilogram of radium, was enormous considering the small weight of matter involved.

Rutherford moved back to England in 1907 to become the Langworthy Professor of Physics at the University of Manchester, succeeding Sir Author Schuster [10]. There, he teamed up with Hans Ludwig Geiger, a German physicist who completed a fellowship with Schuster and stayed on as an assistant, to quantify the alpha particles released by radium. Rutherford had felt from the beginning that the alpha particle was likely a helium atom carrying two-unit charges, but this was not easy to prove experimentally. They developed an electrical method for counting the alpha particles expelled from a radioactive substance. The alpha particles were shot through a tiny opening in a vessel that contained air or other gases at low pressure under the influence of an electric field near the sparking value. There was a sudden deflection of the electrometer needle each time an alpha particle entered the testing vessel (an early version of a Geiger counter). From this sample, they calculated that 3.4×10^{10} alpha particles were emitted per second from 1 g of radium.

Based on these and similar experiments counting the number of scintillations produced on a phosphorescence screen, Rutherford and Geiger calculated the charge of a single alpha particle by dividing the number of alpha particles by the total positive charge carried by all the alpha particles emitted. They found a positive charge of 9.3×10^{-10} electrostatic units per alpha particle, double the 4.65×10^{-10} units of a hydrogen atom. They were convinced that alpha particles carried two-unit charges and if the charge was neutralized, the alpha particles became stable helium atoms. If so, radioactive materials should contain helium and they estimated that 1 g of radium should produce about 160 cubic millimeters of helium per year. Helium was later found in radioactive substances that emitted alpha particles at roughly the rate predicted by their calculations.

Prior to 1910, Rutherford like most scientists of the time, assumed that the positive charge and mass were evenly distributed throughout an atom with tiny negatively charged electrons scattered in between, e.g., the plum pudding model of his former mentor JJ Thomson. But in working with Geiger and his alpha particle counting chamber, he was impressed with the marked variability in the angle of deflection as alpha particles collided with gas molecules in the detection chamber. He had assumed that the alpha particles were too heavy to be so strongly deflected. Rutherford suggested that Geiger along with a young graduate assistant, Ernest Marsden conduct a series of experiments to see how thin foils of different metals and thicknesses deflected alpha particles.

Geiger and Marsden bombarded metal foils with alpha particles from a radium source and used a fluorescent screen to identify the trajectories of the particles. The apparatus was designed to count scintillations from every angle by attaching the microscope and screen to a rotating cylinder that could be moved in a full circle around the foil. They spent hours in the dark counting the tiny flashes of light with a microscope as the alpha particles struck the screen. A variety of metals were used but they preferred gold foil because it was pliable and could be made very thin. If the plum pudding model of atoms was correct, they expected that all the alpha particles would readily pass through the thin gold foil with only a slight amount of deflection.

What they found was surprising. Although nearly all the a-particles passed through the thin gold foil without any apparent deflection, a small number were deflected at large angles and rarely an alpha particle bounced back in the opposite direction as though it had collided with something large and dense in the foil. Furthermore, the amount of deflection increased with increasing thickness of the foil and the atomic mass of the metal.

Based on these findings, Rutherford published his famous paper in 1911 on the structure of the atom that marked the beginning of modern physics [11]. Rather than a diffuse positive charge, he suggested that the positive charge was concentrated in a small dense mass at the center of the atom, the nucleus, and that most of the atom was empty space. If the atom was the size of a large ballroom, the nucleus was so small that it was barely visible. This explained why most of the alpha particles passed through the metal foil with minimal deflection whereas a rare alpha particle struck the positive nucleus and bounced straight back. Negatively charged electrons orbited the positive nucleus to explain the overall neutral charge. As to the composition of the positively charged nucleus, Rutherford speculated that it might be made up of combinations of a-particles (helium nuclei), 49 in the case of gold. This would explain why alpha particles were released by radioactive substances.

A few years later, Rutherford conducted a series of experiments with another young assistant, William Kay in which they studied the effect of bombarding different gasses with alpha particles. When they directed alpha particles into a chamber containing hydrogen, some of the hydrogen atoms were knocked forward and none were deflected backward. Sending a beam of alpha particles into a chamber with nitrogen dislodged hydrogen from the nitrogen atoms suggesting that the basic building block of the nucleus had a single unit of charge (a hydrogen nucleus) and not the double unit of charge in an alpha particle (a helium nucleus). This single unit of charge particle later became known as a proton. Rutherford returned to Cambridge in 1919 where he succeeded JJ Thompson as Professor and head of the Cavendish Laboratory. There he would mentor several future Nobel laureates including JJ Thomson's son GP Thomson.

Discovery of the Neutron

Rutherford had determined that the mass and positive electric charge of an atom were concentrated in a very small central nucleus. By 1920, he and others had noted a disparity between the observed charge units and mass units of the atomic nucleus [12]. The nuclear mass units were about twice the positive electric charge units. Since unstable atomic nuclei like uranium gave off alpha and beta particles, it was already assumed there were electrons in the atomic nucleus. Rutherford proposed that the nucleus had hydrogen nuclei (protons) with about half coupled to electrons so that the nuclear charge (atomic number) was about half the atomic mass. The alpha particle had four hydrogen nuclei, two closely bound to electrons giving a mass of 4 and a charge of +2. In his model, Rutherford placed half the electrons in the nucleus and the other half in orbits around the nucleus to account for the overall neutral charge of the atom. For example, the nucleus of the nitrogen atom had 14 protons and seven electrons with another seven electrons called external electrons orbiting the nucleus.

In 1921, an American chemist, William Harkins named the combined hydrogen nucleus-electron particle a neutron, and, about the same time, the hydrogen nucleus particle was called a proton. Rutherford and a young colleague, James Chadwick, spent much of the 1920s searching for the elusive combined proton-electron particle. But identifying the neutron proved more difficult than expected since the mass of an electron was so much smaller than that of a proton and estimates of the mass of a neutron varied greatly. Furthermore, the spin and magnetic moment of electrons predicted by the rapidly advancing field of quantum physics (see below) were not consistent with them being present in the nucleus. By the end of the 1920s, Rutherford's proton-electron hypothesis was on life support.

In 1930, German physicists, Walther Bothe and Herbert Becker, identified a new highly penetrating radiation when alpha particles from polonium passed through light elements such as beryllium or lithium. Normally alpha particles are deflected by an electric field since they are positively charged. Bothe and Becker, however, noted that the penetrating radiation they observed was not deflected by an electric field, so they speculated that it was gamma rays. But the radiation was more penetrating than known gamma rays. Rutherford and Chadwick at the Cavendish Laboratory were highly skeptical of the Bothe and Becker hypothesis, and Chadwick set out to prove the radiation was not gamma rays.

Chadwick and several colleagues at Cavendish had already been studying the effect of alpha particles on light elements and had developed a way to accurately detect and count protons emitted by the bombardment. Chadwick devised a series of experiments studying the effect of the new radiation by passing it through hydrocarbons and a variety of gases. The radiation consisted of particles with roughly the same mass as protons but without charge, neutrons [13]. Chadwick received the Nobel Prize in physics for the discovery of the neutron in 1935.

Despite Chadwick's discoveries, not everyone was ready to give up on the electron-proton model of a neutron. After all, how did one explain beta radiation emanating from the nucleus of unstable elements. The final piece of the puzzle came in 1934 when the Italian physicist Enrico Fermi described beta decay as a process where a neutron decays to a proton by creating an electron and another particle with no charge and almost no mass, a neutrino [14]. His theory was a landmark being the first to suggest how atomic particles could be created and destroyed. It later became the theory of weak interaction predicting one of the four fundamental forces of nature. The elusive neutrino was not detected until after Fermi's death.

The newly discovered neutron radiation provided a new tool to probe nuclear structure which ultimately led to the discovery of nuclear fission, a type of nuclear transformation where the resulting nuclear fragments are different elements. Liberation of neutrons during the fission process could set off a nuclear chain reaction with the release of massive amounts of energy. This discovery led to the development of the nuclear bomb and nuclear power by the end of the Second World War. Protons and neutrons were thought to be elementary particles until the 1960s when they were shown to be composed of even smaller particles called quarks.

Planck, Einstein, and the Quantum

As Rutherford carried on his groundbreaking discoveries on the structure of the atom, an even more revolutionary breakthrough was occurring in basic physics. Newton and Maxwell's equations defining the nature of matter and electromagnetic waves were being replaced by Max Planck and Albert Einstein's equations based on the concept that all energy was packaged in discrete units, called quanta. The atomic world of Rutherford and the quantum world of Planck and Einstein came together when a young Danish physicist Niels Bohr moved from JJ Thomson's laboratory at Cambridge to work with Rutherford in Manchester. Bohr used the new quantum physics to explain how electrons are distributed in atoms.

At the turn of the twentieth century, there was a great deal of interest in the phenomenon of black body radiation [15]. If one heated a black object, it would glow first red, then yellow, and finally blue. The hotter the body the greater the oscillations in its molecules and the higher the frequency and energy of radiation emitted. Standard thermodynamic equations predicted that the energy increased as frequency and temperature increased. An infinitely hot body was predicted to radiate infinite energy. To deal with this "*calamity,*" the German physicist Max Planck suggested that the energy in black body radiation existed in individual packets called quanta just as matter existed in discrete units. The energy of black body radiation was defined by the equation $E = hv$, where h represented the quantum unit and v the vibration frequency. The quantum dealt with the absurdity of infinite energy at infinitely high frequencies since at extremely high frequencies, the emission of a single quantum would require more energy than was available.

Whether Planck initially understood the full implications of his "*quantum theory*" is doubtful, and it would be 5 years before he published on the subject again. Since data on the relationship between the energy and frequency of radiation was available, he calculated the constant $h = 6.55 \times 10^{-27}$ erg-sec which became known as Planck's constant. In his early lectures, however, Planck hinted that his constant h was a mathematical necessity and should not necessarily be taken too seriously.

According to the physicist and science historian A. Douglas Stone: "*There was nothing innocent about Planck's explanation of the radiation law. If it were the real explanation, it was a time bomb hidden in a thicket of algebra, which would explode with earth-shattering implications. Atoms and molecules were not little Newtonian billiard balls; they obeyed completely different and counterintuitive laws*" [16].

Someone who did take Planck and his constant very seriously was a young relatively unknown physicist working in a Swiss patent office to support his family, Albert Einstein. In a remarkable burst of creative energy in the first half-decade of the twentieth century, Einstein formulated and published his special theory of relativity and his photon theory of light. In his paper entitled *On a heuristic point of view concerning the production and transformation of light* published in 1905, Einstein wrote: "*According to the assumption to be contemplated here, when a light ray is spreading from a point, the energy is not distributed continuously over ever-increasing spaces, but consists of a finite number of energy quanta that are localized in points in space, move without dividing, and can be absorbed or generated only as a whole*" [17]. He suggested that light consisted of a stream of particles (quanta or photons) each with energy hν.

Einstein pointed out that his quantum theory of light not only explained black-body radiation but also phosphorescence (photoluminescence) and the photoelectric effect. When a substance absorbs light and later releases it (phosphorescence), the wavelength of the released light is invariably longer than that of the absorbed light. Einstein speculated that when a molecule absorbs a quantum of light, some of the energy is lost so that the highest frequency of the quantum of light it could release was always equal to or less than that of the quantum of light absorbed.

Most impressively, Einstein's quantum theory of light explained the mysterious photoelectric effect [18]. The observation that light could alter the electrical properties of certain substances dates to Becquerel's father in the mid-nineteenth century (see Chap. 9). In 1887, Heinrich Hertz inadvertently observed the effect when he placed his electromagnetic wave detector, a ring of wire with a spark gap, in a dark box so he could see the spark better. To his surprise, the maximum spark length was decreased in the dark. Light somehow assisted the electrons to jump across the gap. One of the most dramatic demonstrations of the photoelectric effect was seen with a simple gold leaf electrometer. If the electrometer was negatively charged, the negative charge passed on to the gold leaves causing them to separate. If a brief high-frequency light in the blue to ultraviolet range from a spark gap or an arc light was shown onto the cap of the electrometer, the leaves collapsed as electrons were released from the cap. But if lower-frequency light in the red to green range was shown on the cap, the leaves remained separated. It did not matter how intense or how long the lower-frequency light was on, the gold leaves remained separated.

Although this finding was baffling based on classical Maxwellian physics, Einstein's quantum light theory provided an explanation. Each quantum of light interacted with only one electron. To dislodge a negatively charged electron from its positively charged atom, the electron had to absorb just the right amount of energy, the high energy in one quantum of blue or ultraviolet light. There was not enough energy in a quantum of red or green light to dislodge the electron, so there was no photoelectric effect regardless of the intensity of the light (number of quanta). The electrons released by light could be collected and used to create an electric current in a circuit, the basic principle of modern-day solar panels.

So, what was light, a wave or a particle? According to Einstein, it had properties of both. It exhibited refraction and interference like classical waves yet when exchanging energy with matter it acted like a particle. Einstein's genius was that he was willing to accept the apparent contradiction unlike most of his fellow physicists. Electromagnetic waves were a new entity, different from classical waves, and required a new physics. One consequence of this new physics was that there was no longer a need for a medium in which the waves traveled. Ether became superfluous.

Recall that classical physicists from Maxwell to Hertz were convinced of the critical importance of ether for electromagnetic transmission. According to Hertz, "*Take electricity out of our world and light vanishes; take luminiferous ether out of our world and electric and magnetic fields can no longer travel through space*"

[19]. Yet no one had ever found physical evidence of the all-pervasive ether. Einstein concluded: "*one can obtain a satisfactory theory only if one drops the ether hypothesis. In that case electromagnetic fields which constitute the light will no longer appear to be states of a hypothetical medium, but rather independent entities emitted by the source of light*" [20].

Quantum Models of the Atom

A young Danish physicist, Niels Bohr joined Rutherford's laboratory in Manchester in 1912, a year after Rutherford had published his famous paper on the nuclear structure of the atom [21]. Bohr had originally been working with JJ Thomson at Cambridge but found Thomson too conservative for his taste and decided to move on to Manchester where he found Rutherford to be more energetic and inspiring. In late 1911, the Belgian inventor and industrialist Ernest Solvay sponsored a congress in Brussels with the help of Nernst to focus on the revolutionary new quantum physics. It featured Planck and Einstein along with prominent physicists throughout Europe including Rutherford. When Bohr arrived in Manchester, the Solvay Conference was the main topic of discussion in Rutherford's laboratory and Bohr was excited about the possibilities of introducing the quantum into Rutherford's atom.

Bohr immediately saw a problem with Rutherford's atom. The light negatively charged electrons orbited the heavy positively charged nucleus like planets orbiting the sun. The electrons had to be moving since if still they would be pulled into the nucleus. But according to Maxwell's equations, moving electrons radiate energy in the form of electromagnetic waves that would also cause the electrons to collapse into the nucleus. Bohr came to the logical conclusion that classical electrodynamics do not work at the atomic level and as Einstein had concluded before him, it was necessary to introduce Plank's quantum in any discussion at the atomic level. He spent the remainder of his time in Manchester working on a new quantum theory of atomic structure and he published the seminal paper on the new model after returning to Denmark in early 1913.

In a nutshell (pun intended), Bohr proposed that electrons could only revolve in a series of discrete orbits or shells at fixed distances from the atomic nucleus (Fig. 12.1). These so-called stationary shells were at distances given by an integer multiple of Planck's constant nh, where $n = 1,2,3\ldots$for each allowable shell. Electrons in the lowest shell could not get any closer to the nucleus. In these allowable shells, electrons could move without generating radiation or loss of energy. The only way to gain or lose energy was for an electron to absorb or emit electromagnetic radiation by jumping from one allowed shell to another. The frequency (f) of the absorbed or emitted radiation was determined by the energy difference between the orbit levels given by $E_2 - E_1 = hf$, where h is Planck's constant.

Although primarily based on an intuitive application of Planck's quantum, Bohr did have some objective data on which to base the predictions of his atomic model [22]. As discussed in Chap. 8, it had long been known that each element gave off a

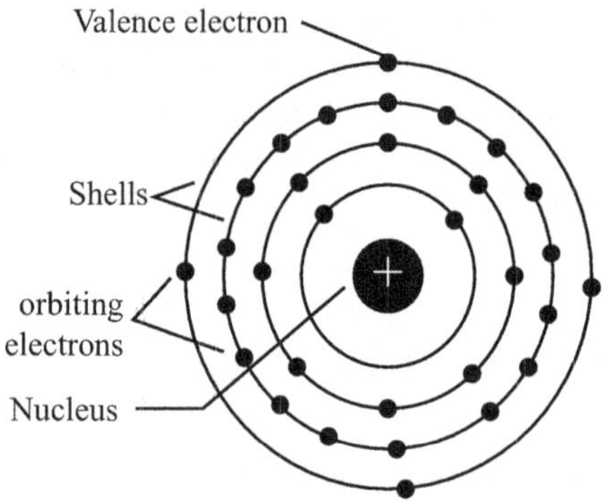

Fig. 12.1 Quantum model of an atom. Electrons orbit the nucleus in a series of discrete shells at a fixed distance from the nucleus based on an integer multiple of Planck's constant. Each shell has a maximum number of electrons (see Table 12.1). Valence electrons are electrons in the outermost shell available to participate in a chemical bond if the outermost shell is not closed

distinct pattern of colored lines (spectral lines) when light was shown upon it. Many of the new elements were discovered using this technique although why the lines were spaced in these patterns was completely unknown. The simplest spectral pattern was associated with the hydrogen atom consisting of a single proton and electron. In 1888, the Swedish physicist Johannes Rydberg described a formula based on the inverse of the square of integers that predicted the spectral lines of hydrogen. Bohr's quantum atomic model predicted the same spectral pattern as Rydberg.

According to classical electrodynamics, light and other electromagnetic waves were produced by the vibration of electrons and the frequency of the radiation reflected the frequency of the vibrating electron. In Bohr's theory, there were two electron frequencies, the frequency of the electron in the initial orbit and the frequency in the final orbit, but neither of these frequencies determined the frequency of the emitted radiation. This was given by the energy difference between the orbit levels. Although Bohr's theory was consistent with many of the features of Einstein's wave-particle duality theory, Einstein initially was highly skeptical questioning whether Bohr's restriction of electrons to a few allowable orbits was arbitrarily designed to fit the hydrogen spectrum. It did not appear to be the new insight that he was searching for.

But over the next several years, Bohr's theory was shown to have surprising predictive power and Einstein jumped on board. The American physicist, Theodore Lyman, discovered a new series of ultraviolet lines in the hydrogen spectrum that were predicted by Bohr's formula. Even more impressive, Bohr's formula explained the so-called Pickering-Fowler spectrum described by astrophysicists E. C. Pickering and Alfred Fowler. Pickering observed a new series of spectral lines in the light from a distant star and Fowler produced the lines with a helium-hydrogen mixture on earth. Bohr was able to produce the series of lines with his formula by assuming they were generated by ionized helium atoms (helium nucleus with only one electron) in the star's photosphere.

Table 12.1 Number of electrons in each shell and subshell

Shell name	Subshell name	Subshell maximum electrons	Shell maximum electrons
K	1s	2	2
L	2s 2p	2 6	8
M	3s 3p 3d	2 6 10	18
N	4s 4p 4d 4f	2 6 10 14	32
O	5s 5p 5d 5f 5g	2 6 10 14 18	50

Years later in review article on the origins of quantum theory published in *Science* in 1951, Einstein praised Bohr's theory: "*All my attempts, however, to adapt the theoretical foundation of physics to this [quantum] knowledge failed completely. It was as if the ground had been pulled out from under one, with no firm foundation to be seen anywhere, upon which one could have built. That this insecure and contradictory foundation was sufficient to enable a man of Bohr's unique instinct and tact to discover the major laws of the spectral lines and of the electron-shells of the atoms, together with their significance for chemistry, appeared to me like a miracle—and appears to me as a miracle even today. This is the highest form of musicality in the sphere of thought*" [23].

But not everyone was as enthusiastic as Einstein with Bohr's electron-shell theory. It worked for hydrogen and a few atoms with small nuclei, but it proved difficult to use with atoms with larger nuclei. Some questioned the basic premise of calculating electron orbits. Into this confusing world of quantum radiation stepped a 26-year-old German physicist, Werner Heisenberg and his famous uncertainty principle [24]. His basic premise was that it is impossible to measure the position and velocity of an atomic particle using classical space-time mechanics so why even try. To predict the future position and velocity of a particle, one must know the current position and velocity. To "see" the particle, one would need to use an electromagnetic wave with short enough wavelength to accurately measure the position. But a quantum of an extremely high-frequency electromagnetic wave will alter the particle changing its velocity in unknown ways. The higher the frequency, the greater the disturbance. The more accurately you measure position, the less accurately you can measure velocity.

After spending a few months in 1924 visiting Niels Bohr in Copenhagen, Heisenberg, joined the physicist Max Born in Gottingen, Germany. Born surrounded himself with an outstanding team of quantum researchers and not only did he win the Nobel Prize but six of his research assistants went on to win the Prize. With his uncertainty principal Heisenberg came to Planck's constant in an entirely different way. He found that the uncertainty of the position times the uncertainty of the velocity times the mass of a particle was never smaller than a constant, Planck's constant. Heisenberg represented the trajectory of a particle with an array of variables in rows and columns with each variable oscillating sinusoidally at a characteristic frequency. In the case of electrons in an atom, the numbers were the frequencies at which the atom absorbed and emitted light.

Within a few months Heisenberg joined with Born and another talented assistant Pascual Jordon to publish a paper describing the mathematics of Heisenberg's new quantum mechanics. Born and Jordan recognized that Heisenberg's number arrays were matrices with well-known mathematical properties. When multiplying matrices, *x* times *y* was often not the same as *y* times *x*, which became a prominent feature of quantum mechanics. As a rule, quantum mechanics does not calculate a specific outcome but rather the statistical likelihood of several different outcomes. It therefore introduces an element of unpredictability and randomness to physics that bothered many physicists of the time including Einstein. Indeed, despite his praise for Bohr's quantum theory of the atom, he stubbornly refused to accept Heisenberg's uncertainty principle epitomized by his famous quote: "*I…am convinced that [God] is not playing dice*" [25].

Dirac and His Beautiful Equations

Over the next few decades, quantum mechanics was formulated and reformulated by a series of young physicists and atomic particles were no longer thought of in terms of position and velocity but in a quantum state, a combination of position and velocity. One of the most enigmatic yet successful contributors to this transformation of quantum mechanics was the English theoretical physicist, Paul Dirac [26]. According to his biographer, Graham Farmelo, Dirac undoubtedly had features of autistic spectrum disorder. Born in Bristol in 1902, he was a mathematical prodigy who eventually made his way to Trinity college at Cambridge. He grew up in a dysfunctional family with a domineering father and he rarely spoke more than a few words to anyone. He later wrote; "*I did not know of anyone who liked someone else—I thought it did not happen outside novels*" [27].

Dirac saw beauty in mathematical equations unrelated to their application to the physical world. After visiting Bohr in Copenhagen and only a few years after he first heard about Maxwell's unified field theory of electricity and magnetism at Cambridge, Dirac set about developing a quantum version of Maxwell's theory. Photons of light were constantly being created and destroyed yet ordinary quantum mechanics did not explain the process. In 1926, Durac developed a mathematical model with a creation operator and an annihilation operator to describe the creation and destruction of photons. As per Maxwell, electromagnetic fields pervade the entire universe, and in Dirac's model, the appearance of a photon represented an excitation of a field at a particular place and time described by the creation operator and the disappearance of a photon, the de-excitation of the field described by the annihilation operator. There was no need to be concerned about whether light was a wave or a particle, it had properties of both.

In Einstein's original description of the photon, he described how photons were emitted and absorbed as electrons jumped from one energy level to another. He also predicted that two photons with same energy level would be released if the stimulating photon had the exact same amount of energy as the difference between the energy levels of the electron—the original photon plus a second photon as the electron returned to the original level. This was a surprising and important prediction that Dirac's mathematical model nicely explained. This phenomenon underlies modern lasers where photons of the same energy level are produced and focused into a beam of monochromatic light. Dirac's paper published in the Proceedings of the Royal Society marked a new era in theoretical physics where mathematical formulation dominated the presentation.

Dirac spent the next several years working on the mathematics of quantum mechanics and he later accepted the Lucasian Professorship of mathematics at Cambridge a position once held by Newton. His "*beautiful*" equations made several seemingly bizarre predictions that were later confirmed. Particles such as electrons and protons had a quality called spin ½ meaning that they did not return to baseline after just one revolution but required two revolutions. The spin, an angular momentum, accounted for the associated magnetic field. Electrons were predicted to have an antielectron, called a positron. If an electron met a positron, both were annihilated. The discovery of the positron in 1932 led to Dirac receiving the Nobel Prize in 1933.

Although many scientists at the time lamented the trend to esoteric mathematical descriptions to explain the actions at the atomic level, it was hard to argue against the new quantum mechanics since it explained so many experimental findings that were otherwise impossible to explain. The theory underlies much of modern technology from ubiquitous lasers in bar code readers and CD players to the transistors and integrated circuits of computers and cell phones. In medicine, it was critical in developing imaging techniques such as CT and MRI and microstimulators that are implanted in patients.

Electron Waves

Recall that JJ Thomson discovered the electron when he showed that cathode rays could be deflected by an electric field indicating that the "rays" were a stream of negatively charged particles (corpuscles). But quantum mechanics predicted that electrons had features of both particles and waves. In a remarkable coincidence, JJ Thomson's son GP Thomson conducted one of the definitive experiments to show that electrons exhibited the property of diffraction just like electromagnetic waves.

One of the first researchers to focus his work on the wave-particle duality of the electron was the young French physicist, Louis de Broglie [28]. Born the youngest of five offspring of Duke Victor de Broglie in 1892, Louis would go on to combine Einstein's theory of relativity with quantum theory into a wave theory of electrons that led to a Nobel Prize in 1929. His much older brother Maurice, a famous quantum physicist in his own right, became a father figure for Louis when the duke died in 1906. Maurice had been a secretary at the 1911 Solvay Conference in Belgium mentioned earlier in the chapter and Louis later recalled: "*I began to think about quanta from the moment that my brother gave me the notes of the Solvay Congress of 1911, probably at the beginning of 1912*" [29].

After completing his undergraduate training, de Broglie was called into military service where he was assigned to a wireless telegraphy unit located under the Eiffel Tower. He completed his service in 1919 and decided to work on quantum physics joining the team of Paul Langevin at the College de France. De Broglie wanted to do theoretical physics and Langevin was one of only a few physicists in France working on quantum theory. On beginning his doctoral thesis, he pondered the wave-particle nature of light and even speculated that a photon of light had an infinitely small mass. He then went on to envision how an electron could be a wave. He had the brilliant insight to combine the two famous equations of Einstein and Planck, $E = mc^2$ and $E = hv$ to relate a wave frequency (v) with a particles mass at rest: $m = hv/c^2$.

De Broglie's equation suggested that every particle including an electron had a baseline oscillation with frequency (v_1) even at rest. When in motion it was associated with a wave of oscillation (v_2) that guided it through space. According to de Broglie, the particle "*glides on its wave, so that the internal vibration of the particle (v_1), is in phase with the vibration of the wave (v_2), at the point where it finds itself*" [30]. As physicist A. Douglas Stone nicely described "*the particle is like a lucky surfer, permanently attached to the crest of the perfect wave*" [30]. De Broglie calculated that the speed of this guiding wave was faster than the speed of light so according to Einstein's theory of relativity it could carry no energy and only guide the particle motion.

As strange as de Broglie's theory appeared, when he applied it to an electron orbiting a nucleus it predicted the same electron shells for hydrogen as predicted by Bohr's theory. As an electron circled the nucleus, it emitted phase waves that speeded ahead lapping the electron circuit. The requirement that the particle oscillations and the wave oscillations had to be in phase limited the allowed orbits to the discrete orbits seen with hydrogen. De Broglie completed his thesis in the spring of 1924 and gave it to Langevin who initially thought it was "far-fetched." But after Einstein read the thesis and gave it a favorable review, Langevin changed his mind, and the thesis committee accepted the thesis.

Meanwhile at Cambridge, one of the few British physicists to take de Broglie's theory seriously was George Paget Thomson (GP), son of JJ Thomson. GP was born in 1892 and he literally grew up at Cambridge University and the Cavendish laboratory [31] . He was home taught by his mother until age 9 and at age 14 he saw his father win the Nobel Prize. JJ guided GP's university curriculum urging him to first focus on the Mathematical Tripos and then move to experimental physics just as he had done. They met regularly for lunch on Sunday followed by a walk in the fens where JJ recommended reading material for the week. It was just a matter of time for GP to join JJ in scientific collaboration at the Cavendish.

JJ Thomson showed little interest in quantum theory and was even slow to accept Rutherford's concept of an atomic nucleus composed of positive particles. He and GP conducted a series of experiments with cathode ray tubes focusing on "positive rays" which suggested that the positive charge of an atom varied from atom to atom unlike the negatively charged electrons. In June of 1922, the position of Professor of Natural Philosophy opened at the University of Aberdeen and GP applied and received the appointment. On his own for the first time, GP used the £1600 allocated for a new laboratory to reproduce his father's laboratory including the cathode ray tube they had been working with.

Over the first few years at Aberdeen, GP continued the work on positive rays started with his father but he was frustrated by his lack of progress. In the summer of 1926, GP attended the annual meeting of the British Association for the Advancement of Science held at Oxford and he was impressed by a presentation of Max Born on quantum wave mechanics. Born mentioned that a young German physicist had tried to detect diffraction of an electron beam passing through a metallic film but was unsuccessful. Diffraction is the bending of waves around an obstacle to produce a geometric shadow of the obstacle. GP discussed the matter with several colleagues and was convinced that the cathode ray tube he had been using at Aberdeen needed only slight modifications to test whether electrons were diffracted when passing through solids.

On returning to Aberdeen, GP Thomson set about modifying his cathode ray tube for the diffraction experiments. He replaced the positive rays with a beam of electrons by simply reversing the current that produced the gaseous discharge. The only other significant change was to add a thin metallic film for the electrons to pass through. For this he relied on a laboratory assistant, C. G. Fraser, who had trained as a watchmaker. But the problem proved much more difficult than he had anticipated. First there were technical problems with the cathode ray tube and then problems with getting the film thin enough. They started with thin celluloid films and then tried a variety of metal films including gold, aluminum, and platinum. It took more than a year but by October 1927 he had high-quality photographs of the diffraction pattern of cathode rays passing through thin metal films showing circular halos typical of the diffraction pattern of electromagnetic waves such as X-rays.

Thomson first published the photographs in December 1927 in *Nature* followed by a series of publications in 1928 and 1929. At the same time, an American physicist, Clinton Davisson, who also attended the presentation of Max Born at Oxford in 1926, teamed up with a colleague at the Bell laboratories, Lester Germer to demonstrate electron diffraction. The three shared the Nobel Prize for their work.

The final and possibly most significant piece of the puzzle regarding the wave nature of electrons was supplied by the Austrian theoretical physicist, Erwin Schrödinger, working at the University of Zurich [32]. Like many other physicists at the time Schrödinger wandered why electrons were restricted to the select shells defined by quantum theory. Bohr had shown that the restricted shells were defined by Planck's constant h but why was this the case? Schrödinger was impressed with de Broglie's description of the wave nature of electrons and thought that the wavelength might determine the shells that are allowed. Only shells where the confined waves fit perfectly can occur (beginning and ending at the exact same spot in the cycle). De Broglie had hinted that this was the case, but Schrödinger went on to use de Broglie's equation to calculate not only the allowed shells but the fact that an electron could only take on certain energy values. Finally, there was an explanation for the quantum that was not arbitrary. Electrons in an atom had quantized energy levels because they were confined to select shells by their wave properties.

Quantum Theory, Semiconductors, and Transistors

As noted earlier, a crystalline solid is made up of atoms, ions, or molecules arranged in a highly ordered structure called a crystal lattice [33]. Most metals are polycrystals made up of many tiny crystals fused together into a single solid. The great majority of electrons in a crystalline solid are bound to the atoms in local shells, but some electrons are free to move about through the crystal. These free electrons, often called "conduction electrons," are most abundant in metals, infrequent in semimetals or semiconductors and rare in insulators. For example, in metals there are one or more free electrons per crystalline unit, in semiconductors about one free electron per one thousand units and in insulators as few as one free electron per trillion units.

In 1927, the German theoretical physicist Alfred Sommerfeld applied quantum mechanics to the free electron model of Paul Drude to create a new model that solved many of the problems of Drude's model [29]. Sommerfeld along with Born were responsible for training most of the young German physicists working on quantum mechanics at the time. According to Sommerfeld, conduction electrons move about a crystalline structure with relative ease since the positive ions left behind occupy only a small fraction of the volume of the crystal and the electrons interact very little with the ions or each other. Applying a voltage to the metal causes the electrons to accelerate and produce an electric current. Electrons in deeper orbits bound to ions through shared covalent bonds move from ion to ion contributing to the overall electric current. Some electrons switch back and forth between the two modes.

An important feature that determines the ability of conduction electrons to move freely within metals and other crystals is a phenomenon called scatter. Electrons do not scatter from ions located in their expected positions in the crystal lattice but if an ion is missing or replaced by an impurity, electrons will scatter from the "defect." Ions are constantly vibrating about their lattice site and occasionally one will be displaced producing a defect from which electrons scatter. At higher temperatures, ion vibration increases producing more defects which increases resistance to current flow due to increased electron scatter. This explains why conduction decreases with increasing temperature and increases with decreasing temperature.

Semiconductor crystals have electrical conduction properties between metals and insulators. Like insulators, semiconductors have no conduction electrons if the crystal is perfect but if there are imperfections in the crystal that contribute conduction electrons, these electrons can move about the semiconductor much easier than in an insulator. At room temperatures, insulators have "traps" where conduction electrons are tightly bound whereas at the same temperature, semiconductors only weakly bind conduction electrons. With a small number of impurities (providing free electrons), many semiconductors can provide exceptionally good electrical conductivity. One can control the conductivity by controlling the type and number of impurities introduced into the semiconductor crystal.

Silicon crystals, commonly used as semiconductors, accept a variety of impurities in a process called doping. Impurities can either provide conduction electrons or "holes" where electrons would normally be in the silicon crystal. Examples of impurities with conduction electrons are arsenic, phosphorus, and antimony and with holes, boron, gallium, and aluminum. The movement of holes in response to an external voltage in a semiconductor is almost as fast as that of conduction electrons. Semiconductors with impurities causing a preponderance of conduction electrons are called n-type (for negative) semiconductors while those with a preponderance of holes are called p-type (for positive). Neither n-type or p-type semiconductors have an electric charge but are ready to emit or receive electrons when a voltage is applied.

If tiny slices of n-type and p-type silicon are joined together with electrodes on each side and a voltage is applied across the electrodes, current will only flow across the junction in one direction from the n-type side to the p-type side. The result is a diode or rectifier to convert alternating to direct current like Fleming's diode vacuum tube but at a molecular scale. Such diodes can also be made to give off light when electricity flows through them producing light-emitting diodes (LEDs). If instead of two, three layers of silicon are joined together (n-p-n or p-n-p) each with separate electrical contacts, the resulting "transistor" can serve as an amplifier or an on-off switch like de Forest's triode vacuum tube or Audion. The third terminal controls the current flow between the other two terminals.

The first patent of the transistor principle was filed by the Austrian-Hungarian physicist Julius Edgar Lilienfeld in Canada in 1925, but Lilienfeld did not follow up with publications and his work was largely ignored. The Americans John Bardeen, Walter Brattain, and William Shockley working at Bell Laboratories in New Jersey, the research division of AT&T, are credited with producing the first working transistors in 1947. Shockley had built an operational version of a transistor from Lilienfeld's patent, but this was never mentioned in any of his publications [34].

The Bell laboratories work on the transistor evolved from war-time work on germanium crystal diodes used in radar receivers [35]. The traditional tube-based circuits did not switch fast enough, so the Bell researchers tried solid-state diodes as a replacement. Shockley, who joined the Bell laboratories in 1932 after obtaining his PhD from MIT, was interested in developing solid-state switches and amplifiers.

He applied for funding and after several failed attempts on his own joined forces with Bardeen and Brattain. Bardeen, a theoretical physicist, would eventually develop a new branch of quantum mechanics to account for the unusual behavior that occurred at the surface of semiconductors.

Although Shockley was their supervisor, the first transistor was produced by Bardeen and Brattain. They used a germanium crystal, a small plastic wedge, and a small strip of gold foil to make the first solid-state junction transistor. Although the device was very delicate, they were able to amplify a signal up to 100 times. It took several more years for engineers at the Bell laboratories to grow germanium crystals that could be used in commercial transistors. In the aftermath of the transistor discovery, a major dispute developed between the three investigators regarding credit. Shockley was insecure since he was not directly involved in the design, yet he felt that it was his idea. Bardeen and Brattain felt that Shockley was trying to steal credit for the invention. Bardeen left the Bell Laboratories to take an academic position at the University of Illinois and Brattain agreed to a transfer to another section. They jointly received the Nobel Prize in1956, but their relationship remained strained.

Quantum Physics and Magnetism

In 1925, a young Austrian theoretical physicist Wolfgang Pauli, working with the Born group in Gottingen, Germany, described the so-called exclusion principle—no two matter particles could exist in the same state [36]. They could not have both the same position and velocity. If two matter particles were near the same position, they would have much different velocities and therefore rapidly separate. The exclusion principle helps explain why electrons align in shells and why certain elements are magnetic. Pauli also was the first to suggest that atomic nuclei had magnetically related spin which proved to be the physical basis of nuclear magnetic resonance (NMR) and magnetic resonance imaging (MRI).

Pauli was an eccentric scientist with little interest in personal credit often not even bothering to publish his findings. He introduced his discoveries in letters to his friends and colleagues such as Niels Bohr, Werner Heisenberg, and Albert Einstein. He received the Nobel Prize in physics in 1945 mainly due to the efforts of Einstein. According to theoretical physicist, Stephen Hawking, Pauli "*was the archetypical theoretical physicist: it was said of him that even his presence in the same town would make experiments go wrong*!" [37]. After suffering a severe mental illness associated with work and a divorce at age 31, Pauli became a patient of Carl Jung who later brought notoriety to Pauli when he published detailed descriptions of his many dreams.

Electrons are in constant motion rotating with angular momentum, so each electron is a tiny magnet with north and south poles [38]. Magnetic materials occur when the individual moments of many electrons are aligned in the same direction. By contrast, in nonmagnetic materials, the electron spins are randomly aligned.

Based on Pauli's exclusion principle, two electrons cannot have the same quantum numbers in a subshell or orbital within an electron shell. When an electron orbital is occupied by two electrons, one must have spin up and the other spin down. The German physicist Friedrich Hund formulated rules for how the electron subshells are filled in a multielectron atom. Rule 1 indicates that electrons arrange themselves in an orbital in a state that maximizes the total spin quantum number.

Electrons in the d- and f-subshells with a capacity of 10 and 14 total electrons respectively, largely determine the magnetic properties of a material (Table 12.1). The d-subshell has five orbital states and ten electrons when full and the f-subshell has seven orbital states and 14 electrons when full. Electrons are added one at a time to each subshell so that all the up-spin positions of the subshell are filled before a down-spin electron is added. For example, when the first electron in d-subshell has spin up, the next four will also have spin up. This is often called the bus seat rule as single bus passengers typically occupy all the double seats before passengers begin to fill the second seat. Elements with many unpaired electrons in the d- and f-subshells are the best magnets. By contrast, elements with all paired electrons in the d-and f-subshells are poor magnets.

In crystals, individual ions cooperatively align their individual angular moments to account for the magnetic properties of the crystal. For example, divalent iron (Fe^{2+}) has six d-electrons arranged to produce maximum spin and magnetic moment. The d-orbitals of neighboring ions overlap into covalent bonds which can align the spin on the two neighbors. The exchange force between neighbors is weak compared to the force within a single ion but it is sufficient to produce ferromagnetism. If a bar of iron is placed within a strong magnetic field, all the magnetic moments become aligned along the external field. When the iron bar is removed from the magnetic field, it will remain magnetized for a long period of time.

Another important factor determining magnetism is the temperature of the ferromagnetic material. There is a critical temperature for each material (the Curie temperature, T_c) above which the magnetic moments of the ions lose their alignment and the crystal loses its magnetism. Of the three most common ferromagnets, nickel, iron, and cobalt, the T_c is 631 K, 1034 K, and 1394 K respectively. If a magnetized ferromagnetic bar is heated above and then cooled to a temperature below T_c, grains within the crystal are magnetic but their moments are randomly oriented so that the bar is not magnetic. The bar can be magnetized again by placing it in another large magnetic field.

Nuclear Magnetic Resonance (NMR)

In the 1930s, several physicists investigated Pauli's idea that atomic nuclei had magnetically related spin [39]. An American physicist working at Columbia University in New York, Isidor Isaac Rabi was the first to discover a way to detect the magnetic spin of different atomic nuclei. Rabi produced a gaseous beam of nuclei from different atoms and sent them past a magnet that deflected the beam before it struck the detector. He added a separate electromagnet whose field strength could be rapidly oscillated to pump electromagnetic wave energy into the beam and noted that at a specific frequency of oscillation, the beam abruptly bent to a new deflection point. Photons in the radio frequency (RF) range changed the direction of nuclear spin.

Rabi received the Nobel Prize for his discovery in 1944 and a nucleus was said to be *in resonance* when it absorbed its unique RF and suddenly changed its magnetic properties. As Pauli had predicted, the protons and neutrons of the atomic nucleus have the intrinsic quantum property of spin. The spinning protons and neutrons generate a magnetic field just as the spinning electrons in their orbitals around the nucleus. Also, like electrons, if the protons and neutrons are paired, the spin is canceled. The nucleus has a magnetic field if the number of protons plus neutrons are an odd number or if the number of protons and neutrons are odd numbers.

When nuclei with gyromagnetic fields are placed in an external magnetic field, the magnetic fields of the individual nuclei align with the external field [40]. Depending on the intrinsic properties of the nuclei, absorption of RF photons with sufficient energy can flip the direction of the magnetic field. The flipping back and forth of the nuclear magnetic field from one energy level to another provides the nuclear magnetic resonance signal.

Other investigators soon expanded nuclear magnetic resonance to use on liquids and solids and nuclear magnetic resonance became broadly used to determine structural and chemical information about molecules. Different atomic nuclei in a molecule resonated at different RFs for the same external magnetic field strength. Thus, new discoveries in quantum physics led to parallel advances in the fields of analytical chemistry, biochemistry, and electronics.

The hydrogen nucleus with its single unpaired proton has the highest net spin and magnetic moment and the strongest nuclear magnetic resonance signal. When hydrogen nuclei are imbedded in molecules, the NMR signal is diminished by the surrounding electron density and binding with other nuclei. For example, the OH bonding and surrounding electrons in water (H_2O) molecules produce a lower magnetic field and lower resonant RF than free hydrogen atoms. Measuring the hydrogen nuclear magnetic resonance in different materials provides key information about the molecular structure of the material.

With a typical NMR machine developed in the 1950s, the material being investigated was placed in a small test tube surrounded by a large, fixed magnet. When an RF pulse at the right energy level was applied to the test tube, protons in the material spun in phase with each other driven by the oscillating waveform. An antenna listened to the emitted oscillating signal from the stimulated protons. After the RF pulse, the proton signal decayed away (relaxation time) as the energy from the pulse dissipated. Two features of the signal decay were typically measured T_1 and T_2 that varied depending on how freely mobile water protons were in the material.

Computerized Tomography (CT)

At the mid-twentieth century, the state of brain imaging was still primitive not much different from the time Roentgen discovered X-rays at the end of the nineteenth century. Routine X-rays of the head visualized the calcified skull bones but not the contents within. Contrast procedures such a pneumoencephalography and angiography provided visualization of specific brain structures but were difficult to perform and not without risk. Pneumoencephalography visualized the fluid-filled ventricles and angiography, the larger arteries; neither provided specific information about brain tissue other than mass shifts.

X-ray tomography of the chest was developed to better visualize the heart and identify abnormal tissue such as tuberculosis and tumors. The basic process was to move the X-ray machine in one direction while the image plate moved in the opposite direction. Structures in the plane of the axis of rotation were in focus while structures outside the plane were blurred. Over time, the process became more complex with the X-ray machine moving in circles and spirals and even rotating the patient so that a series of tomograms could be made. But such a system was of little use for visualizing the brain. What was needed was a shift to computed axial tomography with development of modern computers, a process that was also highly dependent on quantum physics.

The first breakthrough in brain imaging was the computed tomography (CT) scanner first publicly introduced in 1972 followed a few years later by magnetic resonance imaging (MRI) [41]. For CT scans, the X-ray tube rotates around the head while rows of detectors in a gantry measure the attenuated X-ray signals. The massive number of measurements were processed by a computer using tomographic reconstruction algorithms to produce virtual slices of the brain at multiple levels. In 1937, the Polish mathematician, Stefan Kaczmarz developed a method to solve such a large number of algebraic equations and in 1957 the South African physicist, Allan Cormack working at Tufts University in Boston developed the mathematical technique of "line integrals" used in most modern CT scanners.

The basic concept and a prototype CT scanner were developed between 1959 and 1961 by William Oldendorf, a UCLA neurologist, stationed at the VA Hospital in West Los Angeles. Bill, who I trained with during my neurology residency, hated doing pneumoencephalograms and angiograms almost as much as the patients hated undergoing the procedures. He was a kind compassionate man who felt that there must be a less invasive way to visualize the brain.

Oldendorf later recalled that he first got the idea for a CT scanner of the head when he met an engineer at a social event who asked him for advice on making an automated X-ray machine for identifying frostbitten oranges. It immediately occurred to Bill that if it was possible to identify damaged spots inside an orange, it should be possible to identify damaged spots inside the brain. The idea percolated in his brain for years before he finally decided to work on such a device. His goal was to produce a machine that emitted a moving beam of high-energy photons that passed through the head to detectors on the other side. Oldendorf noted: "*The procedure is similar to shadow radiography, but film is replaced by an x-ray detector and the moving narrow beam of x-rays permits a point-by-point plot of the transmission of the object*" [42].

In his prototype device, Oldendorf used a radioactive source of gamma rays instead of X-rays and kept the beam source and detectors stationary while the object representing the head moved. To model the head, he used an outer ring of iron nails simulating the skull and an iron and aluminum nail near the center simulating brain lesions. The device was made entirely from materials found in his garage including an old 16 rpm phonograph turntable and an H-O gauge train flatcar and track from his son's electric train set. The model was mounted on the turntable "*in such a way that the axis of rotation of the turntable intersected the beam of gamma rays. Pulling the flatcar along the track was a radio-receiver-dial string assembly, connected to a pulley that was in turn attached to a discarded spring-driven alarm-clock motor. This apparatus moved the point of intersection of the axis and the beam through the model at 80 millimeters per hour*" [42].

In 1961, Oldendorf published an article describing his device and scraped together $1700 to apply for a patent application. He finally received a US patent in 1963 for a "*radiant energy apparatus for investigating selected areas of interior objects obscured by dense material.*" Oldendorf tried to market his device to several X-ray imaging companies but there was uniform lack of interest. They could see no practical use for his machine, and he ultimately abandoned it and moved on to other areas of research.

Oldendorf's idea of tomographic reconstruction of the brain was taken to another level in the late 1960s by Godfrey Hounsfield, an engineer working at the Central Research Laboratories of EMI (Electrical and Musical Industries) at Hayes, UK. Hounsfield had two major advantages over Oldendorf, a seemingly unlimited supply of research funds from the sale of Beatles' records and a mathematical background with access to the latest developments in digital computers. Furthermore, Hounsfield was not an academic and had no pressure to write grant proposals or publish his results and only applied for a patent after he had a working model of a computerized X-ray tomographic scanner.

Hounsfield's prototype CT scanner used a pencil-thin X-ray beam aimed at a photomultiplier detector that took 160 parallel readings through 180 angles each 1 degree apart. The X-ray beam rotated and translated around the head. The images from these scans took about 2 ½ h to process using algebraic reconstruction on a large computer. The first commercially available CT scanner, the EMI-Scanner, was limited to brain scans and took about 4 min for each brain slice and about 7 min of computation time on a microcomputer for each reconstructed image. Hounsfield initially tried to convince the chief of neuroradiology at the National Neurological Hospital at Queen Square, London, to try his CT scanner, but the neuroradiologist could see no need for such a device. He had better success at the Atkinson Morley Hospital in Wimbledon where the chief neuroradiologist, Jamie Ambrose, had an interest in axial tomography. The first patient brain scan was made on October 1, 1971, at Atkinson Morley but it was not publicly announced until 1972 (Fig. 12.2).

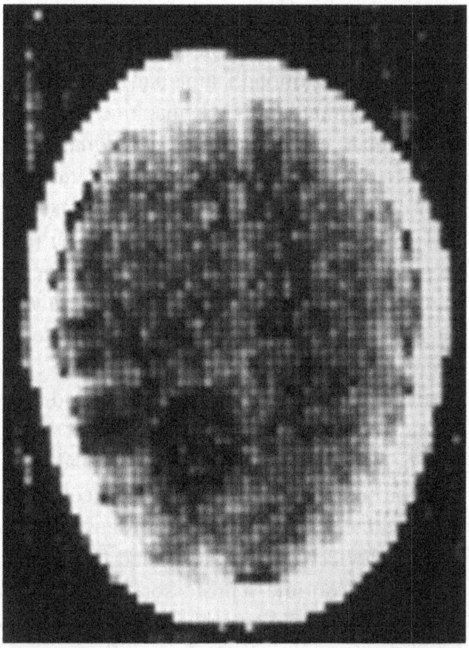

Fig. 12.2 The first CT scan at Atkinson Morley Hospital on October 1, 1971. The patient had a plum sized cyst in the left frontal lobe. (From Higgins ES. Fifty years ago, the first CT scan let doctors see inside a living skull. *Smithsonian Magazine*, October 1, 2021)

When the initial pictures of CT scans of the brain were published, there was world-wide interest and physicians flocked to Wimbledon to see the fantastic machine. EMI was inundated with orders even with a price tag of over $300,000 per scanner. Over the next 10 years, EMI successfully battled to maintain Hounsfield's patent even though the technique was not an invention, and the mathematical calculations were not unique. Interestingly, the one unique feature of the patent was Hounsfield's observation that a low-intensity X-ray beam could distinguish between slight differences in tissue density.

In 1979, Hounsfield and Cormack shared the Nobel Prize for developing CT scanning. Even though Hounsfield acknowledged Oldendorf for his earlier model, the committee felt that Oldendorf's scanner was primitive and lacked the necessary mathematics for proper tomographic reconstruction.

Magnetic Resonance Imaging (MRI)

While Hounsfield was developing and marketing his X-ray CT scanner in England in the early 1970s, another brain imaging technique based on the principles of NMR was developed at the State University of New York (SUNY) by two researchers with completely different backgrounds. This new imaging technique that became known as magnetic resonance imaging (MRI) provided much more detailed images of brain tissue than was possible with X-ray CT scanning and would eventually dominate the field of brain imaging.

In 1970, Raymond Damadian, a research physician at the Brooklyn SUNY campus, used a standard NMR machine to compare T_1 and T_2 relaxation times in tumor tissue and normal tissues and noted that T_2 relaxation was longer in tumor tissue. He published his findings in the journal *Science* in 1971 and suggested that NMR might be a useful tool for identifying the presence of cancer [43]. Damadian concluded that if it was possible to identify a tumor in a piece of tissue in a test tube, it should be possible to identify tumors in a patient. He envisioned a massive NMR machine that could hold a living person. The idea was to have the person standing in a fixed magnetic field with an RF emitter and detector moving in a spiral around the subject from head to foot. Damadian felt that he could record an NMR signal from a pencil vertical column and the subject could be moved in tiny steps to "image" the whole body.

The problem was that Damadian had little understanding of the physics of NMR and the machine that he proposed could not work. It would not make an image because the RF signal was not focused but covered a wide area. Without trying to test the device, Damadian applied for a patent in 1972 and it was granted in 1974.

In the meantime, a professor of chemistry at the Stonybrook branch of SUNY on Long Island, Paul Lauterbur developed a way to make an NMR image that did work. Lauterbur was the CEO of a small company that made and operated NMR equipment and after Damadian's paper in *Science*, Lauterbur used NMR to test rat tissue specimens for tumors. Like Damadian, Lauterbur wandered if it would be possible to use NMR to image the living animal.

Lauterbur later noted that an idea came to him while he was eating a hamburger at a local Bob's Big Boy restaurant, and he immediately jotted it down on his napkin. He would use magnetic gradients to assign different magnetic field strengths to each point (voxel) in the animal's volume. Each voxel would produce a unique identifiable radio frequency (in megahertz) that could be used to produce an NMR image of the animal. For example, the resonant frequency of hydrogen atoms (protons) in water is 200 megahertz for a magnetic field strength of 4.7 Tesla. Slight changes in the magnetic field strength across the gradient cause slight changes in the resonant frequency. The higher the magnetic field strength the greater the changes in resonant frequency and the better image resolution.

When Lauterbur returned to his laboratory, he wrote up his proposal in detail and on September 2, 1971, he had the material witnessed for filing a preliminary patent disclosure. But when it came time to pay for the patent, he decided that it was not worth the money based on the advice of colleagues who saw little practical use for his invention. Lauterbur published the first nuclear magnetic resonance images of a vial filled with water and a clam in 1973 and the first image of a living mouse in 1974. His report on the methodology of MRI was initially rejected by the journal *Nature* when the editor noted the technique was highly specialized and of limited general interest. Lauterbur appealed the rejection, and the paper was eventually published in *Nature* [44].

The third key contributor to magnetic resonance imaging (MRI) was the British physicist, Peter Mansfield [45]. After completing his PhD degree on pulsed NMR spectroscopy at the University of London and postdoctoral research on NMR in the University of Illinois in the United States, Mansfield returned to the University of Nottingham where he was involved in early work on MRI. He received a grant from the Medical Research Council in the early 1970s to develop MRI equipment and improve the mathematical analysis of RF signals for generating images. Probably his greatest contribution to MRI was the discovery of an MRI protocol for rapid imaging called echo-planar imaging. The technique made it possible to obtain T_2 images much faster than previously possible and paved the way for development of functional MRI where images are obtained in real time. Mansfield also developed and tested the first full body MRI machine and he volunteered to be the first person scanned with machine in late 1978.

The controversy regarding the Nobel Prize and patents for the invention of MRI is one of the most bizarre such stories in the history of science. The Prize was given to Paul Lauterbur and Peter Mansfield in 2003 and Raymond Damadian was snubbed despite a media binge by Damadian and his supporters. Damadian was a creationist and he and his supporters were outraged that he was not included in the Prize. They considered it an attack on creationism and took out full page advertisements in newspapers across the country arguing that Lauterbur would not have discovered MRI if it was not for Damadian's earlier discoveries. The Nobel committee did not agree.

Regarding his initial patent, Damadian sued a subsidiary of Johnson and Johnson for patent infringement and was awarded 2.2 million dollars by a jury, but the judge overturned the verdict. He then sued General Electric (GE) and was awarded $100 million dollars only to have that verdict thrown out on appeal. He did finally win a patent infringement suit against GE for a later patent on oblique angle imaging software. Damadian also eventually won an appeal regarding his original patent, but it was based on the finding of prolonged T_2 decay of cancerous tissue and not on imaging capability. Making matters even more confusing, we now know that most cancers do not exhibit prolonged T_2 relaxation as Damadian observed in his original test tube samples.

In over just a few years, neurologists had two new techniques for imaging the parenchyma of the brain in live patients without any discomfort. When I saw my first MRI of the brain in the mid-1970s, I could not believe the detail in brain structure (Fig. 12.3). It was like having a slice of brain tissue to examine. Previous lengthy neurological examinations designed to localize the site of brain pathology were replaced by a short procedure that was much more accurate in localization. Some even questioned whether neurologists would be replaced by the new technology. Fortunately for neurologists, the most common symptoms that bring patients to see them (e.g., back pain, headache, and dizziness) are usually not associated with abnormalities on imaging. The "art" of neurology remains in demand.

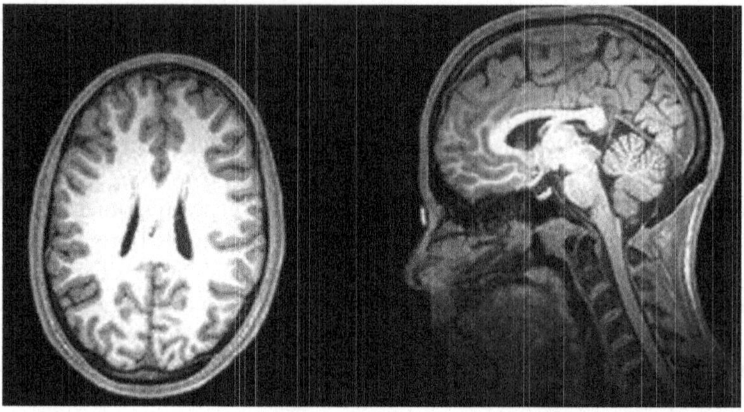

Fig. 12.3 Normal MRI of the brain showing remarkable anatomical detail. Horizontal section (left), sagittal section (right)

Table 12.2 Relationship between photon energy, wavelength, and frequency of electromagnetic waves

Class	Wavelength *Meters*	Frequency *Hz*	Photon energy *Electron volts*
Gamma rays	10^{-12}	3×10^{20}	1.2×10^{6}
X-rays	10^{-9}–10^{-11}	3×10^{17}–3×10^{19}	1.2×10^{3}–1.2×10^{5}
Ultraviolet	10^{-7}–10^{-8}	3×10^{15}–3×10^{16}	12–120
Visible light	10^{-6}	3×10^{14}	1.2
Infrared	10^{-3}–10^{-5}	3×10^{11}–3×10^{13}	1.2×10^{-3}–1.2×10^{-1}
Microwaves	1.0–10^{-2}	3×10^{8}–3×10^{10}	1.2×10^{-6}–1.2×10^{-4}
Radio waves	10^{8}–10	3.0–3×10^{7}	1.2×10^{-14}–1.2×10^{-7}

The boundary between ionizing and nonionizing radiation is about 100 electron volts
Electron volt = energy or work required to move an electron through a potential difference of 1 V. 1.0 Joules = 6.24×10^{18} electron volts

Photon Energy and the Brain

There is a remarkably wide range of electromagnetic radiation with wave lengths of 100 million meters for very low-frequency radio waves to 1 trillionth of a meter for high-frequency gamma rays (Table 12.2) [46]. Recall that the energy (E) of a photon of an electromagnetic wave at a given frequency (f) is given by the formula: $E = hf$, where h is Planck's constant. The relationship between wavelength (λ) and frequency is given by $\lambda = c/f$, where c is the speed of light in a vacuum. The energy in photons with wave lengths shorter than about 10 billionth of a meter in the mid-ultraviolet range (above around 100 electron volts) can damage chemical bonds and strip electrons from molecules. By comparison, the energy in photons of radio waves and microwaves is miniscule. In the higher-frequency end of visible light and the lower end of the ultraviolet spectra, photons have enough energy to cause electron to shift from one shell to another but not enough energy to break chemical bonds.

The frequency in the ultraviolet electromagnetic spectra where the energy reaches the critical level to damage chemical bonds (~10^{15} Hz) marks the junction between so-called nonionizing electromagnetic radiation and ionizing electromagnetic radiation. The boundaries of the different bands of the electromagnetic spectrum overlap and junctional regions have mixed properties of the bands on each side. For example, both visible red light and invisible infrared radiation can power the process of photosynthesis. The separation of the ionizing radiation into X-rays and gamma rays is partly based on the source of the radiation. Gamma rays typically result from nuclear decay while X-rays from accelerated electrons. Gamma rays are usually but not always more energetic than X-rays.

As a rule, nonionizing electromagnetic radiation causes brain damage through the heating of brain tissue whereas ionizing radiation damages the basic building blocks of nerve cells (cell membranes, mitochondria, nuclear DNA) and triggers cell proliferation (cancer) and cell death (apoptosis). However, with the widespread use and exposure to radio frequency particularly microwave frequency waves beginning in the early twentieth century, some scientists speculated that there were direct nonheat effects of these less energetic electromagnetic waves on brain tissue. But these speculations have never been conclusively proven. The picture is complicated by the fact that when the temperature of brain tissue is increased, a wide range of physiological processes are activated in response [47]. For example, heating activates nerve receptors that cause vasodilation and increased blood flow to maintain normal temperature. But heat changes blood brain permeability and metabolic rate that may further increase temperature. Furthermore, heat can damage mitochondria and cell cytoskeleton components. It can be difficult to separate the secondary effects of heat from the primary effects.

A good example of a heat effect of microwaves that some claim to be a direct nonheat effect on the brain is the auditory perception of microwaves, the so-called Frey effect. In the early 1960s, Alan Frey at Cornell University in New York described a series of experiments on normal subjects who reported hearing pulsed microwave radiation (radar) as far as a few hundred feet from the transmitter [48]. The subjects reported hearing buzzing, clicking, hissing, or knocking sounds behind the head regardless of the orientation to the transmitter. The perceived loudness depended on the peak power density of the microsecond pulses which could be as low as 80 mW/cm^2 considered well within a safe range of exposure. Frey noted that placing a 5 × 5 cm metal screen over the temporal lobe just above the ear lobe blocked the auditory perception raising the possibility that the sound resulted from direct brain stimulation.

However, subsequent human, animal, and modeling studies have conclusively proven that the auditory perception of microwaves results from bone conduction of thermally generated sound transients perceived by a normal cochlea [49, 50]. This is not surprising since it was soon shown that pressure waves could be detected in water exposed to pulsed microwave energy [51]. A transient thermal expansion generates stress waves that propagate away from the site of energy absorption. A pulsed thermal expansion of as little as 5×10^{-6} degrees centigrade at the human head is enough to be detected by the remarkably sensitive hair cells in the cochlea. For comparison, the hearing threshold is about a million times lower than the pressure wave required to induce traumatic brain injury. The Frey effect is a normal biological phenomenon without any adverse health effects. Unfortunately, this fact has not prevented some current day scientists from claiming the Frey effect is evidence that microwaves can produce neurological symptoms without overheating the brain (see Chap. 13). Experts agree that protecting against the heat effects of nonionizing radiation is adequate for setting exposure limits.

If the goal is to damage brain tissue such as when treating brain cancer, ionizing radiation is much more effective than nonionizing radiation. Within a year of Roentgens discovery of X-rays, physicians in Europe and America were using X-rays to treat a variety of cancers particularly skin cancers. Treatment of brain cancer with X-rays and gamma rays soon followed but the results were not very encouraging. In some cases, nonionizing radiation was used in conjunction with ionizing radiation since heating the cancer tissue increased susceptibility to the ionizing radiation. Brain cancer provided a particularly difficult challenge, however, since the cancer cells typically infiltrate normal brain tissue often distant from the main tumor mass. Even with the development of better imaging techniques and improved methods for localizing the radiation dose, the results of ionizing radiation treatment for most brain cancers have been disappointing leading to only months of extended survival with potentially significant side effects.

In Brief

JJ Thomson's electron and Henri Becquerel's alpha particle were key to Ernest Rutherford developing his model of atomic structure. Disparities between the charge and mass of the atomic nucleus soon led to the discovery of the neutron. Niels Bohr's application of the new quantum theory of Max Planck and Albert Einstein to atomic structure solved the problem of electrons remaining in orbit and not crashing into the nucleus. Electrons had dual properties of particles and waves. Quantum theory provided new rules for understanding electricity and magnetism that led to a range of modern electronic devices from transistors and lasers to cell phones and modern computers. The quantum prediction that spinning protons and neutrons generate a magnetic field like the spinning electrons in orbit around the nucleus led to the discovery of nuclear magnetic resonance. This in turn led to the discovery of magnetic resonance imaging revolutionizing the medical specialty of neurology.

References

1. Navarro J. The history of the electron. J.J. and G.P. Thomson. Cambridge: Cambridge University Press; 2012. p. 91.
2. Navarro J. The history of the electron. J.J. and G.P. Thomson. Cambridge: Cambridge University Press; 2012. p. 102.
3. Romer A. The discovery of radioactivity and transmutation. New York: Dover; 1964.
4. Becquerel H. On radioactivity, a new property of matter. Nobel Lecture, Dec 11, 1903, Nobelprize.org.
5. Katzir S. The discovery of the piezoelectric effect. Arch Hist Exact Sci. 2003;57:61–91.
6. Reid RW. Marie Curie. New York: Saturday Review Press; 1974.
7. Rutherford E. The chemical nature of the alpha particles from radioactive substances. Nobel Lecture, Dec 11, 1908, nobelprize.org.
8. Rutherford E. Uranium radiation and the electrical conduction produced by it. Phil Mag. 1899;47:116.
9. Rutherford E, Soddy F. Radioactive change. Phil Mag. 1903;5:106.
10. Heilbron JL. Ernest Rutherford: and the explosion of atoms (Oxford portraits in science). Oxford: Oxford University Press; 2003.
11. Rutherford E. The scattering of alpha and beta particles by matter and the structure of the atom. Philos Magaz Ser. 1911;6(21):669–88.
12. Stuewer RH. The nuclear electron hypothesis. In: Shea WR, editor. Otto Hahn and the rise of nuclear physics. Dordrecht: D. Riedel Publishing Company; 1983. p. 19–67.
13. Chadwick J. Existence of a neutron. Proc Royol Soc A. 1932;136:692–708.
14. Wilson FL. Fermi's theory of beta decay. Am J Phys. 1968;36:1150–60.
15. Stone AD. Einstein and the quantum. The quest of the Valiant Swabian. Princeton: Princeton University Press; 2013.
16. Stone AD. Einstein and the quantum. The quest of the Valiant Swabian. Princeton: Princeton University Press; 2013. p. 60–1.
17. Stone AD. Einstein and the quantum. The quest of the Valiant Swabian. Princeton: Princeton University Press; 2013. p. 70.
18. Niaz M, Klassen S, McMillan B, Metz D. Reconstruction of the history of the photoelectric effect and its implications for general physics textbooks. Sci Ed. 2010; https://doi.org/10.1002/sce.20389.
19. Stone AD. Einstein and the quantum. The quest of the Valiant Swabian. Princeton: Princeton University Press; 2013. p. 83.
20. Stone AD. Einstein and the quantum. The quest of the Valiant Swabian. Princeton: Princeton University Press; 2013. p. 85.
21. Cropper W. Great physicists: the life and times of leading physicists fro Galileo to Hawking. Oxford: Oxford University Press; 2004. p. 245.
22. Stone AD. Einstein and the quantum. The quest of the Valiant Swabian. Princeton: Princeton University Press; 2013. p. 177.
23. Einstein A. The advent of quantum theory. Science. 1951;113:82–4.
24. Lindley D. Uncertainty: Einstein, Bohr, and the quantum discontinuity. New York: Doubleday; 2007.
25. Stone AD. Einstein and the quantum. The quest of the Valiant Swabian. Princeton: Princeton University Press; 2013. p. 228.
26. Farmelo G. The strangest man. The hidden life of Paul Dirac, Mystic of the atom. New York: Basic Books; 2009.
27. Farmelo G. The strangest man. The hidden life of Paul Dirac, Mystic of the atom. New York: Basic Books; 2009. p. 13.

28. Stone AD. Einstein and the quantum. The quest of the Valiant Swabian. Princeton: Princeton University Press; 2013. p. 241–53.
29. Stone AD. Einstein and the quantum. The quest of the Valiant Swabian. Princeton: Princeton University Press; 2013. p. 243.
30. Stone AD. Einstein and the quantum. The quest of the Valiant Swabian. Princeton: Princeton University Press; 2013. p. 248.
31. Navarro J. The history of the electron. J.J. and G.P. Thomson. Cambridge: Cambridge University Press; 2012. p. 109–70.
32. Stone AD. Einstein and the quantum. The quest of the Valiant Swabian. Princeton: Princeton University Press; 2013. p. 254–67.
33. Kittel C. Introduction to solid state physics. Wiley: University of Michigan; 1953.
34. Arns RG. The other transistor: early history of the metal-oxide-semiconductor field-effect transistor. Eng Sci Ed J. 1998;7:233–40.
35. Riordan M, Hoddeson L. Crystal fire: the invention of the transistor and the birth of the information age (Sloan Technology series). New York: WW Norton and Company; 1997.
36. Peierls R. Wolfgang Ernst Pauli 1900-1958. biographical memoirs of fellows of the Royal Society. Royol Soc. 1960;6:174–92.
37. Hawking S. A brief history of time. New York: Bantam Books; 2017. p. 70.
38. Blundell SJ. Magnetism: a very short introduction. Oxford: Oxford University Press; 2012.
39. Ramsey NR. Early history of magnetic resonance. Bull Mag Reson. 1985;7:94–9.
40. Rigen JS. Quantum states and precession: the two discoveries of NMR. Rev Mod Phys. 1986;58:433–48.
41. Filler A. The history, development and impact of computed imaging in neurological diagnosis and neurosurgery: CT, MRI, and DTI. Nat Prec. 2009; https://doi.org/10.1038/npre.2009.3267.4.
42. Oldendorf WH. The quest for an image of brain: a brief historical and technical review of brain imaging techniques. Neurology. 1978;28:517–33.
43. Damadian R. Tumor detection by nuclear magnetic resonance. Science. 1971;171:1151–3.
44. Lauterbur PC. Image formation by induced local interactions: examples employing nuclear magnetic resonance. Nature. 1973;242:190–1.
45. Morris PG. Sir Peter Mansfield, 9 October 1933—8 February 2017. Biogr Mem Fellows R Soc. 2021;70:313–34.
46. Barnes FS, Greenebaum B, editors. Biological and medical aspects of electromagnetic fields. 3rd ed. Boca Raton: CRC Press; 2018.
47. Walter EJ, Carraretto M. The neurological and cognitive consequences of hyperthermia. Crit Care. 2016;20:199–06.
48. Frey AH. Auditory system response to radio frequency energy. Aerospace Med. 1961;32:1140–2.
49. Foster KR, Finch ED. Microwave hearing: evidence for thermoacoustic auditory stimulation by pulsed microwaves. Science. 1974;185:256–8.
50. Elder JA, Chou CK. Auditory response to pulsed radiofrequency energy. Bioelectromagnet Suppl. 2003;6:S162–73.
51. White RM. Generation of elastic waves by transient surface heating. J Appl Phys. 1963;34:3559–67.

Chapter 13
Overview

In questions of science, the authority of a thousand is not worth the humble reasoning of a single individual.

Galileo Galilei 1632

Abstract Scientists share a wide range of personal attributes common to humans. They can be biased by their beliefs and prevailing theories, and they can be driven by ambition to distort history and even falsify data. But science is a process designed to deal with human foibles over time. It often appears chaotic in the short term, but history shows us that it works remarkably well in the long term.

Keywords Electromagnetic spectrum · Priority battles · Suggestion · Nocebo effect

Several themes recur throughout this book. Some are common to all areas of science, but others are unique to electricity and neuroscience. A review of these common themes provides a good overview of the book. The reader can check appropriate sections of the book for more details and references.

Serendipity

The history of science is replete with instances of serendipity. A chance observation leads to discovery of a new principal. The key to discovery is that the scientist is primed to recognize the significance of the chance observation. Often the same chance observation occurred before, but the significance was not appreciated.

Luigi Galvani's discovery of animal electricity When sparks from a nearby electric machine leapt to a scalpel touching the sciatic nerve of a dissected frog causing the leg to jump, Galvani suspected that electricity was the mysterious nervous fluid. He immediately devised a series of experiments to test this hypothesis. In the process, he made another chance observation. When he hung a frog leg preparation on his iron balcony with a brass hook through the spinal cord, the frog leg jumped if it was in contact with the iron balcony. Galvani assumed that by joining the two

R. W. Baloh, *Brain Electricity*, https://doi.org/10.1007/978-3-031-62994-5_13

metals, one in the spinal cord and the other touching the leg, he had created an electric circuit causing electricity to flow from the spinal cord to muscle. The muscle contraction did not occur if he placed an insulating material between the two metals or between a cut in the nerve.

Alessandro Volta read Galvani's description of his experiments with great interest, but after experimenting on his own, he soon found that any two metals forming a bimetallic arc could activate a frog nerve-muscle preparation. In fact, he could dispense with the frog leg preparation all together and detect the current from the bimetal arc with a specially designed electrometer. The metals were not closing an electric circuit between nerve and muscle but rather generating external electricity. In other words, the nerve-muscle preparation was acting as a biological electrometer sensing the electricity from the bimetallic arc. Volta's experiments ultimately led to his invention of the Voltaic pile that would revolutionize the fields of electricity and neuroscience. A "friendly" debate developed between Galvani and Volta regarding the existence of animal electricity, but both recognized the importance of the other's findings. Volta introduced the term galvanism and galvanization became the term for the therapeutic use of electricity from Volta's electric pile.

Christian Oersted's discovery of electromagnetism By chance, there was a compass on the laboratory table that Oersted was using to demonstrate to his students that electricity caused a wire attached to a voltaic pile to generate heat and light. When his assistant connected the wire to the pile terminals, the compass needle swung toward the wire and then swung back when the connection was broken. Oersted immediately saw the significance of the observation. He had a long-standing interest in electricity and magnetism and had previously questioned whether electricity might have some effect on magnets.

Oersted moved the compass 360 degrees around the wire and noted that the compass needle maintained a perpendicular position relative to the wire so that when above the wire it pointed in the opposite direction as when it was below the wire. Just as heat and light radiated from a wire carrying electric current, magnetism must also radiate from such a wire. Similar interactions between electricity and magnetism had been observed by several others, but the significance was not previously recognized.

When Oersted published his findings in Latin in a privately printed 4-page pamphlet, it hit the European scientific community like a thunderbolt. Within a few weeks, the pamphlet was translated into English, German, French, Italian, and Dutch. He went on to show that the magnetic force surrounding a conducting wire assumed transverse circles, the planes of which were at right angles to the wire (see Fig. 4.1). The magnitude of the force depended on the distance from the wire, the power of the battery, and the quality of the metal wire. A circular closed circuit carrying an electric current attracted or repelled the pole of a magnet, could be turned by a magnet, and had north and south poles just like a magnet.

Heinrich Hertz's discovery of electromagnetic wave transmission Hertz was preparing an exhibit of Faraday's induction coil for a class he was teaching. To illustrate that electricity was induced in the second coil, he decided to produce sparks across a small gap between two metal balls on the wire terminals from the coil. When the setup was complete and the sparks produced, he was surprised to also see sparks between the ends of a circular piece of wire lying on the table. Hertz immediately recognized the potential significance of the observation since during his training, Hermann Helmholtz encouraged him to try to detect the propagation of electromagnetic waves predicted by Maxwell's equations. The sparks from the induction coil were generating electromagnetic waves that produced enough electromotive force in the wire loop to produce sparks across the gap. The Americans Joseph Henry and Thomas Edison had earlier observed similar phenomena but were not primed to recognize the significance.

Once he was certain that he was detecting electromagnetic waves, Hertz began a series of experiments comparing the detected waves to light waves. He showed that the waves moved at the speed of light and could be polarized and refracted like light waves. In other words, they were the same as light waves but at a lower frequency and longer wavelength than visible light.

Critical Theories by Unlikely People

Unlikely people have made some of the most remarkable scientific discoveries. The most impressive examples are instances where men with humble backgrounds and little or no formal education developed lasting scientific theories.

Stephen Gray's fluid theory of electricity Prior to Gray's experiments, electricity was considered a static phenomenon that could be stored and released. After showing that electricity could flow over long distances in an oily hemp twine, Gray concluded that electricity was "fluid-like." He tested a range of materials for the flow of electricity including live animals even human subjects. His "flying boy" experiments became famous throughout Europe. He suspended a boy with silk cords attached to a horizontal bar supported by two vertical bars anchored to the floor (see Fig. 2.3). When he rubbed his glass tube to generate a static electric charge and placed it near the boy's feet without touching them, the boy's extended hands attracted small pieces of paper just like the electrified tube.

Gray lived a life of poverty despite receiving the prestigious Copely Medal from the Royal Society twice for his work on electricity. After minimal schooling, he was apprenticed first to his father and then to an older brother in the dyeing trade. He was self-taught with a natural curiosity for science. Through friends of the family, he obtained a pensioned position at the London Charthouse, an almshouse for older men and a school for young boys, where he carried out many of his experiments. As in life, he died in poverty buried in an area reserved for pauper pensioners.

Ben Franklin's one-fluid theory of electricity Although Ben Franklin is best known for his kite experiment demonstrating that lightning is electricity, his one-fluid theory of electricity has had the most lasting scientific impact. Based on the observation that chaff was attracted to an electrified amber rod and repelled by an electrified glass rod, the French chemist Charles du Fay concluded that there must be two types of electrical fluid, resinous (amber-like) and vitreous (glass-like). Franklin, on the other hand, concluded that electricity was present in all substances and that electrical charge which could be positive or negative was normally in a balanced or neutral state. Excess or deficiency of positive or negative charge could be produced by friction between substances producing an imbalance in the amount of electric fluid. For example, when a glass tube was rubbed with a silk cloth, the glass became charged with an excess of fluid or a positive charge. By contrast, the opposite occurred when amber was rubbed with fur, the amber developed a deficiency of fluid or a negative charge.

Franklin suggested that electrical fluid would flow from a positively charged substance with excessive fluid to a negatively charged substance with a deficit of fluid. If an uncharged conductor encountered a positively charged glass tube, some of the excess fluid on the tube would flow onto the conductor, whereas, if the same conductor encountered a negatively charged amber rod, electrical fluid would flow from the conductor to the amber to balance the amount of fluid. Franklin's electrical fluid moved from positive to neutral and from neutral to negative. This convention persists to this day even though electrons move in the opposite direction.

Like Gray, Franklin came from a humble background with little formal schooling before he was apprenticed to a printer. He was also self-taught, and the printing shop proved the ideal place for him to learn about the latest scientific discoveries. Franklin soon had his own printing shop and proved to be a successful businessman and entrepreneur. He first became interested in electricity when he attended a public demonstration in Boston like the one Gray conducted in Europe. He acquired an electrical rubbing tube from a colleague in London and later made his own electrical machine with a large wheel and an electrical rubbing glass globe. In his later years, Franklin became a statesman for his newly developing country spending years in London and Paris representing the country's interests. Despite his lack of formal education, he was called "Doctor" Franklin by all and was revered for his work on electricity by the world's scientific community.

Michael Faraday's lines of force theory That the son of an impoverished blacksmith developed a theory that became the foundation of modern physics is one of the most compelling stories in the history of science. With no formal training in physics or mathematics, Faraday conducted a series of ingenious experiments that led to the invention of electric motors, generators, and transformers and his field-of-force theory. After observing how iron filings arranged themselves in a pattern of lines when he sprinkled them on paper held over the poles of a magnet, Faraday saw the lines as an invisible force existing in space. The lines of force connected particles of matter together and electromagnetic waves resulted from vibrations in the lines of force. There was no difference between force and matter. Physical substances could be considered a convergence of forces.

After learning to read and write at a common day school, Faraday was apprenticed for 7 years to a bookmaker where he first had access to the intriguing world of science. His book-binding skills and his habit of taking copious notes led to an offer to work as an assistant in Humphrey Davy's laboratory at the Royal Institute. Faraday rapidly became indispensable to Davy, preparing chemicals for experiments and working on the organization of lectures and presentations. After accompanying Davy on a tour of Europe where he met many famous scientists, Faraday returned to London a confident scientist able to speak French and German. Davy and Faraday had a rocky relationship in later years, but Faraday took over the laboratory when Davy suffered a stroke in 1826.

Less well known, Faraday became interested in animal electricity later in his career and he conducted a series of experiments on the electric eel showing the electricity from the animal's electric organ was no different from inorganic electricity. During these experiments, he worked with a young physician and physiologist, Robert Todd, and Faraday later designed an induction coil for nerve stimulation that Todd used to study electrical stimulation of the rabbit brain. Faraday's induction coil soon became widely used for electrotherapy in Europe and America called faradization.

Priority and Recognition for Scientific Discoveries

Success in science is often based on claims of priority for discoveries during the scientist's career. This usually means being the first to publish a discovery although some scientists are much more aggressive in publishing than others. Many of the early giants in electricity and neuroscience were embroiled in battles over priority while others cared little about priority.

Henry Cavendish and the priority for Ohm's law Cavendish, best known for his discovery of hydrogen and estimating the weight of the earth, performed a series of experiments on electricity in a laboratory in his home that were unknown until more than a century later. Notes from these experiments were collated and published in 1879 by James Maxwell when he became the first Cavendish Professor of Physics and head of the Cavendish Laboratory at Cambridge University.

The son of the politician and amateur scientist Lord Charles Cavendish, Henry spent 3 years at Cambridge University without taking a degree before returning home to work with his father in his own laboratory. Henry was known for his eccentric personality. He spoke little and only to men. He was unbelievably shy around women and communicated with female servants only by written notes. The well-known neurologist and science writer, Oliver Sachs, concluded he probably had Asperger's syndrome.

Using homemade equipment such as a piece of straw with a thin wire running through it and pith balls covered with foil, he discovered the inverse square law of electric force with distance later attributed to Coulomb. He showed that "electrical fluid" exerted a pressure that controlled the flow. This electrical potential that he called the "degree of electrification" was related to the current flow by a simple mathematical relationship that became Ohm's Law. He also developed a mathematical formula for an early unit of capacitance using a plate of glass with metal foil on each side for a Leyden jar.

In a rare publication in 1776, he described an artificial torpedo fish that he constructed out of leather and wood to demonstrate how the fish could produce a strong shock yet no visible spark. He joined together a series of Leyden jars to generate a large current flow at low potential. The electrical potential and not the quantity of electric current determined whether a spark occurred. Cavendish was so excited about his artificial torpedo fish that he broke precedence and invited several scientists (all men of course) to witness a demonstration.

William Crookes, cathode ray tubes, and electrons In contrast to Cavendish, Crookes was a people person who spent much of his career promoting his ideas and discoveries. To him, priority was all important and he became involved in multiple disputes regarding priority. Since he was a journalist, businessman, and scientist, he understood the importance of publicity for achieving recognition. Furthermore, he did not believe in sharing credit for his discoveries, even though in several instances colleagues provided key assistance and ideas.

The story of Crookes cathode ray tube and his idea that cathode rays were moving charged particles nicely illustrates his ability to control the scientific narrative. The vacuum discharge tube was originally developed by the German glass blower and instrument maker Heinrich Geissler in the early 1850s. The German physicist Julius Plücker had Geissler make the vacuum tube for a series of experiments in which he showed that the tube glowed when electricity was passed through the electrodes and that the striated glow could be bent toward a magnet outside the tube. Michael Faraday found a band of darkness between the glow surrounding the negative electrode, the cathode, and the striated glow that filled the remainder of the tube. The band became known as the Faraday Dark Space.

Although Crookes did not invent the cathode ray tube, he and his assistant Gimingham made important modifications including thicker glass and a better vacuum achieved with a new mercury vacuum pump built by Gimingham. Crookes also used a much higher voltage than Geissler to power the tube. German physicists thought that the cathode rays were a form of electromagnetic radiation moving in the ever-present ether. Crookes, on the other hand, was convinced that charged particles of gas were being driven off the cathode by the electric current and that the glow resulted from collisions of these particles with residual gas ions in the evacuated tube. Gradually Geissler's tube became known as Crookes' tube, and it was rare to see a photograph of Crookes without his tube (see Fig. 8.3).

When JJ Thomson later showed that cathode rays were composed of a stream of corpuscles (electrons) with an extremely small mass that could be deflected by an intense electric field, Crookes felt vindicated. He compared Thomson's corpuscles to his radiant matter which he called the fourth state of matter. Of course, he ignored the fact that he previously proposed gas ions the size of molecules and had no concept of Thomson's tiny subatomic particles. Crookes later published a paper based on the new concept of cathode rays in which he used photographic film to detect electrons, positive ions, and X-rays released by cathode ray tubes. He showed that X-rays were best produced when electrons struck a metal surface and that the amount of X-rays produced varied with different metal surfaces.

Carlo Matteucci and Emil du Bois-Reymond and the conflict over priority for the negative variation Sometimes the battle over priority led to major feuds between the competing scientists. The conflict over the priority for the discovery of the negative variation, the forerunner of the action potential, illustrates how the competition for priority can get out of hand.

Matteucci made a key discovery in electrophysiology when he recorded the so-called *demarcation current* between the cut and intact surface of a nerve or muscle. The cut surface had a negative potential compared to the intact surface. In his "*induced twitch*" experiment, he touched the cut end of a nerve in one frog nerve-muscle preparation to the muscle of another preparation and noted that every time the muscle in contact with the cut nerve contracted, it caused the muscle supplied by the cut nerve to contract. The demarcation current between the intact muscle and cut nerve was triggering a negative variation (action potential) that moved down the nerve to trigger a muscle contraction. Matteucci received the Copley Medal for the induced twitch experiment, but he missed the significance of his finding.

Du Bois-Reymond became aware of Matteucci's work when his mentor Johannes Müller gave him a copy of Matteucci's latest paper and suggested that this would be a good subject for his future research. Du Bois-Reymond was initially impressed with Matteucci's findings but soon became frustrated with the experimental technique and interpretation of data. Matteucci was "old school" with a healthy skepticism for technology whereas du Bois-Reymond was of the "new breed" that considered instrumentation the key to successful research. He made his own sensitive galvanometer and modified Faraday's induction coil stimulator with a moveable outer coil so the voltage could be easily adjusted.

With these and other technical improvements, du Bois-Reymond was able to record the electrical events associated with nerve excitation of muscle that he called the negative Schwankung (the negative variation), which represented the first instrumental recording of what later became known as the action potential. He argued that his experimental technique was much more sensitive than Matteucci's and that it should be possible to measure the speed of conduction of the negative variation as it moved down a nerve. In the meantime, his German friend and colleague, Herman Helmholtz, used a much simpler technique, measuring the brief time interval between nerve stimulation and muscle contraction, to show that the speed of nerve conduction was much slower than anyone anticipated (about 30 m/s).

In the conflict over priority for discovery of the negative variation, du Bois-Reymond was much the aggressor from the start. He was young and eager to make a name for himself and he felt that he was not getting proper respect from colleagues, particularly in France and England. Du Bois-Reymond accused Matteucci of conducting sloppy experiments, constantly changing his opinions, and publishing multiple versions of the same experiments. Matteucci responded that the "*young physiologist from Berlin*" was trying to destroy his scientific reputation. In the end, du Bois-Reymond received credit for the discovery of the negative variation.

David Ferrier and electrical stimulation of the cerebral cortex In their competition for priority, some researchers simply ignored the reported findings of earlier investigators. Although Ferrier made many original observations and was the first to study the primate cerebral cortex, he consistently failed to credit earlier researchers in the field. The most glaring example was his failure to recognize the work of Robert Todd even though there was a statue of Todd at the entrance of King's College Hospital in London where Ferrier conducted his experiments on primates. A quarter of a century earlier, Todd had used Faraday's induction coil to electrically stimulate the cortex of rabbits producing localized motor responses and inducing focal and generalized seizures.

Ferrier began his electrical stimulation studies in West Riding, Yorkshire on the brains of rabbits, cats, and dogs in 1873. A few years earlier, German physicians Gustav Fritsch and Eduard Hitzig used galvanic stimulation of the exposed cortex of dogs and identified what appeared to be a topographic map of body movements on the contralateral frontal cortex. Using faradization with high-frequency pulsed current, Ferrier found many more excitable areas of the cortex than the Germans reported and by increasing the duration of stimulation to more than 5 s, he could trigger a focal motor seizure that would move over the motor cortex just as Hughlings Jackson predicted. When he submitted his work for publication, it was rejected because the reviewers felt that he had not given proper credit to Fritsch and Hitzig.

In the meantime, Ferrier returned to King's College Hospital to begin a similar series of studies of the cerebral cortex of monkeys. In the process, he found that animals higher on the phylogenetic scale had a greater degree of localization than animals lower on the scale. The next year he decided to separately write up his work on monkeys which he saw as a way around the priority issue with Fritsch and Hitzig. In 1876, he published a book summarizing his work on all animals, *The Functions of the Brain.* The book contained much of the work included in the previously rejected paper and although he still did not mention Todd, he did give Fritsch and Hitzig some credit for their work. In his book, Ferrier was so confident that the cortical localization findings in the monkey were applicable to humans that he included a figure of a human brain with a functional map of the monkey cortex superimposed on the human cortex (see Fig. 5.4).

Interestingly, Ferrier later became involved in what was publicized as the last great battle between the generalizers and the localizers of cerebral cortical function. At the 1881 Seventh International Medical Conference in London, the section on physiology included a session on the contrasting views about cortical localization of David Ferrier and Fredrich Goltz. Goltz was a German physiologist who claimed he had removed large parts of the cerebral cortex in dogs and never produced paralysis, deafness, or blindness. Ferrier pointed out that dogs have a relatively small cerebral cortex compared to primates and that Goltz probably did not destroy as much of the cortex as he thought. Ferrier resected the motor cortex in a monkey that he had electrically stimulated, and the monkey developed a profound paralysis like a human after a stroke.

To illustrate his point, Goltz brought a dog that had most of his cerebral cortex resected yet the dog ran around and appeared to have normal hearing and vision. Ferrier brought the monkey that was paralyzed on the right side after having the left motor cortex removed. The chairman asked the men whether they were willing to have a postmortem examination of the dog and monkey to see if the lesions were as presented and both agreed. After the animals were sacrificed, the brains were examined by a panel of experts. In the case of Ferrier's paralyzed monkey, the damage was localized to the left frontoparietal region involving the motor cortex as predicted, but in the case of Goltz's dog, much more motor and sensory cortex remained intact than he had thought. The localizers claimed victory and Ferrier was widely acclaimed.

Marie Curie and ionizing radiation Marie Curie had problems with priority and recognition simply because she was a woman. Even though she initiated the research and made several of the initial discoveries on her own, the scientific community assumed it was her husband Pierre Curie who was behind the research that they jointly published.

When Marie Sklodowska, a young Polish graduate student at the Sorbonne in Paris, read about Becquerel's exciting findings on ionizing radiation, she decided to study the subject for her doctoral thesis. She met her future husband, Pierre Curie, at the Sorbonne. After marriage, she followed Pierre to the Municipal School of Industrial Physics and Chemistry where he set up a make-shift laboratory for her to do her research.

Her studies of the uranium minerals, pitchblende and torbernite indicated that pitchblende had four times the radioactivity as uranium alone and torbernite twice the amount. This suggested that these minerals likely contained other elements more radioactive than uranium. She began a search for these substances and by early 1898 she isolated the element thorium which also spontaneously released charged particles. Pierre was so impressed with her work that he decided to set aside his work on crystals and join her on radioactivity. By the end of 1898, Marie and Pierre Curie published papers announcing the discovery of two other radioactive elements in pitchblende, polonium, and radium. They then began the tedious process of extracting appreciable quantities of these new elements from pitchblende.

In 1903, the Nobel committee decided to award the Prize to Becquerel and Pierre Curie for their work on radioactivity. It was only after Pierre complained to the committee that Marie was added to the award becoming the first woman to win the Prize. After the tragic accidental death of Pierre in 1906 despite being the first woman to become a professor at the University of Paris, she failed to be elected to membership in the French Academy of Sciences.

The story surrounding her receiving a second Nobel Prize in 1911 is even more bizarre. At the time Marie was involved in a romantic relationship with the physicist Paul Langevin who although married was separated from his wife. The tabloids accused her of home wrecking and preying on the younger Langevin, 5 years younger than Marie. A mob formed outside her home, and she had to flee with her two daughters to the home of a friend. The chair of the Nobel Committee, Svante Arrhenius, suggested that Marie should not attend the Nobel ceremony because of questionable morals, but she ignored his comments and made her Nobel presentation.

Scientific Fraud

In the history of electricity and neuroscience as in all fields of science, scientific discovery runs hand in hand with scientific fraud. Although in some instances, the deceit is intentional, in most cases it is just bad science.

Andrew Crosse generates life with electricity In the late eighteenth and early nineteenth centuries, "gentlemen scientists" in the mold of Ben Franklin were flourishing throughout Europe. In England, electrical enthusiasts got together to form the London Electrical Society, open to all comers regardless of background or training. At one of the early Society meetings, a squire from rural Somerset, Andrew Crosse, read a report claiming that he had generated insects from volcanic rock with electricity.

Crosse, known as the "thunder and lightning man" by the locals, had a lifelong interest in electricity. The trees of his estate were covered with wires which he used to collect electricity for his experiments. The wires were connected to a battery of 50 Leyden jars kept in his library. When conditions were right such as during a dense fog the battery would discharge with the sound of a cannon fire. When voltaic batteries became available, Crosse used the batteries to provide electrotherapy for the local poor with rheumatism and paralysis.

Also interested in geology. Crosse also had an interest in geology and he was convinced that electricity was the driving force behind the generation of metal lodes and crystals. He subjected different solutions including mineral water from a local cavern to electric current from a voltaic battery for weeks to months carefully watching for developing crystals. To produce crystals of silica, Crosse steadily dripped a solution of silicate of potash and muriatic acid onto a piece of volcanic rock that was electrified with batteries. After a few weeks he noticed tiny "nipples" on the surface of the stone and these nipples gradually increased in size. By day 26, he observed a fully formed insect and over the next 2 weeks hundreds of insects were moving about over the rock. Furthermore, Crosse reported he was able to repeat the experiment a second time.

As news of Crosse's discovery leaked out to the media, public reactions varied from amazement to incredulity. To a public enthralled with electricity, Crosse's claim was not so far-fetched. Some considered his research blasphemy. The religious community condemned Crosse for interfering with God's laws and there was a rumor that a local vicar performed an exorcism at Crosse's laboratory. Another society member reported at a subsequent meeting that he was able to repeat Crosse's experiment. Crosse sent samples to Michael Faraday, but Faraday refused to enter the fray ignoring the ludicrous claims.

Reviving dead bodies with electricity With Volta's electric pile, researchers had a tool to study Galvani's theory that "nervous fluid" was electricity that originated in the brain and flowed in nerves to activate muscles. Foremost among these researchers was Galvani's nephew, Giovani Aldini, who took over Galvani's laboratory and performed a series of experiments on the effects of galvanic stimulation of the nervous system in dead animals and humans.

In January and February of 1802, with the help of several physicians, Aldini galvanized the heads of decapitated criminals in a large public square near the Bologna Palace of Justice. He found that the best responses were obtained within about 3 h of death. After filling the ears of a fresh head with salt water, Aldini inserted the terminal wires from a voltaic pile in each ear. All the muscles of the face contracted violently producing frightful expressions. He noted that the eyelids became particularly animated and even speculated that the movements might have reflected the person's premorbid temperament. Many of the spectators were sure that the criminals transiently returned to life.

In late 1802, Aldini toured Europe performing similar public demonstrations in London and Paris. In London, electrical stimulation of 26-year-old George Foster, hanged for the murder of his wife and child, was a spectacular event. As a large crowd watched with nervous anticipation, stimulating rods were applied to Forster's ear and mouth. Foster's jaw quivered, his facial muscles contorted into a frightful expression, his left eye opened like a wink, and he even appeared to breath. Sensational details of the exhibition were reported in all the London newspapers including *The Times*. Aldini was portrayed as a new Dr. Faust probing the forces that govern human existence.

In 1818, at the University of Glascow, another sensational public display of the life-giving power of electricity was conducted by James Jeffrey, a professor of anatomy, and Andrew Ure, professor of natural philosophy. Mathew Clydesdale had been convicted of a brutal murder, savagely beating an old man with an axe, and the judge sentenced him to be hanged and delivered to Jeffery for public dissection. Jeffrey dissected and Ure galvanized the body using a powerful battery with 270 pairs of metal plates.

With Jeffery and Ure standing over Clydesdale's body, Jeffery made careful incisions to expose nerves for Ure to galvanize. One wire from the battery was placed in the spinal cord and the other used to stimulate different nerves. When stimulating individual nerves, Ure described purposeful movements such as pointing at spectators or like playing a violin. When Ure stimulated the phrenic nerve in the neck, he claimed that Clydesdale began to breath. Ure was a showman, and his goal was to convince his audience that electricity brought Clydesdale back to life.

In London a year later, academics at Cambridge were highly skeptical of Ure's sensational report. A young professor of Chemistry, James Cummings decided to reproduce Ure's galvanization experiments to see the results for himself. His subject was Tom Weems, an illiterate mill-hand hanged for killing his young wife.

Cummings was determined to follow Ure's experimental protocol as close as possible so that he could expose the Glasgow exhibit as a flight of fantasy. He used a similar battery and stimulated the same nerves in the same fashion described by Ure. He tried to get Weems to breathe, to move his arms and legs and to produce facial expressions just as Ure reported. But instead of organized breathing, Cummings produced only slight shuddering and instead of purposeful movements only brief convulsive twitching of muscles.

Karl Weinhold and the brain as a battery with attached wires (nerves) The public who witnessed or read about the demonstrations that electricity could activate intentional movements in dead bodies were convinced that the bodies were transiently revived. The brain was simply a voltaic pile passing electricity throughout the body via nerves acting as wires.

The German physician, Karl August Weinhold set out to demonstrate that the brain worked like a battery with attached wires. After obtaining his medical degree in Wittenberg in 1805, Weinhold traveled through Europe studying medicine and electrotherapy before holding several medical appointments throughout Germany. He became a professor of medicine and surgery in Halle where he carried out his infamous experiments claiming he could revive kittens after the brain and spinal cord were removed by filling the cavities with bimetal mixtures.

Weinhold described how he scooped out the brain with a small spoon through an opening in the back of the skull. After the animal lost all vegetative function, he filled the cavity with the "amalgam" of zinc and silver. Weinhold reported that the animal returned to life for almost 20 min, raising its head, opening its eyes, and crawling and hopping about. The heartbeat and respiration returned and persisted for 15 min after the animal "sank down exhausted." Weinhold believed that he was not simply stimulating nerves with his bimetals but recreating "physical life" in the kittens. In addition to rejuvenating circulation and heart function, he claimed that the treated kittens responded to sensory stimuli such as loud sounds and bright lights.

Electrotherapy and Suggestion

Most of the founding fathers of the new medical specialty of neurology that developed in the nineteenth century were practicing electrotherapists. Electrotherapy was used for a wide range of neurological conditions from paralysis to epilepsy, but it was found most effective for treating hysteria and neurasthenia. The role of suggestion in treating hysteria and neurasthenia was recognized by several of these neurologists, but the concept of a placebo and its importance in all medical treatments awaited the twentieth century.

Jean-Martin Charcot and hysteria Often considered the father of modern neurology, Charcot was the Director of the medical service at the Salpêtriére, a public hospital for destitute women on the outskirts of Paris. Charcot applied a revolutionary new approach of clinical-pathological correlation to study neurological diseases. The most common diagnosis at the Salpêtriére was hysteria and Charcot was convinced that it was an organic disease of the brain even though he was unable to find brain abnormalities on postmortem examination. An inherent weakness predisposed patients to triggers such as physical or emotional trauma or even the presence of another person with hysteria. Patients were highly susceptible to suggestion. If patients with hysteria were housed on the same ward with epileptic patients, the patients with hysteria developed epileptic-like fits.

Charcot became convinced that electrotherapy was the best treatment for hysteria. He used an induction coil like the one developed by du Bois-Reymond to faradize patients in and between attacks of hysteria. He made the bizarre discovery that if he delivered electrical stimulation to the face of a patient with hysteria under hypnosis, the facial grimace became fixed and remained even after electrical stimulation was stopped. At the time photographs required a prolonged exposure and Charcot noted that the prolonged facial expressions induced in hysterical patients were useful for obtaining good photographs of otherwise fleeting expressions. His most famous patient, often called the queen of hysterics, Blanche Wittman, posed for a series of photographs showing persistent facial expressions after faradization.

George Beard and neurasthenia Beard was an early American electrotherapist and neurologist practicing in New York. He became interested in electrotherapy when in his first year of medical school, he developed symptoms of neurasthenia that was cured with electrotherapy. The term neurasthenia had been used to describe a general physical and mental exhaustion, but Beard would later redefine the syndrome to include a range of symptoms due to lack of "nervous force" resulting from subtle changes in the brain. He felt that electricity could restore the nervous force. Beard saw neurasthenia from a unique American perspective. He felt that the American drive to be successful and the pace of modern life provoked nervous disorders. In 1872, Beard teamed up with his partner in practice A.D. Rockwell to publish what was considered the definitive textbook on electrotherapy in America, *The Medical and Surgical Uses of Electricity*.

Beard made a chance observation later in his career that forced him to consider psychogenic factors in the cause of neurasthenia. He was about to treat a patient with electrotherapy when he noticed that his battery was dead. On a whim, Beard decided to go ahead with the treatment, and to his surprise the patient had a good response. Beard even went so far as to test the effect of telling patients that a given symptom would disappear at a certain time and day and surprisingly the symptom disappeared. Beard presented his observations at a meeting of the newly formed American Neurological Association, but the other 12 members were not impressed some even criticizing his experiment as immoral.

Paul Julius Möbius and the suggestive power of electrotherapy Möbius began the practice of neurology and electrotherapy after obtaining his medical degree at the University of Leipzig in 1876. He specialized in using electrotherapy for treating neurasthenics and for pain relief in patients with neuralgias. But he later became disillusioned with the claims of dramatic cures with electrotherapy. By 1891, Möbius was convinced that in four out of five cases, the positive results of electrotherapy were suggestive rather than physical in nature. For example, he noted that there had never been a well-documented case of a cure for paralysis with electrotherapy. Even in the one out of five who responded to electrotherapy such as patients with neurasthenia and pain syndromes, the effects could still have been due to suggestion.

Most of Möbius' fellow German neurologists were appalled with his attack on electrotherapy. They considered the curative effects of electrotherapy for a wide variety of neurological diseases as an "empirical fact." But he did receive some support. Ewald Hecker felt that the large number of symptoms that occurred with functional diseases were due to autosuggestion and could be cured by reassurance from a caring physician without the need for more aggressive intervention. Electrotherapy was simply acting as a "carrier of suggestion" or as German neuropsychiatrist, Emil Kraepelin, later called "hypnotizing patients through galvanization of their head."

Möbius considered suggestion an unconscious mental process, that could cause or eliminate symptoms both with organic and functional diseases. Suggestibility was a basic human attribute based on beliefs. Hypnosis was a rare specific case of it. Möbius felt that it was impossible to exclude suggestion as part of the efficacy of any type of medical treatment. Möbius was anticipating our modern concept of the placebo. His skepticism marked the beginning of the end of electrotherapy which finally occurred in the first half of twentieth century. Interestingly, there has been a resurgence of claims for transcranial electrical stimulation of the brain for a variety of neurological conditions in the early twenty-first century supported by the same type of evidence that Möbius questioned in the late nineteenth century.

Fear of Electrocution

Fear of electricity in ancient people was mainly due to the mystery associated with it. That rubbing a piece of amber could attract light objects and produce mild shocks seemed like a supernatural phenomenon. As it became possible to accumulate large quantities of electricity in the eighteenth century, it was apparent that electricity was potentially dangerous and could kill people. When electricity became widely distributed to the public in the late nineteenth century, fear of electrocution became an important factor in determining how electricity was generated and distributed.

Georg Richmann's fatal experiment With Ben Franklin's invention of the lightning rod, it became possible to harness massive quantities of electricity. Richmann, a Baltic German physicist working in St. Petersburg, Russia, decided to use a lightning rod to measure atmospheric electric charge with a new electrometer he invented. He made the fatal mistake of not grounding the insulated iron rod connected to a wire that ran to the top of his roof. During a thunderstorm in August

1753, Richmann leaned over near the rod to monitor his electrometer and a ball of lightning jumped from the iron bar to Richmann's head knocking him unconscious.

Efforts to resuscitate him were futile and he was pronounced dead. Postmortem examination revealed that the electricity had entered the body through the head and exited from a foot, shredding his shoe. There was discoloration of the skin where the electricity entered the head and bleeding into the lungs but no obvious damage to the brain or heart as the electricity passed from head to foot.

Franklin lamented that if only Richmann had grounded his lightning rod, the accident would not have happened. Franklin personally had suffered an accidental shock to the head that knocked him out but without any residual effect. He described a dramatic experiment where he lined up six men in a column each holding his hand on the head of the man next to him. The man at the end of the line was connected to the outside of two fully charged Leyden jars. When he brought a discharging rod connected to the inside of the Leyden jars to the head of the man at the beginning of the line, all six suddenly lost consciousness and dropped to the floor. Franklin reported that on recovering consciousness all had retrograde amnesia for the event and could not recall feeling anything or how they fell.

Thomas Edison stokes the fear of high-voltage alternating current When Edison finally developed a practical long-lasting incandescent light bulb, he chose a direct current (DC) system to supply electricity to homes and businesses using the bulbs. He had multiple reasons for the choice including the proven safety of low-voltage DC current. Two major limitations of a DC system were the short range of distribution (about a half mile) and the need for massive amounts of expensive copper wire. Every square mile supplied with electricity required an electric generating power plant with its own thick copper wires. When George Westinghouse decided that high-voltage alternating current (AC) was a more efficient and cost-effective system for distributing electricity, Edison latched onto the safety issue as his best weapon for attacking Westinghouse.

At the time in New York City where Edison was installing his first DC system, the city was already inundated with overhead wires mostly carrying DC current for telegraph and alarm systems but a few with AC current for newly popular arc lights for street lighting. Several well publicized accidental electrocutions by high-voltage AC power lines provided Edison the ammunition he needed. He embarked on a campaign to influence public opinion by exaggerating the danger of electrocution from high-voltage alternating current systems. He bought advertisement in leading newspapers emphasizing the risks of alternating current and lobbied powerful political friends against the use of high-voltage alternating current.

Edison needed someone, preferably not involved with his company, to take up the torch and lead the crusade against alternating current. In stepped a little-known New York engineer and electrical consultant, Harold Brown. After one of the accidental electrocutions, Brown wrote a letter to the New York Evening Post condemning companies using alternating current for lighting. Brown ranted about the dangers of high-tension alternating current and suggested safety rules that included outlawing alternating current above 300 V. He wanted to legislate alternating current out of business. He soon teamed up with Edison who provided financial support for a series of public demonstrations during which Brown electrocuted a variety of animals with AC current particularly dogs and horses. These exhibitions had mixed results in that they did induce fear in the public but also outrage for the brutal killing of innocent animals.

Edison saw another opportunity to publicize the dangers of high-voltage AC when he received a query from the New York State Death Commission regarding electrocuting condemned criminals. After initial hesitation, he wrote back that electricity was ideal for producing a rapid painless death. Furthermore, he recommended using an alternating current dynamo, like the ones manufactured in America by George Westinghouse. Even the slightest contact with Westinghouse's high-voltage AC current would lead to instantaneous death. With Edison's endorsement, the New York State Legislature passed a law establishing electrocution as the means of capital punishment starting January 1, 1889. In the end, despite Edison's persistent efforts, the AC system won out simply because it was a better more cost-effective way of distributing electricity.

William Hedley and the mechanism of electrocution As the risks of injury and death from electricity became a broad public concern, it was only natural that electrotherapists would be called upon for their expertise in treating patients after electrocution. Articles on the mechanism of tissue damage and cause of death from electric shock were common in medical journals in the 1890s. In an article published in *Lancet* in 1894, Hedley, the director of the electrical department at the London Hospital, summarized medical understanding of electrocution at the time.

Hedley noted that death from electrocution resulted from either tissue destruction due to the heat generated by the electricity or from arrest of respiration, asphyxia. If the tissue was heated by the passage of electric current to more than 45 degrees Celsius irreparable tissue damage occurred, and nothing could be done. In the case of asphyxia, the patient could sometimes be saved with artificial respiration. Hedley suggested that electricity caused asphyxia by direct or indirect effects on the lungs, heart, or nervous system. Stimulation of afferent nerves such as the vagus nerve could inhibit brainstem respiratory centers or have direct effects on the lung and heart muscle. He pointed out that applying electricity to a dog's heart could produce an arrhythmic fibrillation called "delirium cordis."

Tissue damage from electrocution was obviously related to the type, quantity, and duration of the current flow. There was a consensus that alternating current was more dangerous than direct current. The current pathway was important in determining the degree of injury. Current through the thorax could produce cardiac and/or respiratory arrest whereas current through the brain could cause seizures and paralysis in addition to respiratory arrest. Hedley later wrote that the medulla oblongata of the brainstem was particularly vulnerable because of its control of vital functions such as breathing and blood circulation.

Fear of Electromagnetic Waves

Unlike electricity with its associated sparks and shocks, people exposed to electromagnetic waves are unaware of the exposure unless there is enough energy to cause a sensation of heat. In fact, other than the narrow band of electromagnetic waves in the range of visible light, people were unaware that electromagnetic waves existed until the pioneering work of James Maxwell in the late nineteenth century.

Electromagnetic waves in the radio frequency (RF) range From the time of Marconi's earliest wireless telegraphy transmissions, health concerns were raised. Marconi's company received several letters from people who claimed they were made ill by the wireless waves passing through their bodies. One well-publicized incident occurred at the Wimereux station in France built to receive the first wireless transmission across the English Channel in 1899. A gunman burst into the wireless room complaining that the wireless waves were causing him intense body pains and that he would shoot the engineer if he did not shut down the station. The quick-thinking engineer managed to diffuse the problem getting the man to empty his pocket of metal coins and lay down the gun.

Around the same time, the French physiologist Arsenne d'Arsonval and the Serbian American inventor Nikola Tesla were experimenting with high-voltage, high-frequency alternating current on the human body. To generate the current, they used a high-voltage primary transformer, capacitor, and air core induction like the wireless transmitter developed by Hertz and modified by Marconi. Clearly, they were generating electromagnetic waves in the RF range, up to 10 kHz by d'Arsonval and up to 20 kHz by Tesla. Both men experimented on themselves, on normal human subjects, and on patients. Both noted that the most consistent effect was a sensation of warmth followed by perspiration.

d'Arsonval noted that the skin flushed after a period of sweating which he attributed to vasodilation. He suggested that high-frequency therapy could be helpful to improve circulation and kidney output. Tesla regularly treated himself since he was convinced of the long-term health benefits. Within minutes of exposure, Tesla described feeling very warm and he began to perspire. After an hour or so, he described an "immense fatigue" like the sensation experienced after extreme exertion. He reported that it was difficult to keep his eyes open. Fortunately, Tesla had the common sense to limit his exposure when he felt the early symptoms of heat stroke. If blood circulation and other compensatory mechanisms fail and heat buildup persists, heat stroke can occur with permanent brain damage.

In the twentieth century as radio transmission became widespread, it became apparent that the only consistent effect of electromagnetic waves in the RF range on the human body was to heat tissue with the amount of heat proportional to the energy in the electromagnetic waves and inversely proportional to the distance from the source (inverse square law). In usual circumstances where people were distant from the transmitting antenna, there was no risk at all. In the rare occurrence where people were accidently too close to the source, they initially felt extreme heat and perspired diffusely in most cases causing them to flee the area.

The eyes and testicles are the organs at greatest risk to heat damage since they have the least shielding and relatively low blood flow. Cataracts and transient impotence have resulted from accidental overexposure to RF electromagnetic waves. Regarding long-term risk, probably the two human subjects with the greatest lifetime exposure to RF electromagnetic waves were d'Arsonval and Tesla. Unlike the pioneers who worked with ionizing radiation, d'Arsonval and Tesla lived well into their 80s without any ill effects from long-term exposure to RF electromagnetic waves.

Electromagnetic waves in the microwave range With the development of microwaves for diathermy and cooking in the mid-twentieth century, there were renewed concerns about the dangers of these high-frequency radio waves on the brain. There is significant overlap between radio waves and microwaves. Typically, microwaves are defined as frequencies in the low end of the RF range between roughly 300 MHz to 30 GHz (see Table 12.2). A few victims developed skin burns and nerve damage in the hands after accidental exposure to high-energy microwaves from leaking microwave ovens. The skin and the skull are major barriers to these relatively low-penetrating electromagnetic waves, however. The skin would be burned before the brain is severely overheated. As with other RF electromagnetic waves, accidental overexposure to microwaves causes initial symptoms of heat sensation and perspiration.

One of the most bizarre episodes of human exposure to microwaves occurred at the US embassy in Moscow from the early 1950s to the late 1970s. Embassy staff and their families complained of a wide variety of neurological symptoms including headaches, sleep disturbances, memory and concentration difficulties, and dizziness which they and some government officials blamed on microwaves beamed at the embassy by the Soviets. In 1952, the Soviet government asked the US to move their embassy from the Kremlin to a residential area several miles away where they renovated an empty apartment building. In the process, the Soviets implanted several listening devices that were activated by microwave signal from a nearby apartment building. In the spring of 1953, routine surveillance identified the microwave beam aimed at the embassy and the staff was informed of the findings.

Because of continued symptoms among the staff and their families and concerns about the exposure to microwaves, the US Department of State commissioned an epidemiological study by researchers at Johns Hopkins University in 1976. The basic design was to compare health records and health questionnaires of the Moscow staff and families with those of the staff and families of other Eastern European embassies and consulates. Regarding the Moscow microwave exposure, the frequency range was 2.5–4.0 GHz with a maximum exposure of 5 $\mu W/cm^2$ when standing undressed by an open window on the side of the building facing the beam but much less and even miniscule away from the window and in other rooms. None of the staff reported heat-related symptoms typical of microwave overexposure.

The details of the study were not published at the time, but it was finally published in review form in 2012. The main finding was that there were no significant differences between symptoms, morbidity, or mortality of the Moscow staff, and families versus the staff and families at the other Eastern European embassies. No adverse health effects due to low-level microwave exposure were identified.

Remarkably, a similar phenomenon occurred in the early twenty-first century. US staff and their families at the Cuban Embassy complained of similar nonspecific neurological symptoms that were initially attributed to some type of sonic weapon since many reported hearing strange sounds before developing the symptoms. Once the impracticality of a sonic weapon became apparent, others including a panel of "experts" from the National Academy of science concluded that a microwave weapon was most likely the culprit. The panel attributed the sounds heard by the victims to the "Frey effect" of microwaves (see Chap. 12). The so-called Havana syndrome became a national media sensation and soon spread to US embassies around the world including the Moscow embassy. Someone was attacking US embassies worldwide, but it was unclear who or why.

Unlike the earlier Moscow embassy episode, despite extensive searches by several US intelligence agencies, no evidence of microwaves or any weapon was ever found. After 5 years of investigation, a joint committee of the US intelligence agencies concluded that in the great majority of patients with Havana syndrome, the symptoms could be explained by either preexisting illnesses such as migraine or do to stress, a type of mass psychogenic illness.

Electromagnetic hypersensitivity Electromagnetic hypersensitivity is a condition in which people attribute a variety of nonspecific neurological symptoms to exposure to man-made electromagnetic fields. The symptoms range from sleep disturbances, headaches, dizziness, memory and concentration difficulties, and mood disorders, and the electromagnetic field sources include power lines and power transformers, home appliances, mobile phones and base stations, computers, Wi-Fi routers, and even fluorescent and halogen lighting. It is predominantly a twenty-first-century phenomenon, although early cases were described in Sweden in the 1980s and there are roots in the 1950s with similar complaints from radio and radar technicians in the Soviet Union. It is estimated to affect about 5% of the populations of developed countries. People take extreme measures to avoid electromagnetic field exposure from wearing protective clothing and getting rid of electronic devices to moving away from society to live the woods or caves.

Modern double-blind provocative studies have shown that people who claim they are sensitive to electromagnetic fields are unable to detect the presence of electromagnetic fields and sham exposure to electromagnetic fields causes symptoms as frequently as real exposure. Studies that assess people's expectations prior to being exposed to sham or real electromagnetic fields show that if people expect to have symptoms, they are much more likely to have them. This so-called nocebo effect is the opposite of the placebo effect. People with preexisting anxiety are more likely to develop symptoms than those without anxiety. Men and women are equally affected. Not surprisingly, the media plays an important role in people's expectations. Media warnings about the adverse effects of modern life may be self-fulfilling.

Thinking Outside the Box

Major advances in science often require the ability to ignore accepted doctrine and interpret unexpected experimental findings in a completely new way. It requires an open mind willing to suffer through the inevitable backlash from traditional thinking colleagues. The book is replete with such examples, but a few stand out because of the magnitude and implications of the discoveries.

James Maxwell and modern electromagnetic theory Maxwell was one of the few scientists who took Faraday's lines of force theory seriously and was determined to express the theory in mathematical terms. He began with a simple concept of *flux* to describe the amount of force acting through any cross section in space. A high density of flux corresponded to a high concentration of Faraday's lines of force, the higher the density of flux the stronger the electric or magnetic force. To manage the fluid's flow through space, Maxwell imagined tubes for the fluid to flow in, all fitting together leaving no gaps between. The fluid moved fast where the diameter of the tubes narrowed and slow where they widened. By analogy, the electric and magnetic lines of force were strong where the tubes were narrow, and weak where the tubes were wide.

Maxwell next tackled the difficult problem of explaining the relationship between a moving electric current and the surrounding electromagnetic field and electromagnetic induction. He came up with an ingenious thought model consisting of rows of tiny rotating spheres ("particles of electricity") surrounded by large circular vortices representing the magnetic field (see Fig. 7.2). When revolving in place, the tiny spheres were just stationary electric charges but under the influence of an electromotive force, they would translate along the rows causing the surrounding vortices to rotate producing magnetic flux. To explain induction, he showed how the rotating vortices (magnetic field) could transiently set in countermotion neighboring rows of electrical particles and the on through the next layer of vortices to other more distant rows of electrical particles.

Maxwell's imaginary spheres and vortices were everywhere, in the ether of space and in any matter that occupied space. In conducting materials, like a copper wire the spheres were relatively free to move along their row whereas in insulating materials, the spheres were relatively fixed. With an applied electromotive force, electrical current rapidly moved along a conductor, but in an insulator (dielectric), there was little moving electrical current generated, what Maxwell called a *displacement current*. Maxwell gave the spheres elastic walls so that even though they did not move they could distort in one direction, the beginning of a current. With removal of the electromotive force, the distorted spheres would snap back producing a transient current in the opposite direction. Furthermore, if the spheres were set into elastic motion by an oscillating electric field, there would be a wave-like displacement of current along with transverse waves of magnetism.

There were eight equations in Maxwell's Dynamical Theory of the Electromagnetic Field. The equations described the behavior of electromagnetism in terms of the electric field, the magnetic field, and the currents. He soon left his mechanical model of spinning spheres and vortices behind and replaced it with intricately linked quantities that varied in space and time represented by abstract mathematical symbols. Many of his colleagues were frustrated with the lack of a mechanical explanation and most were overwhelmed by the mathematics of the new theory. They just were not prepared for this paradigm shift in the world of physics that would become the norm in the twentieth century.

Maxwell emphasized that he used mechanical models to get at the relevant mathematical equations to describe the relationship between electricity and magnetism. They did not necessarily represent nature's actual mechanism. But the mathematical equations predicted outcomes that were not previously identified by experiment or common logic. Modern scientists working on cell phone technology are still using them.

Albert Einstein's photons As revolutionary as Maxwell's equations describing electromagnetic waves were in the later nineteenth century, the modern world of physics experienced an even greater revolution with quantum theory at the turn of the twentieth century. To deal with the dilemma of infinite energy at infinitely high frequencies of black body radiation, the German physicist Max Planck suggested that the energy in black body radiation existed in individual packets called quanta just as matter existed in discrete units. The energy of black body radiation was defined by the equation $E = hv$, where h represented the quantum unit (Planck's constant) and v the vibration frequency. A little-known physicist working in a Swiss patent office, Einstein immediately recognized the implications of Planck's suggestion and developed his photon theory of light.

Instead of being a wave that spread out in all directions from a source, light consisted of a finite number of energy quanta that could be absorbed or generated only as a whole. Light was a stream of particles (photons) localized in space each with energy hv. Einstein's photon theory of light explained the otherwise mysterious phenomenon of phosphorescence. When a substance absorbs light and later releases it (phosphorescence), the wavelength of the released light is invariably longer than that of the absorbed light. Einstein speculated that when a molecule absorbs a photon of light, some of the energy is lost so that the highest frequency of the photon of light it could release was always equal to or less than that of the photon of light absorbed.

Equally mysterious was the so-called photoelectric effect. If a gold leaf electrometer (see Fig. 3.3) is negatively charged so that the leaves are widely apart, shining light in the blue to ultraviolet range onto the cap causes the leaves to collapse as electrons are released. But if lower-frequency light in the red to green range is shown on the cap, the leaves remain separated. No matter how intense or how long the lower-frequency light is on, the gold leaves remained separated. According to Einstein's theory, each photon of light interacts with only one electron. To dislodge an electron, it must absorb just the right amount of energy, the high energy in one photon of blue or ultraviolet light. There is not enough energy in a photon of red or green light to dislodge the electron, so there is no photoelectric effect regardless of the intensity of the light (number of photons). The electrons released by light can be collected and used to create an electric current in a circuit, the basis of modern-day solar panels.

According to Einstein, light is both a wave and a particle. It exhibits refraction and interference like classical waves, yet when exchanging energy with matter it acted like a particle. Einstein's genius was that he was willing to accept the apparent contradiction unlike many of his fellow physicists. Electromagnetic waves were a new entity, different from classical waves requiring a new physics. One consequence of this new physics was that there was no longer a need for a medium in which the waves traveled. The ether that was so important to Maxwell and scientists dating back to antiquity might be superfluous.

Robert Heath and deep brain stimulation for psychiatric disorders Sometimes thinking outside the box can have unintended consequences. When Robert Heath was recruited from Columbia University in New York to head psychiatry at Tulane university in New Orleans in 1949, there were no proven treatments for psychiatric disorders. Frontal lobotomy was commonly used to treat schizophrenia, but it benefited care givers more than patients. Heath, who trained in neurology and psychiatry, was convinced that the frontal lobes were not the problem in schizophrenia but rather a small region deep below the frontal lobes just above the midbrain called the septal region. Heath believed this region, the focus of emotions such as pain and pleasure, was malfunctioning in patients with schizophrenia. With the rapid advances in stereotactic neurosurgery, Heath was convinced that it was possible to electrically stimulate the region without damaging it.

After some preliminary experiments in cats and a few monkeys, Heath and neurosurgeon Francisco Garcia who followed him from Columbia, began performing deep brain electrical stimulation in psychiatric patients within a year. After drilling a hole in the skull, Garcia, passed a thin silver wire electrode deep through the frontal lobe near the base of the brain into the septal region. He guided the electrode using a series of anatomical markers outlined with pneumoencephalography. For chronic recordings, the wire was held in place by a bone plug covering the hole with the skin pulled tight around it. Typically, patients underwent multiple stimulations over several days with a range of stimulus parameters from 2 to 4 mA at 4–8 V. The patients remained fully alert during the stimulations so they could respond to questions and report any discomfort.

By 1952, Heath had stimulation data from 22 patients, the majority with a diagnosis of schizophrenia. Overall, the results of the stimulation were mixed with only a few having a dramatic response. A 26-year-old woman with catatonic schizophrenia who had responded to electroconvulsive therapy in the past began to speak after electrical stimulation and upon returning to the ward she talked with the nurses for the first time in months. A few others showed good improvement, but four patients had no response, and one had a worsening of symptoms after stimulation.

Three of the patients were considered controls since they had no mental illness. The patients agreed to the procedure since they suffered from medically intractable pain, one from severe arthritis and the other two from terminal cancer. Heath felt that stimulating the "pleasure" centers in the septal region might block the sensation of pain. The team reported that deep brain stimulation suppressed the pain in all three patients. One of the cancer victims who lived for 7 months after surgery reported that a 15-min stimulation session relieved her pain for a week.

When Heath's team presented their results, there was overall skepticism. The operation and the interest and attention shown the patients before and after the surgery were a strong placebo. Some suggested the zeal exhibited by Heath and his team may have influenced interpretation of the data to conform to their hypothesis.

The 1950s was the "wild west" for deep brain stimulation. There was no such thing as "informed consent" and few raised ethical concerns. Robert Heath not only implanted electrodes in the septal region but in several patients, he implanted electrodes throughout the brain to assess the response to septal stimulation. It was hard to justify these additional electrodes on a therapeutic basis. By the 1970s, there was a backlash in public and physician opinion regarding the ethics of deep brain stimulation for psychiatric illness, but there has been a resurgence of interest in the early twenty-first century, and preliminary studies suggest it may be helpful for treating patients with post-traumatic stress disorder (PTSD).

In Brief

Scientists share a wide range of personal attributes common to humans. They can be biased by their beliefs and prevailing theories, and they can be driven by ambition to distort history and even falsify data. But science is a process designed to deal with human foibles over time. It often appears chaotic in the short term, but history shows us that it works remarkably well in the long term.

Index

R. W. Baloh, *Brain Electricity*, https://doi.org/10.1007/978-3-031-62994-5

The manufacturer's authorised representative in the EU is Springer Nature Customer Service Centre GmbH, Europaplatz 3, 69115 Heidelberg, Germany. If you have any concerns regarding our products, please contact ProductSafety@springernature.com

Printed and bound by CPI Group (UK) Ltd, Croydon, CR0 4YY
28/09/2026
02224065-0007